# Landscape
# of the World

Patrick E.F. O'Dwyer

Gill & Macmillan

Gill & Macmillan Ltd
Hume Avenue, Park West
Dublin 12
with associated companies throughout the world
www.gillmacmillan.ie


0 7171 2819 9
Index compiled by Patrick Gillan
Print origination in Ireland by Design Image, Dublin
Colour separations by Typeform Repro, Dublin

*The paper used in this book is made from the wood pulp of managed forests. For every tree felled, at least one tree is planted, thereby renewing natural resources.*

*This book is dedicated to the memory of my parents,*
*Constance and Edward O'Dwyer*

# Contents

# Preface

The purpose of this book is to contribute to an understanding of some of the physical landscapes of the world. To this end the author attempts to develop an awareness of some of the natural forces that shape our landscapes and in some instances shows how people interact with these processes.

While it is essential that the processes are clearly explained it is equally important, I believe, that a textbook should be attractive and user-friendly for both student and teacher.

Recent scientific research has provided a window into the forces that drive the earth's crustal plates. These processes, such as plate tectonics, earthquakes and volcanic eruptions, directly affect the lives of people on all of the earth's continents. The inclusion of a detailed but hopefully a simple examination of plate tectonics is directly related to a change in the phrasing of recent examination questions and to increased media coverage of processes in recent years that have caused devastation and death in so many places.

Photographs, Ordnance Survey maps and block diagrams are used lavishly throughout the text to develop facility in map-reading and sketching and knowledge of and training in interpreting the landscape both physical and human.

There are many activities throughout each chapter. Some of these may be simple exercises while others may require deeper scrutiny. One purpose of these more difficult activities is to create discussion within the classroom while at the same time develop skills by more detailed examination of both maps and photographs. Some questions do not require a single correct answer. Varying answers may be acceptable as long as students can support their views with evidence from either map or photograph.

When using this text it is important that teachers and students should focus their attention only on the areas in the book that interest them while at the same time focusing on recent examination questions at senior level. It is the author's wish that they also find this work easy to follow, well structured, focused on examination questions and, by no means least, enjoyable.

P.E.F.O'D.

# Acknowledgments

Writing this book, while challenging and enjoyable, was also a major undertaking and its completion would have been impossible without the help of a number of individuals. Their interest, support and advice deserve special mention and I acknowledge them here, with sincere thanks.

Special thanks are due to Hubert Mahony, Editorial Director, managing editor Gabrielle Noble and the staff at Gill & Macmillan who worked tirelessly behind the scenes to make this book a reality. Angela Rohan, a skilled editor, sharpened, pruned and fine-tuned the script. Her attention to detail was most impressive and nothing was too insignificant to escape her scrutiny.

The artistic, creative and elegant design and illustrations of this book reflect the talents of Dara O'Doherty and her team at Design Image in Dublin. Their ability to understand and produce the countless and often complex illustrations in such a limited time has to be admired.

Sincere thanks to Dr Mary Knightly, whose encouragement and advice was invaluable.

The author also wishes to thank the following: Leonardo Fusciardi, chief geologist at Lisheen mines, for his helpful comments on part of the script; my colleagues in St Joseph's Secondary School, Doon, James Hayes, Michael Quirke and Linda Fitzpatrick, for their advice day and night on my many queries; my many friends in the Cork Geological Association and Pascal Gray in Nenagh CBS, whose companionship at lectures and on field trips made the study of geology so enjoyable.

Special thanks to Dr John Feehan, whose company, knowledge and teaching expertise I enjoyed and admired so much on his many outings.

**Maps** in this book are reproduced by permission of Ordnance Survey Ireland and Ordnance Survey of Northern Ireland.

**Photos** are reproduced by permission of the following:

J. Allan Cash Photo Library (Figs. 2.3, 2.6, 2.12, 2.13, 2.14, 4.12, 4.13, 9.13, 10.16, 10.32, 11.19, 13.6, 14.27);

Kevin Dwyer (Ireland) Limited (Figs. 2.11, 7.17, 7.20, 7.38, 7.39, 11.39, 11.40, 11.60, 13.18, 13.43, 13.63, 13.65, 13.66, 13.88, 16.6, 16.9, 16.16, 16.22, 16.23, 16.24, 17.5, 17.22, 17.23, 17.30, 17.35, 17.40, 17.42, 17.51b, 17.52, 17.53, 17.55, 17.56, 17.57, 17.59, 17.97, 19.21, 20.3);

Ordnance Survey Ireland (Figs. 2.21, 7.2, 10.37, 11.41, 12.51, 17.4, 17.12, 17.13, 17.25, 17.88, 18.6, 18.15, 18.16, 18.17, 18.18, 18.20, 20.1, 20.7, 20.12, 20.14, 20.16);

Marie Tharp: Oceanographic Cartographer (Figs. 3.2, 3.24, 3.42);

Topham Picture Point (Figs. 4.6, 11.25, 13.48, 19.15);

Barnaby's Picture Library (Figs. 4.16a, 12.23, 13.59b, 13.75, 14.20);

Science Photo Library (Figs. 4.32, 12.1, 12.5);

Camera Press (Figs. 4.34, 13.3, 14.13, 19.14a, 19.14b);

Finbarr O'Connell (Figs. 7.4, 17.15, 18.4, 18.5);

Rex Roberts (Figs. 7.5, 13.70);

Gillian Barrett (Fig. 7.29);

Aerofilms (Figs. 7.35, 12.45, 16.4, 16.7, 16.15, 17.2, 17.6, 17.19, 17.24, 17.33, 17.43);

Planet Earth Pictures (Figs. 8.1, 11.1, 11.10, 12.24, 13.1, 14.1, 14.8a, 14.16b, 14.22, 14.25, 19.1, 19.5);

Patrick O'Dwyer (Figs. 9.9, 9.26c, 11.12, 12.37, 12.38, 12.60, 12.61, 13.5, 13.10, 13.11, 13.15, 13.16, 13.37, 13.54, 13.76, 13.86);

University of Cambridge: Committee for Aerial Photography (Figs. 9.15, 12.16, 12.26);

Frank Spooner Pictures (Figs. 9.19, 9.20, 9.24, 13.19, 13.57, 13.59a);

Associated Press (Fig. 9.23);

Sygma (Figs. 9.25, 19.4a, 19.4b);

US Geological Survey (Figs. 10.12b, 12.53);

Woodfall Wild Images (Figs. 10.19, 10.33, 14.6);

Robert Harding Photo Library (Figs. 10.25, 13.81, 14.8b, 19.13);

Magnum Photos (Figs. 10.30, 13.60);

D.D.C. Pochin Mould (Figs. 10.36, 12.21, 15.11, 15.13, 17.16, 17.17, 17.21);

*The Irish Times* (Fig. 11.61);

Lonnie G. Thompson (Fig. 12.55);

George Westerman (Fig. 13.38);

Ordnance Survey of Northern Ireland (Figs. 12.59, 13.49, 13.55, 15.1, 16.5, 17.11, 17.29, 18.19, 20.8, 20.10);

The Travel Library (Fig. 13.46);

Aerocamera Hofmeester (Fig. 13.72);

Impact Photos (Figs. 13.83, 14.16a);

Dúchas, The Heritage Service (Figs. 15.8, 16.17);

Dublin Corporation (Fig. 18.1);

Don Sutton International Photo Library (Fig. 18.2);

National Roads Authority (Fig. 18.3);

Network Photographers (Figs. 19.15a, 19.17).

**Illustrations**

Ordnance Survey Ireland (Fig. 13.84);

Department of the Environment (Figs. 18.7-18.12);

Cork University Press, for the following illustrations reproduced from *Atlas of the Irish Rural Landscape* (Aalen, Whelan, Stout) –

Matthew Stout (Figs. 6.5, 10.34, 12.43, 16.1, 17.8, 17.38, 17.75, 17.76, 17.77, 18.13);

Ordnance Survey of Ireland (Figs. 17.9, 17.20);

Ordnance Survey of Northern Ireland (Figs. 17.27, 17.45).

# Planet Earth

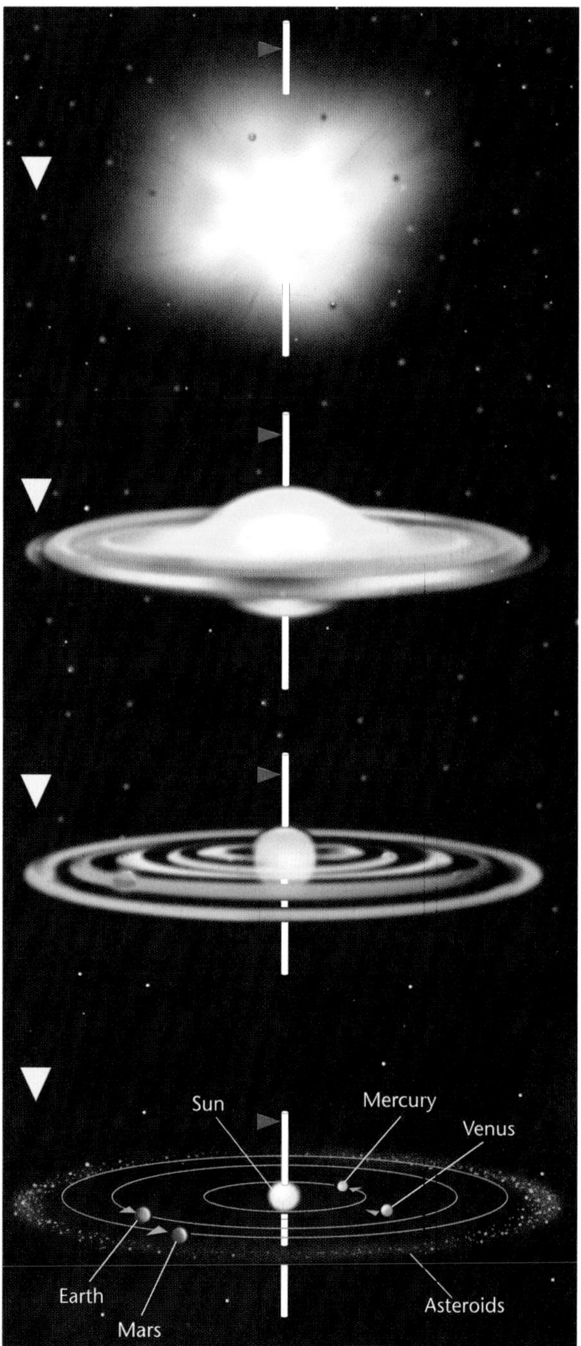


Fig. 1.1
The gathering of atoms (the tiniest particles of matter) in space created a rotating cloud of dense gas. The centre of the gas cloud eventually became the sun; the planets formed by condensation of the outer portions of the gas cloud.

## Solar System

Scientists believe that the solar system developed from a huge nebula (cloud of gas and dust) that once swirled around the sun. The sun itself may have been formed from the central part of this nebula. As the nebula whirled around the sun, it slowly flattened out. Sections of the cloud began to spin like eddies (whirlpools) in a stream. Gas and cloud collected near the centres of these eddies. The collections of gas and dust grew by attracting nearby particles of matter. They slowly developed into the spinning planets that now travel around the sun. The sun is a star, one of billions of stars that make up a galaxy called the Milky Way. The Milky Way and billions of other galaxies make up the universe (Fig. 1.1).

## The Planet Earth

The planet earth is only a tiny part of the universe.

Some 4.6 billion years ago a cloud of dust condensed into planet earth, which soon turned molten from meteorite impacts and radioactive decay. As it cooled, heavier materials sank, forming a layered globe 12,740 kilometres in diameter. Still cooling, the earth brings molten rock from its core to the crust. This molten rock brings heat to the surface for release. This **convection** (current) creates earthquakes and volcanoes and mountain ranges that shape our life-bearing lands and seas. The currents driving these vast movements sweep slowly through the underlying **mantle** rock. The idea that solid rock can flow is hard to grasp because rock seems unchanging in the human time-frame. Yet over millions of years, rock can flow like glacier ice.

## The Earth's Crust

### The Earth's Crust Has Two Layers

Continents have an upper layer of granite-like rocks called sial. The lower layer, called sima, extends under the oceans and consists of materials similar to hardened lava. Continents appear to 'float' like icebergs in the heavier rock of the mantle.

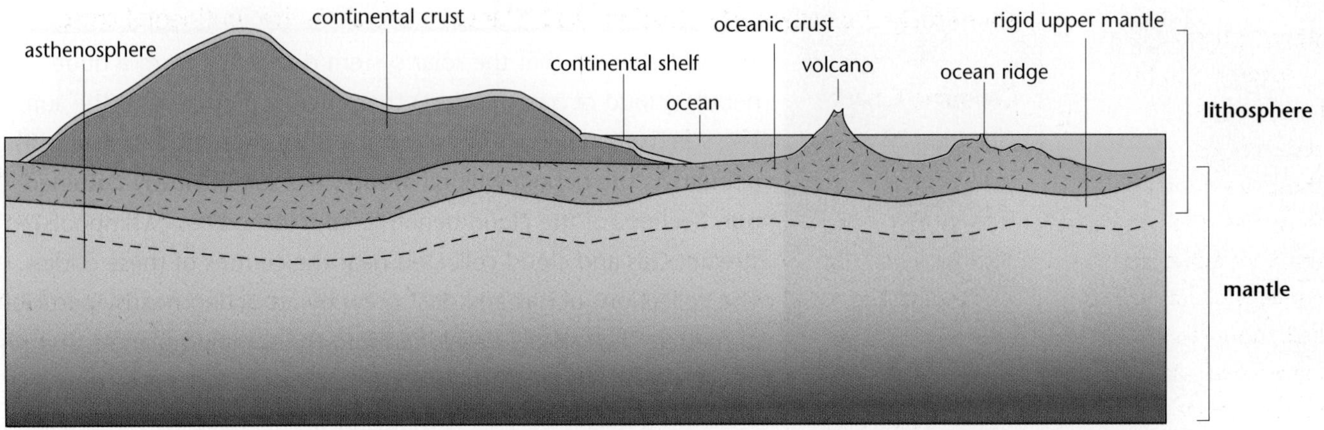


Fig. 1.2

## The Earth's Interior

### Section across the Earth at Latitude 20°N

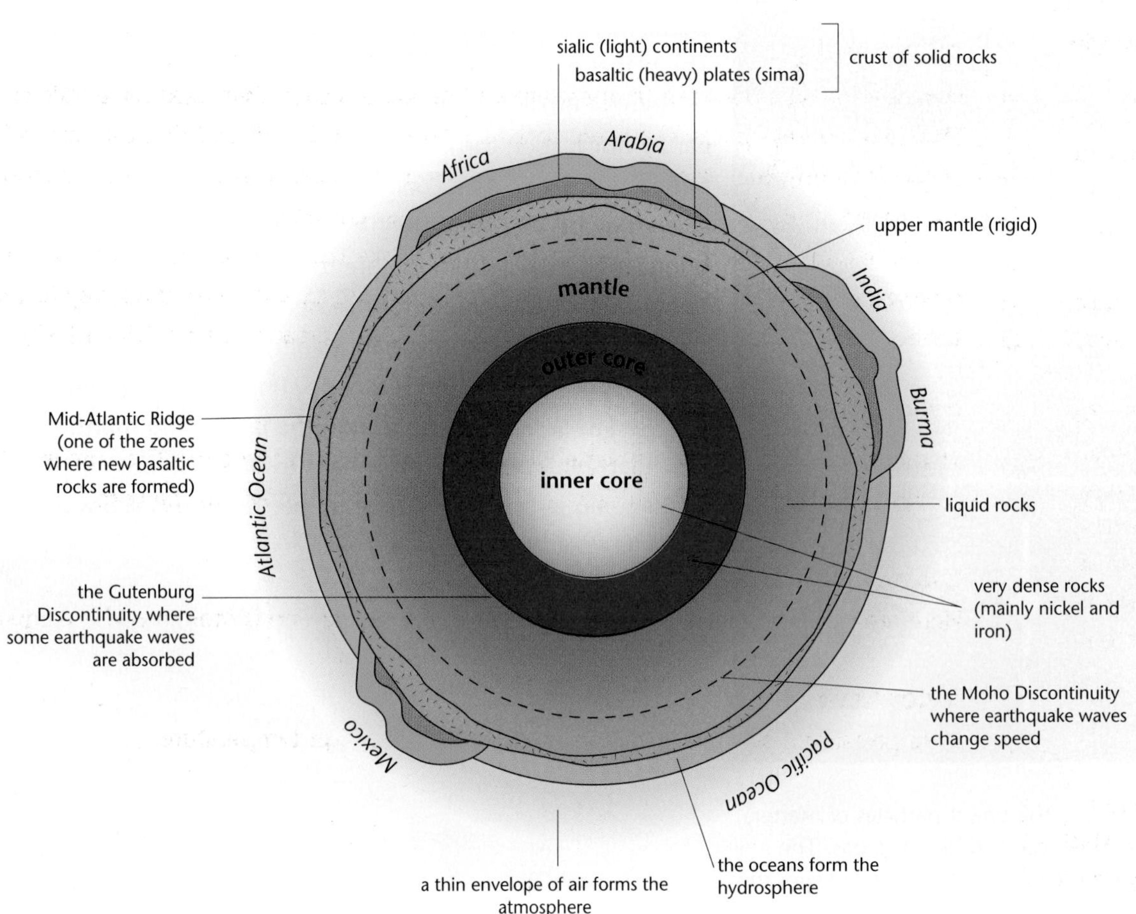


Fig. 1.3

**In the Lithosphere**
Rocks are cooler, stronger and more rigid in the lithosphere than in the asthenosphere.

According to the theory of plate tectonics, the lithosphere is broken into plates. The plates, which are much like segments of cracked shell, move relative to one another, sliding on the underlying asthenosphere.

**Moho Line**
In 1909 a Yugoslavian seismologist, Andrija Mohorovicic, presented the first convincing evidence for layering within the earth. By studying earthquake waves he found that the speed of waves increased abruptly below a depth of 50 km. This boundary line which separates the crust from the underlying mantle is known as the **Mohorovicic Line**, shortened to **Moho Line**.

**Note:**
Basalt represents a group of rocks which cooled quickly on the surface.

Lithosphere

## Crust 0–50 km

### Continental Crust

The crust of the continents is **thick** compared to that of the ocean floor. It has an average thickness of 45 km and is up to 70 km in thickness under the mountain ranges. The rocks which form the continental crust are **light** in weight. They are rich in **si**lica and **al**umina. These rocks are collectively called the **sial**. The sial is therefore the continental crust.

### Oceanic Crust

The crust of the ocean floor is **thin**; it averages 8 km in thickness and may be as thin as 3 km in places. The rocks forming the oceanic crust are mostly **basalt** and are **heavy** and rich in **si**lica and **ma**gnesium. These rocks are collectively called the **sima**. This sima layer also underlies the continental crust.

**Rocks of the sial are much lighter than those of the sima and hence the continents can be regarded as floating on the sima.**

## Rigid Upper Mantle 50–100 km

The upper mantle is made up of rigid rocks of a mixed nature which are of even greater weight than those of the sima. These rocks plus the sima and the sial are known as the **lithosphere** or the earth's crust.

Convecting Mantle

## Asthenosphere and Lower Mantle 100–200 km

Here the balance between temperature and pressure is such that rocks have little strength. This zone consists partly of melted rock (10 per cent). It is believed that the 'plastic' rock of the asthenosphere moves about as fast as fingernails grow. The plates of the lithosphere are moved about on these slow-moving convection currents.

Some geologists believe that convection **pipe-like currents** of hot plastic rock, having received heat from the outer core, **rise as mantle plumes** and make their way to the surface, where they erupt to form hot spots and lava plateaus (see pp. 38 and 52).

## Core

At the centre of the earth is the densest of the three layers, the **core**. The core is composed mainly of **iron** with some **nickel**. There are two parts to the core.

### Outer Core

Here temperature and pressure are so balanced that the iron is molten and is a liquid.

### Inner Core

Here pressure is so great that iron is solid despite its high temperature.

# The Rock Cycle

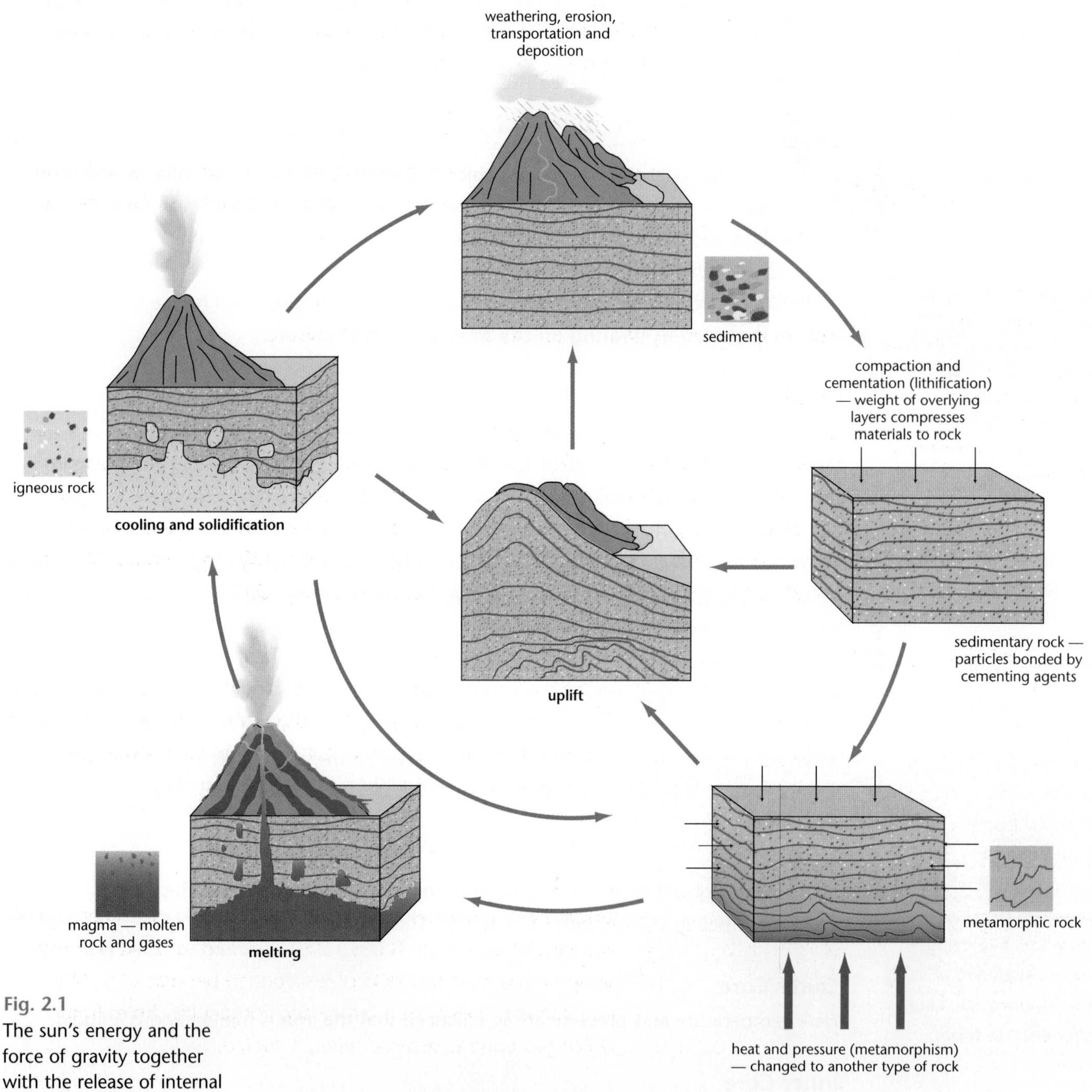

**Fig. 2.1**
The sun's energy and the force of gravity together with the release of internal heat of the earth power the rock cycle. This has been in operation for at least four billion years and is likely to continue for as long as there is energy available to drive it.

The earth has a highly varied and ever-changing surface. Minerals and rocks change as well. Like with rivers, the earth is always trying to achieve a state of equilibrium where everything is in perfect balance. If this were to happen, the earth would be a boring place to live. Nothing would change. However, this is not the case. The internal and external forces of the earth's heat-engine continue to interact, creating constant change.

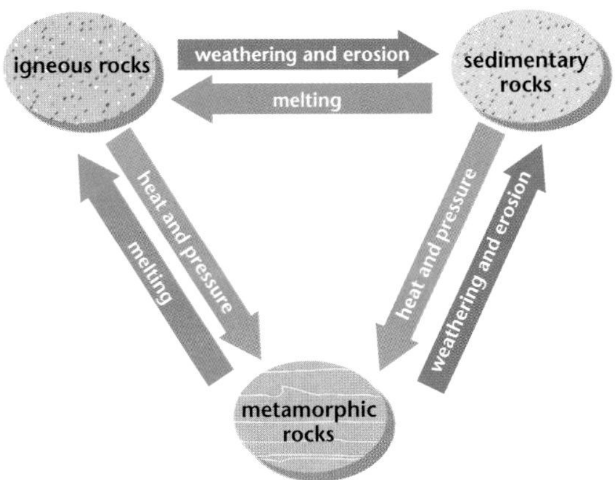


Fig. 2.2
The cycle of rock change. In this simplified version of the cycle of rock change, the three classes of rock are transformed into one another by weathering and erosion, melting, and exposure to heat and pressure.

## Classification of Rocks

The rocks of the earth's surface may be classified into three types according to the way in which they were formed. These three types are **igneous**, **sedimentary** and **metamorphic**. Each of these rocks may form at the expense of another.

### Igneous Rocks

These rocks were formed from magma either inside the earth's crust (called **intrusive** or plutonic rocks) or on the earth's surface (called **extrusive** rocks).

#### Intrusive Rocks

Most of these are crystalline, i.e. they are composed of crystals.

#### Granite Rocks

Some of these intrusive (plutonic) rocks **cooled very slowly** deep in the earth (Pluto was the Roman god of the Underworld) and are composed of large crystals. These rocks are called granite. Granite rocks are composed of three minerals, mica, feldspar and quartz. The percentages of each of these vary, thus creating granites with a range of colours. Some may be black and white, much like a firelighter; others may range from pink to grey. Therefore the term granite relates to a group of rocks. (See p. 50.)

Fig. 2.3
Pink granite. The colour of granite varies according to the percentages of the various minerals that form the rock.

#### Hypabyssal Rocks

Magma which solidified **in** the crust slowly but nevertheless faster than granite is called hypabyssal rock. This may have been near to the surface or may have formed in a narrow sill or dyke. (See 'Hypabyssal Rock — An Intermediate Rock', p. 51.)

#### Extrusive Rocks (Volcanic Rocks)

#### Basalt

Magma which reaches the surface **cools quickly** into small crystals and forms basalt rock. This is a dark rock and some surface rock samples may have tiny dots of rust formed because of its high iron content. (See p. 52.) It may also contain tiny holes from which gas bubbles escaped when it cooled.

#### Pyroclasts

A small group of rocks are formed when they are ejected from a volcano. They include solid lava, cinders, ash and dust. As a group they are classified as tephra. (See 'Tephra', p. 46.)

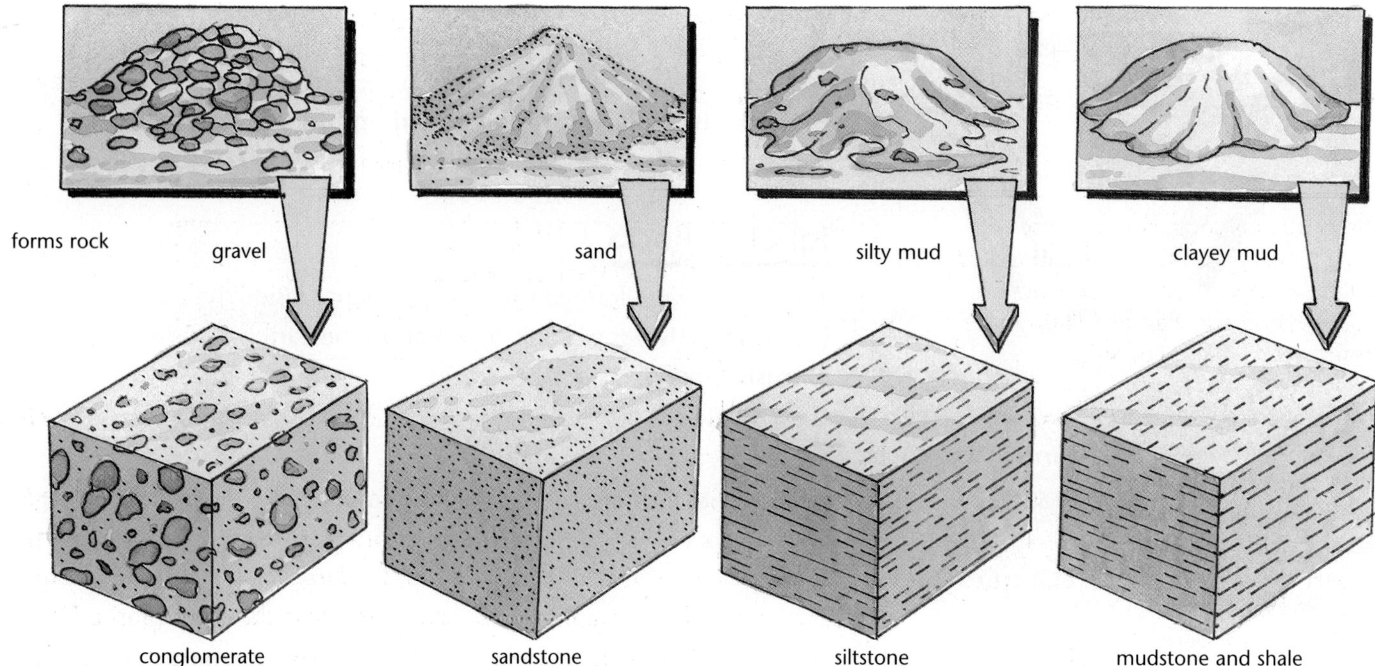


Fig. 2.4
Principal kinds of eroded (clastic) sediments and sedimentary rocks formed from them

**intrusive rocks**
Rocks formed from magma within the earth. These are also called plutonic rocks.

**extrusive rocks**
Rocks formed by magma on the earth's surface.

**hypabyssal rocks**
Rocks formed within the earth close to the surface.

**pyroclasts**
Rock formed from materials ejected from a volcano.

**mafic rock**
Mafic refers to rocks which are rich in iron and magnesium.

**ultramafic**
This refers to rock which is higher in iron and magnesium content than mafic rocks and so is heavier.

## Sedimentary Rocks

Sedimentary rocks are formed from particles of all other rocks, including other sedimentary rocks. Once exposed at the surface of the earth, rocks are subjected to the processes of denudation, i.e. the processes of weathering and erosion. (See p. 113.) Weathered and eroded rock particles are transported downslope and are deposited in layers at lower altitudes, generally lowland, valley floors, deserts or in fresh or salt water. The line of division between each layer is called a bedding plane.

Sedimentary rocks may be classified according to either their **mode of formation** or their **composition**. Under the first classification, they may be described as

1. mechanically formed
2. organically formed and
3. chemically formed

Under the second classification, they may be divided into groups such as conglomerates and sandstones. It is convenient to combine the two classifications when one is studying sedimentary rocks.

## 1. Mechanically Formed Sedimentary Rocks

When igneous and metamorphic rocks are subjected to weathering and erosion, rock particles are transported by wind, water or ice and deposited elsewhere. Where these particles accumulate they become compacted by the weight of the overhead particles.

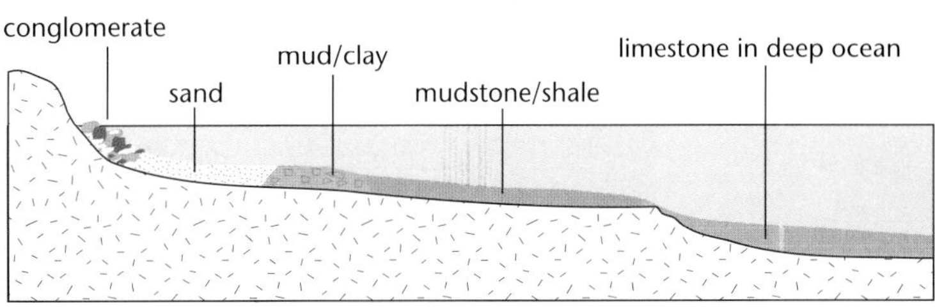


Fig. 2.5
The location of sediments in the sea varies, depending on the size and weight of the particles. The coarse particles are found near to the coast and finest particles far out into the ocean.

Fig. 2.6
Fossiliferous limestone (limestone with fossils)

This reduces the spaces between them. Over time, cementing agents, such as silica, bind the particles together.

The term **clastic** is often used to refer to these rocks as they are formed from other rock particles ('clastic' derives from the Greek word for broken).

When sediments are carried to the sea, the larger particles are deposited near the coast as they are heavy. Medium-sized particles are deposited further out and fine particles are carried far from the coast.

Coarse particles form **conglomerate** rock. Such rocks are composed of large pebbles, small pebbles and sand grains which are all cemented together. They may have formed near the coast or other places of deposition, such as an alluvial fan or river-bed. During times of flash floods in desert areas, pebbles and stones may be carried long distances. As flash floods may stop as quickly as they began, deposits many metres in thickness may form one individual layer (stratum). The Devil's Bit in Co. Tipperary in Ireland is formed of conglomerate and was formed in Devonian times (Old Red Sandstone conglomerate).

Fine particles of silt and mud are deposited furthest from the coast. These create shale and mudstone.

## 2. Organically Formed Rocks

These rocks are composed of the remains of once living organisms, the hard parts of which have accumulated over long periods of time. The most widespread and varied group is limestones.

We think of limestone as a single rock type. But there are many different types of limestone, e.g. those with fossils and those without fossils, and dolomite (magnesium carbonate) as distinct from 'ordinary' limestone, which is calcium carbonate. The best way to observe limestone is along a cliff face or by examining a core sample extracted from a quarry or mine. The latter clearly shows the broken shells and sediment much as a soil analysis does in the classroom.

Limestone is a rock made up mostly of calcium carbonate. It is usually a greyish colour, but all colours of limestone from white or pink to black are found. Scientists test natural rock to see if it is limestone by pouring cold diluted hydrochloric or sulphuric acid on it. If it **fizzes** it is limestone.

## Peat, Lignite and Coal

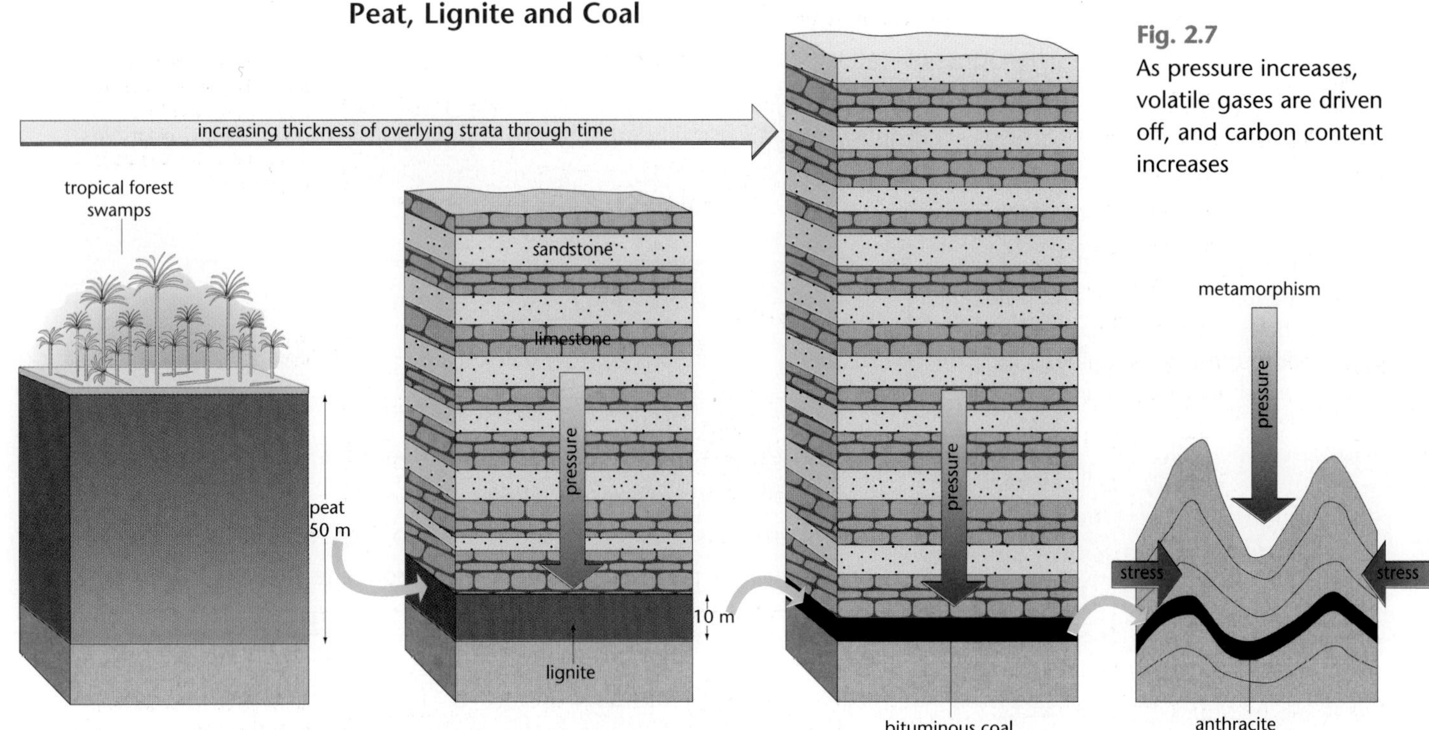


**Fig. 2.7**
As pressure increases, volatile gases are driven off, and carbon content increases

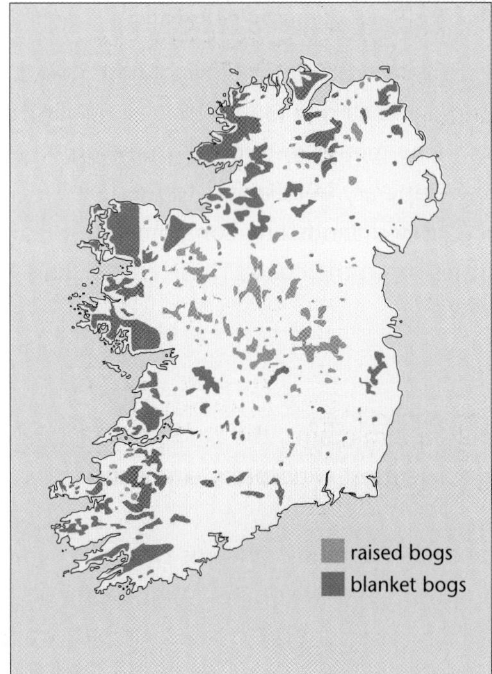


**Fig. 2.8**
Distribution of bog types in Ireland

### Peatlands

Peatlands cover 16.2 per cent of the landmass of the island of Ireland and 17.2 per cent of the area of the Republic. This percentage is exceeded in global terms by only three countries, namely Finland, Canada and Indonesia. See Figs. 2.8 and 2.9.

### What is Peat?

Undrained peat consists of 95 per cent water and 5 per cent solids material. The solids are composed of the partly decayed remains of a variety of plants. The plant remains include roots, stems, leaves, flowers and seeds. In some locations the solids material is made up of a mixture of partly decayed remains of plants and trees. In such cases the residues of trunks and roots of trees such as yew, oak, pine and birch are found.

### Peatland (or Bog) Formation

Peat forms in situations where a number of conditions exist over long periods of time. These are:

1. the continuous annual growth of vegetation
2. the presence of moderate to relatively high levels of rainfall — Ireland's maritime climate ensures rain throughout the year
3. the existence of poor drainage which leads to waterlogging of the soil surface
4. the very low level — or absence — of oxygen in the waterlogged soil (i.e. anaerobic conditions prevail)

THE ROCK CYCLE

## Types of Peatland

Three types occur in Ireland:

**Fens:** Fens are bogs which form in shallow lakes from vegetation which is fed from ground waters rich in nutrients.

**Raised bogs:** These bogs formed on fens. They have a dome-shaped surface and are very deep (up to thirteen metres). The plants which grew to form these get their nutrients and water supply from rainfall.

**Blanket bogs:** These bogs formed on hilltops and hillsides where vegetation grew in oxygen-free saturated soils. They are thin (only one to three metres deep).

**(a) Early stage**

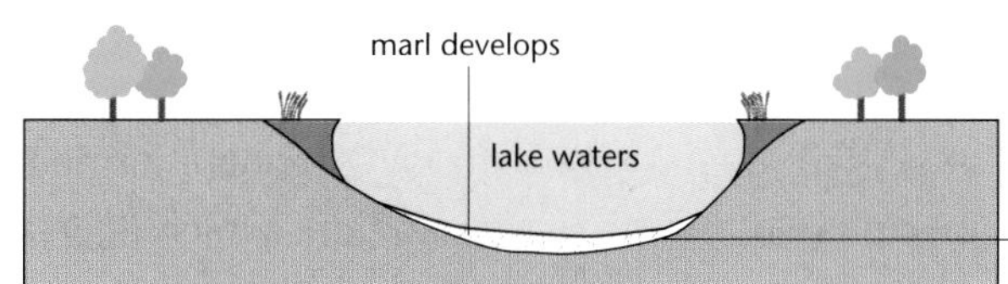
marl develops
lake waters
when rivers which flow across limestone areas flow into shallow lakes they deposit their dissolved limestone on the lake-beds to form marl

**(b) Fully developed fen**

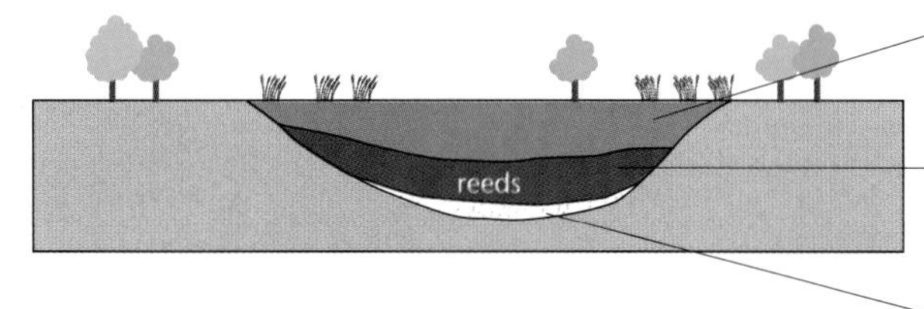
reeds
peat develops and gets its moisture from below
reeds grow getting their water and minerals from the shallow lake and lake-bed until the lake fills in
calcium carbonate is precipitated to form marl

**(c) Fully developed raised bog**

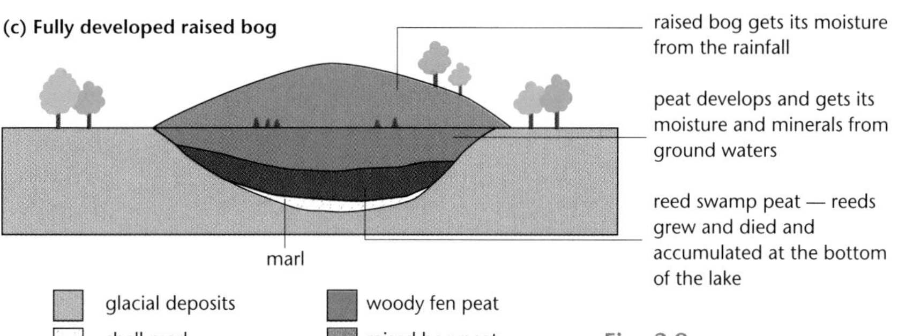
raised bog gets its moisture from the rainfall
peat develops and gets its moisture and minerals from ground waters
reed swamp peat — reeds grew and died and accumulated at the bottom of the lake
marl
glacial deposits
shell marl
reed swamp peat
woody fen peat
raised bog peat

Fig. 2.9
Stages in the development of a raised bog

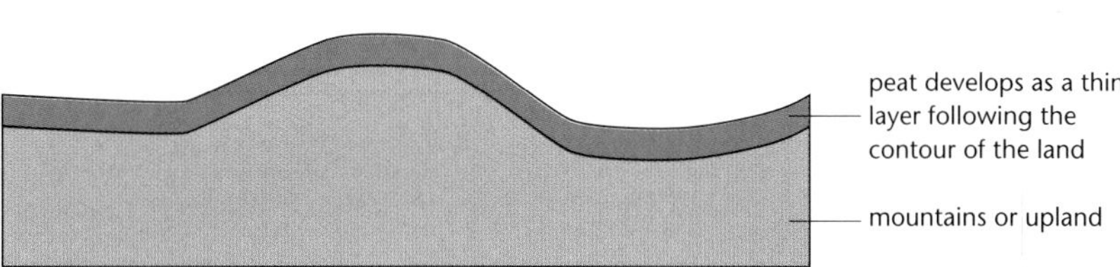
peat develops as a thin layer following the contour of the land
mountains or upland

Fig. 2.10
Blanket bog

## Influence of Ice Age

Any discussion on the formation of the peatlands should make reference to the last glaciation period, or ice age, because of the major influence this had on the evolution of peatlands, especially in the central plain in Ireland. About 13,000 years ago, the ice fields had melted and retreated northward, due to a slight increase in temperature, leaving behind glacial deposits (i.e. eskers, moraines, drumlins) which resulted in very irregular topography or landscape. These glacial deposits prevented normal drainage and so surface

waters were impounded to form lakes. It was in these shallow lakes that bogs such as those in the central plain of Ireland were formed.

Underneath the raised bogs in the central plain is a white layer of soft rock called 'marl'. This is a layer of calcium carbonate formed from dissolved limestone and deposited (precipitated) by the surface streams which flowed into these lakes initially.

**Fig. 2.11**

Bogs formed in shallow glacial lakes which were impounded (dammed) by eskers and moraines in the Irish midlands, e.g. near Clonmacnoise, in Co. Offaly in Ireland

Study the photograph of the Clonmacnoise area in the Irish midlands (Fig. 2.11, p. 10). Then do the following:

1. Draw a sketch map of the photograph and on it mark and label:
   (a) the River Shannon
   (b) the bogland areas (dark brown areas in background)
   (c) the roads
   (d) the early Christian settlement of Clonmacnoise
   (e) esker and glacial deposits
2. Carefully examine the undulations (ups and downs) on the photograph and then with the aid of drawings use evidence to account for the formation of the boglands in this area. See also Fig. 2.8, p. 8 and Fig. 2.9, p. 9.

## Chemically Formed Rocks

Among the most common chemical rocks are rock salt and gypsum, each of which is formed by evaporation of sea water or lake water and has economic value. Gypsum is used as plasterboard (ceiling slabs) in the construction industry. Rock salt is used as a cow-lick.

Fig. 2.12
Stockpiles of raw salt at salt processing plant, Great Salt Lake, Utah, USA

### Mineral Deposits in Chemical Sedimentary Rock

Sedimentary rocks provide employment for most of the world's geologists. Vast accumulations of sedimentary minerals, such as petroleum and coal, are trapped in sedimentary rocks. These two minerals drive our modern civilisation. This became quite apparent to all during the oil crisis in 1973: European industry and transport ground to a halt when Middle Eastern countries, e.g. Kuwait, reduced their oil output.

### Banded Iron Deposits

Modern society is more dependent on iron than on any other industrial metal. Many of the world's most important deposits of iron are found in sedimentary rocks billions of years old in Brazil, Canada, the former USSR and Australia.

### Evaporites

An evaporite is created by the evaporation of sea water or lake water.

Some lake water evaporite minerals are sodium carbonate, sodium sulphate and borax. These and other salts have many uses, including the manufacture of soap, paper, detergents, antiseptics and chemicals for tanning and dyeing.

Sea water evaporites include gypsum for plaster in construction, rock salt for cattle, and potassium for fertilisers.

## Metamorphic Rocks

Igneous and sedimentary rocks may change physically and chemically as a consequence of heat or pressure or both. Some metamorphic rocks are

- marble — formed from limestone
- slate — formed from shale or clay
- anthracite — formed from bituminous coal
- quartzite — formed from sandstone

Earth movements subject rocks to great pressure. When these occur they are usually on a large scale and the results are therefore widespread.

Other changes may be brought about as a consequence of heat. A rise in temperature generally occurs because of a mass of hot igneous rock nearby. A zone of contact alteration, called an aureole, can sometimes be distinguished. Such a zone may be seen at the Grotto in Pallas Green in Co. Limerick, Ireland, where basalt meets a limestone layer.

In granite, where such a contact zone exists, a gradual transition in the character of the rocks can be traced as new minerals appear.

Metamorphic rocks are, in general, compact and resistant to erosion. They tend to form masses within areas involved in mountain building. The counties of Galway, Mayo, Donegal, Down and Wicklow in Ireland, much of the highlands of Scotland, the Isle of Anglesey, Scandinavia, much of Africa and western Australia are composed of metamorphic rocks.

Fig. 2.13
Marble forms from limestone. Some 'marble' is not real marble but polished rock created for household use.

Fig. 2.14
Slate forms from shale

## The Age of Present-day Land Surfaces

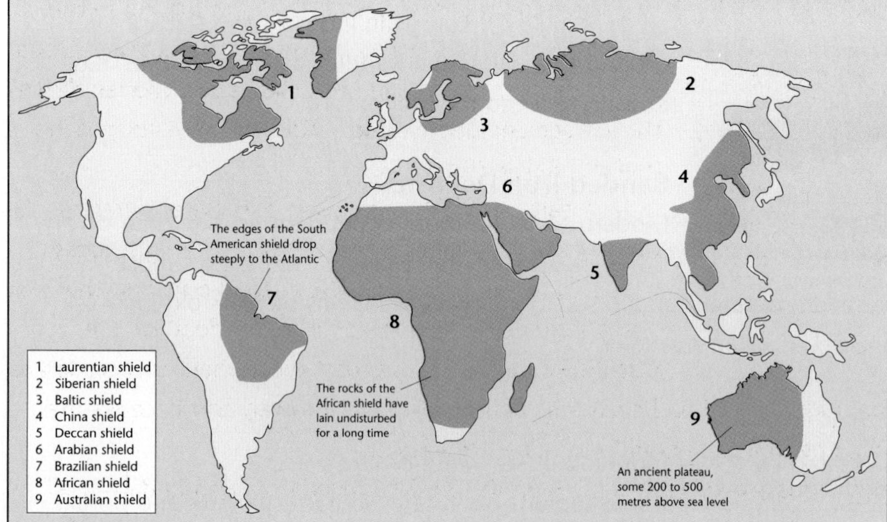


Fig. 2.15
The shields are ancient areas of stable rock. Onto these masses have been 'welded' mountain systems of various ages and around and upon their margins have grown sheets of sediments which cover much of the world's lowland areas.

The distribution of land surfaces has not always been as it is today. There are large areas of land composed of sedimentary rocks which contain shells and the remains of sea organisms. Such rocks must have been laid down in water, and these regions at one time must have formed the floors of seas. There are, however, many regions composed of very old crystalline rocks which must have been land for long periods of time. These regions are called **stable blocks** or **shields**. The shields are frequently rich in mineral deposits such as **gold**, **silver**, **lead**, **cobalt** and **nickel** and other ore bodies, such as iron ore.

## Isostatic Change

The theory of isostasy has been accepted for a long time to explain why vertical and horizontal movements take place in the earth's crust. In this theory we know that the force of gravity must play an important role in determining the elevation of the land.

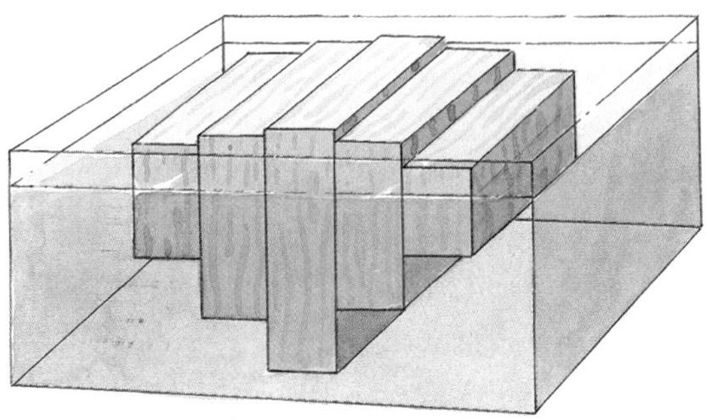

Fig. 2.16
This drawing illustrates how wooden blocks of different thicknesses float in water. In a similar manner, thick sections of crustal material float higher than thinner crustal slabs, and they also extend deeper into the mantle.

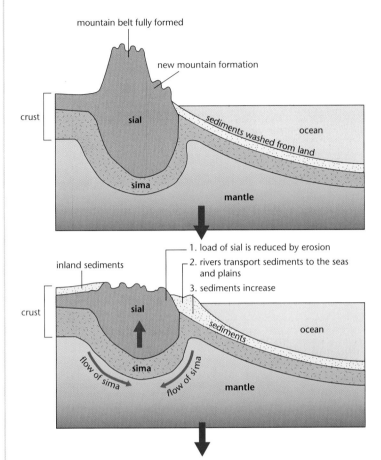

The theory of isostasy compares the continental blocks with a series of wooden blocks of different heights floating in water. Notice that the **thicker** wooden blocks **float higher** in the water than the thinner blocks. Their **bases** also **extend deeper** into the water than those of the thinner blocks. In this way, the wooden blocks and the water are in perfect balance. Any reduction in the timber blocks will cause a change in the water level. In a similar manner, mountain belts (sial) stand higher above the surface and also extend further into the supporting material below (sima).

The earth's crust is composed of rocks of differing densities. The continents are composed of light rocks which are classified as **sial**. The ocean floors and the heavy materials which the continents 'float' on are classified as **sima**. The sial of the continent is regarded as 'floating' on the underlying sima because its density (weight) is lower, and because the sima is thought to be in a semi-liquid state (like thick jelly).

When the highlands are lowered by erosion the eroded material is deposited on the surrounding lowlands and in the seas under the pull of gravity. As weight is removed from the highlands the remaining sial underneath rises. As weight is added to the lowlands the sial underneath the lowlands sinks. Sima underneath the lowland sial then flows towards the base of the rising mountains. This adjustment of balance is reflected on the surface by a levelling of the landscape.

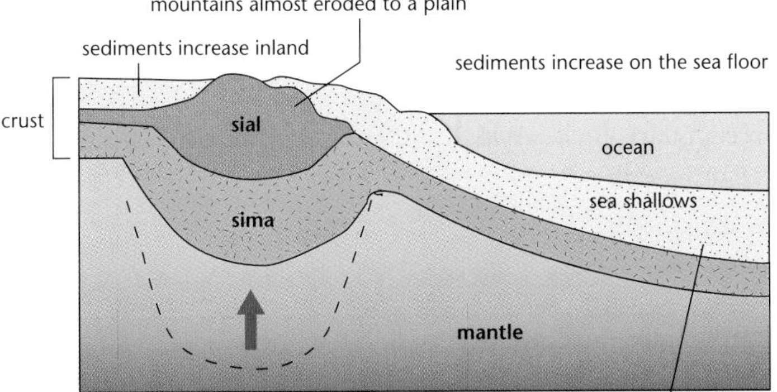


Fig. 2.17
Isostatic change

# The Value of Rocks to People

## Building Construction

Most dwellings in Western Europe are built from rock, such as limestone or sandstone, or rock compounds such as concrete blocks or clay bricks. Incorporated into some rock compounds are metals such as iron and steel to reinforce buildings or precast concrete units. Stone Age people — palaeolithic, mesolithic or neolithic as we now classify them — were so called because tools used by them were made from stone or flint, and Iron Age people used stone as a foundation for their lakeside crannog settlements. Early Christian monks built round towers to protect them from attacks by the Vikings, and corbelled cells and churches. The Normans built huge castles and defended settlements to protect their conquered lands.

Today in domestic house construction, stone is used for filling, foundations (concrete), blocks for walls, concrete tiles for roofing, and gypsum for plaster in ceilings and walls.

## Iron and Steel Production

Vehicles, such as cars, excavators, cranes, farmyard machinery and trains are all made from iron and steel. In the processes of iron and steel manufacture, coal and petroleum are used as energy sources in the smelting of the metal. In the same process, limestone is used as a flux, i.e. it is added to the heated iron ore to remove impurities by reducing their melting points. The impurities rise to the surface of the melt and are skimmed off.

## Transport

Roads are constructed from rock. Gravel and sand are used for road foundations. Rock chips and tar are used for surface dressing, while tarmacadam and bitumen are used for road surfaces.

Trucks and ships are made from iron and steel, while aeroplanes are made from steel and alloys (mineral compounds). Their electric wiring is made from copper, extracted from copper ore.

## Fertiliser

Limestone is broken into powder and sprayed on farmland to improve soil quality. It also helps in drainage; lime causes soil particles to cluster and this creates more pore spaces for groundwater to drain away to trenches.

## Ornaments

Gold has long been prized for its beauty. In Celtic times, gold bracelets, collars, chalices and shrines were used. Today, gold forms the raw security for nations' wealth. Diamond, the hardest rock, is used for engagement rings and also for industrial drilling. Other coloured stones such as rubies and emeralds are used for rings and necklaces. Less valuable rocks such as marble are used for fireplaces, flooring and sporting trophies.

## Hydrocarbon Deposits in Rocks

### Processes, Oil, Natural Gas and Plate Tectonics

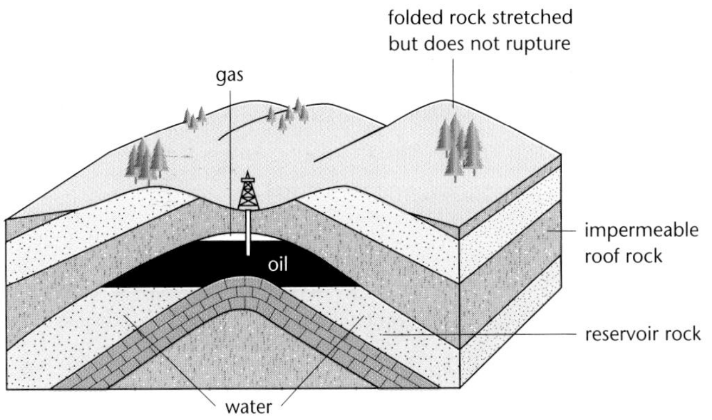


Fig. 2.18
Structural traps

Most of the world's oil 'wells' exist because of the way rocks deform. An oil well is not a well in the usual sense of a pool. The term actually refers to layers (strata) of rock in which oil occupies all the pore spaces (saturated).

The great oil wells of the Middle East, such as those in Saudi Arabia, Kuwait and Iraq, are in 'structural traps' (see Fig. 2.18). These traps were formed by faulting or folding as a result of earth movements created when the African plate collided with the Eurasian plate.

Such mineral deposits are regarded as having huge economic and strategic value. Iraq invaded Kuwait in 1990 in order to control its oil wells, and the United Nations went to war the following year to expel the Iraqi army.

### Case Study: Uraninite (Uranium Ore Rock)

Uraninite is one of the world's most valuable ores. It contains both radium, which can be used to cure cancer, and uranium, which is a source of atomic power. Both of these minerals are radioactive.

## Disadvantages of Some Rocks to People

### Radon Gas

Radon gas is a naturally occurring gas from decaying radium. It rises to the surface and becomes trapped within houses. As a consequence of domestic house design, we sometimes create an artificial climate indoors that is not always healthy.

**Home Activity**

Find out:

- which regions are more prone to the effects of radon gas
- which types of rock produce radon

Examine Fig. 2.19 and then explain what other sources of pollution regularly occur in family homes.

Fig. 2.19
Pollution levels indoors, especially in buildings that have been 'tightened' for energy conservation, can be ten times higher than those outdoors.

Radon, a colourless, odourless, naturally occurring radioactive gas, seeps into basements from decaying uranium in soil. A million families in the USA alone are thought to be exposed to radiation levels higher than those faced by uranium miners.

LANDSCAPES OF THE WORLD

## Uranium

The processing of uranium produces radioactive waste, some of which is released into waterbodies due to accidental or intentional discharges. This waste pollutes seas and rivers. The Irish Sea is regarded as the most radioactive sea in the world. This is due to discharges from the Sellafield plant on the western coast of England.

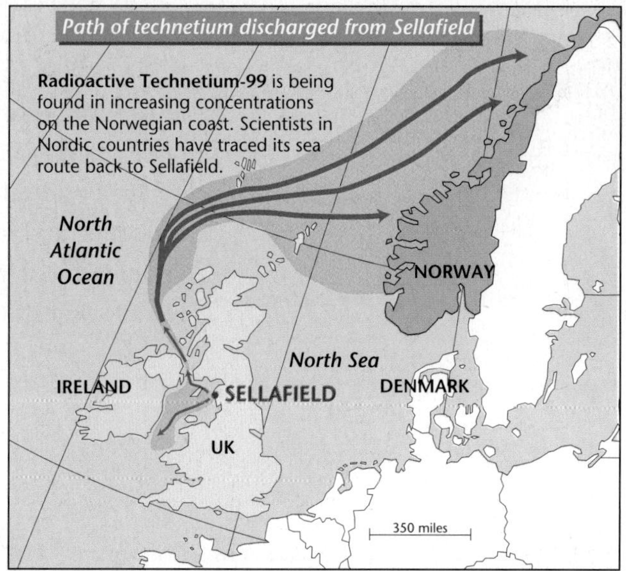
Path of technetium discharged from Sellafield
**Radioactive Technetium-99** is being found in increasing concentrations on the Norwegian coast. Scientists in Nordic countries have traced its sea route back to Sellafield.
North Atlantic Ocean
NORWAY
North Sea
IRELAND
SELLAFIELD
DENMARK
UK
350 miles

**Fig. 2.20**
Discharges from Sellafield

As technology improves, the waste from nuclear power stations changes. Both gas and liquid discharges are released from these plants. At present radioactive technetium (Tc-99) discharges from Sellafield are causing concern to countries such as Norway, Ireland and Iceland, who as a consequence are demanding the closure of the Sellafield plant.

Carefully examine the photograph of the Roadstone quarry at Belgard in Co. Dublin (Fig. 2.21, p. 17). Then answer the following:

1. What type of raw material is mined in this quarry? Use evidence on the photograph to support your answer.
2. What finished products are/could be produced here?
3. The finished products produced here are used as raw materials in what other economic/infrastructural activities? Use these headings: roads, housing, farms.
4. What disadvantages may be experienced by local people/communities near the quarry?
5. Explain the purpose/meaning of truck wash, settling pond, quarry face, stockpiles.
6. Suggest the purpose of the two buildings at the bottom of the photograph.

Fig. 2.21
Roadstone quarry at Belgard in Co. Dublin

# Plate Tectonics

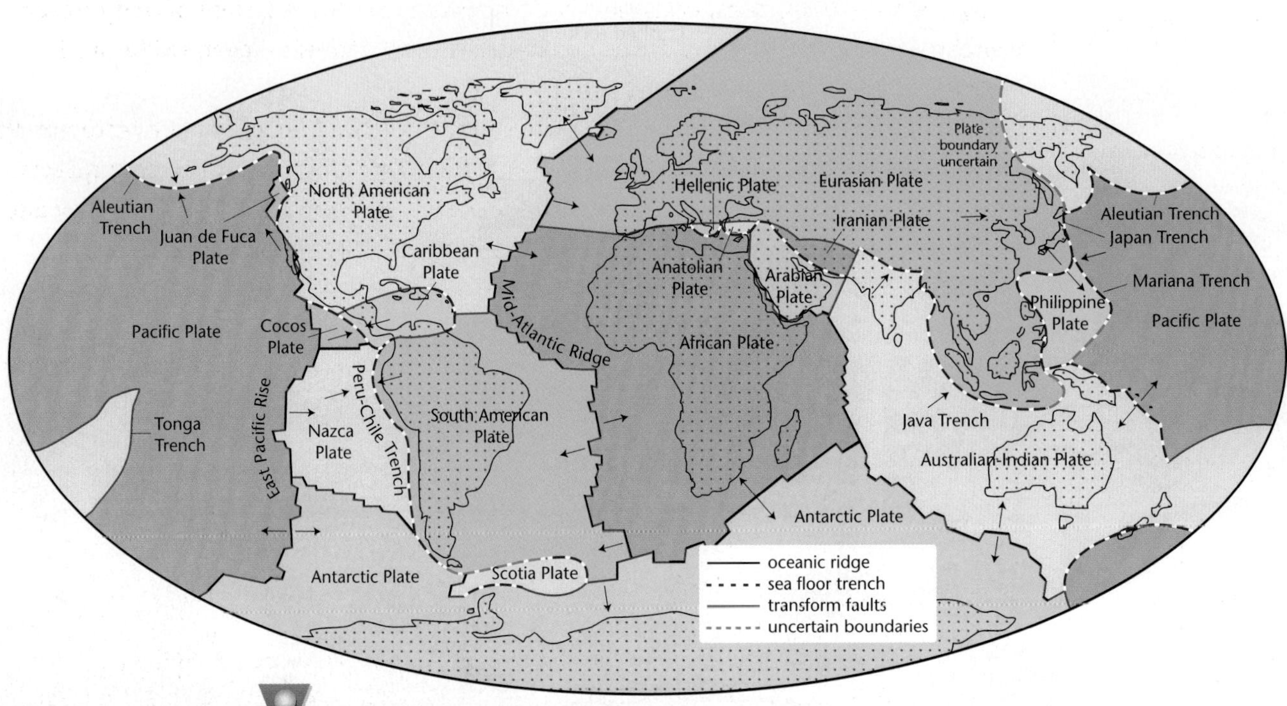


Fig. 3.1
Six large plates of lithosphere and several smaller ones cover the earth's surface and move steadily in the directions shown by the arrows

**tectonic**
Belonging to the structure of the earth's crust and to general changes in it such as folding, faulting, etc.

**ocean trench**
A narrow scar thousands of metres deep along a destructive plate boundary on the ocean floor.

**subducted**
At a destructive plate boundary where two plates come together, the heavier ocean plate sinks and is destroyed beneath the lighter continental plate.

**asthenosphere**
A zone within the upper mantle between the depths of 100 and 700 km in the earth's crust.

**lithosphere**
The rigid outer layer of the earth, including the crust and upper mantle.

**sea floor spreading**
The theory that new oceanic crust is created at the mid-ocean ridges, spreads laterally (sideways), and descends back into the mantle at the deep ocean trenches.

The theory of plate tectonics suggests that the earth's crust is divided into huge slabs of creeping rock that move slowly across the surface of the globe. The slabs or **plates** are driven by enormous **convection currents** (see Fig. 3.4, p. 20) within the earth's core and mantle. **Alfred Wegener**, a German meteorologist, proposed this idea in 1915 in a book called *The Origin of Continents and Oceans*, and today it is accepted as one of the most important theories of earth science. Simply, he proposed that **all the continents of the modern world had drifted apart from an original 'supercontinent', which he called Pangaea** (meaning 'all land'). Others before him, intrigued by the 'jigsaw puzzle' shape of the continents, had suggested a similar hypothesis, but Wegener was the first to support it with evidence, systematically collected and analysed.

Wegener based his proposals on

1. the distribution of identical fossils on continents separated by thousands of miles of ocean
2. the matching shapes of the African and American coastlines
3. the presence of marine sediments on high mountains

**island arc**
A curved chain of volcanic islands near deep ocean trenches.

**mantle**
A layer of hot rock which lies between the earth's core and the crust.

**mid-ocean ridge**
A volcanically active ridge of mountains that runs continuously on the ocean floor between the continents.

He also argued that no evidence existed for most of the temporary land bridges that scientists of the time proposed as an explanation for the migration of human and animal species from one continent to another.

By the mid-1950s, oceanographers had mapped a worldwide system of **deep ocean trenches, island arcs** (curved chains of volcanic islands) and **mid-ocean ridges and mountains** on the sea floor. (See Fig. 3.2.)

They noted that all of these landforms were related to the theory of plate tectonics. For instance, seismologists (scientists who study earthquakes) noted that many earthquakes which occurred deep in the earth's crust were located beneath ocean trenches. They also observed that volcanic activity is concentrated along the **mid-ocean ridges** and **subduction zones** that correspond to plate boundaries.

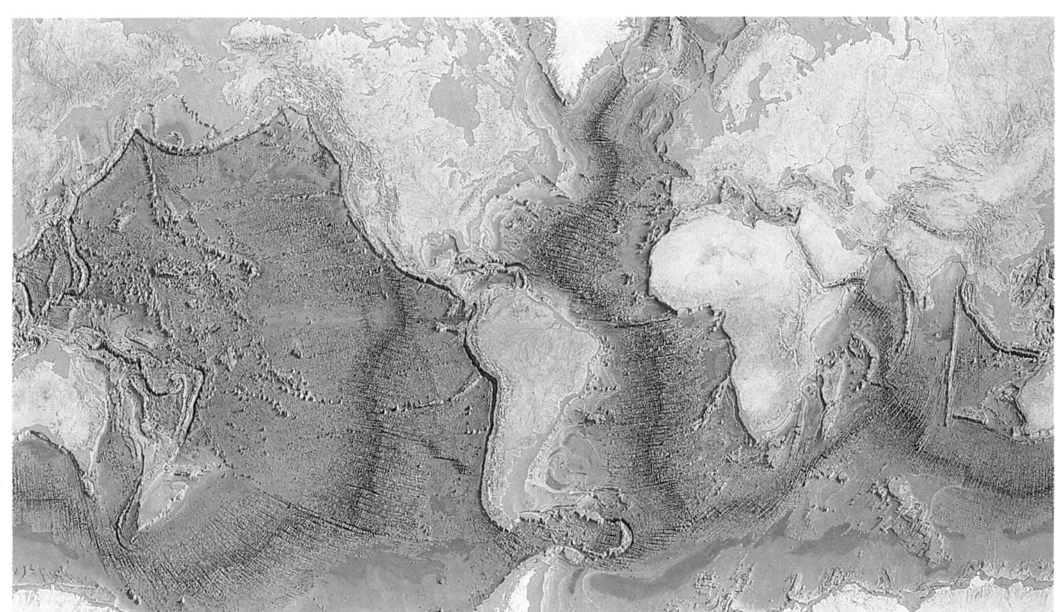

Fig. 3.2
The continents and sea floors of the world

**convection current**
A process which occurs because hot, less dense (lighter) materials rise and are replaced by cold, heavier, down-flowing materials to create a current.

**subduction**
When two plates of the earth's crust come together, the heavier ocean plate sinks and is destroyed beneath the lighter continental plate.

## Processes of Plate Tectonics

The crust of the earth can be compared to the hard shell of an egg. However, unlike an egg, the entire crust has many cracks separating the various pieces, which are called **plates**.

These huge plates of rock **float** on heavier, semi-molten rock called the **mantle**. The **plates move** relative to each other, carried along by **convection currents** within the mantle. To us, this movement is very slow but in terms of geological time it is fast. Over time, they separate, collide and in some places slide past each other. These movements cause folding, volcanic activity and earthquakes. This entire process repeats itself in an endless cycle over hundreds of millions of years.

# Convection Currents within the Earth

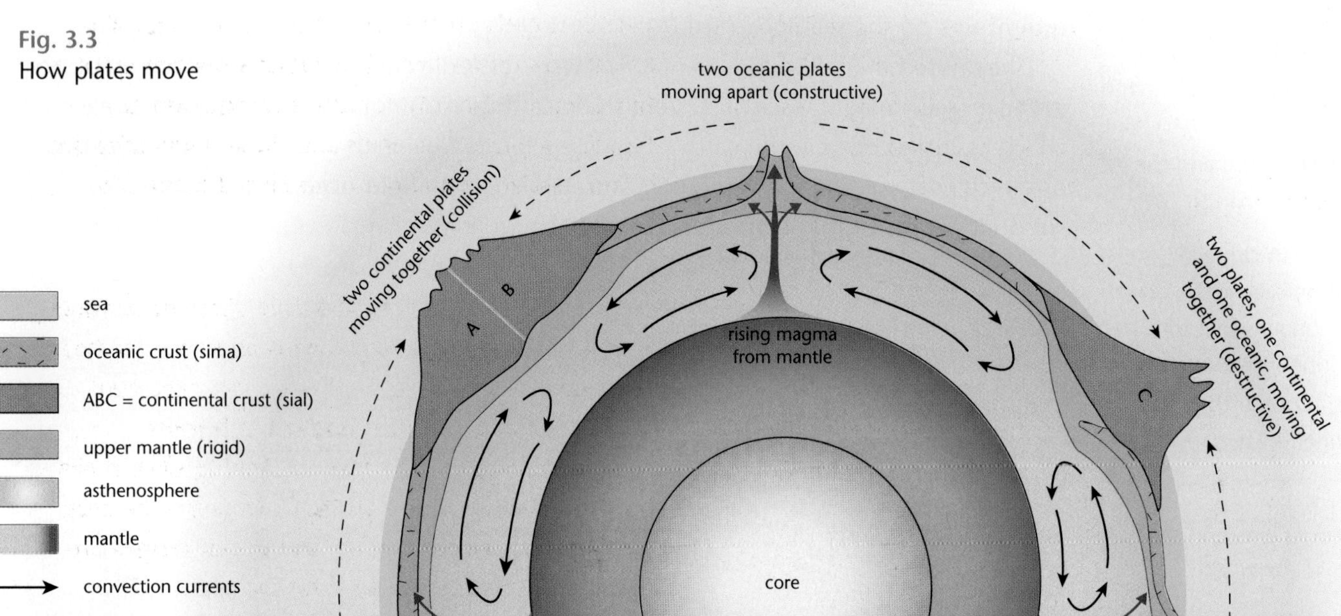


Fig. 3.3
How plates move

| | Continental crust (sial) | Oceanic crust (sima) |
|---|---|---|
| **Thickness** | 35–40 km on average, reaching 60–70 km under mountain chains | 6–10 km on average |
| **Age of rocks** | very old, mainly over 1,500 million years | very young, mainly under 200 million years |
| **Weight of rocks** | lighter, with an average density of 2.6 | heavier, with an average density of 3.0 |
| **Nature of rocks** | light in colour; many contain silica and aluminium; numerous types, granite is the most common | dark in colour; many contain silica and magnesium; few types, mainly basalt |

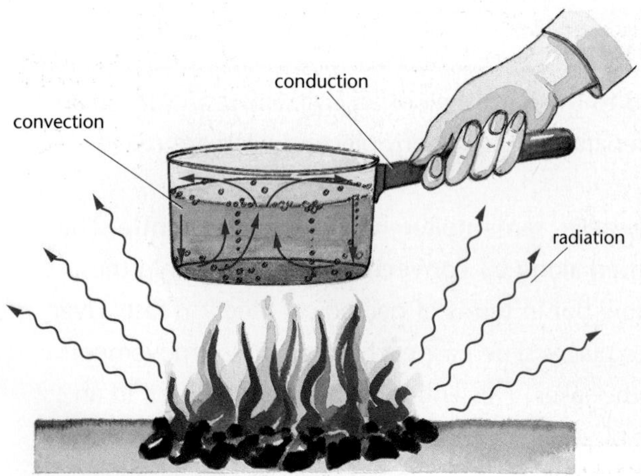


Fig. 3.4
Illustration of conduction, convection and radiation

Water that is heated in a saucepan expands and rises. As it rises, it starts to cool, flows sideways and sinks, eventually to be reheated and to pass again through the convection cell.

Though much slower than convection in a saucepan, the principle is the same here. Hot rock rises slowly from deep inside the earth. It cools, flows sideways and sinks. The rising hot rock and sideways flow are believed to be the factors that control the positions of ocean basins and continents.

## Tectonic Plate Movements on the Earth's Surface

### Where Plates Slide Past Each Other

In some places, convection currents cause plates to slide past each other horizontally along a fault-line called a transform fault. This process **neither creates nor destroys** the earth's crust. **Earthquakes** are associated with these areas and these boundaries are called **passive boundaries**. See Figs. 3.5 and 3.6.

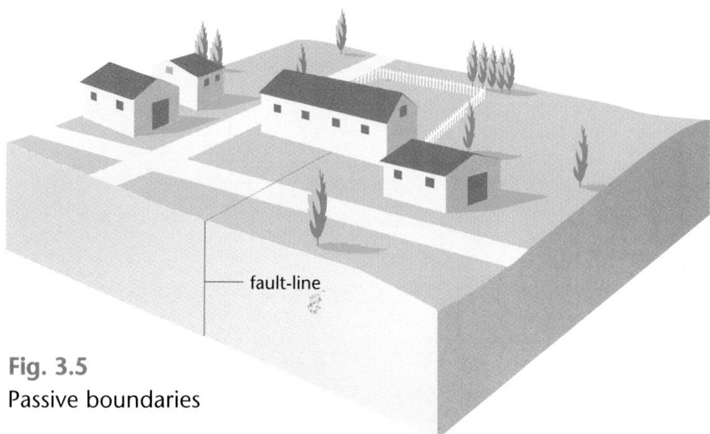
fault-line

Fig. 3.5
Passive boundaries

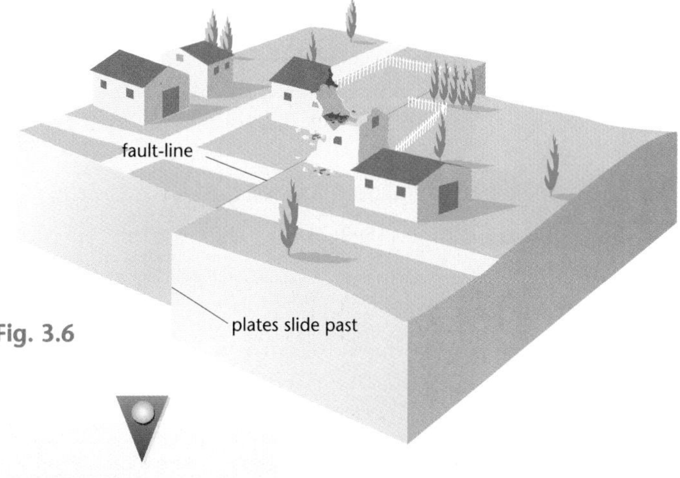
fault-line
plates slide past

Fig. 3.6

### Where Plates Meet

The plates of the earth's crust have different densities (weights). Some, generally the ocean floors, are heavy while continents are light. Where they meet, **the heavier ocean crust dips under the lighter continental plates**. This heavier dipping plate melts as it slides down into the hot mantle beneath the crust. **Earthquakes, volcanoes** and **mountain ranges** are associated with these areas. This process is called **subduction** and these areas are called **destructive boundaries**. (See Fig. 3.7.)

**magma**
Melted hot rock within the earth's crust. It is in a plastic-like state.

**mantle plume**
A pipe-shaped volume of hot plastic-like soft rock which rises from the mantle towards the crust.

**earth's crust**
The outer layer of the earth, which is divided into two types of material, sial and sima.

**Note:**
What is subduction?
See definition, p. 19.

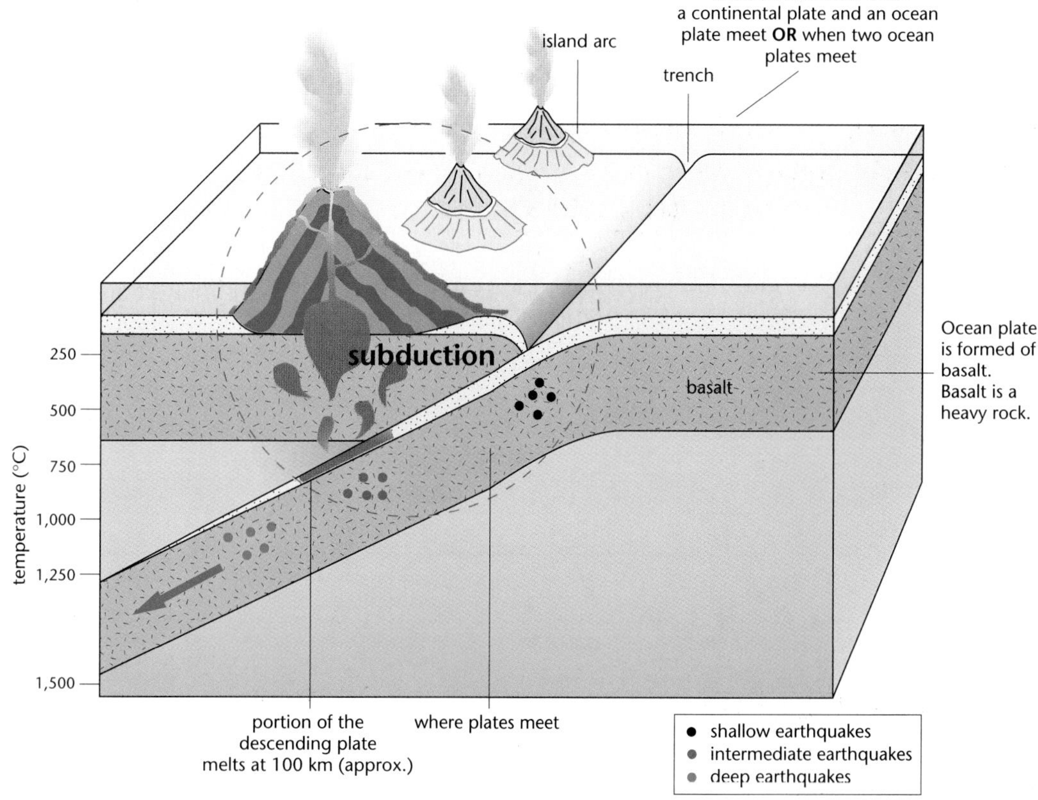
subduction occurs when a continental plate and an ocean plate meet OR when two ocean plates meet
island arc
trench
subduction
basalt
Ocean plate is formed of basalt. Basalt is a heavy rock.
temperature (°C)
250
500
750
1,000
1,250
1,500
portion of the descending plate melts at 100 km (approx.)
where plates meet
shallow earthquakes
intermediate earthquakes
deep earthquakes

Fig. 3.7
Subduction zones

## Where Plates Separate

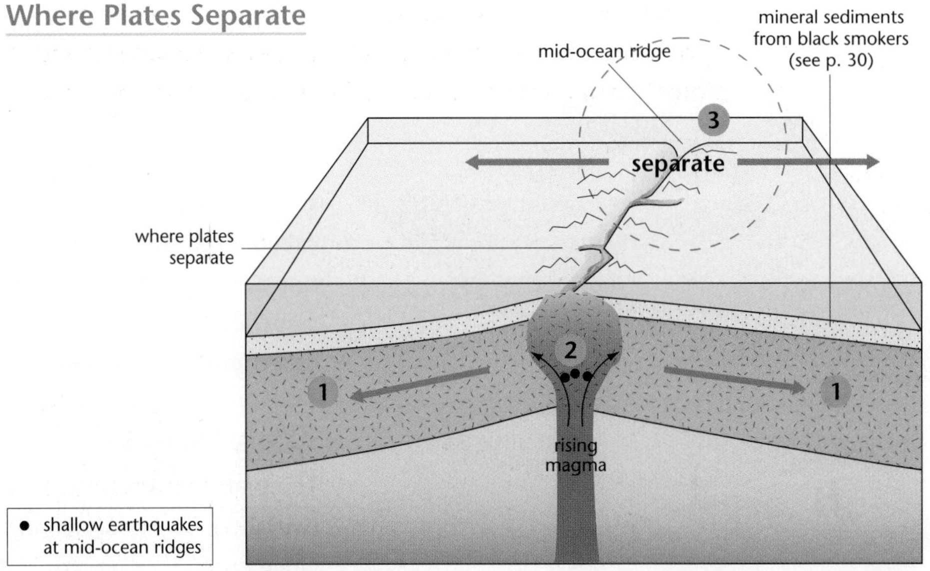


Fig. 3.8
Zones of divergence

Where plates separate, hot liquid **magma** rises between the plates at the earth's surface and creates new igneous rock. The plates separate because of some or all of the following three processes:

1. convection currents carry the plates away from each other
2. the rising magma pushes the plates apart
3. the plates are pulled apart because the older end sags and contracts as it cools

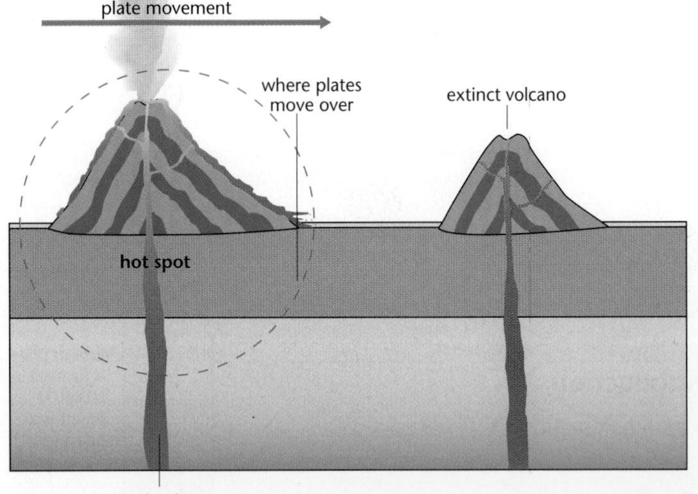


Fig. 3.9
Hot spots

**Earthquakes**, **volcanoes** and submerged **mountain ranges** are associated with these zones of divergence. These areas are called **constructive plate boundaries**.

### Where Plates Pass Over

Hot spots are places on the earth's surface where magma plumes rise from the earth's core or mantle. These confined areas have very active volcanoes with basic (fluid) lavas (not so dangerous).

As plates pass over these hot spots, those 'spots' directly above the **magma plumes** become active. As they move, this activity stops.

## The Theory of Continental Drift

The theory of continental drift suggests that the continents have moved great distances on the earth's surface and are still moving today. According to the theory, the continents once formed part of a single landmass, called Pangaea, which was surrounded by the world's single ocean, called Panthalassa. About 200 million years ago, Pangaea began to break apart. It split into two landmasses, called Gondwanaland and Laurasia.

Gondwanaland then broke apart, forming Africa, Antarctica, Australia, South America and North America. The formation of the present continents and their drifting into their present positions took place gradually over millions of years.

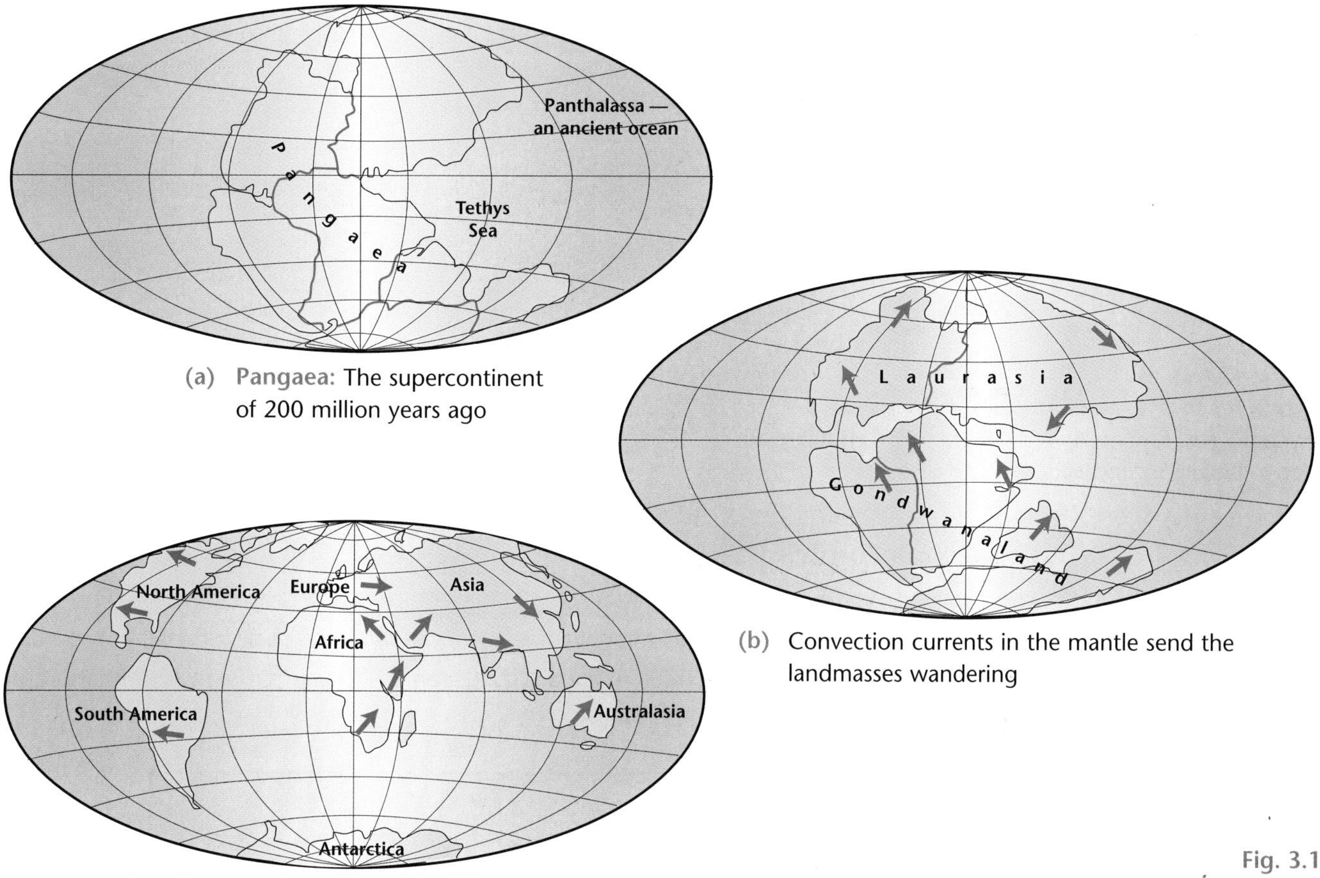

(a) Pangaea: The supercontinent of 200 million years ago

(b) Convection currents in the mantle send the landmasses wandering

(c) Tomorrow's world — 50 million years hence

Fig. 3.10

**Precambrian time**
All geological time before 570 million years ago — before fossils and life.

Although Pangaea split up 200 million years ago, the continents were moving much earlier; Pangaea itself was formed by the collision of many small continents. Recent work shows that continents have been in motion for the past two billion years (some geologists say four billion), well back into **Precambrian** time. For half or more of earth's history, the continents appear to have collided, welded together, then split and drifted apart, only to collide again, over and over, in an endless, slow dance.

## Proofs of Continental Drift

### 1. Matching Mountain Range across the North Atlantic

Geological studies of ancient mountain systems show a connection between the continents. These studies suggest that the Appalachian Mountains of the eastern United States extend through Newfoundland. The Appalachians may have been connected to the Caledonian mountain system, which runs through the north of Ireland, Scotland and Scandinavia.

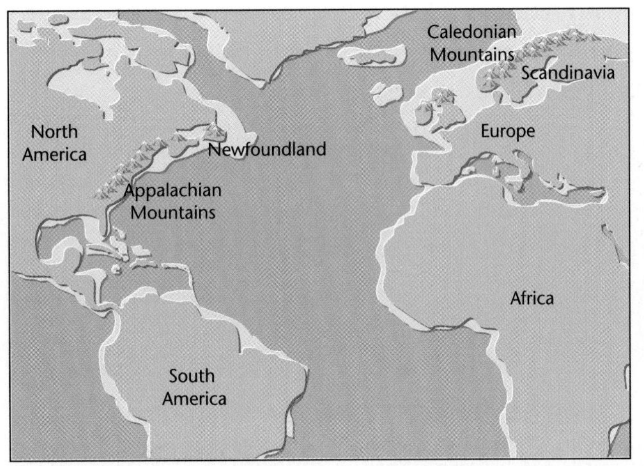


Fig. 3.11
The Appalachian Mountains trend along the eastern edge of North America and disappear off the coast of Newfoundland. Mountains of similar age and structure are found in Britain, Ireland and Scandinavia.

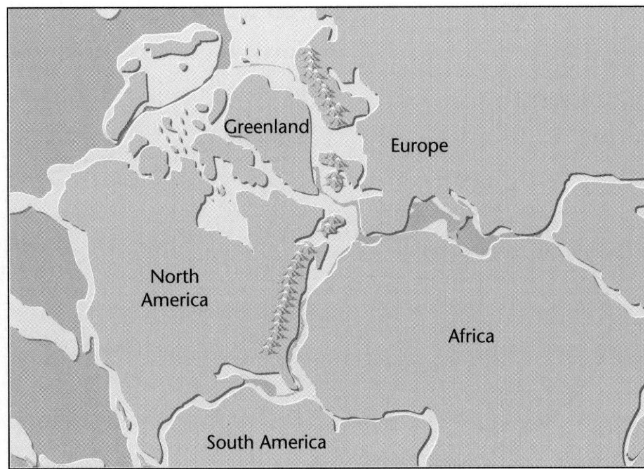


Fig. 3.12
When these landmasses are placed in their pre-drift locations, these ancient mountain chains form a nearly continuous belt. These folded mountain belts formed roughly 300 million years ago as the landmasses collided during the formation of the supercontinent of Pangaea.

## 2 Matching Fossils

Other evidence for such a connection comes from palaeontologists (scientists who study fossils). They have found fossils of similar land mammals in the rocks of Asia, Europe and North America that are 100 million years old. It seems unlikely that similar animals could develop on widely separated continents.

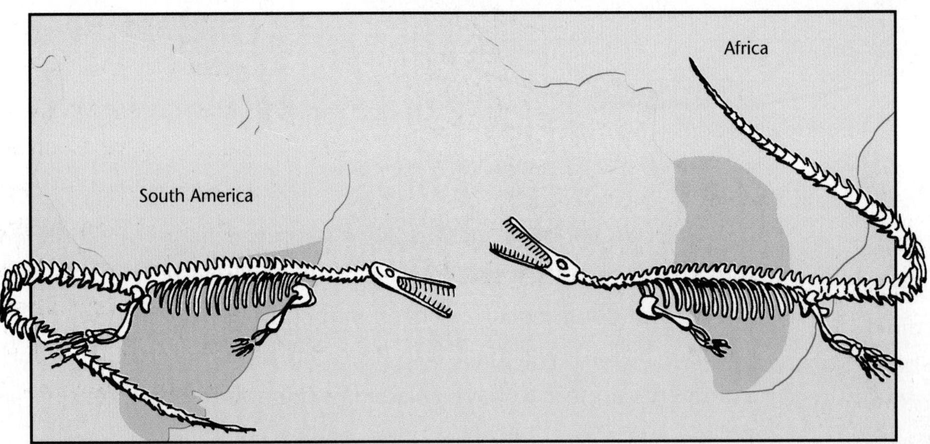


Fig. 3.13
Fossils of *Mesosaurus* have been found on both sides of the South Atlantic and nowhere else in the world. Fossil remains of this and other organisms on the continents of Africa and South America appear to link these landmasses before they separated.

## Age of the Ocean Floor

If Europe and North America are moving away from each other, the distance measured between the two continents will be greater from one year to the next. Only recently have we developed the technology (such as satellite measurements) to measure accurately distances between continents. So in the 1960s other methods of proof had to be tried. One prediction was that new rock at the mid-ocean ridge will be the youngest on the ocean floor and the rocks of the oceanic crust will be progressively older the further they are from the crest of the ridge.

**asthenosphere**
A zone within the upper mantle between the depths of 100 and 700 km in the earth's crust.

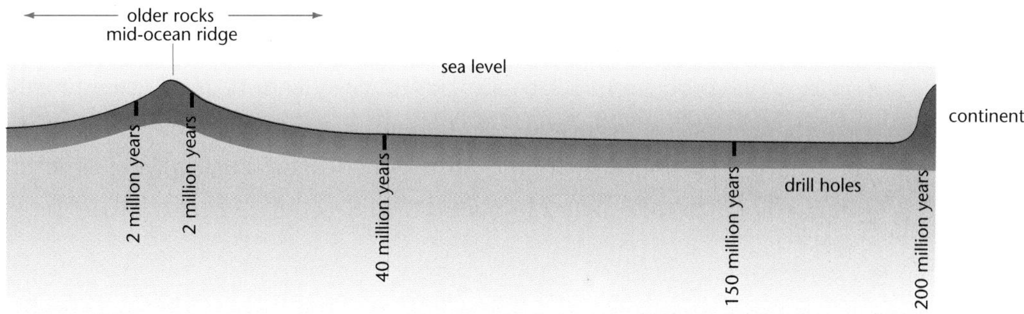


Fig. 3.14

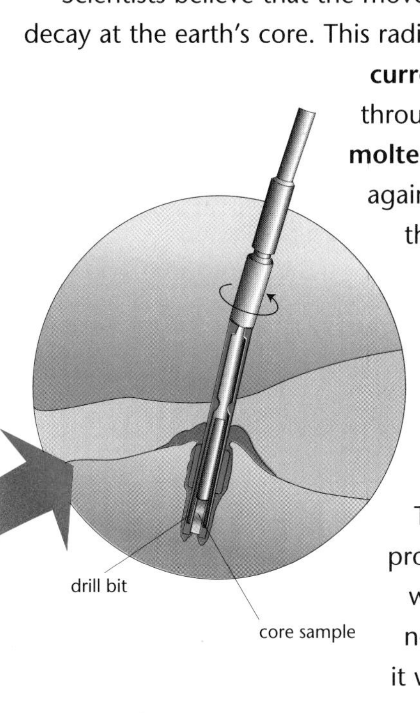


Fig. 3.15
Drilling rock samples on the ocean floor

Holes were drilled in the deep sea floor from a specially designed ship. Rocks and sediment were collected from these holes and the ages of these materials were found. As predicted, the youngest sea floor is near the mid-oceanic ridges, whereas the oldest sea floor (up to about 200 million years old) is furthest from the ridges.

Scientists believe that the movements of the crustal plates are due to radioactive decay at the earth's core. This radioactive process produces heat, forming **convection currents** that rise from the core. These currents circulate through the soft, hot material of the mantle. They carry **molten rock** up from the **asthenosphere**, pushing it against the **crust** and forcing it into large cracks where the plates separate. Molten magma wells up along the length of these cracks. As the molten rock meets the cold ocean water, it hardens, forming new rock and high mountain ridges. This continuous addition of new rock along the ridges pushes the ocean floor and the continents away from them, forcing the sea floor to spread apart. This is known as a **spreading zone**, and apparently provides the power that drives the plates. In other words, the convection currents in the rock carry the newly formed crustal plate away from the ridge as if it were riding on a conveyor belt.

## Ancient Ice Age

During the late Carboniferous period, about 300 million years ago, a continental ice sheet covered parts of South America, southern Africa, India and southern Australia. (See Fig. 3.16a.) However, if 300 million years ago the continents had been in the positions they occupy today, an ice sheet would have had to cover all the southern oceans and in places would have had to cross the equator. Such a huge ice sheet could mean only that the

world climate was exceedingly cold. Yet at this time no evidence of glaciation of this type exists in North America and Eurasia in the northern hemisphere. Here warm climates produced thick coal deposits. This dilemma is explained neatly by continental drift: 300 million years ago, the regions covered by ice were close to the south pole. (See Fig. 3.16b.) At that time North America and Eurasia were close to the equator. No landmass covered the north pole so there was no northern ice sheet.

Fig. 3.16
Ancient ice age

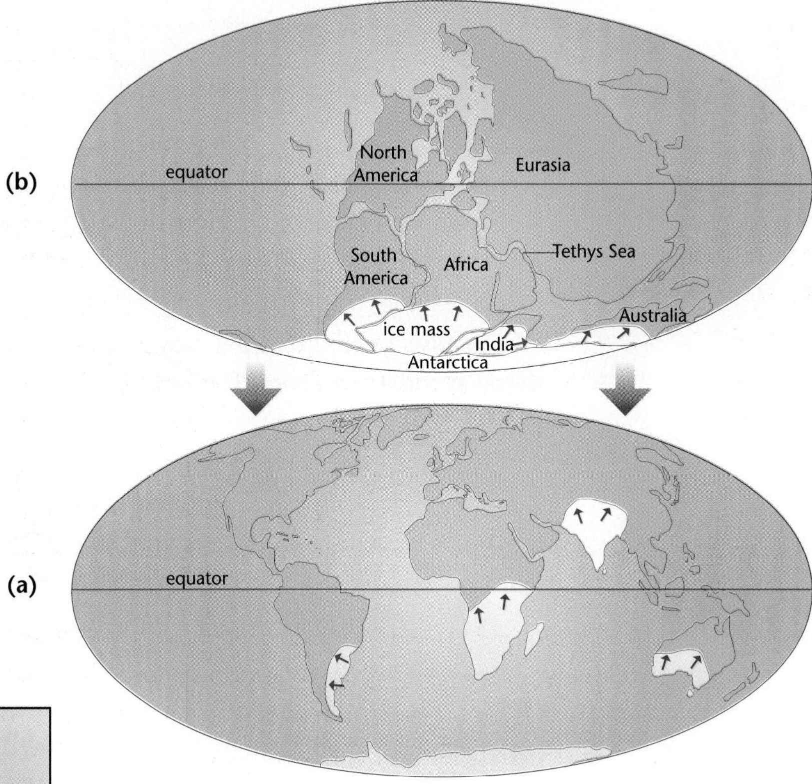


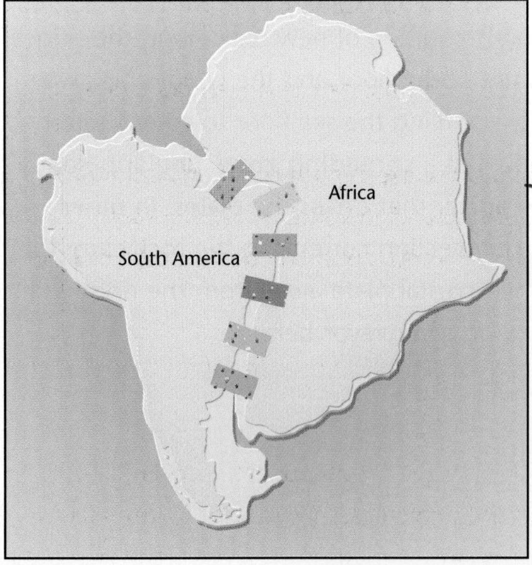


Fig. 3.17
This shows the **best fit** of South America and Africa along the continental slope at a **depth of 500 fathoms** (about 900 metres). The areas where continental rocks match appear in different colours.

## Matching Continents

Convincing evidence for continental drift came from exact matches between now separated continents. If continents are fitted together like pieces of a jigsaw puzzle, the 'picture' should match from piece to piece.

When the edges of the **continental shelves** of Africa and South America are **fitted together**, the rock types match on each side. Areas of distinctive rock extend out to sea to the edge of the continental shelf off the coast of Africa. The identical rock types are found in precisely the right position on the shore of South America. (See Fig. 3.17.) The ages of rocks also match between these continents. For example, detailed matches have been made between rocks in Brazil, in South America, and Gabon in Africa. These rocks are similar in type, structure, sequence, fossils, ages and texture.

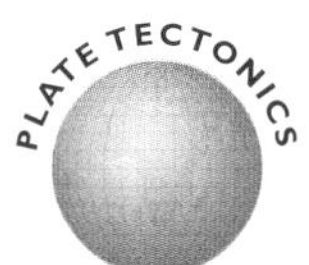


## Fossil Magnetism

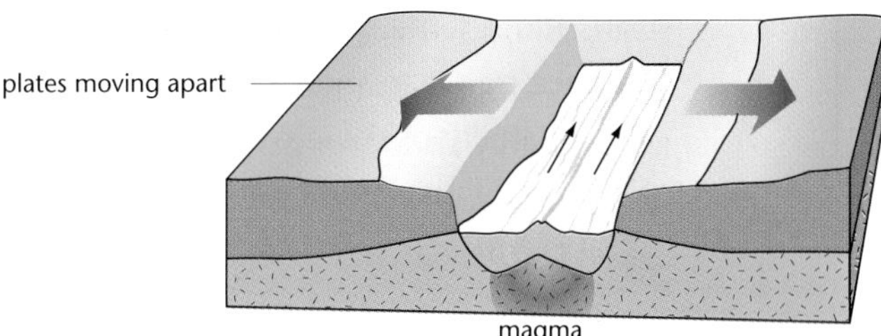


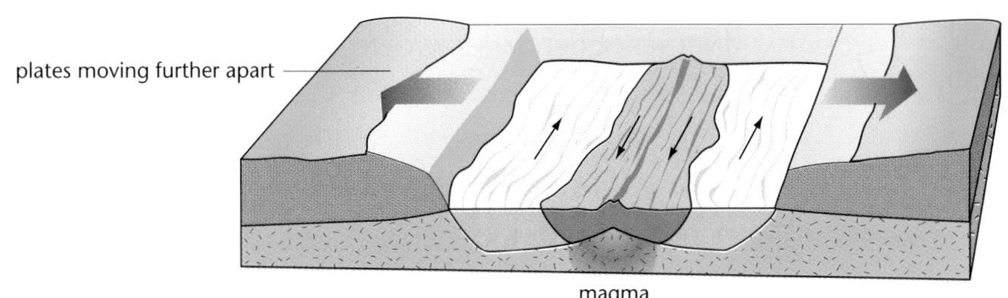


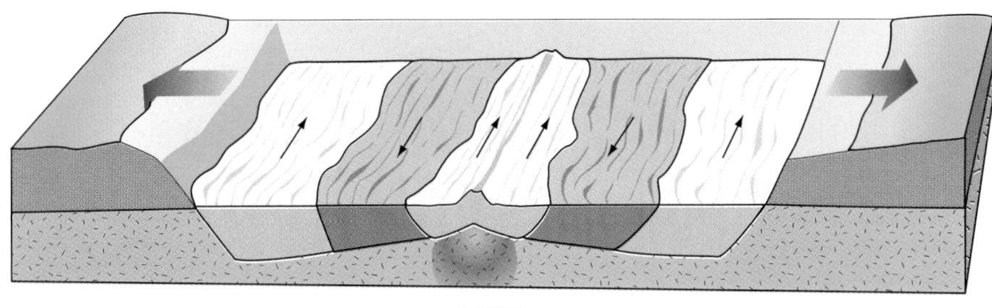


Fig. 3.18
Magnetic reversal

**magnetic reversal**
A switch in the direction of the earth's magnetic field such that a compass needle that today points north would, at the time of reversal, have pointed south.

In 1963 British scientists used old magnetic measurements near the ocean ridges to prove the sea floor spreading theory. They based their experiments on two facts.

1. Magnetic particles in the sea floor rocks recorded the direction of the earth's magnetic field when the rock hardened.
2. The direction of the field reversed itself from time to time as the ocean floor was formed.

If the sea floor spreads, the pattern of normal and reversed magnetism should match on both sides of the ridge. Experiments found these matching patterns. Today many scientists accept the basic idea that the continents have moved in the past as part of large rigid plates.

## Measurement by Satellite and Radio Telescopes

### Satellite Laser Ranging (SLR)

This system uses ground-based stations to bounce laser pulses off specific satellites. Precision timing of returning pulses allows scientists to calculate the precise locations of the ground stations in relation to each other over time.

**Terms to remember**
**SLR:** the Satellite Laser Ranging System.
**VLBI:** the Very Long Baseline Interferometry.

### Radio Telescopes (VLBI)

Large radio telescopes at ground stations send radio signals to distant objects called quasars which lie billions of light years from earth. As these objects are so far away from the earth they act as stationary reference points. The millisecond differences in the arrival

**sial (light)**
This rock forms the continents. It is composed of silica and alumina.

**sima (heavy)**
This rock forms the ocean floor. It is composed of silica, magnesium and iron.

time of the same signal at different ground stations provide a means of measuring the distance between them.

The ability of these two systems to measure with unprecedented accuracy proves that plate motion has been detected.

## Close-up on Plate Boundaries

There are three types of plate boundaries:

1. divergent boundaries, where the plates are created and spread away from a mid-ocean ridge
2. transform boundaries, where the plates slide by one another horizontally
3. convergent boundaries, where the plates collide, causing one to descend into the mantle beneath the deep sea trench

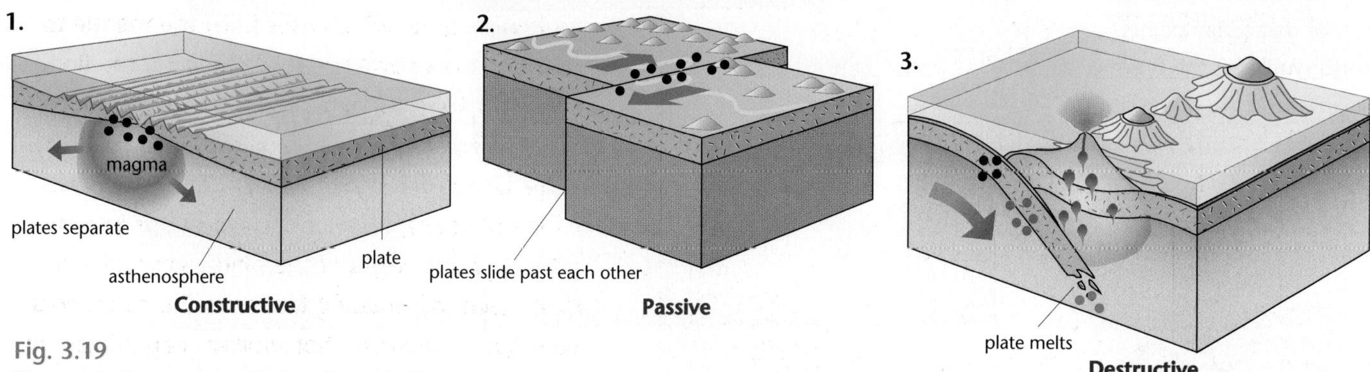


Fig. 3.19
There are three types of plate boundaries:
1. constructive boundaries, i.e. mid-ocean ridges (divergent)
2. passive boundaries, i.e. transform faults (conservative)
3. destructive boundaries, i.e. (a) oceanic-oceanic, (b) oceanic-continental, (c) continental-continental

| Type of plate boundary | Description of changes | Examples |
|---|---|---|
| A **Constructive margins** (spreading or diverging plates) (divergent) | two plates move away from each other; new oceanic crust appears forming mid-ocean ridges with volcanoes | Mid-Atlantic Ridge (Americas moving away from Eurasia-Africa)<br>East Pacific Rise (Nazca and Pacific Plates moving apart) |
| B **Destructive margins** (subduction zones) | oceanic crust moves towards continental crust but being heavier sinks and is destroyed, forming deep sea trenches and island arcs with volcanoes | Nazca sinks under South American Plate (Andes)<br>Juan de Fuca sinks under North American Plate (Rockies)<br>island arcs of the West Indies and Japan |
| **Collision zones** (convergent) | two continental crusts collide and, as neither can sink, are forced up into fold mountains | Indian collided with Eurasian forming Himalayas<br>African collided with Eurasian forming Alps |
| C **Conservative or passive margins** (transform faults) | two plates move sideways past each other — land is neither formed nor destroyed | San Andreas Fault in California |

**Simplified and labelled diagrams**

Practise and memorise these sketches before you begin this.

1.

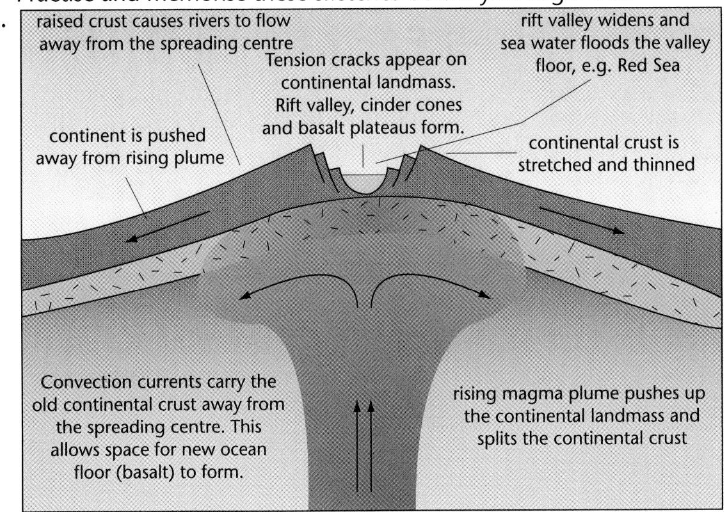
raised crust causes rivers to flow away from the spreading centre
rift valley widens and sea water floods the valley floor, e.g. Red Sea
Tension cracks appear on continental landmass. Rift valley, cinder cones and basalt plateaus form.
continent is pushed away from rising plume
continental crust is stretched and thinned
Convection currents carry the old continental crust away from the spreading centre. This allows space for new ocean floor (basalt) to form.
rising magma plume pushes up the continental landmass and splits the continental crust

2.

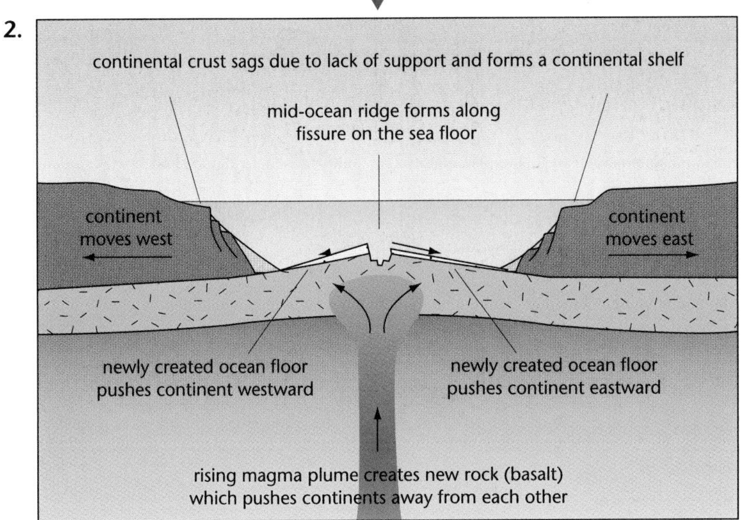
continental crust sags due to lack of support and forms a continental shelf
mid-ocean ridge forms along fissure on the sea floor
continent moves west
continent moves east
newly created ocean floor pushes continent westward
newly created ocean floor pushes continent eastward
rising magma plume creates new rock (basalt) which pushes continents away from each other

3.

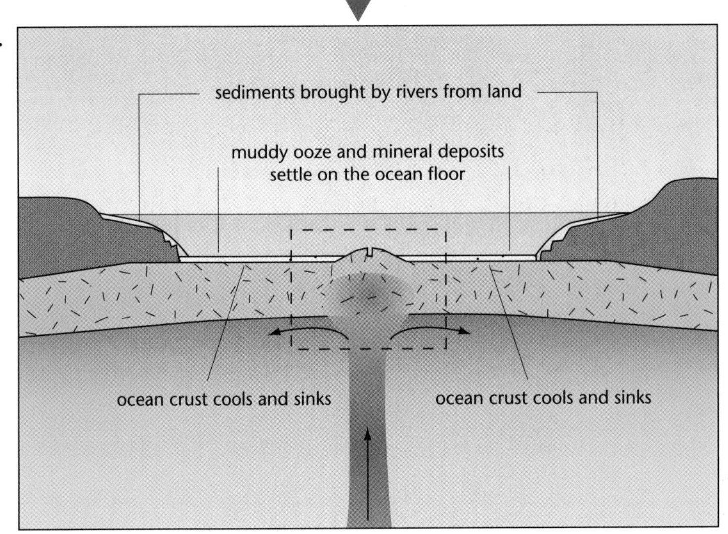
sediments brought by rivers from land
muddy ooze and mineral deposits settle on the ocean floor
ocean crust cools and sinks
ocean crust cools and sinks

Fig. 3.20

Stages in the formation of a new sea, a new ocean floor and a mid-ocean ridge

## Constructive Plate Boundaries

### The Process of Ocean and Mid-ocean Ridge Creation

#### Formation

1. A hot mushroom-shaped 'plume' of magma rises towards the surface, from the mantle, and pushes the continental plate upward. The crust is stretched and thinned. Numerous faults break the crust along the line of separation. Some crust drops down to form a rift valley. This process is called **rifting**.

   The faults (cracks) act as pathways for molten magma, which rises from the mantle to form volcanoes and basalt plateaus on the floor of the rift valley. Shallow earthquakes also occur.

   An example of a spreading boundary at this stage is the **African Rift Valley** in eastern Africa.

2. As the plates separate, sea water floods into the rift valley. The edges of the plates sag and slip down (slump), creating faults. As the continents continue to separate, hot molten magma rises in the middle from the mantle and cools rapidly on meeting the cold sea water to form basalt. This basalt rock forms the new sea floor (oceanic crust), **sima**, between the two continents.

3. The plates continue to separate, widening the sea. The light, hot rock of the magma chamber uplifts the new crust and creates a mid-ocean ridge along the plate boundary.

   As the newly formed rock at the mid-ocean ridge moves away from the plate boundary, it cools, contracts and sinks, thus deepening the ocean away from the ridge.

   The **edges of the continents** also **subside** as their **support is removed** and as the hot rock beneath them cools. Subsidence continues until the edges of the continents are under water. Many layers of sediment derived from river and coastal erosion accumulate on these continental margins to form shallow **continental shelves**.

   The Atlantic Ocean is at present at this stage of separation.

## Natural Processes at Mid-ocean Ridges on the Ocean Floor

4. Molten rock (magma) is forced up through cracks to form new basalt rock.
5. This magma cools quickly on contact with the cold ocean water and forms pillow lavas (like pillow shapes) of basalt.
6. **Black Smokers**
   This is the name given to hot liquids and gases which issue from fractures on the ocean floor at mid-ocean ridges. Black smokers **emit** dissolved **minerals**, which **settle** on and in the **ocean floor** to form mineral ores.

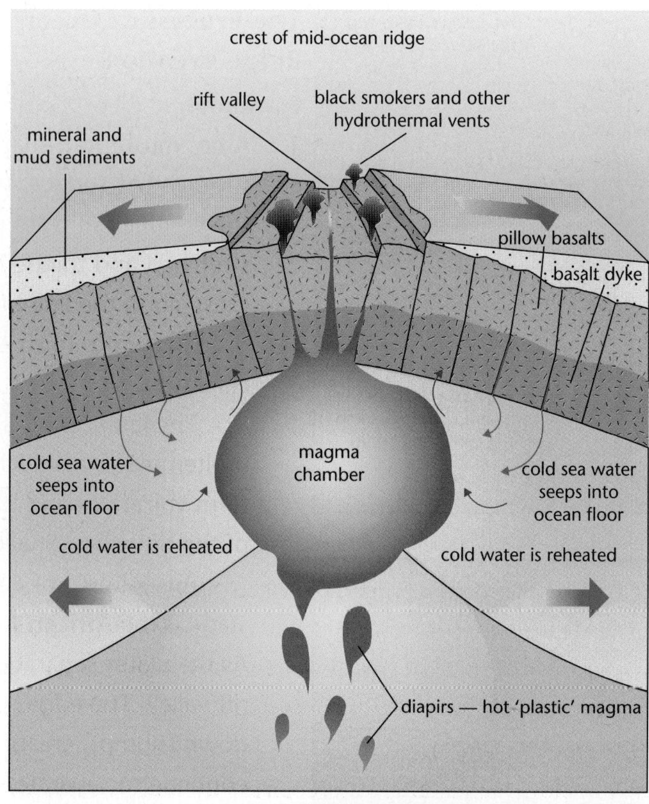


3. Hot liquid magma is lighter than surrounding cold rock. This magma pushes up the ocean floor to form a mid-ocean mountain range.
2. As magma rises, its pressure reduces and this allows it to become liquid.
1. Small hot 'plastic' masses of magma, called diapirs, melt their way towards the surface of the ocean floor. As this magma rises, it expands, due to reduced confining pressure. (See 'Confining Pressure', p. 44.)

Fig. 3.21

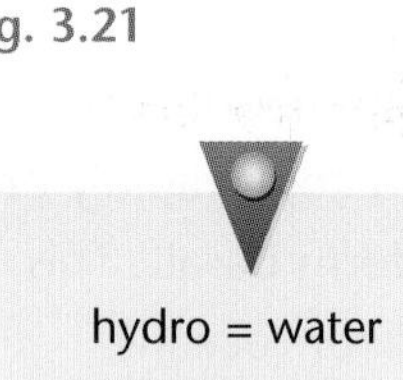

hydro = water
thermal = heat

Fig. 3.22

## Hydrothermal Vents on the Ocean Floor

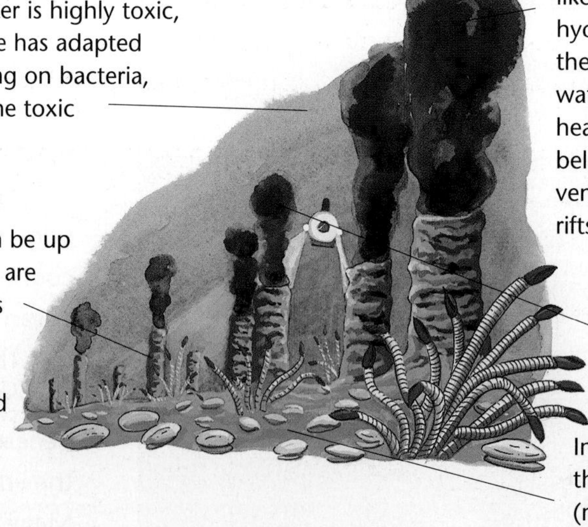

**Poisonous jets**
The sulphurous water is highly toxic, but vent-site wildlife has adapted uniquely by surviving on bacteria, which flourish on the toxic compounds.

**Moulded mounds**
Smokers, which can be up to 10 m (33 ft) tall, are built up by minerals dissolved in the gushing water. The minerals solidify and accumulate as the water cools.

**'Black Smokers'**
These strange features perch like chimneys on top of some hydrothermal vents, cracks in the ocean floor where sea water gushes up, having been heated by hot volcanic rocks below. First found in 1977, vents occur close to volcanic rifts in the seabed.

Super-hot water cools as it emerges.

In and around the tube-worm thickets lie giant molluscs (mussels and clams) up to 30 cm (1 ft) long.

**mid-ocean ridge**
A mid-ocean ridge is not straight. Transform faults cause an offset pattern.

**hydrothermal vent**
This is an opening on the earth's surface, either on land or on the sea floor, through which super-hot water with dissolved minerals gushes from the mantle.

### Minerals and Black Smokers

- Mineral particles 'settle out' and form part of the 'mud' sediment on the ocean floor.
- Mineral deposits are also deposited in cracks in the newly formed rock to form mineral veins.

## Passive Boundaries

### Transform Boundaries (Faults)

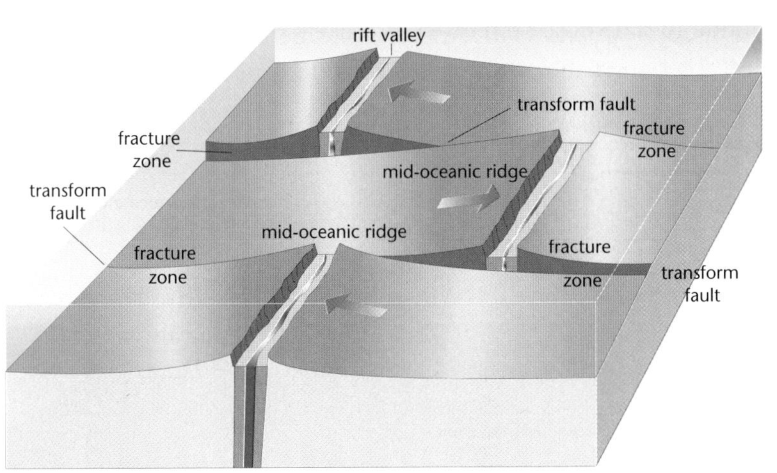


Fig. 3.23
Transform faults allow the earth to retain its spherical shape

Fig. 3.24
A mid-ocean ridge and numerous transform faults in the Indian Ocean

Transform boundaries (faults) occur where plates slide past each other without the creation or destruction of crust. The line along which the plates slide is called a fault-line.

Transform faults provide the means by which the ocean crust created at a mid-ocean ridge can be carried to its site of destruction at a converging boundary (subduction zone/deep ocean trench).

Most transform faults are located under the oceans, but a few, including the San Andreas Fault in California, are situated within continents.

### The San Andreas Fault

The San Andreas Fault is the boundary between the North American and Pacific Plates. Both plates are moving in a north-west direction. However, the Pacific Plate moves faster than the North American Plate, giving the illusion that they are moving in opposite directions.

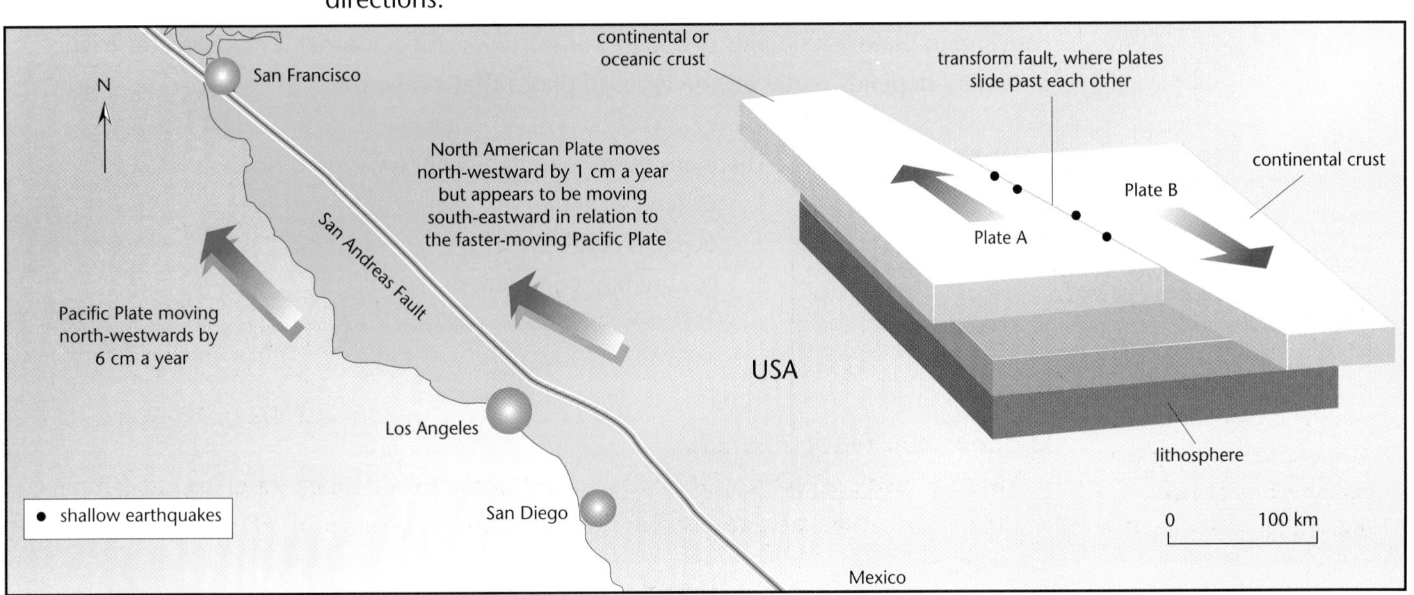


Fig. 3.25
San Andreas Fault — a transform boundary

The Pacific Plate moves about six centimetres each year. Sometimes it jams until pressure builds up, allowing it to jerk forward. The last major movement resulted in the Los Angeles earthquake of 1994. (See p. 64.)

Should these plates continue to slide past each other, it is likely that Los Angeles will eventually be on an island off the Canadian coast.

## Destructive Plate Boundaries

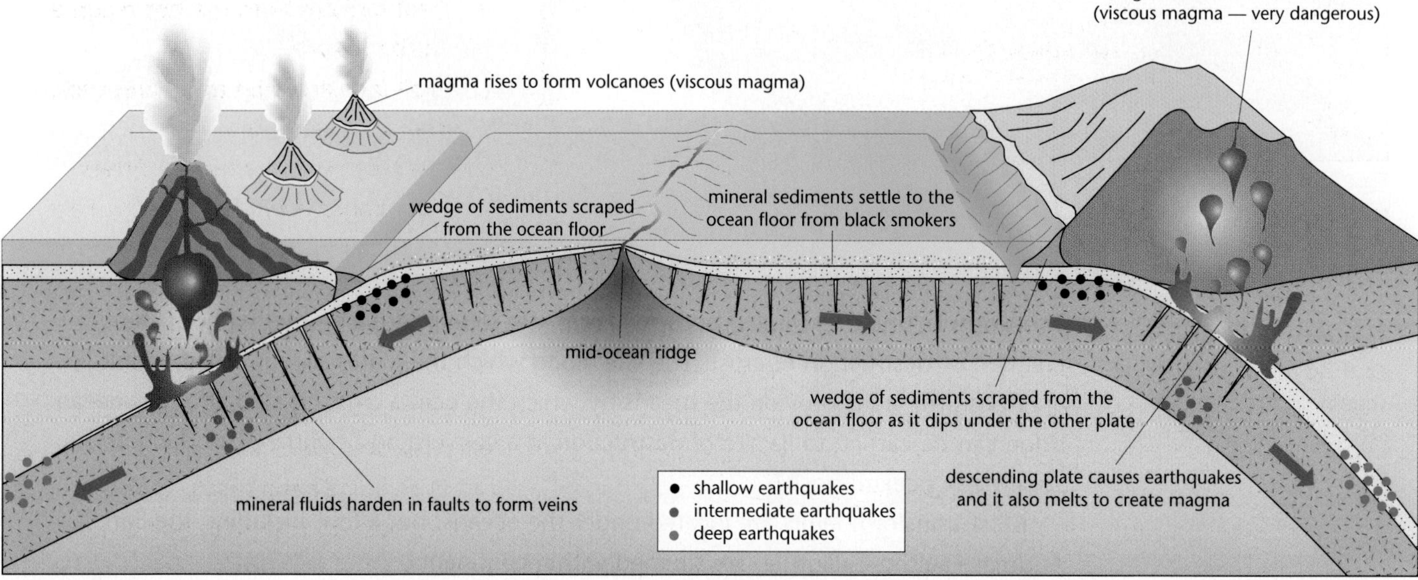


Fig. 3.26
Constructive and destructive plate boundaries

### Converging Boundaries

There is good evidence that the earth is not expanding. So there must be a 'global conveyor belt system' linking zones of creation (diverging plate boundaries) to zones of destruction. Thus a second simultaneous requirement for continental drift is a zone of **subduction**, where a moving plate is pulled down into the mantle and destroyed. At converging plate boundaries two plates move towards each other. The character of the boundary depends partly on the types of plates that converge.

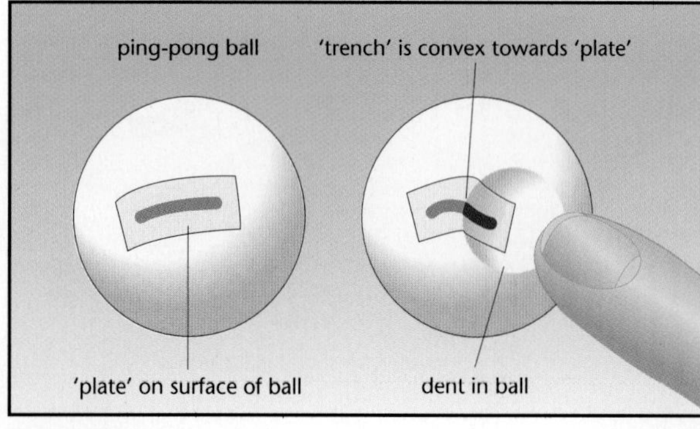


Fig. 3.27
Why ocean trenches and island arcs are curved

There are three main types of converging boundaries:

1. oceanic-oceanic
2. oceanic-continental
3. continental-continental

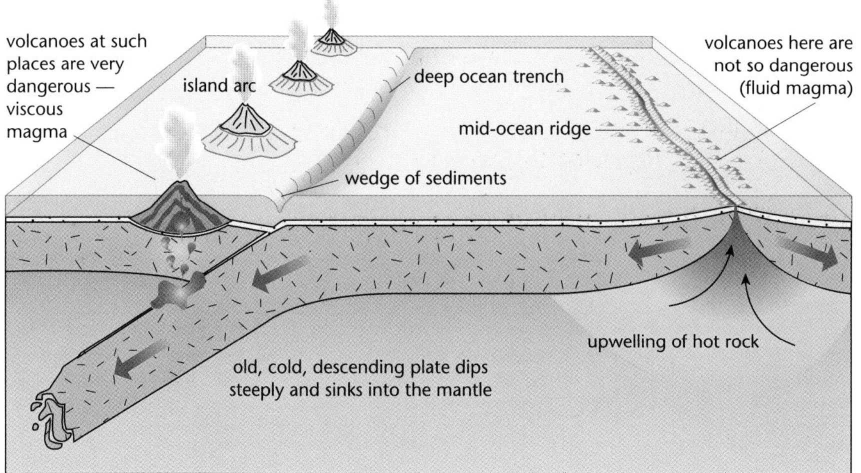
volcanoes at such places are very dangerous — viscous magma
island arc
deep ocean trench
volcanoes here are not so dangerous (fluid magma)
mid-ocean ridge
wedge of sediments
upwelling of hot rock
old, cold, descending plate dips steeply and sinks into the mantle

**Fig. 3.28**
Island arc begins

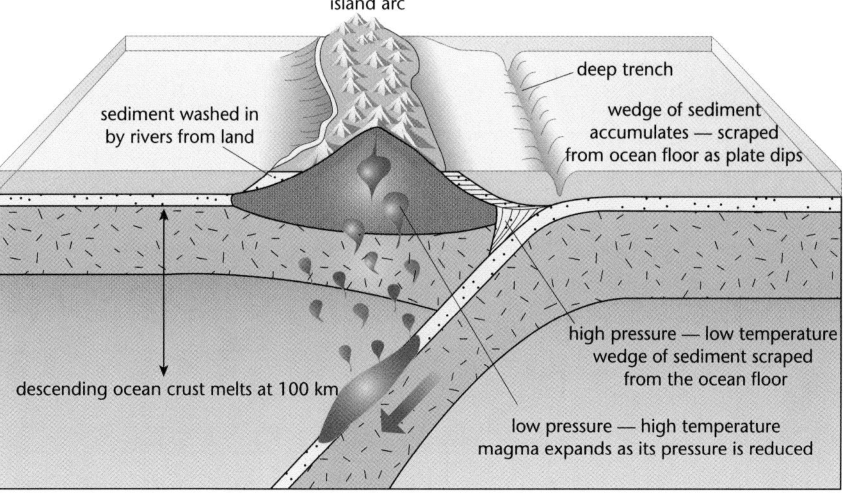
island arc
deep trench
sediment washed in by rivers from land
wedge of sediment accumulates — scraped from ocean floor as plate dips
high pressure — low temperature wedge of sediment scraped from the ocean floor
descending ocean crust melts at 100 km
low pressure — high temperature magma expands as its pressure is reduced

**Fig. 3.29**
Mature island arc

**Ocean-ocean convergence**
Features of this type of plate boundary include:

1. a deep sea trench where the denser plate is subducted
2. a wedge of sediment scraped from the descending plate
3. a volcanic island arc fed by magma created by the subducted (descending) plate

## Ocean-Ocean Convergence

Where two plates capped by oceanic crust converge, one plate subducts (is pulled) under the other. The subducting plate bends downward forming a deep curved ocean trench.

As the subducted plate descends, it melts, because:

- heat radiates from the hot magma in the mantle
- heat increases due to compression (being squeezed)
- heat is created by friction (press your finger down hard on the floor and then push it along the surface)

As the plates meet, the plate of the ocean floor, composed of denser (heavy) rock containing **si**lica and **i**ron and **ma**gnesium (sima), sinks or is pulled under the continental plate of lighter rock containing only **si**lica and **al**umina (sial).

The boundary where the plates meet is marked by a deep, curved trench on the ocean floor. These trenches form the deepest parts of the ocean, e.g. the Mindanao Trench off the Philippine Islands and the Japan Trench off the coast of Japan.

**Case Study:** Japan, An Island Arc

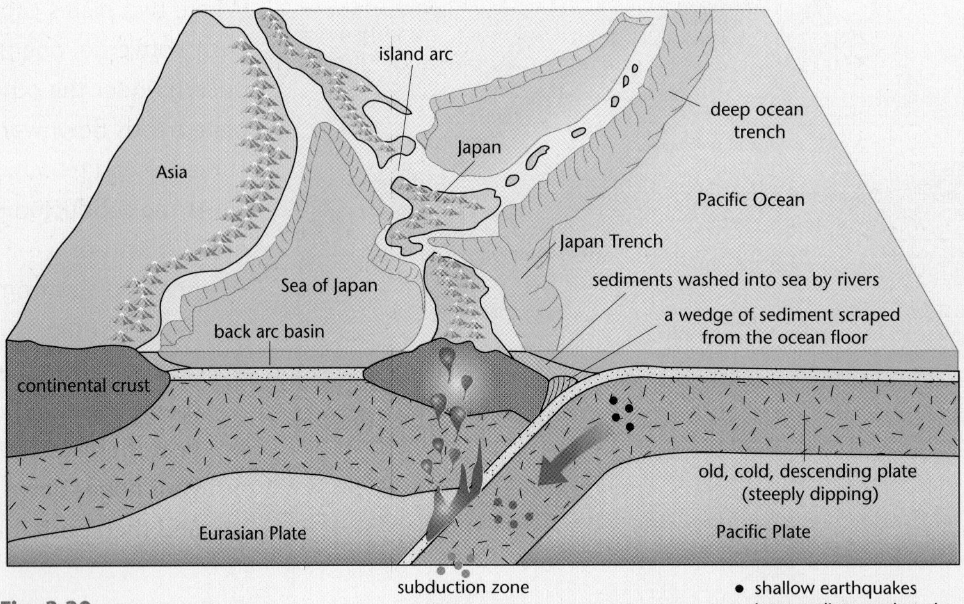


**Fig. 3.30**
Japan, a zone of collision/subduction

As an ocean plate is subducted beneath another, the descending plate melts at a depth of about 100 km. Magma rises and forms volcanic cones on the ocean floor. Dry land eventually emerges from the ocean depths to form an island arc, a curved string of islands parallel to the ocean trench. These volcanoes are **highly explosive** because the magma which creates them is **high in silica (viscous/thick)**. This prevents gases from escaping freely. The **trapped gases expand** as they rise towards the surface. Thus acid lava is created which regularly results in dramatic and sometimes **deadly violent eruptions**. See pp. 45 and 55.

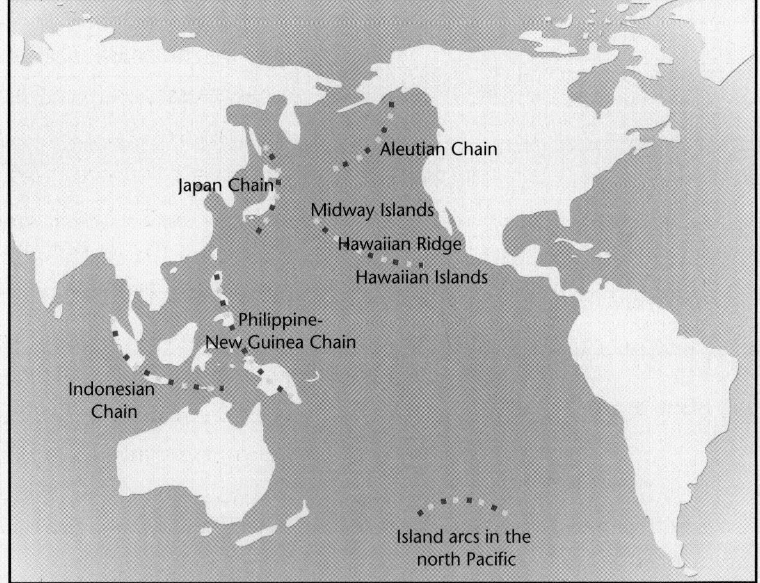


**Fig. 3.31**
North Pacific Ocean island arcs

## Ocean-Continental Convergence

A mountain belt, bounded on its seaward edge by an ocean trench, is formed when a plate capped by oceanic crust is subducted under a continental plate. The heavy oceanic crust sinks into the asthenosphere, where it melts at a depth of approximately 100 km. The hot magma rises and the presence of water vapour contributes to the high gas content and explosive nature of this magma. Volcanoes form on the continental edge parallel to the subducted plate.

Fig. 3.32
A destructive boundary

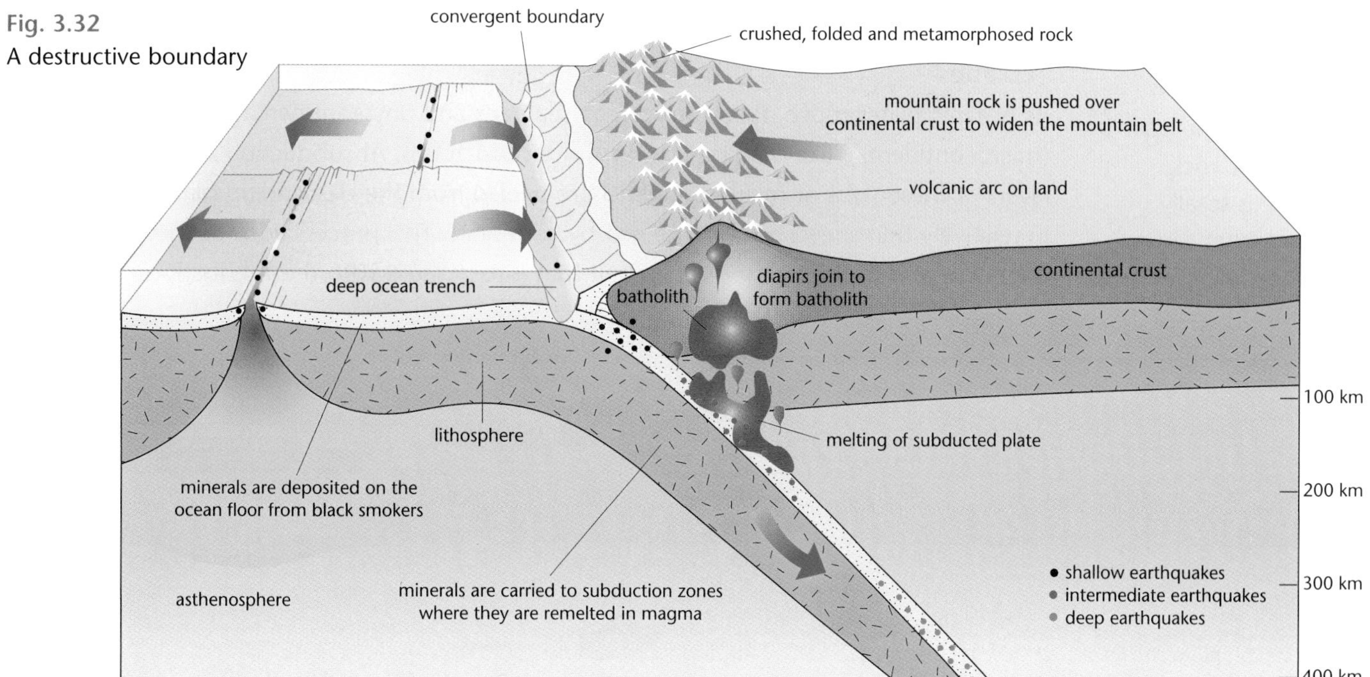


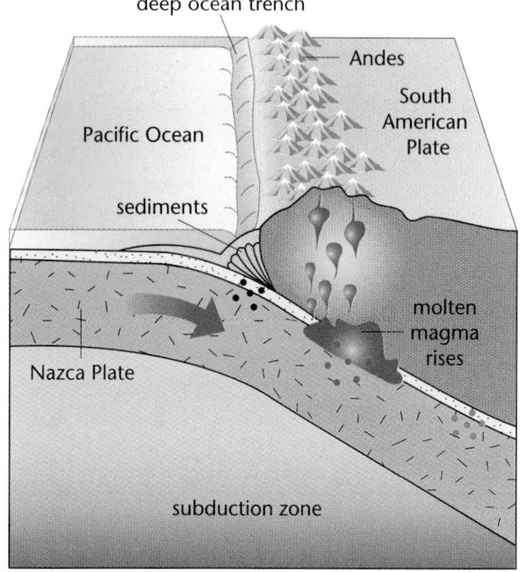


- shallow earthquakes
- intermediate earthquakes
- deep earthquakes

Fig. 3.33
Formation of the Andes

Beneath the volcanoes are large masses of plastic-like magma which thicken the crust, causing uplift. After millions of years of severe erosion, these masses are exposed as batholiths of granite.

Compression caused by the colliding plates causes faulting and folding on the continental side of the volcanic belt. This pressure also pushes some of the mountain rock landward over the continental surface, thereby increasing the width and the mass of the mountain belt. The Rockies of North America and the Andes of South America were formed in this way.

All major mountain belts, including the Alps, Urals, Himalayas, Appalachians and Caledonian mountains of Scotland and Ireland, are **complex fold mountains**. Although major mountains differ from one another in particular details, all possess the same basic structures.

Mountain chains generally consist of roughly parallel ridges of folded and faulted sedimentary rock, volcanic rock and often active, dormant or extinct volcanoes.

Minerals that formed within the crust at the mid-ocean ridges along with some minerals on the plate surface are subducted where they are remelted and concentrated in the various magmas and rocks along plate boundaries.

Some of the minerals are carried by rising magma or super-hot fluids towards the surface to form mineral deposits and precious stones such as gold, silver, copper and lead. Other minerals such as diamonds in South Africa and silver in Bolivia in the Andes mountains in South America are also found in igneous rock.

Different magmas and temperatures create different mineral deposits.

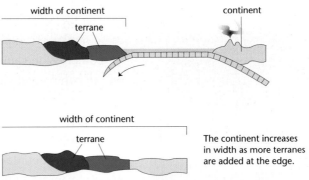

**terrane**
A section of the earth's crust whose rock structure is totally different from those on either side of it.

## How Fold Mountains Increase in Width

### Terranes

As oceanic plates move, they carry ocean plateaus and tiny continental slabs (microcontinents), like Madagascar, to subduction zones. At subduction zones, the upper layers of these slabs or microcontinents are peeled from the descending plate and pushed in relatively thin sheets against the nearby continent. This process adds to the width of the continent and to the fold mountain belt that is associated with the subduction zone.

Fig. 3.34



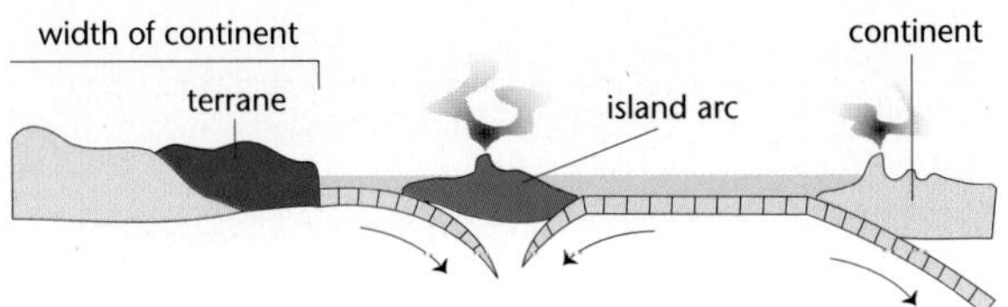




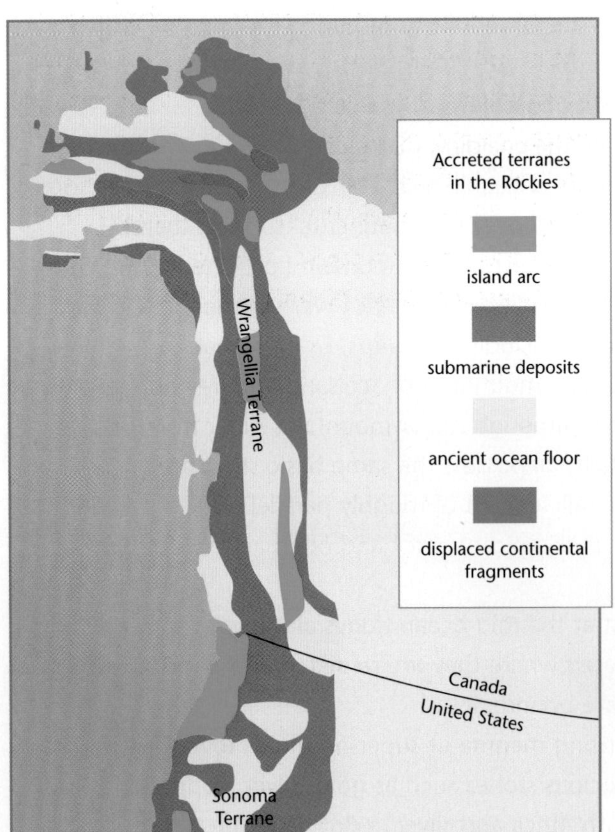


Fig. 3.35
Terranes in the Rocky Mountains in the United States of America. Terranes add to the width and the mass of a complex fold mountain belt.

These added crustal blocks are called 'terranes'. Terranes vary greatly in size. Some are no longer than volcanic islands on the ocean floor. Others, such as the entire Indian subcontinent, are huge.

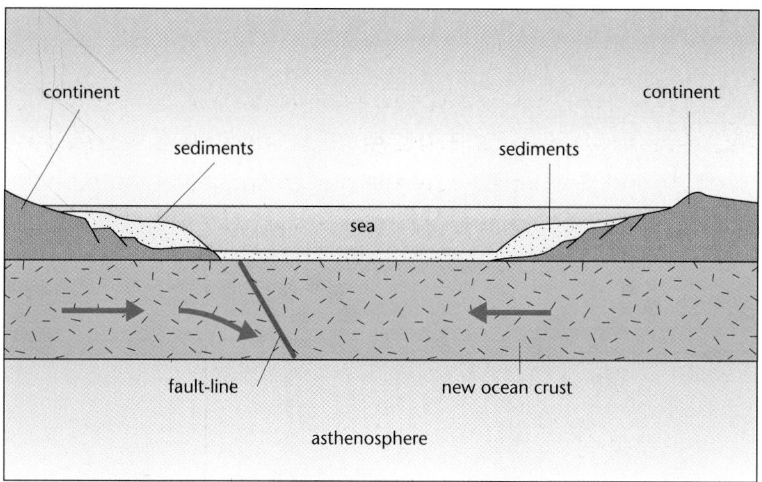

Fig. 3.36 Stage 1

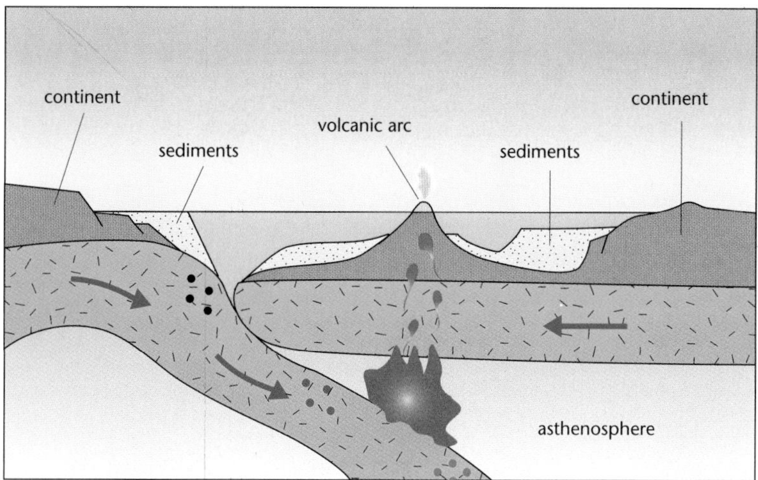

Fig. 3.36 Stage 2

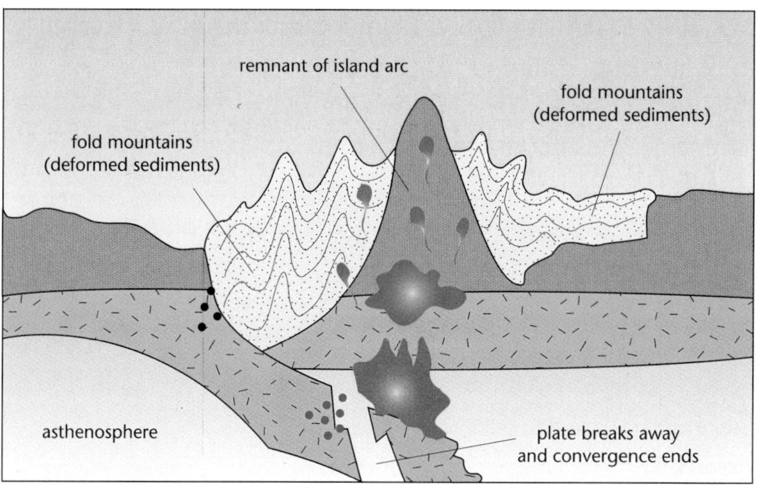

Fig. 3.36 Stage 3

## Continental-Continental Convergence

If two continental plates approach each other and collide they will form a high complex fold mountain system.

After the break-up of a continental landmass sea water will pour in to fill the area between the new continental margins to form an ocean. When this happens a thick wedge of sediments is deposited along these margins thereby increasing the size of the new continents.

For reasons not yet understood, at some stage the ocean basin begins to close and the continents begin to converge. Plate convergence results in the subduction of the intervening ocean slab and starts a long period of volcanic activity.

Eventually, the continental blocks collide. This event, which often involves further igneous activity, severely deforms and metamorphoses the trapped sediments. Continental convergence continues for a long time. The sediments, the volcanic arc and faulted rock shorten and thicken, producing an elevated fold mountain system. For example, the Himalayan mountain system formed as a consequence of Continental-Continental collision of India and Asia.

# Hot Spots

## Distribution of Hot Spots

Fig. 3.37
Long-lived hot spots at the earth's surface, each a centre of volcanic activity

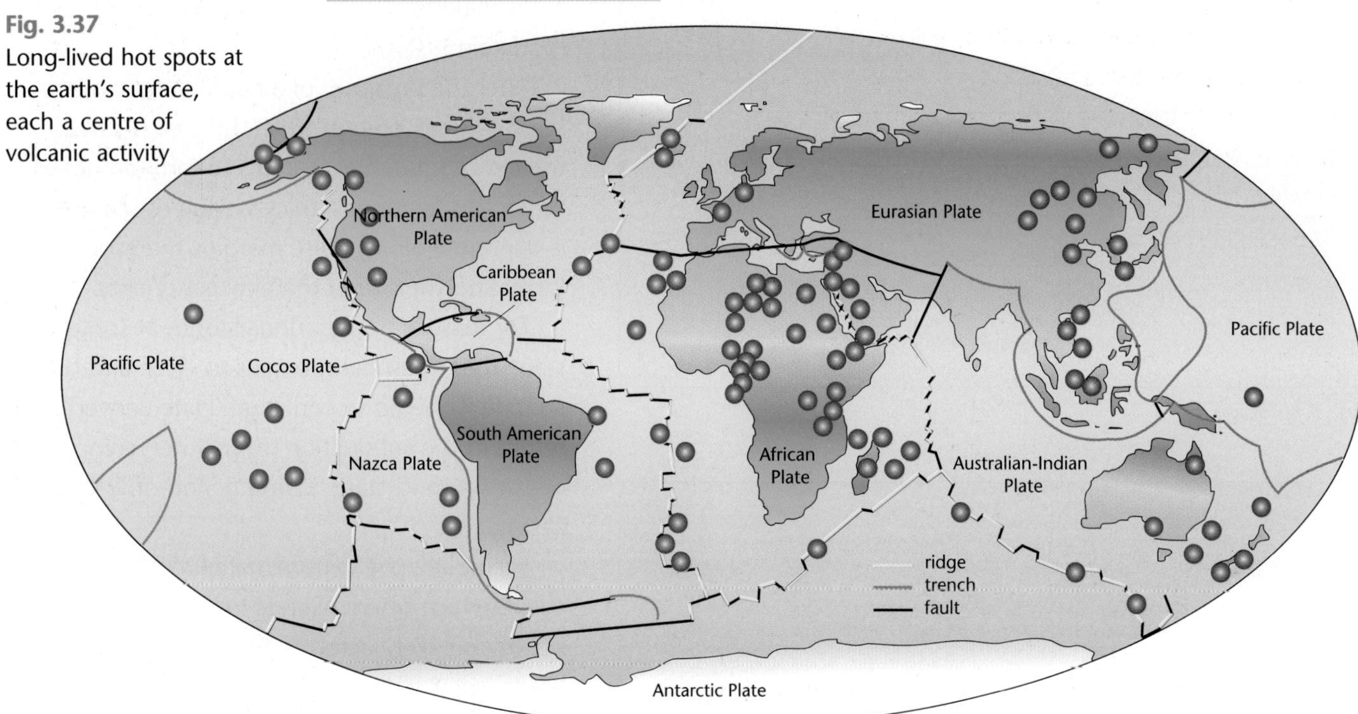


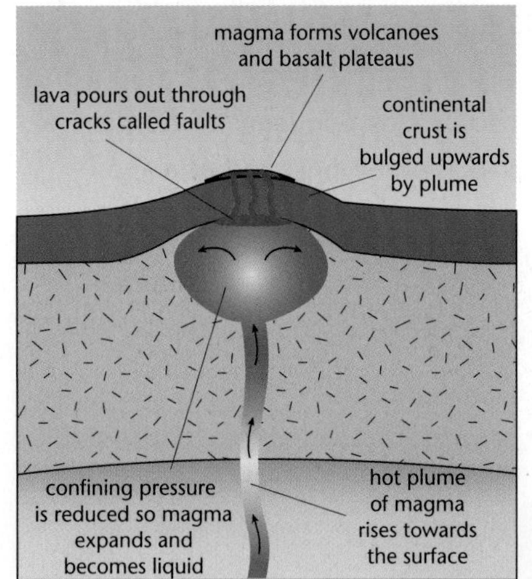


Fig. 3.38
Plume at hot spot

Most evidence indicates that hot spots remain stationary. Only about 20 of the 120 hot spots that are believed to exist are near plate boundaries.

### What are Hot Spots?

Hot spots are unusually warm areas found deep within the earth's mantle. Here high temperatures produce a rising plume of molten rock, which frequently starts volcanic activity at the earth's surface.

Some geologists believe that narrow columns of hot mantle rock called **'mantle plumes'** and **'super plumes'** rise through the mantle, much as smoke rises through a chimney. These plumes are believed to have large 'mushroom-shaped' heads above narrow tails.

The rising plumes cause uplift and stretching of the crust. The tail or funnel of magma that follows produces a localised spot of volcanic activity, much smaller than the head.

A hot spot beneath Iceland is thought to be responsible for the unusually large accumulation of lava found in that portion of the mid-Atlantic ridge. Another hot spot is believed to exist beneath Yellowstone National Park in the United States and to be responsible for geysers such as 'Old Faithful'.

**super-plumes**
Plumes that rise from the earth's core. These create huge plateaus of basalt on the floors of the oceans.

**plumes (magma)**
Plumes that rise from the earth's mantle.

## How a Hot Spot May Divide a Continent

### Case Study: Africa, A Divided Continent

A mantle plume rising beneath a continent should heat the land and bulge it upward to form a dome marked by volcanic eruptions. As the dome forms, the stretched crust typically fractures in a three-pronged or Y pattern. Continued radial flow outward from the hot spot eventually separates the crust along two of the three fractures, which become new continental edges separated by a new sea. The third fracture is left inactive as a failed rift that fills with sediment.

An example of this type of fracturing may exist along the Red Sea. The Red Sea and the Gulf of Aden are active diverging boundaries along which the Arabian peninsula is being separated from north-eastern Africa. The third, inactive, rift is the northernmost African Rift Valley. (See Fig. 3.41.)

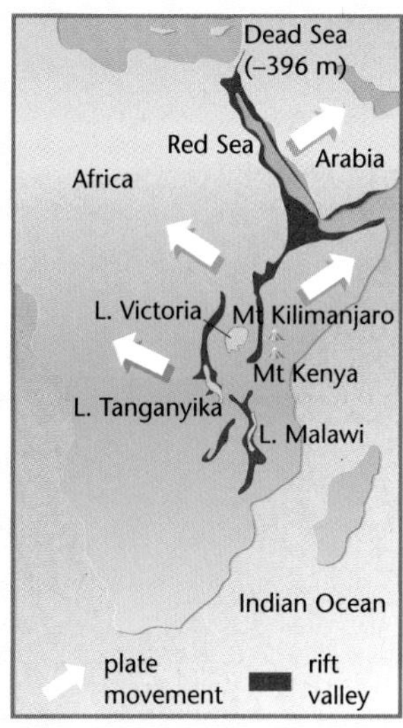


Fig. 3.39

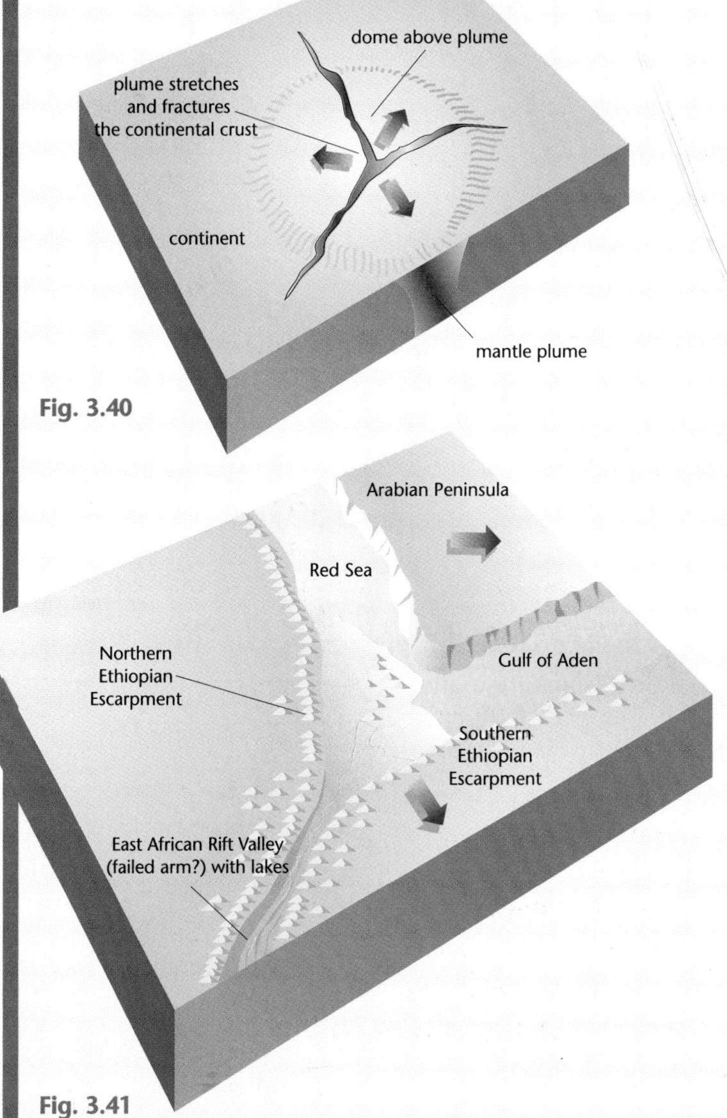


Fig. 3.40

Fig. 3.41

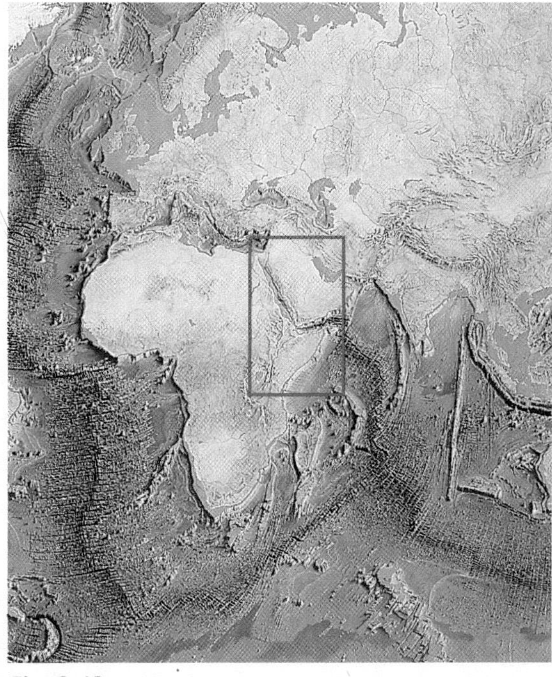

Fig. 3.42

Theory of how a divergent plate boundary may develop

(a) A rising mantle plume beneath a continental crust forms a dome with three branching features (a triple junction).

(b) The fractures spread apart, forming narrow rift valleys that widen and deepen far enough for magma to well up from the mantle and form oceanic crust. The third rift does not completely separate. It remains a failed arm.

(c) This detail shows the triple-junction relationships among the Red Sea, the Gulf of Aden and the East African Rift Valley.

## Hot Spots and Ocean Floors

### Case Study: Hawaiian Islands

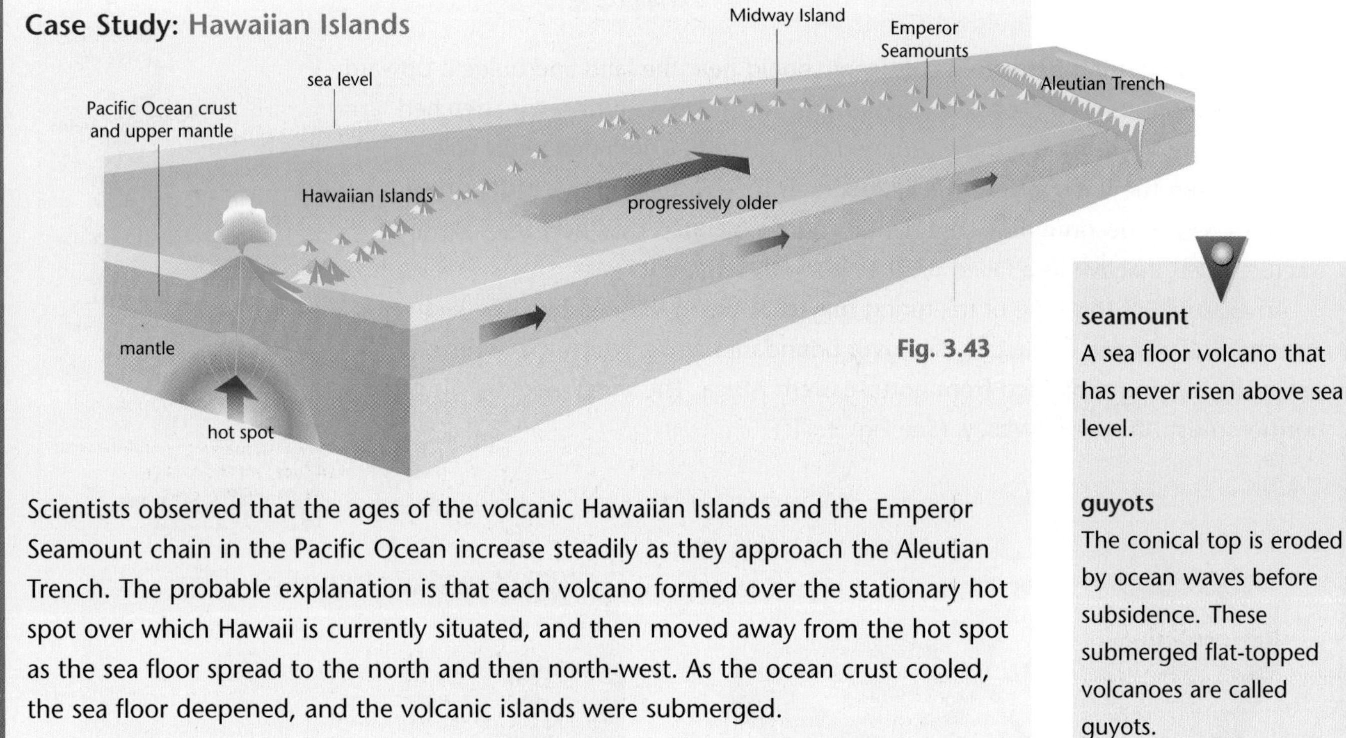


Fig. 3.43

Scientists observed that the ages of the volcanic Hawaiian Islands and the Emperor Seamount chain in the Pacific Ocean increase steadily as they approach the Aleutian Trench. The probable explanation is that each volcano formed over the stationary hot spot over which Hawaii is currently situated, and then moved away from the hot spot as the sea floor spread to the north and then north-west. As the ocean crust cooled, the sea floor deepened, and the volcanic islands were submerged.

**seamount**
A sea floor volcano that has never risen above sea level.

**guyots**
The conical top is eroded by ocean waves before subsidence. These submerged flat-topped volcanoes are called guyots.

**Crystal Formation**
When minerals solidify they form particular angular shapes called crystals. The kind of crystals that form in a magma is determined by
(a) the composition of the magma
(b) the length of time a magma stays at a particular temperature. If the temperature falls, crystals of a new mineral form. If the temperature rises, existing crystals may melt if the temperature is sufficiently high.

## Mineral Deposits and Plate Tectonics

Since the earliest times, people have depended for their existence on the presence of mineral deposits. Palaeolithic people used broken rock edges as their earliest tools. Mesolithic people used flint and chert, which when 'worked' produced razor-sharp edges, as scrapers, arrowheads and axes, efficient as daily tools.

With the discovery of smelting, copper, tin and later bronze and iron were moulded into tools, cooking vessels, arms and ornaments. Even at that time, some thousands of years ago, gold ornaments were displayed as jewellery and as a sign of wealth and status.

Today, as never before, we depend on mineral deposits to support our culture and daily existence. Cars, dishwashers, shower heads, iron and steel products for the construction industry, televisions, radios, aeroplanes, space shuttles and a myriad of other objects are all manufactured from mineral products.

The theory of plate tectonics allows us to get a 'window' into the processes that form some of the many minerals. A somewhat simplified account of the cycle of mineral formation is given below.

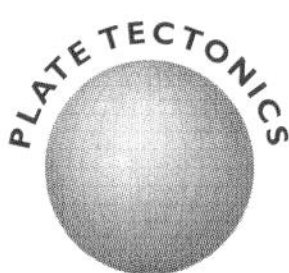

## Cycle of Mineral Formation under the Oceans

Fig. 3.44

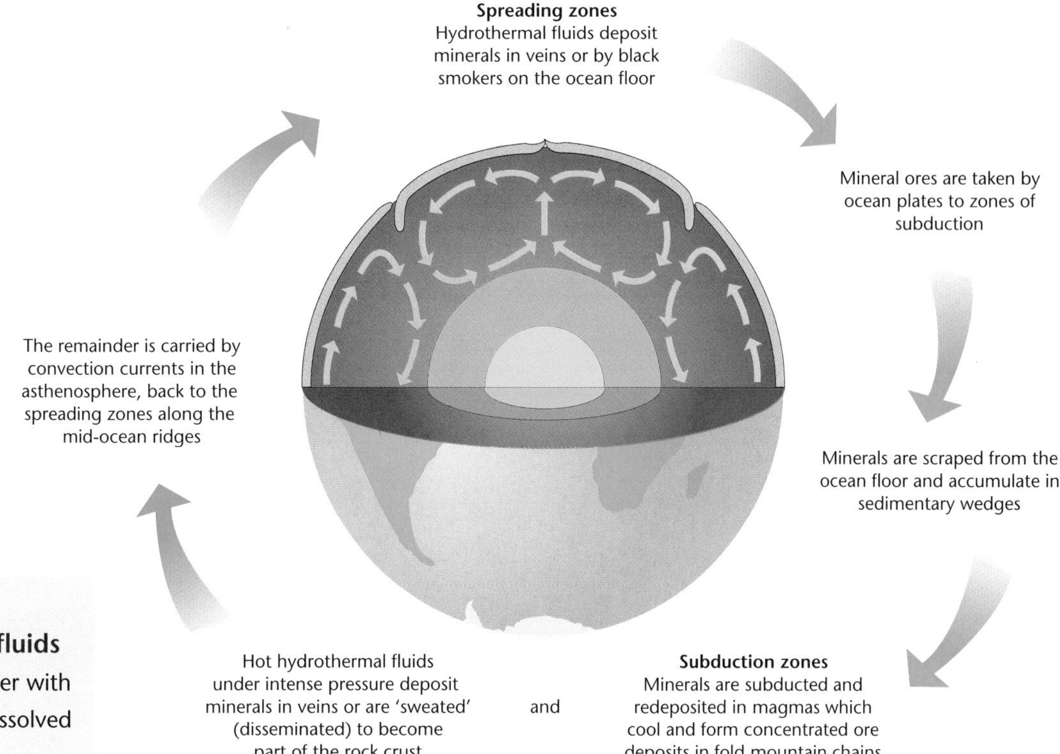
**Spreading zones**
Hydrothermal fluids deposit minerals in veins or by black smokers on the ocean floor
Mineral ores are taken by ocean plates to zones of subduction
Minerals are scraped from the ocean floor and accumulate in sedimentary wedges
**Subduction zones**
Minerals are subducted and redeposited in magmas which cool and form concentrated ore deposits in fold mountain chains
and
Hot hydrothermal fluids under intense pressure deposit minerals in veins or are 'sweated' (disseminated) to become part of the rock crust
The remainder is carried by convection currents in the asthenosphere, back to the spreading zones along the mid-ocean ridges

**hydrothermal fluids**
Intensely hot water with many minerals dissolved in solution.

## Mineral Formation at Subduction Zones

### Processes at Work

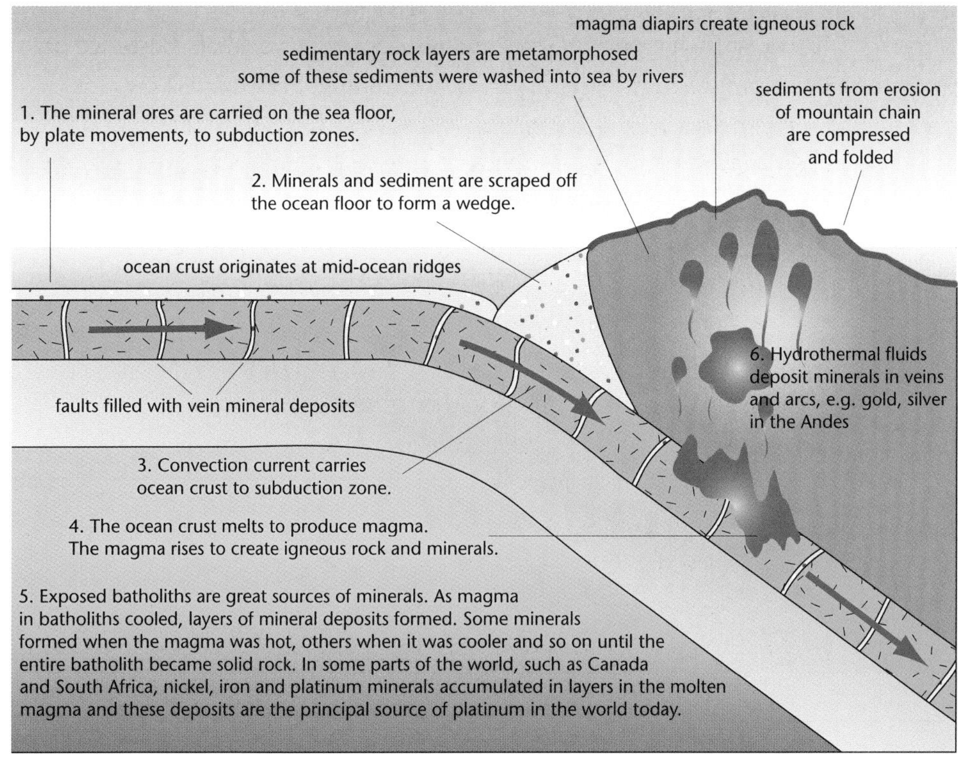
magma diapirs create igneous rock
sedimentary rock layers are metamorphosed
some of these sediments were washed into sea by rivers
sediments from erosion of mountain chain are compressed and folded
1. The mineral ores are carried on the sea floor, by plate movements, to subduction zones.
2. Minerals and sediment are scraped off the ocean floor to form a wedge.
ocean crust originates at mid-ocean ridges
faults filled with vein mineral deposits
6. Hydrothermal fluids deposit minerals in veins and arcs, e.g. gold, silver in the Andes
3. Convection current carries ocean crust to subduction zone.
4. The ocean crust melts to produce magma. The magma rises to create igneous rock and minerals.
5. Exposed batholiths are great sources of minerals. As magma in batholiths cooled, layers of mineral deposits formed. Some minerals formed when the magma was hot, others when it was cooler and so on until the entire batholith became solid rock. In some parts of the world, such as Canada and South Africa, nickel, iron and platinum minerals accumulated in layers in the molten magma and these deposits are the principal source of platinum in the world today.

Fig. 3.45
Mineral formation at subduction zones

Sea water and minerals in the subducted plate are heated and vaporised to create hot metal-rich fluids, called hydrothermal solutions.

In other areas, such as in the centre of large continental plates, heat builds up over many millions of years. This can cause molten magma to rush from great depths to the surface. These hot magmas may trap diamonds which form under great heat and pressure and rush them to the surface, after which the rock cools. Such deposits form the diamond mines of South Africa, Australia, Canada and northern Russia.

Avoca copper mine deposits in Co. Wicklow in Ireland probably formed in this way.

In other instances, rather than being concentrated in narrow veins, these ores are distributed as minute masses (sweated into rock) throughout the entire rock mass. Much of the world's copper ore deposits are formed in this way.

## Mineral Formation at Mid-ocean Ridges

### Processes at Work

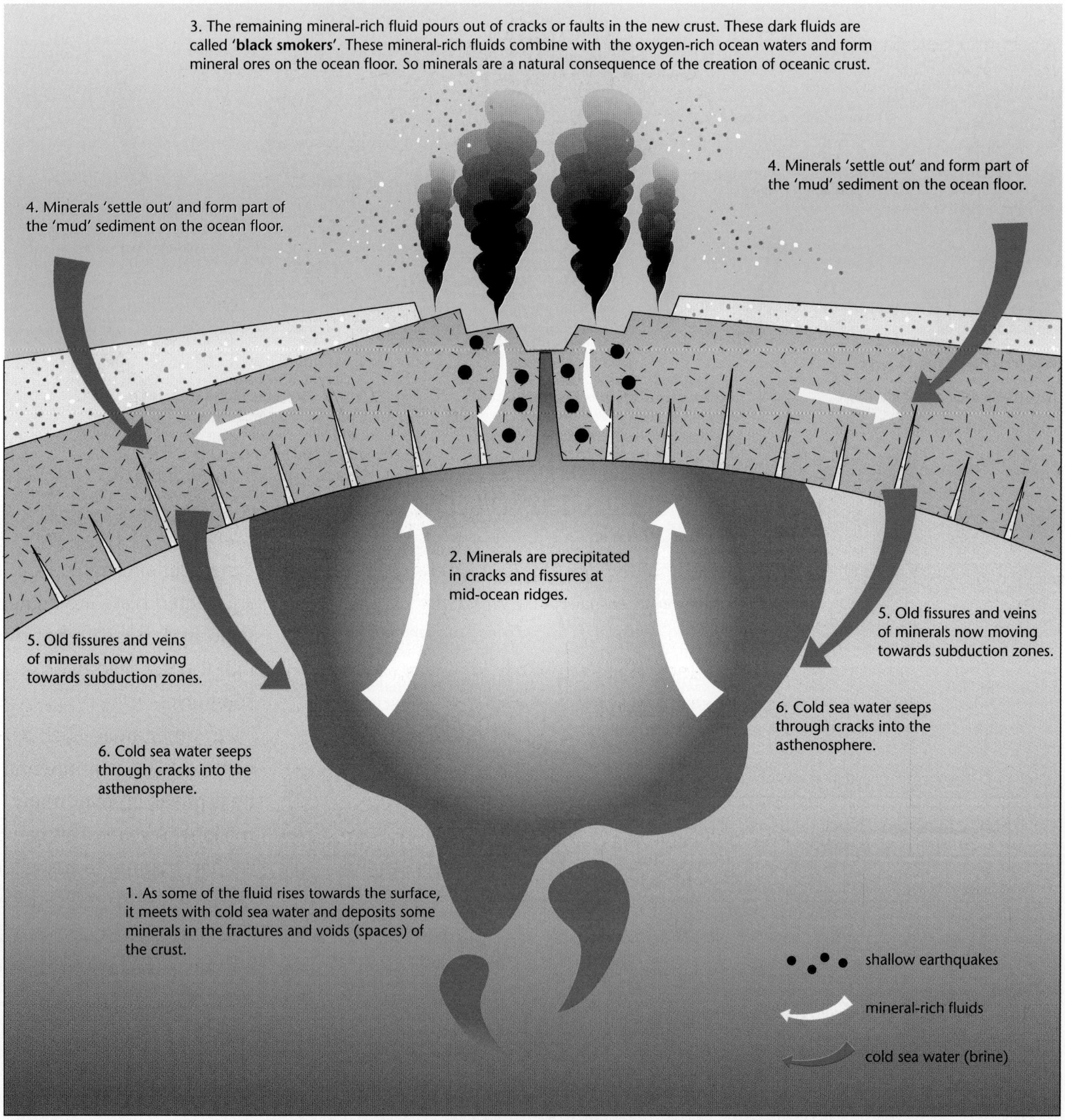

3. The remaining mineral-rich fluid pours out of cracks or faults in the new crust. These dark fluids are called '**black smokers**'. These mineral-rich fluids combine with the oxygen-rich ocean waters and form mineral ores on the ocean floor. So minerals are a natural consequence of the creation of oceanic crust.

4. Minerals 'settle out' and form part of the 'mud' sediment on the ocean floor.

4. Minerals 'settle out' and form part of the 'mud' sediment on the ocean floor.

2. Minerals are precipitated in cracks and fissures at mid-ocean ridges.

5. Old fissures and veins of minerals now moving towards subduction zones.

5. Old fissures and veins of minerals now moving towards subduction zones.

6. Cold sea water seeps through cracks into the asthenosphere.

6. Cold sea water seeps through cracks into the asthenosphere.

1. As some of the fluid rises towards the surface, it meets with cold sea water and deposits some minerals in the fractures and voids (spaces) of the crust.

shallow earthquakes

mineral-rich fluids

cold sea water (brine)

Fig. 3.46

# Volcanism

## Volcanoes

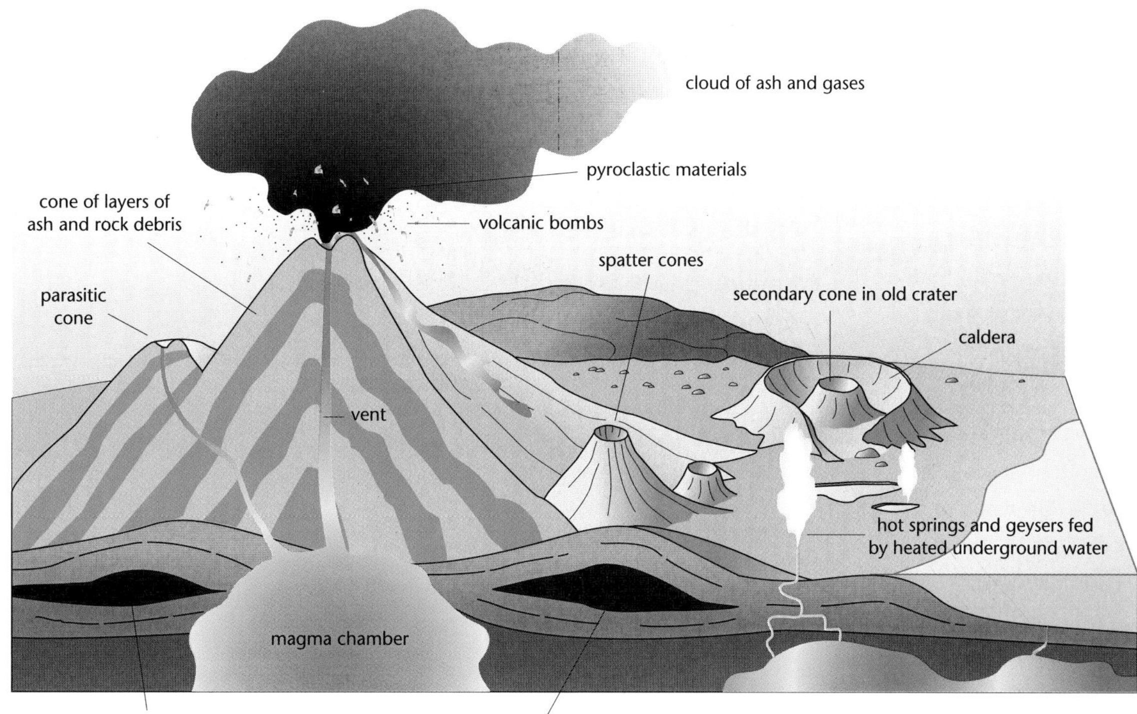


Fig. 4.1
A volcanic landscape (not to scale), showing some of the major features commonly associated with volcanic activity

For thousands of years, volcanoes have struck terror and wonder into the heart of man. They moved the Romans and the American Indians to worship. The term volcano stems from **Vulcan**, the **Roman god of fire**, and the name was first used for Vulcano, one of the Lipari islands in the Mediterranean Sea where the god was thought to live. Today, volcano refers to a pipe-like outlet, called a **vent**, through which molten rock, gases, rock fragments and dust erupt and form conical landforms constructed of these materials.

Rock which lies deep within the earth's crust has a very high temperature. Under intense pressure within the earth, this rock remains in a **semi-solid state**. However, in some places some of the rock becomes liquid. This liquid rock is called **magma**. Rock becomes liquid within the earth for one or a combination of the following reasons:

- earth movements such as folding and faulting
- the presence of water vapour
- radioactive decay
- friction created by the movement of tectonic plates
- a fall in pressure (see 'Confining Pressure', p. 44)

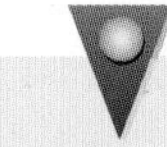

**igneous rock**
Rock which has cooled from a hot liquid state and has its origin in magma either inside the earth's crust (called **intrusive** rocks) or on the earth's surface (called **extrusive** rocks).

**magma**
Molten (liquid) rock material, generated within the earth, that forms igneous rocks when it hardens.

**lava**
Magma that flows out onto the surface of the earth; also refers to the rock formed after the magma cooled.

**volcano**
A vent in the surface of the earth through which magma, gases and rock fragments erupt; also the term for the landform that develops around the vent.

Due to its being of lower density (lighter) than surrounding rocks, the magma works its way towards the earth's surface and on occasion breaks through, producing volcanic activity. Once the magma pours out onto the surface it is called lava.

## Life Cycle of a Volcano

Volcanoes usually pass through three stages in their life cycle. In the beginning, eruptions are frequent and the volcano is **active**. Later, eruptions become so infrequent that the volcano is said to be **dormant** (sleeping). This is followed by a long period of inactivity. Volcanoes which have not erupted in historical times are said to be **extinct**.

# Magma

## Key Process, Key Idea

**As magma rises towards the earth's surface, trapped gases, such as water vapour, expand dramatically.**

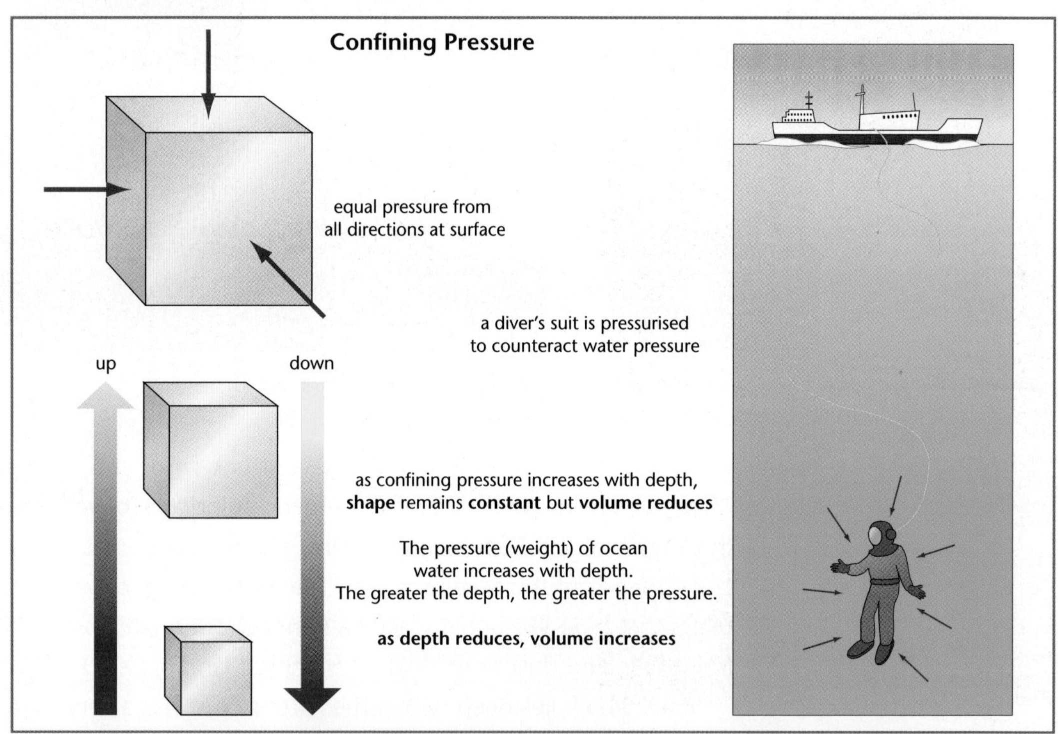


Fig. 4.3

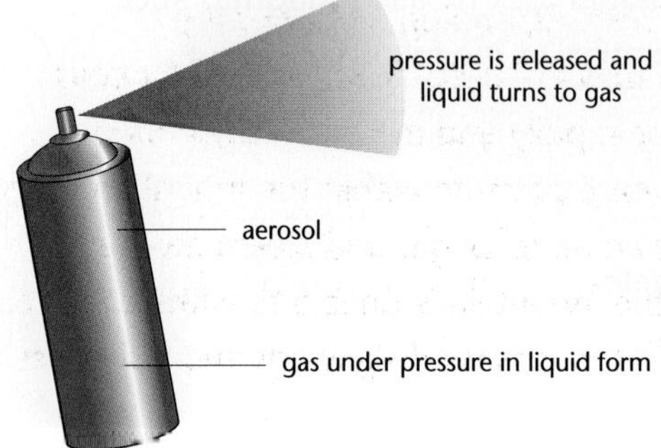


Fig. 4.2

## Confining Pressure

Confining pressure results when pressure is applied equally in all directions. This type of stress or pressure occurs in the earth's mantle under many kilometres of rock, and leads to a reduction in volume. So, the greater the depth, the greater is the confining pressure and the smaller the volume. Conversely, the lesser the depth, the lower is the confining pressure and the greater the volume.

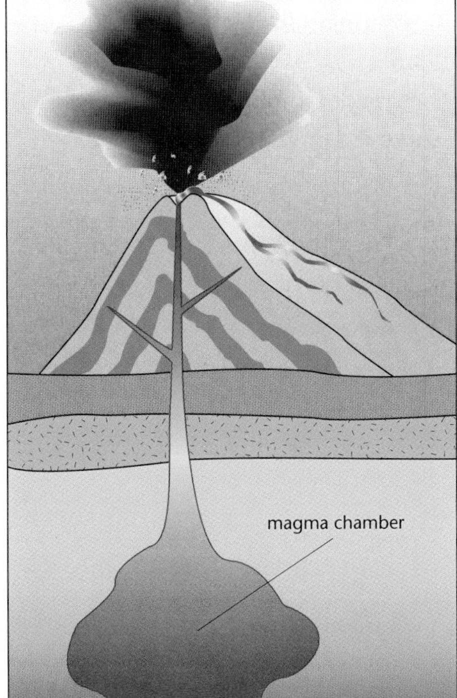

1. Molten rock (magma) from the earth's mantle forms a chamber as much as ten kilometres (six miles) below the surface. As the magma rises, gases dissolved in the molten material expand and bubble off.
2. The resulting froth exerts tremendous outward pressure that propels the magma upward. As the molten material comes into contact with groundwater, the volcano becomes a pressure cooker.
3. The froth and compressed gases push through cracks in the volcano. When they reach the surface, the pressures are suddenly released, the bubbles expand dramatically, and the volcano erupts in an explosion of ash and molten rock. Warning signs of an eruption include rising levels of carbon dioxide, sulphur dioxide and other gases, swarms of miniquakes, rising ground and water temperatures and small changes in the mountain's shape.
magma chamber


Fig. 4.4
Under the volcano

Put simply, when gases are located deep in the earth's mantle, their confining pressure is great and correspondingly their volume reduces. When these same gases move up towards the surface of the earth's crust, their confining pressure reduces and their volume increases. These gases include water vapour, carbon dioxide, nitrogen and sulphur compounds.

## Key Process: Rising Magma and Eruption

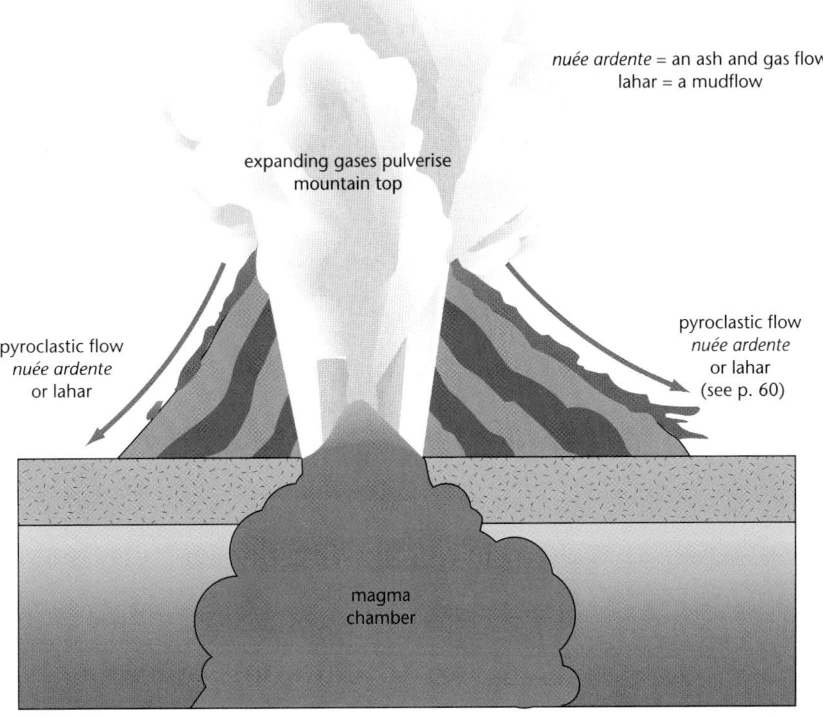


Fig. 4.5

When magma rises from deep underground, temperatures of 1,000 °C (1,830 °F) and low near-surface pressures allow the gases to expand. These expanding gases occupy hundreds of times their original volume than when they are deep down. In acid (viscous, thick) lavas the high content of **silica prevents** these rapidly and massively expanding gases from **escaping**. This leads to a build-up of intense pressure within the mountain. The volcanic mountain bulges and some very small amounts of gas and steam escape. These symptoms indicate extreme danger. Finally, the mountain is unable to withstand the increasing internal pressure and the entire mountain top is blasted skywards and pulverises its rock to dust and ash.

Fig. 4.6
Ejected, red-hot pyroclasts form a lava fountain

## Pyroclastic Rock

Pyroclastic rock is ejected from all kinds of volcanoes.

During an eruption, particles of rock and lava are ejected into the air. Some particles fall near the vent and form a cone structure. Others such as volcanic ash and pulverised rock may be carried great distances by the force of the eruption or by the wind or both. The particles produced by these processes are called **pyroclasts**, or as a group **'tephra'**. The ejected lava fragments range in size from very fine dust to sand-sized volcanic ash to large volcanic bombs and blocks.

The following are examples of pyroclastic rock.

- Pyroclastics the size of **peas** are called **cinders**. They contain numerous voids, caused when ejected lava bombs are pulverised by escaping gases.
- Pyroclastics the size of **walnuts** are called **lapilli**.
- Particles **larger** than **lapilli** are called
  - — **blocks**, when they are made of **hardened lava** and
  - — **bombs**, when they are ejected as **red-hot lava**.

Some pyroclastic rock may have some large crystals, while others may be very small and not visible to the naked eye. Such a rock is said to have a porphyritic texture. The large crystals are called **phenocrysts**.

Some pyroclasts have no crystals at all. This rock is called glass. Obsidian is a glassy black rock.

## World Distribution of Volcanoes

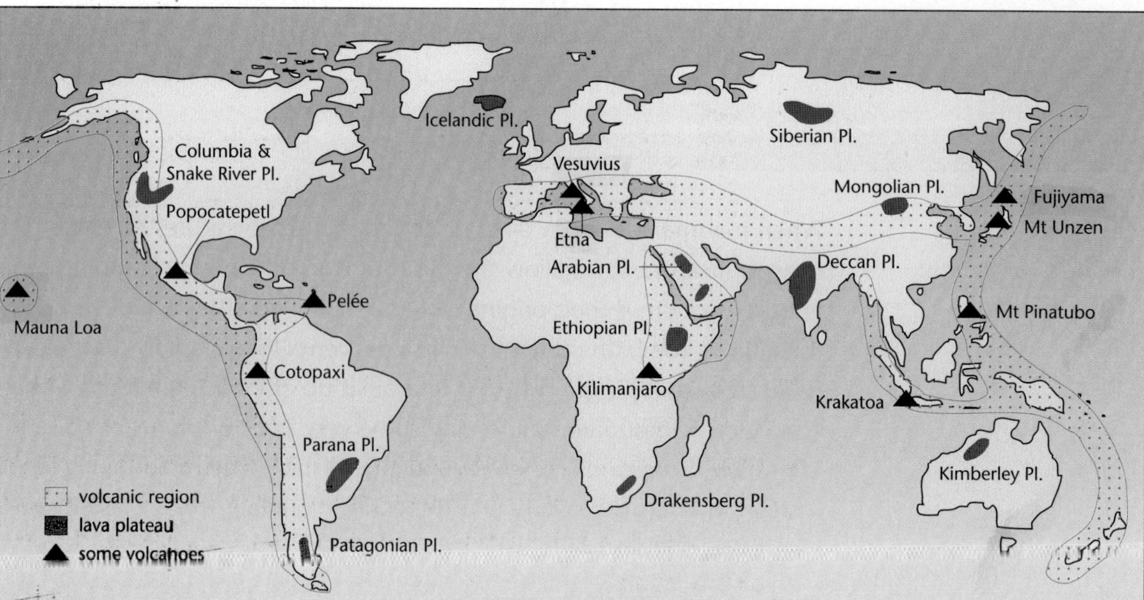

Fig. 4.7
World distribution of volcanoes and lava plateaus

**intrusive**
'Intrude' — enter by force. Formed within the earth's crust.

**granite**
A coarse-grained intrusive igneous rock which cooled slowly deep in the earth's crust.

**volcanic crater**
A circular hollow at the summit of a volcano.

**Benioff Zone**
A distinct earthquake zone that begins at an ocean trench and slopes landward and downward into the earth at an angle that varies from thirty to sixty degrees.

**Moho Line**
The boundary separating the crust from the mantle beneath it. The full title is **Mohorovicic Discontinuity**, named after the man who first discovered the boundary.

Volcanoes are found at the following locations:

1. **converging plate boundaries**
2. **diverging plate boundaries**
3. **hot spots**

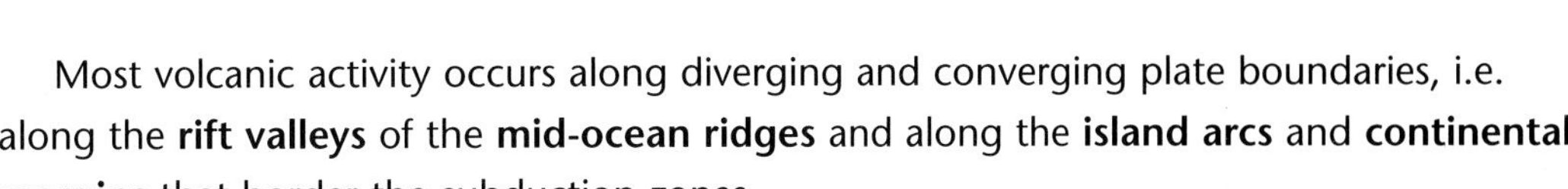

Most volcanic activity occurs along diverging and converging plate boundaries, i.e. along the **rift valleys** of the **mid-ocean ridges** and along the **island arcs** and **continental margins** that border the subduction zones.

## The Formation of Igneous Rock

Igneous rocks form when magma/lava cools and solidifies. Scientists divide igneous rocks into two groups:

- intrusive rocks (see p. 50)
- extrusive rocks (see p. 52)

## Some Intrusive and Extrusive Landforms

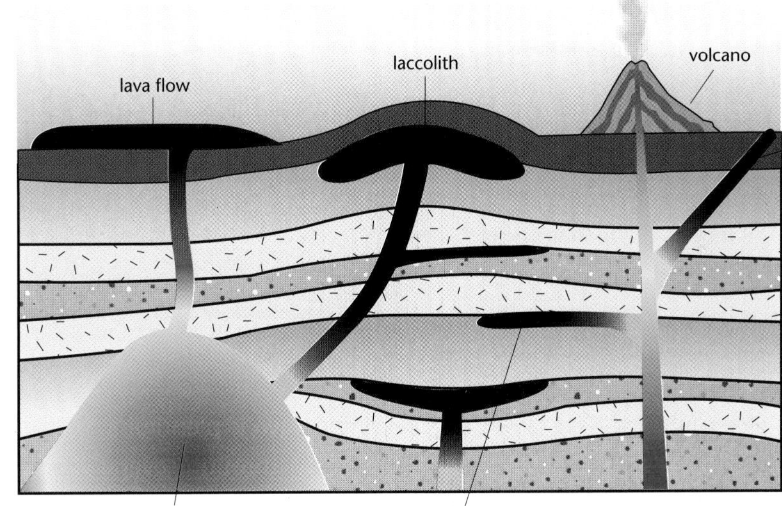


**Dyke**
A vertical or sloping intrusion with horizontal cooling cracks. Cools rapidly on contact with surrounding colder rock. Cuts **across bedding planes**.

**Batholith**
Deep-seated and surrounded by rock. The magma cools slowly so that:
(a) large crystals form
(b) rocks in contact with batholith are **metamorphosed** (changed to other rocks due to extreme heat).

**Sill**
A horizontal or sloping intrusion **along** (between) **bedding planes** with vertical cooling cracks. Cools rapidly on its outside, on contact with surrounding rocks — squeezes **between rock layers**.

Fig. 4.8
Some intrusive and extrusive landforms

## Lava and Magma

When magma reaches the surface it is called lava. Lava is of two types:

1. **Basic lava.** This lava is low in silica (45 to 55 per cent). It is very fluid (mobile). Eruptions take place quietly; flows have been measured at 30 km per hour on steep slopes but speeds of 10 to 300 m per hour are more common. Gases escape freely and in doing so aid the movement of the lava. Basic lava forms gently sloping volcanic cones.

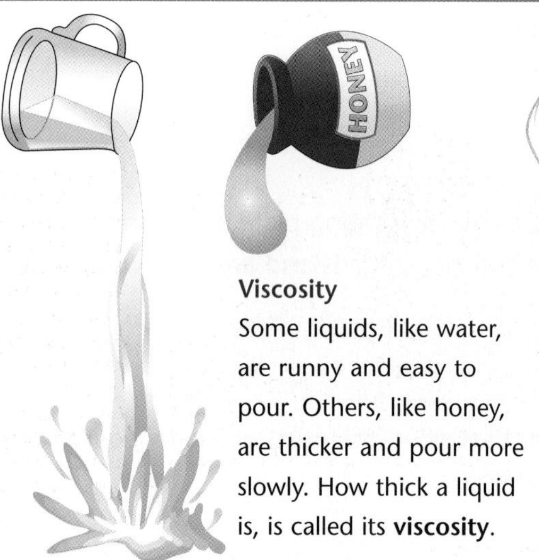

**Viscosity**

Some liquids, like water, are runny and easy to pour. Others, like honey, are thicker and pour more slowly. How thick a liquid is, is called its **viscosity**.

Try pouring clear honey after leaving it in the fridge for a few hours. Liquids become thicker, or more viscous, the colder they get. As they warm up, they become less viscous.

**What is Silica?**

Silica is a white or colourless substance and is the most abundant solid constituent of our globe. It is a mineral formed of **silicon** and **oxygen**.

Pure quartz is composed of silica (silicon oxide $SiO_2$). It is the main constituent of beach sand.

Magma with a high content of silica is thick and traps gases. Such magma is viscous and forms **acid lava**.

Magma with a low content of silica is fluid and gases can escape freely. Such magma forms **basic lava**.

**Did You Know?**

Glass is not a solid, but a liquid. You cannot see it flowing because it is very viscous. Very old windows are thicker at the bottom because the glass has flowed downward over the years.

Fig. 4.10

2. **Acid lava.** This lava is high in silica (over 70 per cent). It is **not fluid, so it is viscous.** It contains many gases which may build up within the magma to make it explosive and so cause violent eruptions, such as Mount Vesuvius in Italy. Viscous lavas give rise to steeply sloping cones.

## Basic Magma

Magma that is low in silica is 'runny', it pours out quietly and gases escape freely from the vent. Fluid lava forms a low, broad cone. Lava fountains throw liquid magma into the air. This activity attracts many tourists to volcanoes such as Mount Etna in Sicily and Mauna Loa in Hawaii.

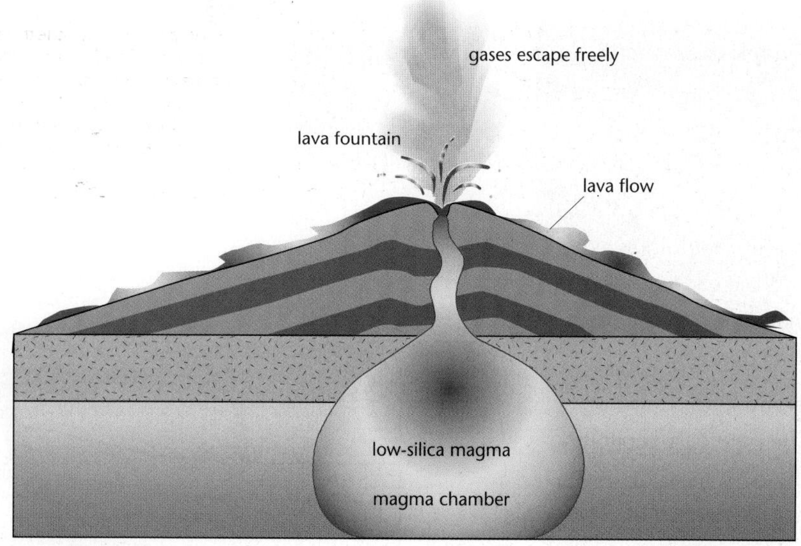


Fig. 4.9

Where lava is plentiful and eruptions frequent — through several vents — the result is likely to be a **shield volcano**, a large structure up to tens of kilometres across and with gentle slopes constructed from hundreds to thousands of successive lava flows. Mauna Loa is a shield volcano.

**pyroclastic material**
Relating to rock material formed by an explosive ejection from a volcanic vent.

## Acid Magma

Highly viscous magma (high in silica) retards the expansion and passage of gas through the magma, so that pressure builds up within the volcano. When the volcano does erupt, it is rapid, violent and explosive. The debris from explosive eruptions falls near the vent, forming a **steep** cone.

Fig. 4.11

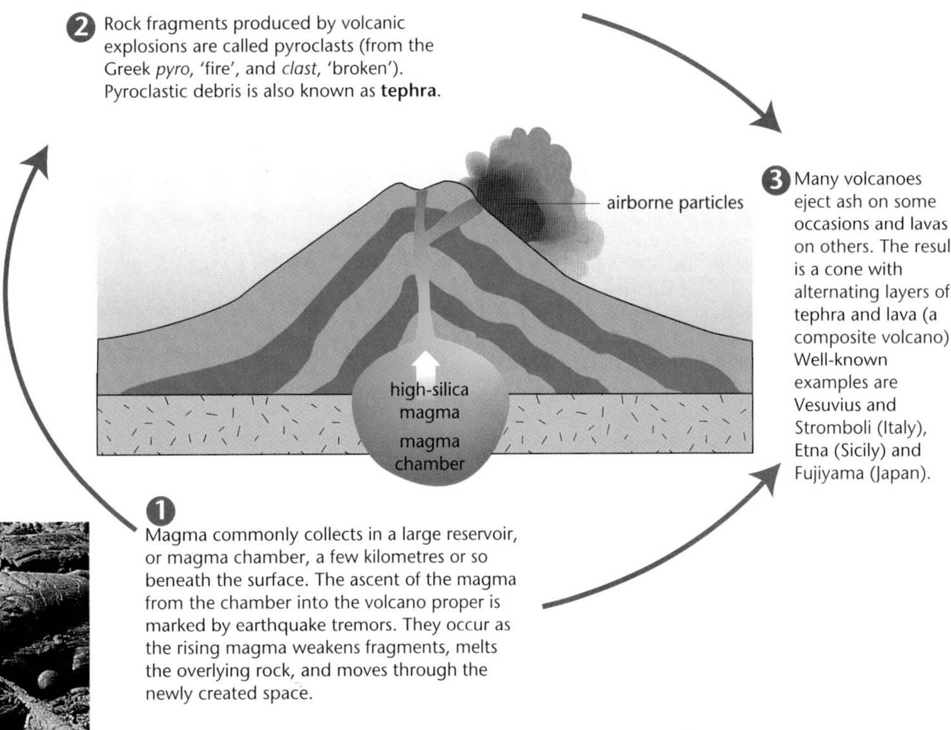

2 Rock fragments produced by volcanic explosions are called pyroclasts (from the Greek *pyro*, 'fire', and *clast*, 'broken'). Pyroclastic debris is also known as **tephra**.
airborne particles
3 Many volcanoes eject ash on some occasions and lavas on others. The result is a cone with alternating layers of tephra and lava (a composite volcano). Well-known examples are Vesuvius and Stromboli (Italy), Etna (Sicily) and Fujiyama (Japan).
high-silica magma
magma chamber
1 Magma commonly collects in a large reservoir, or magma chamber, a few kilometres or so beneath the surface. The ascent of the magma from the chamber into the volcano proper is marked by earthquake tremors. They occur as the rising magma weakens fragments, melts the overlying rock, and moves through the newly created space.

Fig. 4.12
Pahoehoe flow

Fig. 4.13
Aa flow

## Mobility of Lava

The mobility and cooling of lava depends on the amount of gases dissolved within the lava. In Hawaii, lava flows containing lots of gas have remained mobile until they cooled at 850 °C. Other lavas with less gas have cooled at 1,200 °C. Thus gas helps lava flow.

When lava cools, it may take a number of forms. Some of these forms have been given Hawaiian names as a consequence of research carried out there. **There are three main forms of lava.**

- **Pahoehoe flows** (pronounced pa-ho-e-ho-e)
  Pahoehoe lava has a 'wrinkled' or 'ropy' or 'corded' surface. This type of lava forms where the lava is at a high temperature, but the gases escape quietly and the flow congeals smoothly.
- **Aa flows** (pronounced ahh-ah)
  Aa flows solidify into irregular block-like masses of a jagged, angular, clinkery appearance. This type of lava forms when the flow is partially cooled and the front is shoved forward as a pile of rubble.
- **Pillow lavas**
  When lava oozes out from the ocean floor like toothpaste from a tube, it solidifies to form rounded blobs. As other blobs are forced out from the same source, a pile of flattened pillow-shaped masses is formed (Fig. 4.32, p. 58).

## Intrusive Volcanic Rocks and Landforms

Intrusive igneous rocks form from magma that fails to reach the earth's surface. It wells up, cools and solidifies within the crust to form a variety of features (see Fig. 4.8, p. 47).

Intrusive rocks are of two types:

1. plutonic rock
2. hypabyssal rock

### 1. Plutonic Rock

When magma cools in large masses deep in the earth's crust, the process is slow and the resulting rocks are coarse in texture and large-crystalled. Such masses of rock are called 'batholiths'.

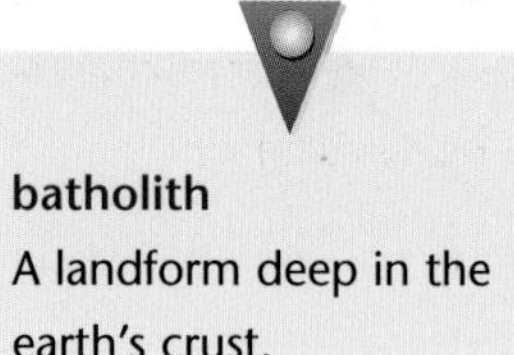

**batholith**
A landform deep in the earth's crust.

A **batholith** is a large mass of magma which has cooled slowly deep underground to form a large dome-shaped mass of granite. Large masses of such granite rock occur in the hearts of mountain ranges, such as the Wicklow Mountains in Ireland. They were formed by deep-seated earth movements on an enormous scale. The surrounding rocks with which the hot magma came into contact are often metamorphosed (changed in their nature) due to the intense heat of the magma. Eventually, due to prolonged denudation (weathering and erosion) of the overlying rock layers, the granite cores are exposed on the surface.

Fig. 4.14
Domestic hot water tank

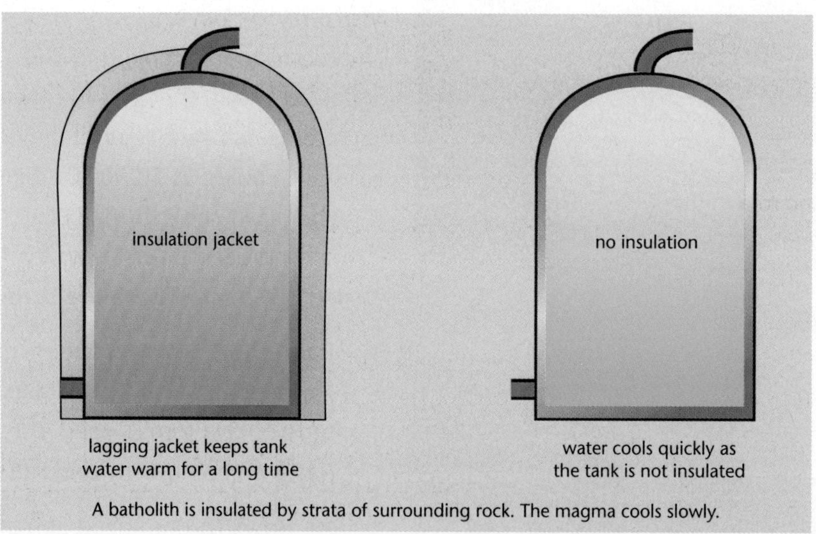
insulation jacket
no insulation
lagging jacket keeps tank water warm for a long time
water cools quickly as the tank is not insulated
A batholith is insulated by strata of surrounding rock. The magma cools slowly.

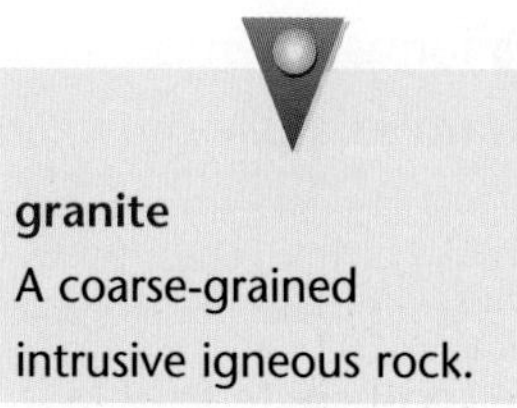

**granite**
A coarse-grained intrusive igneous rock.

Granite is the name given to a **group of rocks** that cooled slowly deep in the earth's crust. Granite consists mainly of **fel**dspar and **si**lica (quartz) as well as some mica and hornblende crystals. Consequently, the term **felsic** is applied to granite. As the magma cooled, crystals were given time to develop and grow, resulting in the formation of **large crystals**. Therefore granite is a coarse-grained and large-crystalled rock. The quartz grains are clear and glassy, the feldspars vary from white to pink and crystals of mica and hornblende are black. This rock occurs in great masses or batholiths and forms the core of most of our fold mountain ranges.

The colour of granite varies, depending on the percentages of the various minerals which form the rock.

**porphyry**
A group of rocks showing a combination of large and small crystals.

## 2. Hypabyssal Rock — An Intermediate Rock

**Formation:** Where magma moves along cracks and lines of weakness in rock the cooling is more rapid than in a large mass but slower than on the surface. This produces an intermediate category of rocks called hypabyssal rocks.

**Laccolith:** A laccolith is a small dome-shaped mass of magma close to the surface. It forms when a tongue-shaped mass of magma forces the overlying strata into a dome, producing a hill directly above it. (See Fig. 4.15.)

**Sill:** When a sheet of magma **lies along the bedding plane** it is called a sill. Some sills form horizontal layers while others are inclined due to a tilt in the bedding planes.

**Dyke:** When a sheet of magma **cuts across the bedding planes** and forms a 'wall' of rock it is called a dyke. Dykes are formed when magma rises through vertical or near-vertical fissures and cools to form igneous rock. Thus dykes may be vertical or inclined.

**volcanic ash**
Pyroclastic material less than two millimetres in diameter.

**volcanic breccia**
A pyroclastic rock composed of angular fragments that are larger than sixty-four millimetres.

***nuée ardente***
A very hot ground-hugging gaseous cloud erupted from a volcano which 'flows' down the volcanic cone at great speed.

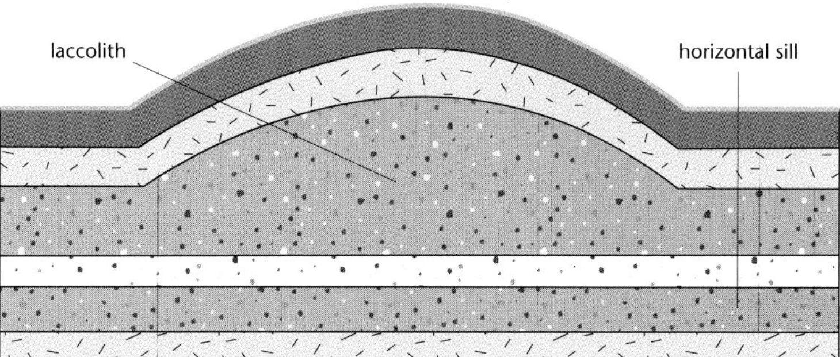


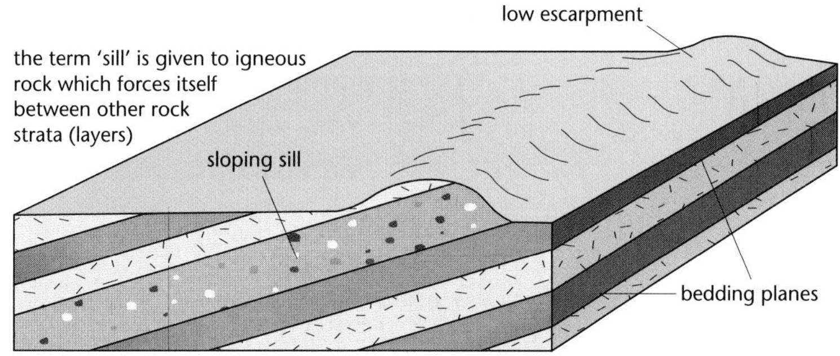


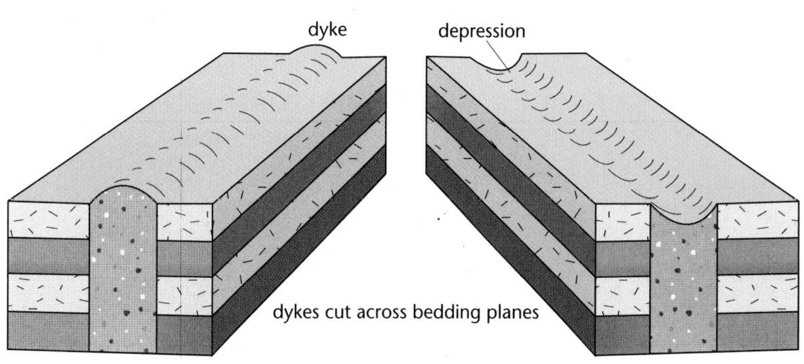


Fig. 4.15

## Extrusive Rocks (Volcanic) and Landforms

Extrusive rocks are of two types:

1. basalt
2. pyroclastics (see p. 46)

**basalt**
A small-grained crystalline rock.

### Fissure Eruptions

### Formation of Plateaus

When magma flows out through long cracks in the earth's surface, it does so quietly, without much violence, and spreads out evenly. The flows are five to six metres thick and often occur at long intervals.

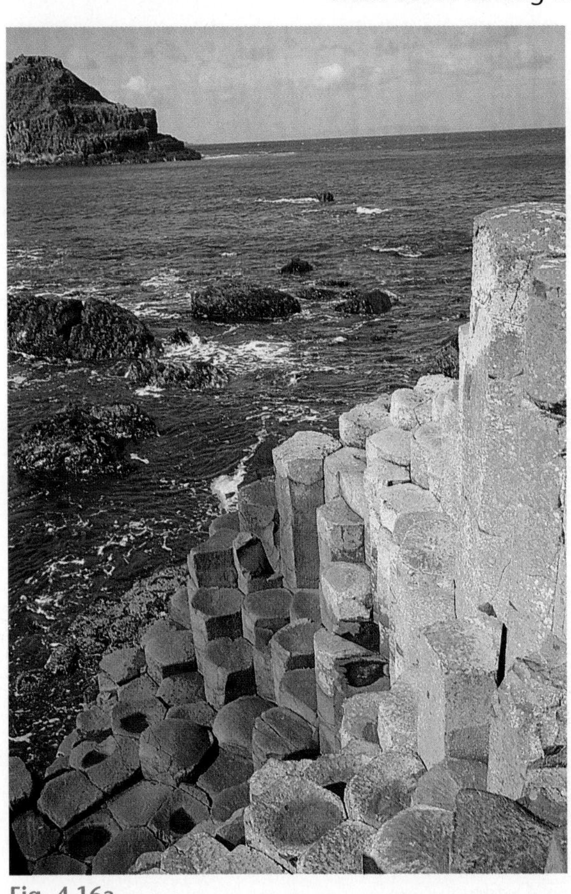

Fig. 4.16a
The Giant's Causeway. Columns of basalt are exposed on the Antrim coastline in Ireland.

Normally, the lava 'wells out' quietly smothering the pre-existing relief under a sheet of basalt. These sheets have no obvious relation to craters, but occur in areas where crustal tension (pulling apart) produces the fissures (cracks) through which the lava extrudes (pours out). The eventual product is a plateau, often with stepped sides as the later flows fail to extend as far as the earlier ones.

Lava flows through surface cracks (fissures) gave rise to the present Antrim Plateau in north-eastern Ireland. Other plateaus to form in this way are the Deccan Plateau in India and the plateau of western Arabia.

Basalt is the name given to a group of rocks that cooled and **crystallised quickly** on or very close to the earth's surface. They have a high iron (Fe) and magnesium (Mg) content and are low in silica. The term **mafic** is given to this rock group. Basalt is a dark green to black, heavy, hard rock composed of tiny crystals. It is formed from crystallised lava which once poured red-hot and liquid from surface fissures or volcanoes. Regularly, when lava cools to form basalt, it splits into columns of four, five or six sides. It forms great plateaus such as the Antrim Plateau in Ireland, which displays perfect hexagonal columns of basalt at the Giant's Causeway. Columns of basalt are also to be seen at Pallas Green, Kilteely and Knockroe in east Limerick, in Ireland.

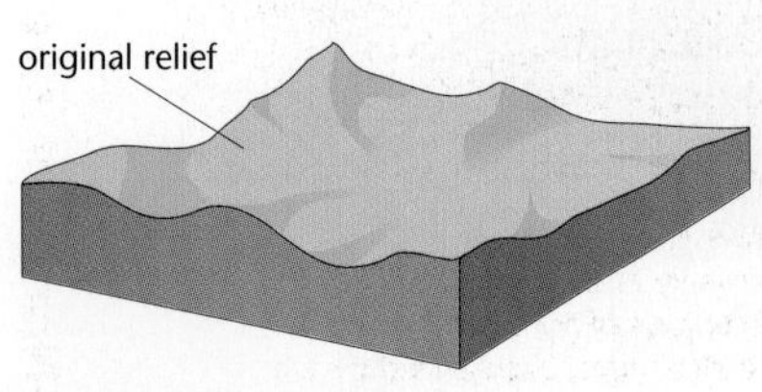


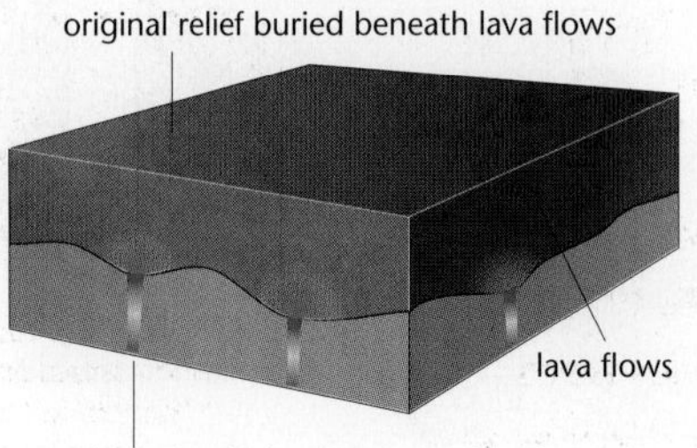


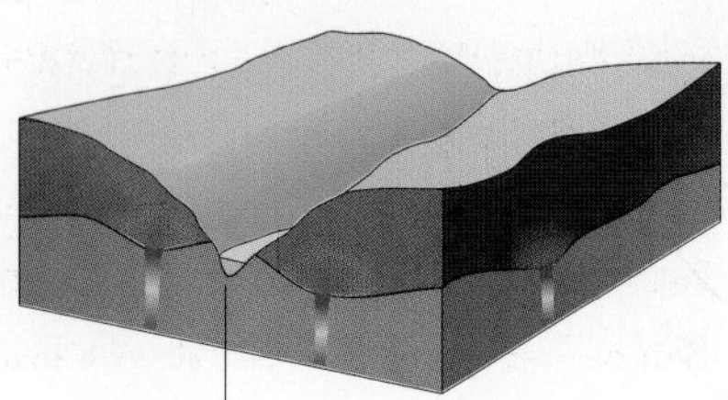


Fig. 4.16b
Stages in the formation of a lava plateau

## Types of Volcanic Cones

**Ash (cinder cone):** During a violent eruption lava is blown high into the sky. Rocks, ash, cinders and volcanic bombs fall down around the vent and begin to build up into a steep-sided volcanic mountain, e.g. Paricutin in Mexico. The gradient of the volcano sides is thirty to forty degrees (Fig. 4.17).

**Composite cones:** Composite cones are produced when viscous (high silica content— thick) lavas are produced over a long time and quite suddenly the eruption style changes and the volcano violently ejects pyroclastic material. In time, this material is covered by lava to repeat the process. Occasionally, both activities occur simultaneously. These volcanoes emit acid lavas as well as basic lavas at different times.

Composite cones are the earth's most picturesque and interesting volcanoes. Most are located in a narrow zone called the 'Ring of Fire' that encircles the Pacific Ocean (see Fig. 4.18 and Fig. 4.7, p. 46).

Fujiyama in Japan, Mount Mayon in the Philippines and Mount St Helens and Mount Shasta in the Cascade Range in the Rocky Mountains in the north-western United States are all composite cones.

The slopes are generally steep near the crest at thirty degrees, easing off to five degrees near the foot.

**Dome cones:** Acidic lava is high in silica. It forms a thick syrup-like lava containing large amounts of gases which burst through the vent. It flows only a short distance and cools quickly. As a result, the cone has a 'blown-out' appearance, like the puys of the Massif Central in France (Fig. 4.19).

**Shield cones:** Basic lava, which is low in silica and allows expanding gases to escape freely, is extremely fluid. The freely escaping gases propel the magma and cause it to travel long distances to form a cone with gentle slopes.

In Hawaii the active volcano Mauna Loa is further increased in size by smaller secondary cones on its sides. The gradient of the volcano sides varies from about ten degrees near the summit to two degrees at the foot of the volcano (Fig. 4.20).

Explosive Type

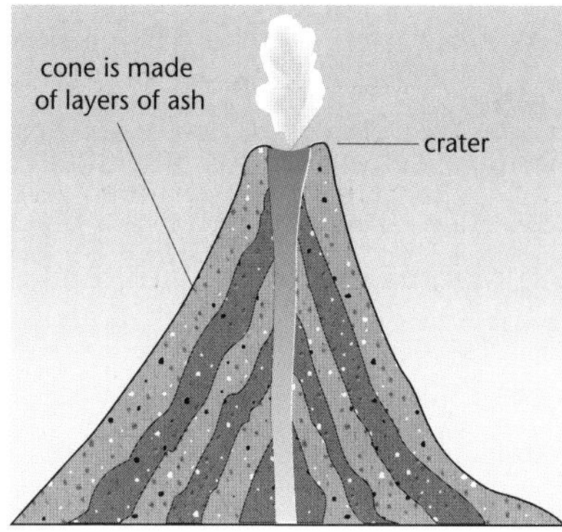


Fig. 4.17

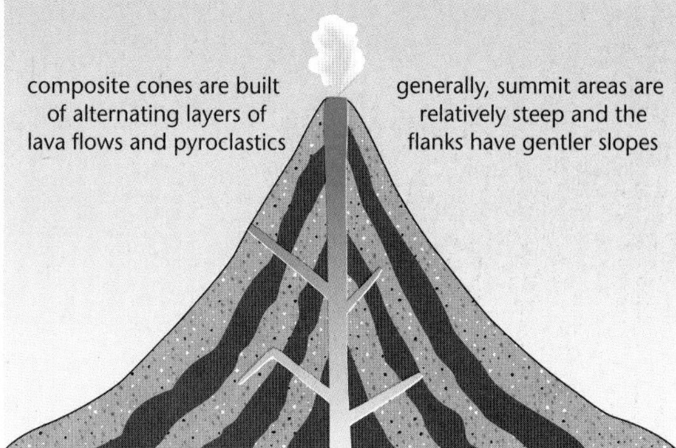


Fig. 4.18

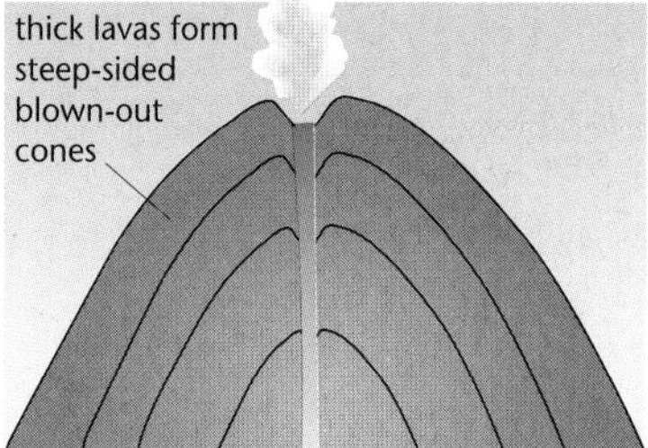


Fig. 4.19

Non-explosive Type

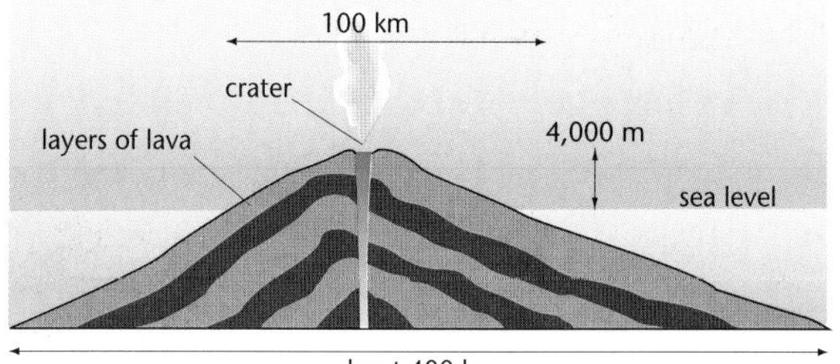


Shield volcanoes are built primarily of fluid basaltic lava flows and contain only a small percentage of pyroclastic materials. These broad, slightly domed structures, exemplified by the Hawaiian Islands, are the largest volcanoes on earth.

Fig. 4.20

## Case Study: Mount Mazama and Crater Lake in Oregon, USA, 6,600 years ago

### Formation of a Caldera

Some eruptions violently eject enormous amounts of volcanic ash (up to seventy cubic kilometres) and partially empty the underlying magma chamber. With the loss of support from within, the whole top of the volcano sinks into the empty magma chamber beneath the vent. This explosion creates a much larger crater than before. This larger crater is called a **caldera**. Over time, rainwater may fill the caldera to form a crater lake. Subsequent eruptions may produce another 'secondary cone' within the caldera.

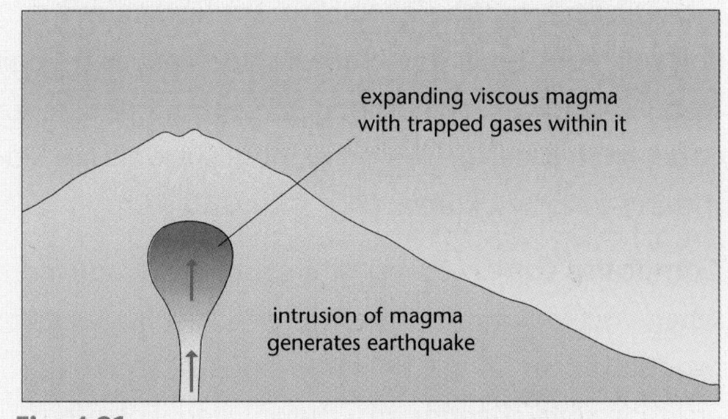


Fig. 4.21
Stage I

puffs of steam and ash
bulge
bulge
Massively expanding gases build up intense pressure. Some poisonous gases and steam and ash seep out through fissures.
continued intrusion of magma produces bulge

Fig. 4.22
Stage II

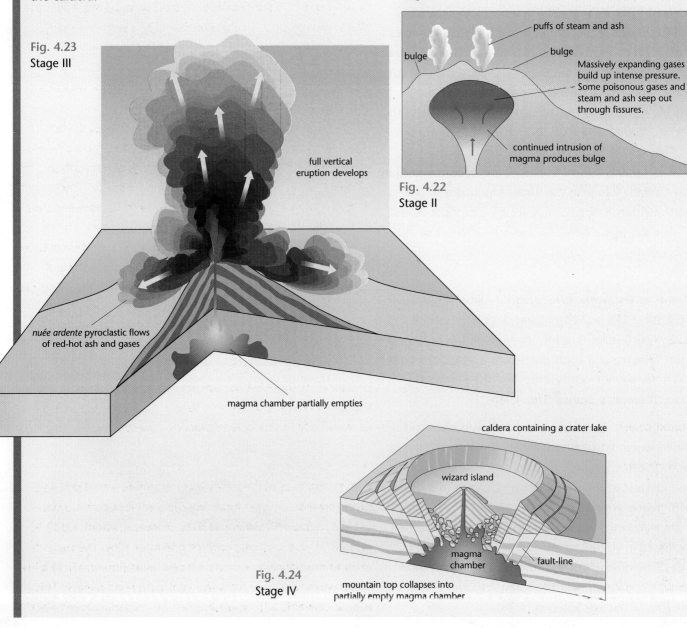

Fig. 4.23
Stage III

full vertical eruption develops
*nuée ardente* pyroclastic flows of red-hot ash and gases
magma chamber partially empties

caldera containing a crater lake
wizard island
magma chamber
fault-line
mountain top collapses into partially empty magma chamber

Fig. 4.24
Stage IV

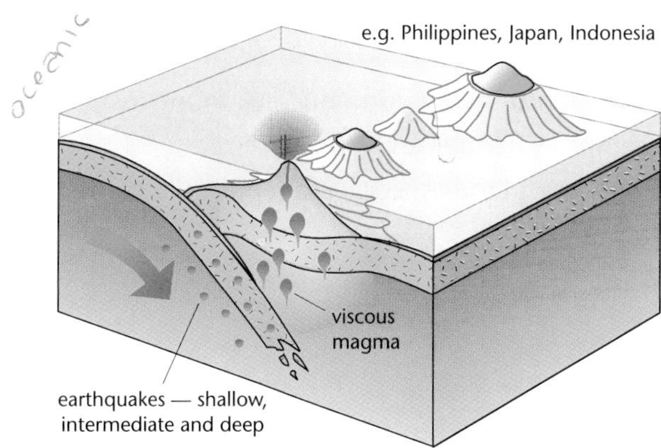


Fig. 4.25
Ocean-ocean

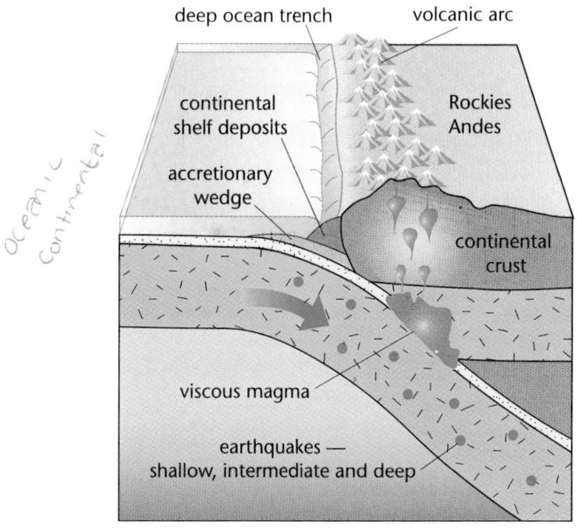


Fig. 4.26
Ocean-continent

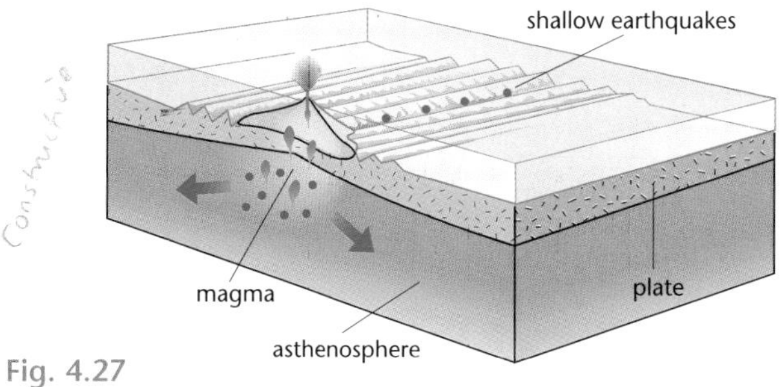


Fig. 4.27
Mid-ocean ridge

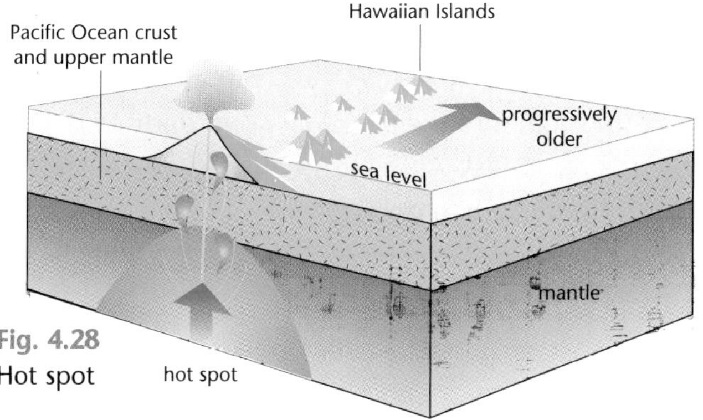


Fig. 4.28
Hot spot

## Volcanoes at Subduction Zones

Volcanoes that are formed at subduction zones are among the most violent and potentially dangerous forces on earth.

### Why are these volcanoes the most dangerous?

The high percentage of silica in the magma of these volcanoes is the key factor which determines their explosive nature. Silica prevents expanding gases from escaping through the magma. As magma rises to the surface, confining pressure is reduced and gases expand, occupying hundreds of times their original volume.

So there is a 'build-up' of pressure which finally explodes, causing death and destruction to local inhabitants; it can even affect climates on the opposite side of the earth for many years.

Pyroclastic minerals such as ash, bombs, lapilli and *nuée ardente* are common in such volcanoes. Lahars or mudflows may also occur. (See p. 59.)

The high silica content of these magmas reduces their ability to 'flow' and their ability to allow gases, such as water vapour and sulphur gases, to escape into the atmosphere. This 'stiffness' or resistance to 'flow' is called 'viscosity'. The magma gathers under the volcanoes, the mountain expands and finally explodes under intense pressure.

These magmas with a high silica content are referred to as 'viscous magmas', 'acid lavas' or 'andesitic lavas': this last name derives from the Andes mountains on the Pacific ring of fire, where many of these viscous volcanic magmas exist.

## Volcanoes at Diverging Plate Boundaries (Mid-ocean Ridges) and Hot Spots

Volcanoes at these centres are very dramatic, throwing hot spurting magma into the air. However, they are less dangerous than those at subduction zones.

### Why are these volcanoes less dangerous?

The gas content of magma affects its movement. **Escaping gases** provide enough force to propel molten rock from a volcanic vent. When magma rises, temperatures of 1,000 °C (1,830 °F) and low, near-surface pressures allow the gases to expand dramatically. Very fluid **basic lavas** (low in silica) allow these gases to escape from the vent with relative ease. This release prevents a build-up of pressure. Escaping gases eject

Did you know?

**Record breaker**
In 1904 the Waimangu geyser (now inert) in New Zealand erupted to an amazing 457 m (1,500 ft), higher than the Sears Tower in the US city of Chicago, which is 445 m (1,460 ft) tall.

Fig. 4.29

red-hot pyroclastic materials hundreds of metres into the air, forming 'lava fountains'. Such magmas appear very dramatic, but are not as dangerous as they seem.

Lava fountains, such as those in Hawaii or Mount Etna in Sicily, are regularly included in tourist itineraries.

(3) Hot spots pg 38

## Other Forms of Volcanic Activity

### Hydrothermal Areas

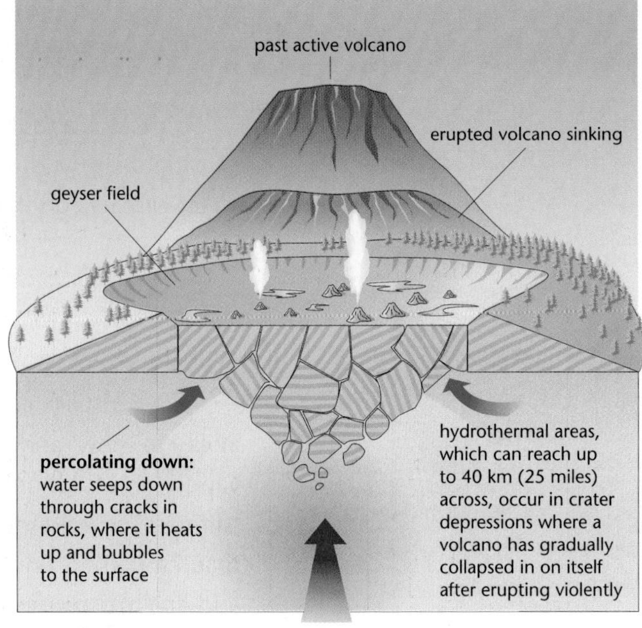
past active volcano
erupted volcano sinking
geyser field
percolating down: water seeps down through cracks in rocks, where it heats up and bubbles to the surface
hydrothermal areas, which can reach up to 40 km (25 miles) across, occur in crater depressions where a volcano has gradually collapsed in on itself after erupting violently

what is left of old magma chamber still produces a lot of energy

Fig. 4.30
Geyser field formation

**sinter terrace**
A stepped hard crust of mineral deposit on rocks caused by the evaporation of water issuing from a number of mineral springs.

**Geyser** fields, otherwise called **hydrothermal** areas, include such phenomena as boiling hot springs, vents which emit jets of sulphur gases, sinter terraces and boiling mud pools. Geysers are, however, the most spectacular of these.

Hydrothermal areas occur at sites of past volcanic activity where moisture is trapped and heated by the cooling (underground) volcanic rocks.

1. **Geysers**

   High-pressure jets of hot water and steam shoot into the air (often rising thirty to sixty metres) at regular intervals. Hot water is ejected first. After the jet of water ceases, a column of steam rushes out, usually with a thundering noise. The most famous geyser in the world is 'Old Faithful' in Yellowstone National Park in the USA, which faithfully erupts about once each hour. Geysers are common in Iceland and New Zealand.

## 2. Hot springs

Mineral explorations throughout the world have shown that temperatures in deep mines and oil wells usually rise with an increase in depth below the surface at a rate of about 2 °C per 1,000 metres. Therefore, when groundwater circulates at great depths, it becomes heated, and if it rises to the surface, the water may emerge quietly as a hot spring. Hot springs are common in Iceland and New Zealand.

## 3. Fumaroles and Solfataras

(a) Some secondary cones on volcanic mountains emit only steam or gases. These are called **fumaroles**.

(b) If the steam contains much dissolved sulphurous gas, the vent becomes surrounded by yellow sulphur deposits. This type is known as a **solfatara**.

## Ocean Depths and Plate Tectonics

The ocean floors are still largely unexplored. From their shallow edges, or continental shelves, they descend steeply, levelling off into ocean basins that may be six kilometres below the surface. Ocean trenches plunge even deeper, up to eleven kilometres (seven miles). This undersea world contains huge plains called **abyssal plains**, **mountain ranges**, **volcanoes** and **canyons** that dwarf anything on land. Whatever life that exists here has adapted to two extremes:

1. the enormous pressure exerted by the weight of water above (called confining pressure, see p. 44)
2. the total absence of light.

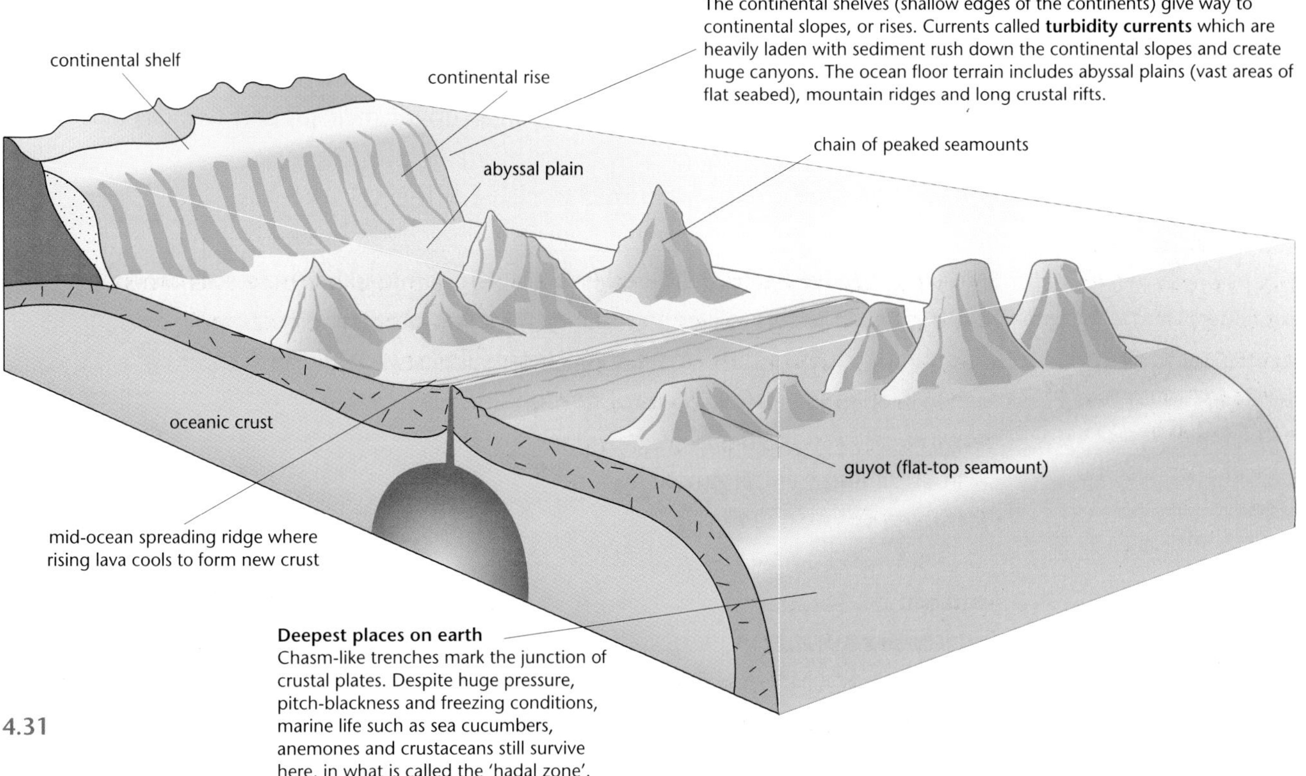

The continental shelves (shallow edges of the continents) give way to continental slopes, or rises. Currents called **turbidity currents** which are heavily laden with sediment rush down the continental slopes and create huge canyons. The ocean floor terrain includes abyssal plains (vast areas of flat seabed), mountain ridges and long crustal rifts.
continental shelf
continental rise
abyssal plain
chain of peaked seamounts
oceanic crust
guyot (flat-top seamount)
mid-ocean spreading ridge where rising lava cools to form new crust
**Deepest places on earth**
Chasm-like trenches mark the junction of crustal plates. Despite huge pressure, pitch-blackness and freezing conditions, marine life such as sea cucumbers, anemones and crustaceans still survive here, in what is called the 'hadal zone'.

Fig. 4.31

**fissure eruption**
A volcanic eruption (generally non-violent) through a long fracture in the earth's crust.

**lava plateau**
An elevated, flat-topped area composed of thick horizontal layers of lava flows.

**obsidian**
A dark-coloured glass formed from hot liquid lava which cooled (instantly) when ejected from a volcano.

Fig. 4.32
Pillow lavas at mid-ocean ridges
Lying near to the deep sea spreading ridges are fields of 'pillow lava', molten rock that has oozed up through fissures in the ocean floor and then cooled.

The continental shelves that are submerged beneath the seas are the edges of the plates that separated following the breaking up of Pangaea. Their gentle slopes indicate the sagging of the plate edges following break-up, as their support was taken away.

## People and Vulcanicity — Some Positive Effects

### Soils

***terra rossa***
Red soils, called laterites, formed due to the weathering of iron-rich soils within or near the tropics.

Soils which derive from weathered lava are rich in minerals. These soils are used to grow fruit trees such as figs, olives and oranges. Market gardening is carried on on land surrounding cities to provide fresh produce, such as vegetables for urban markets. Cities such as Naples have small areas of surrounding precious fertile level land for vegetable growing.

In Brazil and Central American countries, cash crops such as coffee are produced on weathered volcanic *terra rossa* soils. Brazil is the largest exporter of coffee in the world.

### Tourism

Tourism is a major industry based directly on volcanic structures such as craters. Volcanic mountains such as Vesuvius attract hundreds of thousands of tourists each year who pay a fee to climb and view its crater. Nearby archaeological sites such as Pompeii and

Herculaneum which were devastated by Vesuvius in AD 79 also attract people in vast numbers. Local spin-off industries such as the hotel and catering industry, souvenir trade and local guides and shops all benefit from such tourism.

## Geothermal Energy

Geothermal energy is the using of natural steam to generate electricity. This type of energy source is available in areas where subsurface temperatures are high due to relatively recent volcanic activity.

Icelanders have perfected the use of geothermal energy. They have found many clever ways to harness volcanic heat in order to combat the cold climate. Using water warmed by volcanic rocks, they heat their houses, grow tomatoes in hothouses and swim year-round in naturally heated pools, even though the temperatures outside may be below freezing. Icelanders also generate most of the electricity they need by using volcanically produced steam.

## Geothermal Energy from Dry Rocks

Cold water is pumped through fractures at the bottom of a deep well, becomes heated, then flows back up to geothermal power plant, where heat energy is converted to electricity.

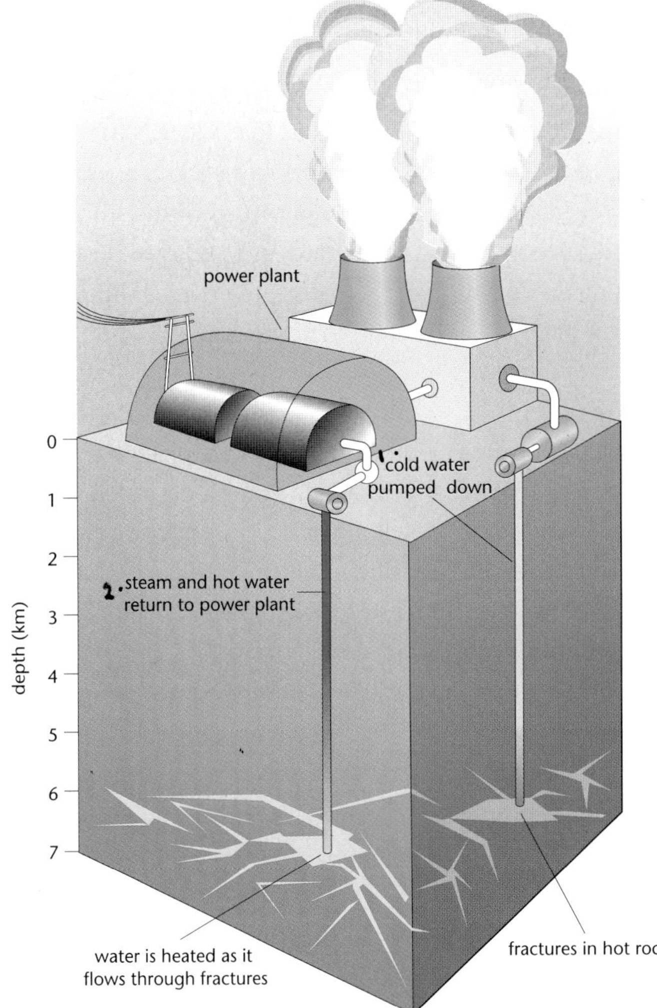


Fig. 4.33
Geothermal energy from dry rocks

## Geothermal Energy from Wet Rocks

Geothermal energy from wet rocks is produced when this process occurs naturally.

## People and Vulcanicity — Some Negative Effects

Large composite cones often generate a type of mudflow known as a lahar. Such mudflows are created when:

1. deposits of volcanic ash and debris are saturated with rainwater and flow down steep volcanic slopes, generally following river valleys or channels
2. large volumes of ice and snow melt on ice-capped volcanic peaks during an eruption. This melting is caused when heat is radiated to the surface from within the volcano

   or

   when hot gases and near-molten materials fall on the ice and snow during an eruption.

Fig. 4.34
A *nuée ardente*, with its poisonous gases, speeds downhill in the Philippines in 1991

### Lahars and People

Lahars can bring death and destruction to people and property in the vicinity of volcanic cones. In November 1985, Nevado de Ruiz, a 5,300-metre (17,400-foot) volcano in the Andes in Colombia, erupted and generated a lahar that killed 20,000 people in the town of Almero. The eruption melted the snow and ice that covered the uppermost 600 metres of the summit. This produced torrents of hot, thick mud, ash and pyroclastic debris which raced down the valleys of nearby rivers that radiate from the volcano, devastating everything in its way.

### *Nuée Ardente*

When expanding hot gases with glowing ash are ejected from a volcano, they may produce a heavy fiery grey cloud called a *nuée ardente.* Also referred to as glowing avalanches, these devastating clouds of steam, gases and ash flows, which are heavier than air, race down steep volcanic slopes at speeds up to 200 kilometres (125 miles) per hour.

These ground-hugging pyroclastic flows with poisonous gases regularly include both vast amounts and larger rock particles, which are suspended from the ground by the hot, expanding gases much as a hovercraft travels over land or water. Some of their deposits travel more than 100 kilometres (60 miles) from the source.

Some *nuée ardentes* happen so quickly that local populations are given no warning whatsoever and can be wiped out in moments.

In 1902 a *nuée ardente* from Mount Pelee in Martinique in the Caribbean killed 28,000 people in a nearby port town. In 1991 a pyroclastic flow at Japan's Mount Unzen killed thirty-one people, including three geologists.

### Volcanic Activity and Climate

Explosive volcanic eruptions emit huge quantities of gases and dust into the atmosphere directly above the explosion. From here, gases and small amounts of the lightest particles may spread around the globe and may remain suspended for many months, even years. However, most dust particles 'settle out' (return to the earth's surface) in a relatively short time and have little or no negative effect.

### *Jokulhlaups*

Spectacular floods known as *Jokulhlaups*, or glacier bursts, can result from volcanic action under glaciers. Huge meltwater lakes formed by a volcano's heat can suddenly burst from under the glacier in one of nature's most awesome displays of power. In 1918 Iceland's Myrdalsjokull glacier produced a *Jokulhlaup* with waterflow estimated at three times that at the mouth of the Amazon River. In 1996 the Loki volcano under Vatnajokull ice cap erupted. The hot magma created an underground lake which later burst through from under the ice and created a *Jokulhlaup*.

# Earthquakes

## What is an Earthquake?

An earthquake is a shaking, rolling or sudden shock of the earth's surface. There may be as many as one million earthquakes in a single year. Most of them take place beneath the surface of the sea, and few of these cause any damage. But earthquakes that occur near large cities can cause much damage and loss of life, especially if the cities rest on soft ground, e.g. Mexico City.

**Earthquakes create a vibration** of the earth produced by the rapid release of energy. This energy radiates in all directions from its source, the **focus**, in the form of waves, like those produced when a stone is dropped into a calm pond. Just as the impact of the stone sets water waves in motion, an earthquake generates **seismic waves** that radiate throughout the earth. Even though this energy decreases rapidly with increasing distance from the focus, instruments located throughout the world record the event.

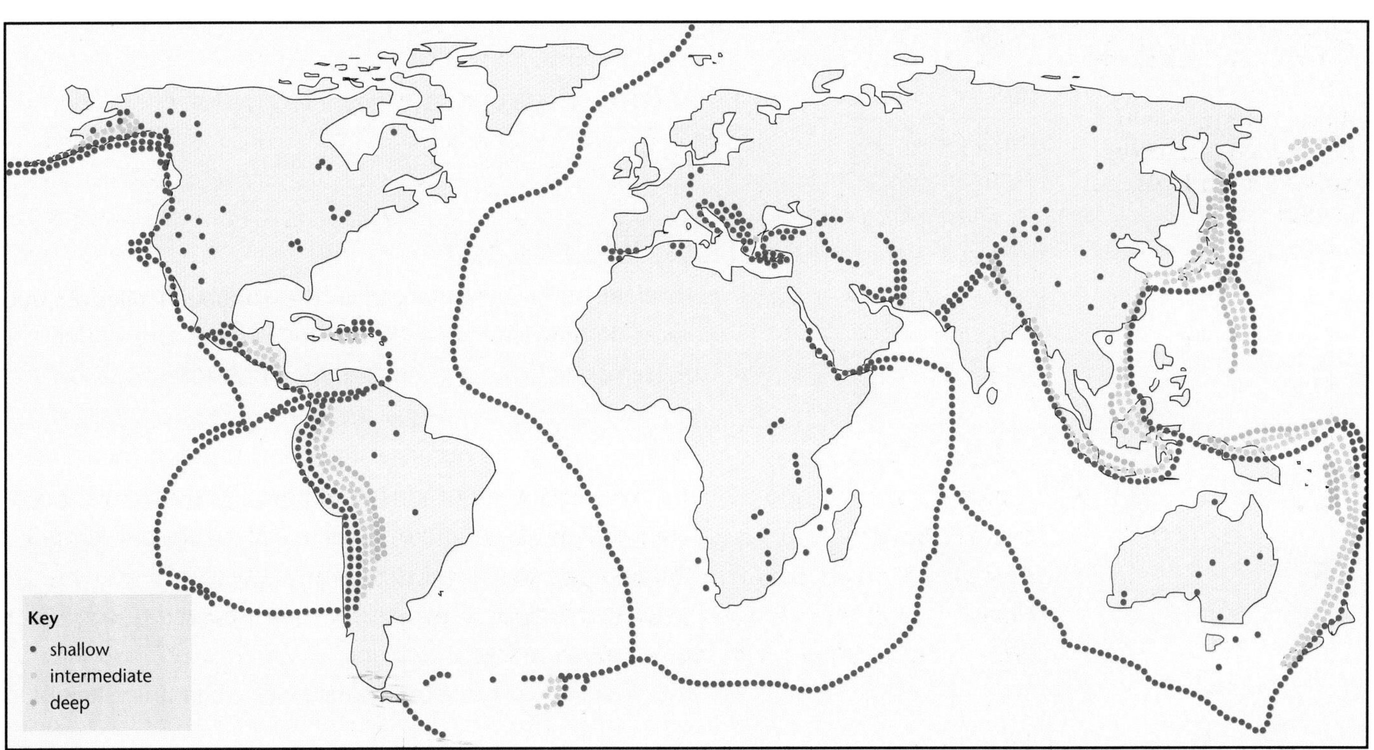


Fig. 5.1
World distribution of earthquakes. The epicentres of over 99 per cent of the earthquakes that occur each year are confined to the boundaries of the earth's crustal plates.

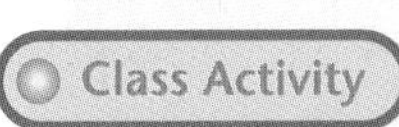


Examine the world distribution map of earthquakes, Fig. 5.1, the world distribution map of volcanoes, Fig. 4.7, p. 46, and the plates of the earth's surface, Fig. 3.1, p. 18. What similarities, if any, do you recognise on each of these maps?

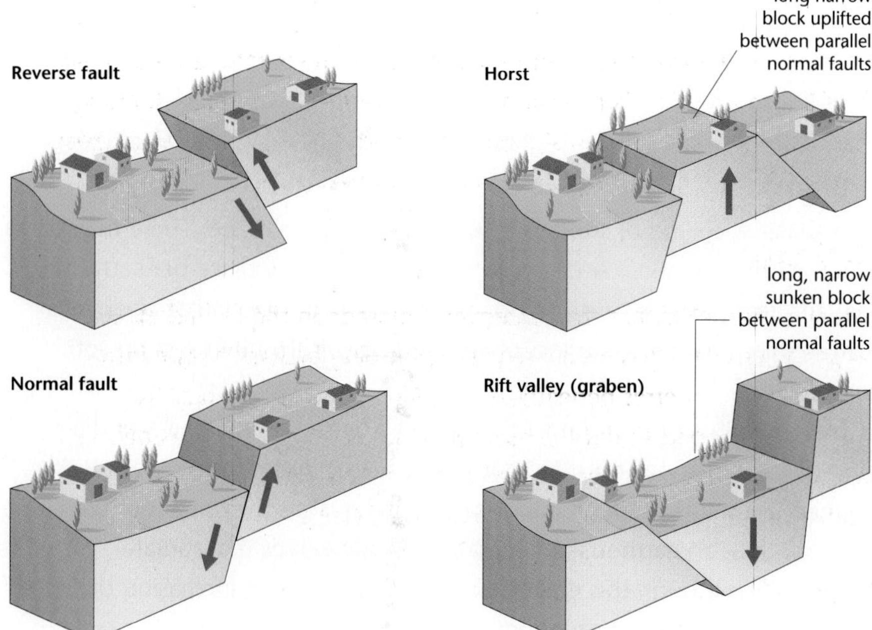


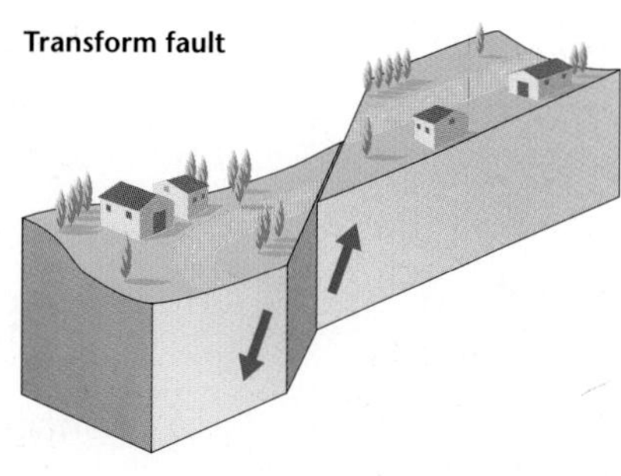


Fig. 5.2
Faults caused by earthquakes

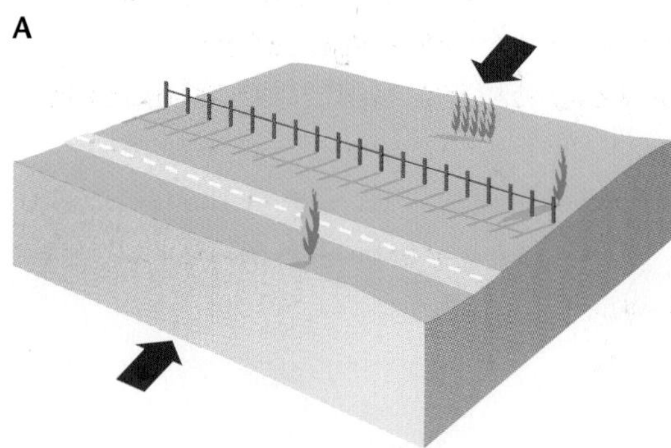


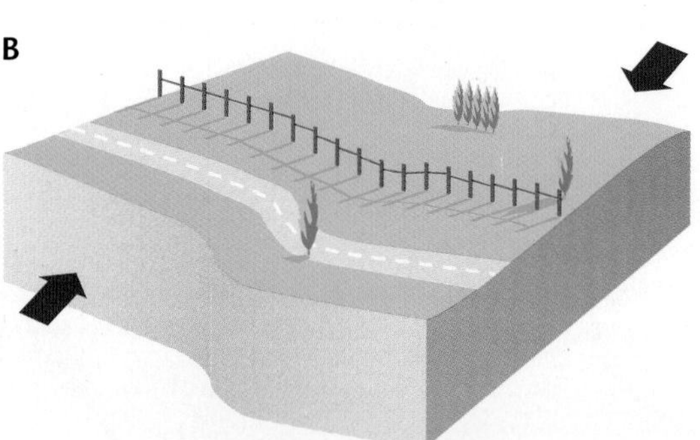


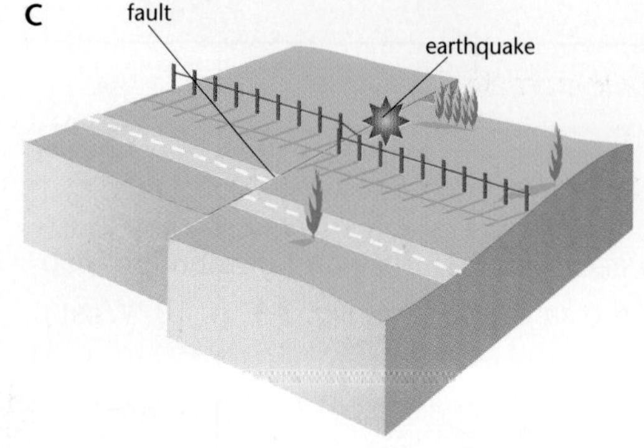


## What Causes an Earthquake?

Earthquakes occur for a number of reasons.

### 1. Elastic Rebound

As plates move ever so slowly, they deform the crustal rocks on both sides of a fault-line. (See Fig. 5.3.) Under these conditions, rocks are bending and storing 'elastic energy' the way a wooden stick would if bent. Eventually, slippage of a plate occurs at the weakest point (the focus). This displacement or slippage will exert stress further along the fault-line where more slippage will occur until most of the built-up strain is released. This slippage allows the deformed rock to 'snap back' to an unstrained position in a process called 'elastic rebound'. Rock behaves elastically much as a stretched elastic band does when it is released.

Fig. 5.3
The Elastic Rebound Theory of the Cause of Earthquakes
(A) Rock with stress acting on it.
(B) Stress has caused strain in the rock. Strain builds up over a long period of time.
(C) Rock breaks suddenly, releasing energy, with rock movement along a fault. Horizontal motion is shown; rocks can also move vertically.

**earthquakes**
Sudden tremors or vibrations in the earth's crust.

**focus**
The place directly beneath the earth's surface where an earthquake occurs.

**epicentre**
The spot on the surface directly above the focus.

**fault-lines**
The lines along which plates meet, e.g. the San Andreas Fault in California.

**Pacific Ring of Fire**
The largest earthquake and volcano zone, which lies around the edge of the Pacific Ocean.

**shock waves**
These may be compared to the ripples in a pool when a stone is thrown in.

## 2. Ice Age

Rebound effect may also be associated with the melting of the great ice sheets that covered much of North America, Europe and Asia thousands of years ago. Melting relieved the load that depressed the crust, and the hypothesis is that stress has been transferred to old faults as the crust slowly rises to its original level.

## 3. Subsurface Water

The series of small earthquakes that struck Denver, Colorado in the United States in 1964 were found to correspond precisely with the pumping of liquid waste into faulted rocks, thousands of metres beneath the outskirts of the city. When the pumping of the waste stopped so did the earthquakes. Some seismologists believe that some shallow earthquakes (earthquakes near the earth's surface) may be prevented or controlled by pumping water in and out of faults.

## 4. Ancient Faults

Fewer than 1 per cent of all earthquakes occur away from plate boundaries but some of these quakes have caused rampant destruction to populated regions. The magnitude 7.6 earthquake that hit the heavily populated industrial city of Tangshan in China in 1976 killed at least 250,000 people (the official figure); some estimate that up to 650,000 died.

Some believe that these devastating earthquakes may be related to reactivation of ancient faults, mainly deeply buried in the continental crust.

## Categories of Earthquakes

There are three categories of earthquakes:

1. shallow earthquakes
2. intermediate earthquakes
3. deep earthquakes

### Shallow-focus Earthquakes

Shallow-focus earthquakes occur at:

#### Rift Valleys

(a) Rift valleys which occupy a central line along mid-oceanic ridges.
(b) Rift valleys on land, e.g. the Rift Valley in Africa.

In both of these areas the crust is new and thin and splits easily and so is not capable of storing the strain energy necessary to generate large or deep earthquakes.

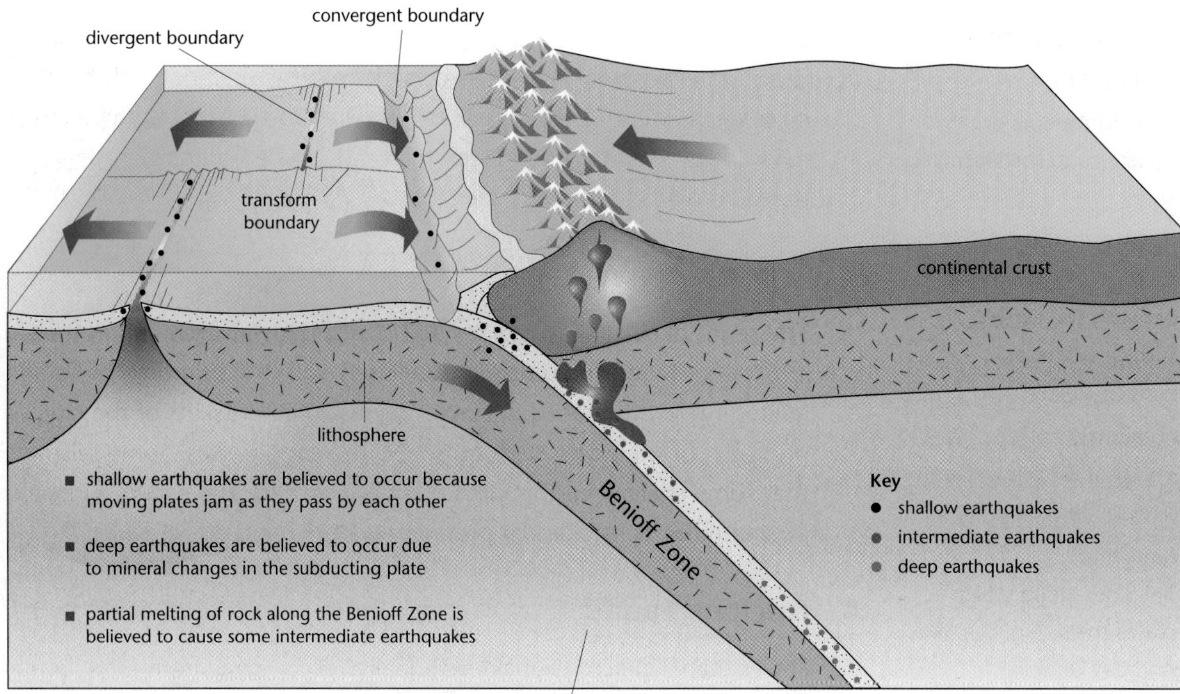


**Fig. 5.4**
Three categories of earthquakes — shallow, intermediate and deep — occur at various depths as an oceanic plate is subducted beneath a continental plate. These zones of earthquake activity are called Benioff Zones after the man who first recognised them. Such zones are found around the rim of the Pacific Ocean, the island arcs of south-east Asia and the eastern Mediterranean.

### Transform Boundaries

As plates slide past each other strain energy is built up. Rocks bend and store elastic energy until finally slippage occurs at the weakest point, called the focus. San Francisco and Los Angeles lie on a transform boundary, namely the San Andreas Fault.

### Subduction Zones

Shallow earthquakes also occur along the surface of zones of subduction. These include such areas as the deep ocean trenches that exist off the South American west coast and off the coast of Japan and the Philippines.

## Intermediate and Deep-focus Earthquakes

Virtually all the earth's deep-focus earthquakes are associated with subduction zones. The pattern of intermediate and deep-focus earthquakes forms continuous lines within the ocean crust that trace the paths of subducting plates as they descend into the mantle.

**Foreshocks** usually precede the main event as the rocks in the vicinity of the focus begin to crack. The main earthquake then arrives and can last from a few seconds to a few minutes.

**Aftershocks** follow as the rocks along the fault readjust and strain is transferred to the next jamming point, called the **asperity**. Then there is a period of quiet while strain accumulates again.

## Case Study: Japan

### The Distribution of Earthquakes with Depth

A Benioff Zone of earthquakes begins at an ocean trench and dips under a continent (such as South America). Detailed study of earthquakes at such destructive boundaries and associated deep ocean trenches shows that the foci (focuses) follow a well-defined pathway called a **Benioff Zone**, named after the scientist who first recognised this phenomenon.

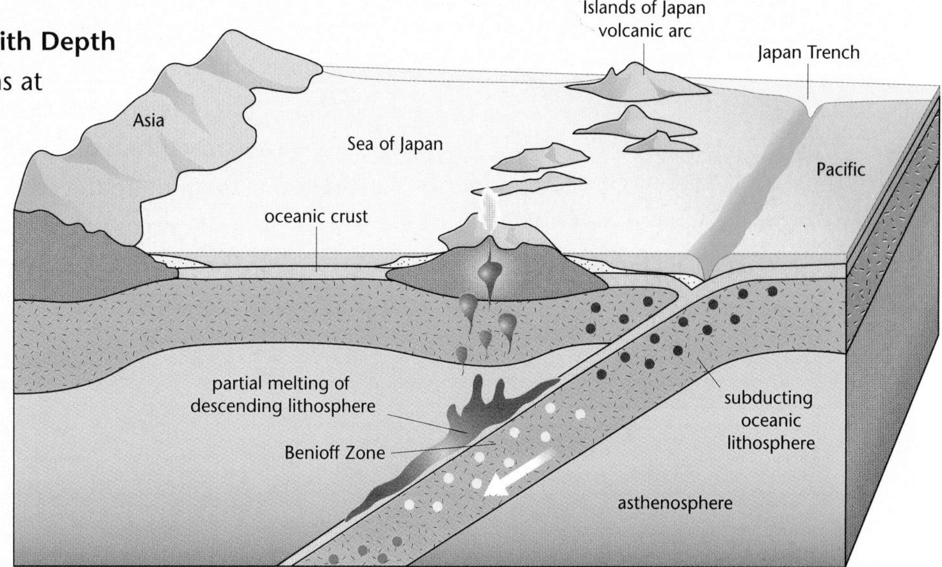


Fig. 5.5
Distribution of earthquake foci in the vicinity of the Japan Trench. Note that intermediate and deep-focus earthquakes occur only within the sinking slab of oceanic plate.

Class Activity

Study Figs. 5.5 and 5.6 and then answer the following:

1. Which earthquake type is nearest to the Japan Trench?
2. Which earthquake type is furthest from the Japan Trench?
3. Explain the pattern in the distribution of the various types of earthquakes.

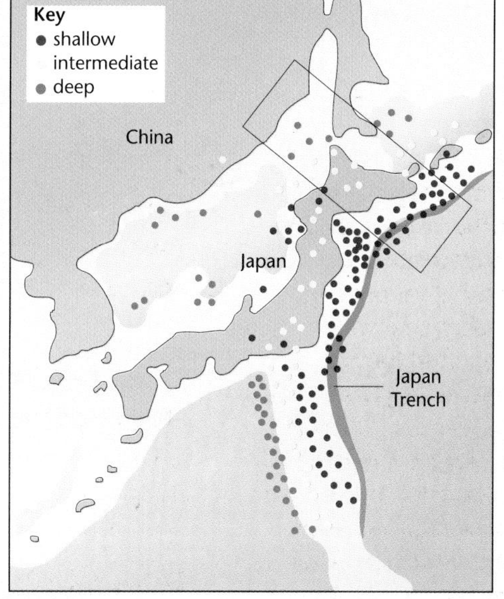


Fig. 5.6
Distribution pattern of earthquakes near Japan

## Recording and Locating Earthquakes

### Seismic Waves

An earthquake releases two classes of seismic waves:

1. body waves that travel through the interior of the earth
2. surface waves that travel mainly near the surface.

The body waves that travel through the interior of the earth are in turn divided into two types:

(a) the faster-travelling primary waves (P)
(b) the slower secondary waves (S).

## Body or Interior Waves

P-waves travel through liquids, solids and gas. They alternately compress and expand the rocks they pass through in an accordion-like motion. These are the first waves to arrive at a recording station.

S-waves travel through solids only (they cannot travel through liquids or gas) and are the second to arrive. They cause the rock to vibrate at right angles to the direction of the wave-path, much like what happens when you shake the free end of a rope tied to a pole.

## Surface Waves

Surface waves are last to arrive as they take the longest route along the surface.

During a powerful earthquake, the ground rises and falls and at the same time it sways from side to side, which is one reason for damage to foundations, sewers and pipelines.

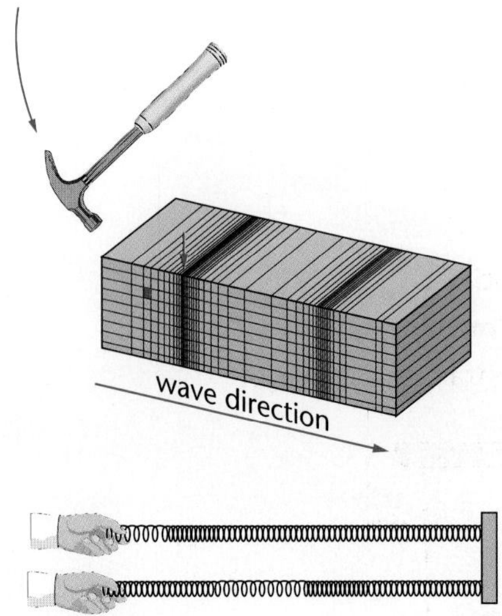


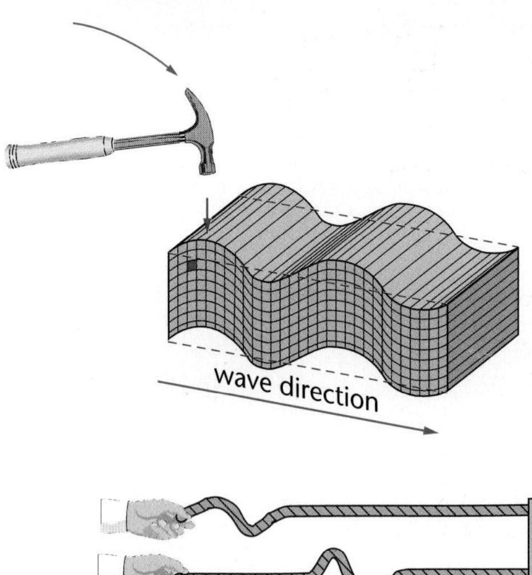


**Fig. 5.7**
Wave direction

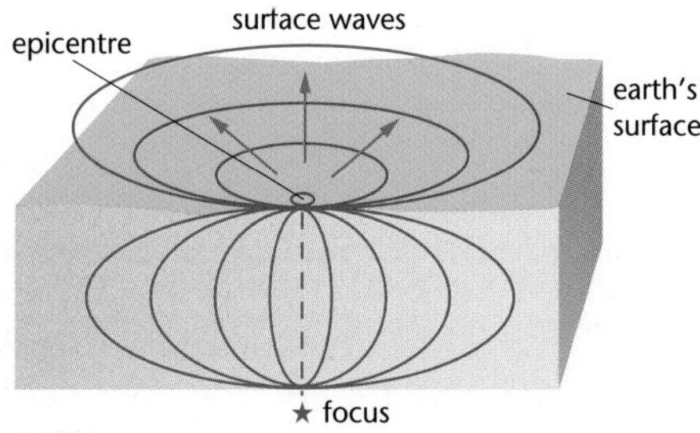


**Fig. 5.8**
The focus is the spot under the earth's crust where the earthquake originates. The epicentre is the spot on the earth's surface that is directly above the focus.

**foreshocks**
Less powerful earthquakes that may directly precede a higher-magnitude earthquake.

**aftershocks**
Less powerful earthquakes that may directly follow a higher-magnitude earthquake.

**elastic rebound theory**
The concept that earthquakes are generated by the sudden slippage of rocks on either side of a fault plane. In the process, the rocks release gradually accumulated strain energy and are returned to an unstrained condition.

## Recording and Locating Earthquakes

Earthquakes are recorded on a **seismograph**.

### Types of Seismographs

Figure 5.9 shows three types of seismograph:

1. a **rotating drum** which **measures vertical motion**
2. a **rotating drum** which **measures horizontal motion**
3. a **glass tube** which **records strain** — called a strain seismograph

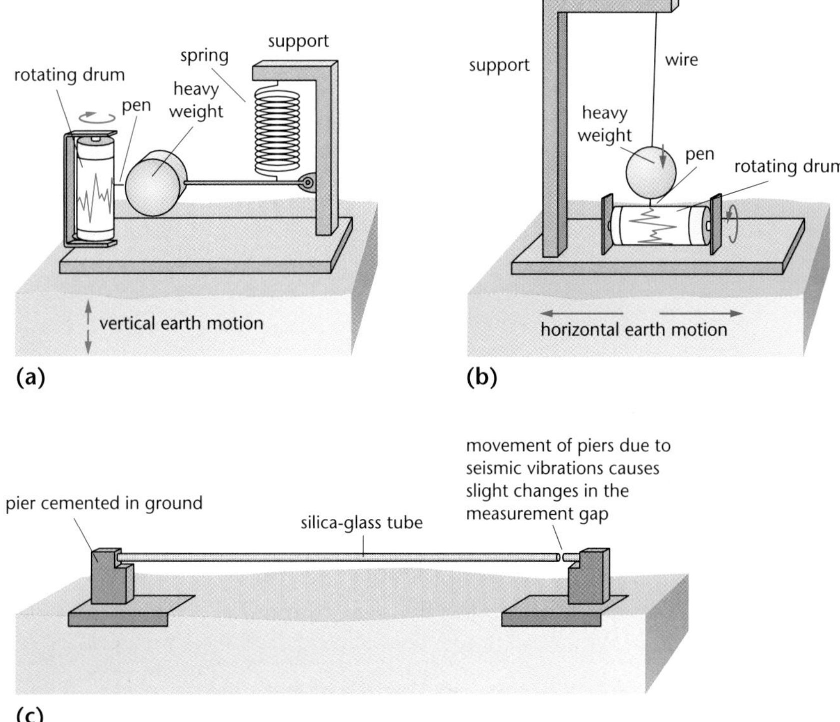


Fig. 5.9
Simple seismographs

## Locating the Epicentre

The epicentre of an earthquake can be located by examining the arrival times of the P-, S- and surface waves from at least three recording stations. Their arrival times indicate the distance the epicentre is from the recording station. Intersecting circles from these stations then locate the epicentre.

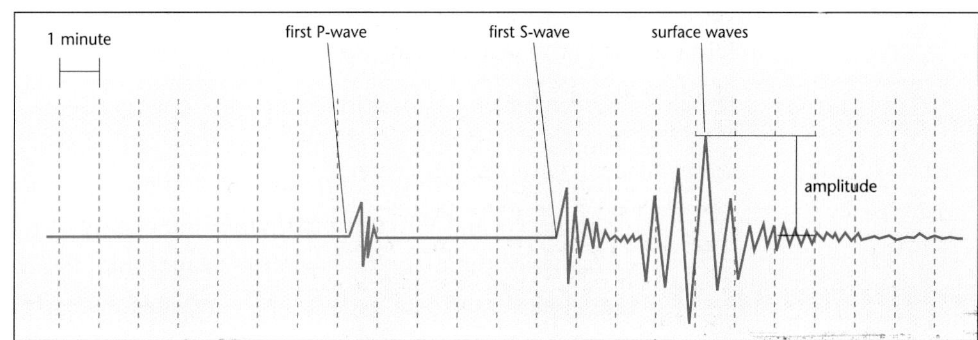


Fig. 5.10
Typical seismic trace. Note the time interval between the arrival of each wave type.

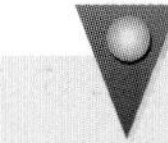

**seismologist**
A person who studies earthquake movements.

**Mercalli intensity scale**
A twelve-point scale that measures earthquake severity in terms of the damage inflicted.

**liquefaction**
The transformation of normal solid sediment or soil to liquid when ground shaking causes the particles to lose contact with each other.

## Locating the Focus

The principle used in locating the focus of an earthquake, which lies directly beneath the epicentre, is much the same as above. The greater the difference in arrival times of the shock waves, the greater is the depth of the focus.

### Earthquakes' Magnitude

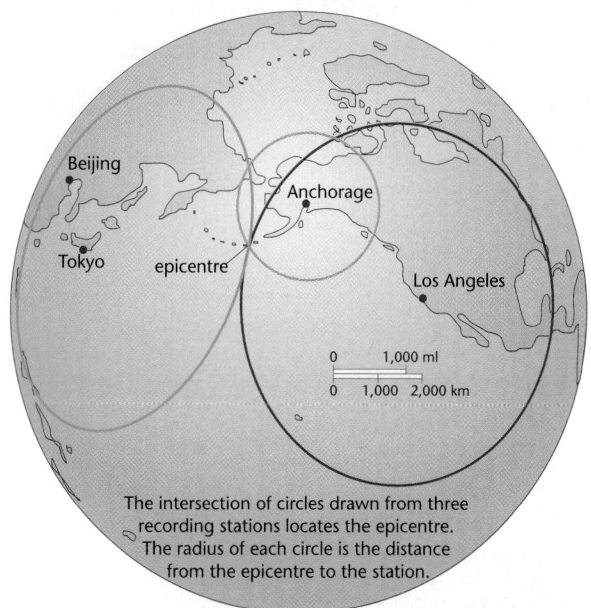
Beijing
Tokyo
epicentre
Anchorage
Los Angeles
0 1,000 ml
0 1,000 2,000 km
The intersection of circles drawn from three recording stations locates the epicentre. The radius of each circle is the distance from the epicentre to the station.

Fig. 5.11
Location of earthquake focus

Earthquakes vary enormously in strength. Great earthquakes produce traces having wave amplitudes (see Fig. 5.10) that are thousands of times larger than those generated by weak tremors.

### Magnitude on the Richter Scale

On the Richter scale an earthquake of magnitude 5 is ten times the amplitude (strength) of a magnitude 4 earthquake. Earthquakes of magnitude 8 and above, which occur once every few years, are classified as great earthquakes.

A magnitude 2 earthquake, although ten times the amplitude of magnitude 1, is still a very mild quiver undetectable by all but the most sensitive instruments. By contrast, a magnitude 8 earthquake is an enormous one. Magnitude measurement differs from energy release (see below).

### Energy Released by Earthquakes

Researchers have established that the energy release of an earthquake increases roughly thirty times for each magnitude number on the Richter scale. Thus a magnitude 7 earthquake releases thirty times the energy of a magnitude 6 earthquake and 900 (30 x 30) times the energy of a magnitude 5 earthquake. Because of the magnitude–energy relationship, the energy released during a great earthquake far exceeds the combined energy of the hundreds of thousands of small earthquakes that occur each year. At magnitude 8.9, the energy release is nearly one million times greater than from a single atomic bomb.

### Modern Seismographs

Modern seismographs are placed in the bedrock beneath the earth's surface. These ultrasensitive instruments can record the vibrations created by running cattle and horses some fields away.

**tsunami**
The Japanese word for a large deadly sea wave set in motion by an undersea earthquake or landslide.

Fig. 5.12

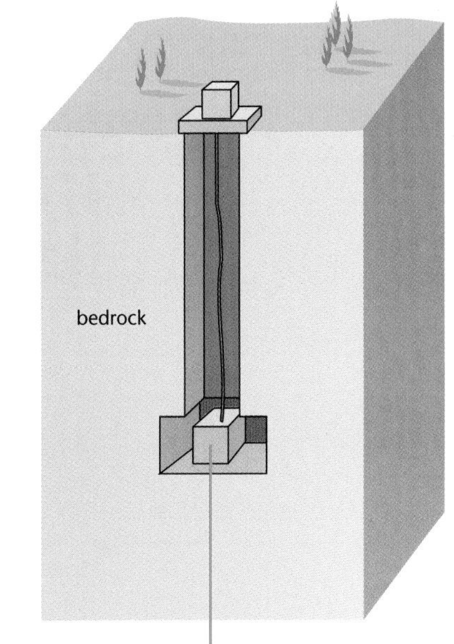


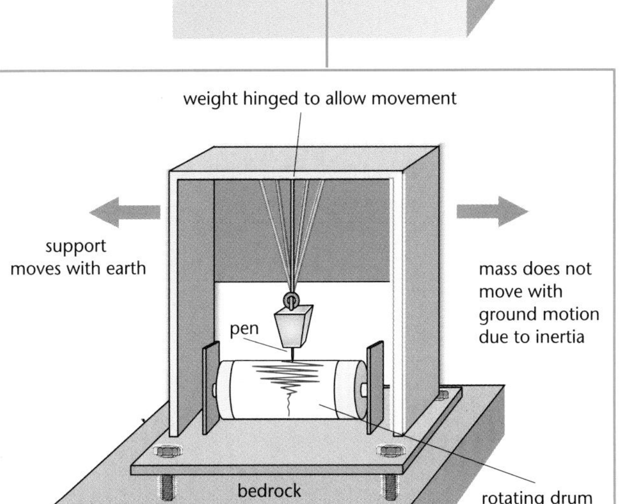


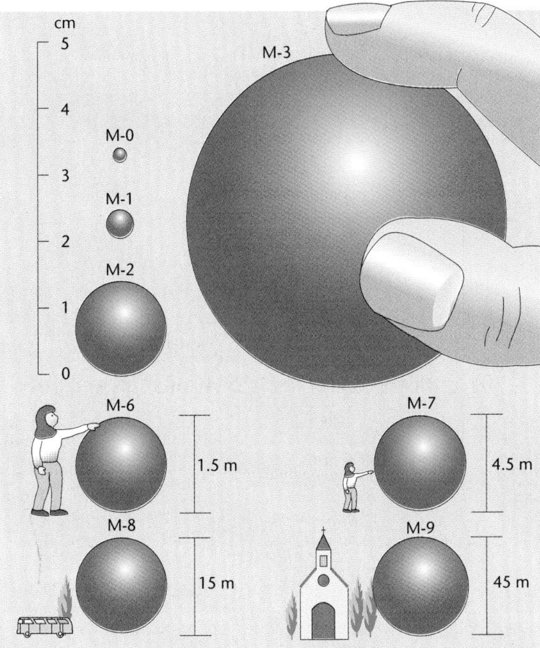


Fig. 5.13
Energy release compared for various Richter magnitudes in terms of the relative volumes of spheres

## Earthquakes: Processes and Damage

### Damage Caused by Earthquakes

The extent of the damage and loss of life caused by an individual earthquake depends on a number of factors:

1. magnitude of earthquake
2. depth of focus

3. the types of rock and soil through which the waves travel
4. the proximity of the epicentre to population centres
5. the buildings and utilities (water supply, gas pipes, etc.) affected
6. the time of occurrence, i.e. day or night, rush hour, etc.

**A Simplified Mercalli Intensity Scale**

| Number on scale | Type | Effect of earthquake on skeleton Mercalli Scale |
|---|---|---|
| 1 | Negligible | Detected by seismographs only |
| 3 | Slight | Vibration like the passing of a heavy truck; generally felt by many people indoors |
| 5 | Not very strong | Felt by all except heavy sleepers; plaster cracks, windows break, unstable objects are overturned |
| 8 | Destructive | Panic; chimneys, factory stacks and monuments fall; slight damage to well-designed buildings but heavy in poorly built structures |
| 10 | Disastrous | Panic, many buildings collapse; large landslides; ground badly cracked; rails bent |
| 12 | Superpanic | Damage nearly total: waves seen on the ground; objects flung into the air |

Fig. 5.14

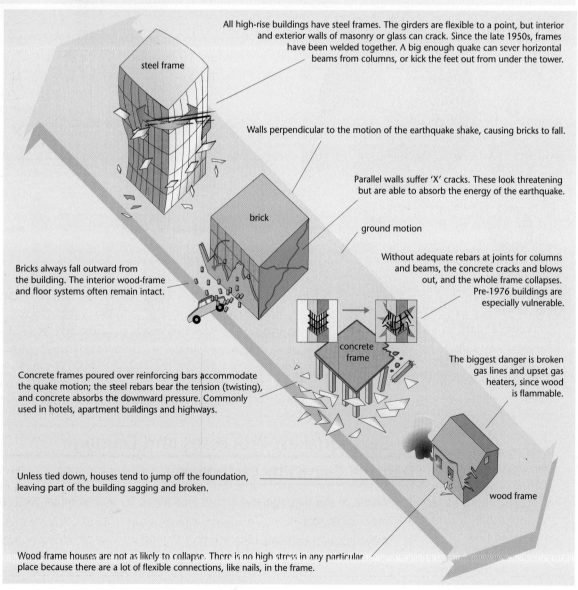

All high-rise buildings have steel frames. The girders are flexible to a point, but interior and exterior walls of masonry or glass can crack. Since the late 1950s, frames have been welded together. A big enough quake can sever horizontal beams from columns, or kick the feet out from under the tower.

steel frame

Walls perpendicular to the motion of the earthquake shake, causing bricks to fall.

Parallel walls suffer 'X' cracks. These look threatening but are able to absorb the energy of the earthquake.

brick

ground motion

Bricks always fall outward from the building. The interior wood-frame and floor systems often remain intact.

Without adequate rebars at joints for columns and beams, the concrete cracks and blows out, and the whole frame collapses. Pre-1976 buildings are especially vulnerable.

concrete frame

Concrete frames poured over reinforcing bars accommodate the quake motion; the steel rebars bear the tension (twisting), and concrete absorbs the downward pressure. Commonly used in hotels, apartment buildings and highways.

The biggest danger is broken gas lines and upset gas heaters, since wood is flammable.

Unless tied down, houses tend to jump off the foundation, leaving part of the building sagging and broken.

wood frame

Wood frame houses are not as likely to collapse. There is no high stress in any particular place because there are a lot of flexible connections, like nails, in the frame.

## Liquefaction

Where unconsolidated materials, such as silt or sand, are saturated with water, earthquakes can generate a phenomenon known as liquefaction. Under these conditions, what had been a stable soil turns into a fluid that is no longer capable of supporting buildings or other structures.

As a consequence, underground objects, such as storage tanks and sewer pipes, may float towards the surface of their newly liquefied environment. Buildings may sink or collapse.

## Tectonic Plates, Cities and People

1997 (ii).

# Shifting plates put cities on the edge of disaster

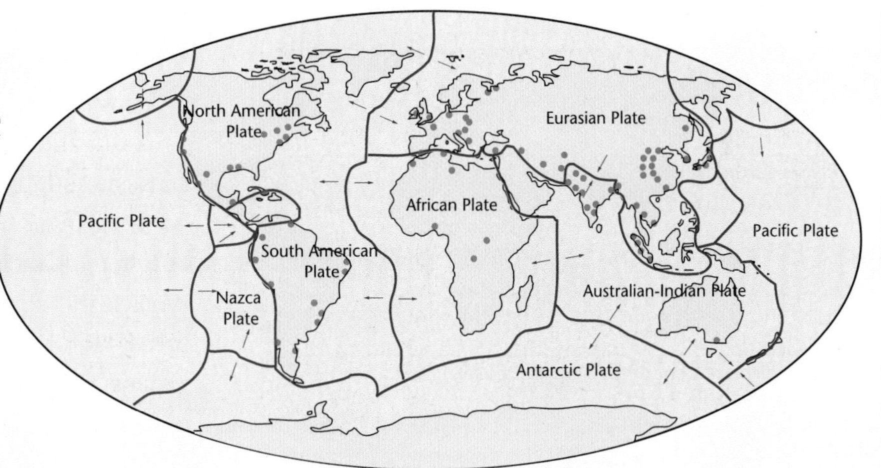


Fig. 5.15
Major cities of the world (•) in relation to tectonic plates

The earthquake in Kobe, Japan on 17 January 1995 captured media headlines and evoked worldwide sympathy for the victims and survivors.

Horrific earthquakes have become more common because of where many of the large and fast-growing cities are located.

A recent survey shows that about 40 per cent of the world's largest cities lie either close to a tectonic plate boundary or near where an earthquake has occurred. Each of these cities will have over two million inhabitants by 2000 and more than 600 million people will live in these cities by 2035.

Since they lie in areas of higher seismic risk, the stage may be set for unprecedented disasters.

Figure 5.15 shows the world's plate boundaries in relation to major cities.

Most earthquake deaths are caused by falling buildings. It follows, therefore, that the construction of buildings resistant to earthquakes should lessen the loss of life. It is widely acknowledged that much of the damage in a recent Armenian earthquake resulted from the collapse of poorly constructed buildings.

In recent years, many new multi-storey apartment blocks were built in the region to cope with the growing population. These were often built with low-quality concrete and badly designed joists, and lacked steel reinforcements. Similarly, many fast-growing cities in regions of high seismic risk contain only basic shelter for people and not earthquake-resistant buildings.

While much of the damage from future earthquakes will probably occur in the Third World, cities in the developed world are certainly not earthquake-proof. Tokyo is probably the most earthquake-conscious city in the world. It was destroyed by an earthquake in 1923 which killed more than 100,000 people.

Elaborate precautions have been incorporated into today's city to limit the damage that a future earthquake might cause.

In spite of this, a recent study estimates that an earthquake measuring 9.5 on the Richter scale occurring just east of Japan (similar to the 1923 earthquake) would destroy about 2.5 million buildings, and cause about 150,000 deaths and 200,000 injuries.

LANDSCAPES OF THE WORLD

Read the article 'Shifting plates put cities on the edge of disaster', then answer the following questions:

1. Why may horrific earthquakes become more common in the future?
2. Identify ten major cities of the world that lie on or close to areas of high earthquake risk.
3. Why are these areas of the earth's crust regarded as zones of high earthquake risk? Explain fully, using labelled diagrams and examples.
4. Why do some earthquakes cause more deaths than others, apart from the fact that they are more powerful? According to the article, what can be done to lessen the number of deaths in zones of high earthquake risk?

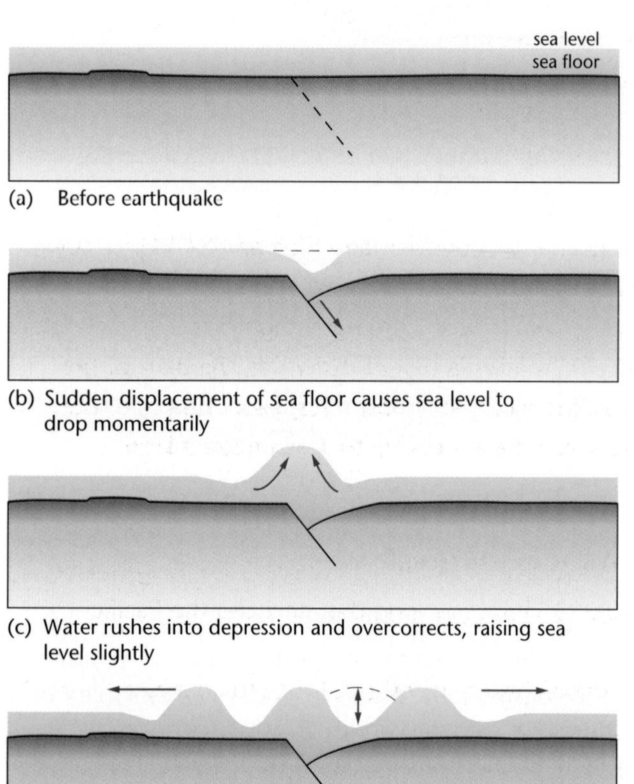
sea level
sea floor
(a) Before earthquake
(b) Sudden displacement of sea floor causes sea level to drop momentarily
(c) Water rushes into depression and overcorrects, raising sea level slightly
(d) Sea level oscillates before coming to rest; long, low waves (tsunamis) are sent out over sea surface

**Fig. 5.16**
How tidal waves form

## Plate Tectonics, Tidal Waves and People

**Tsunami**, a Japanese word, is the correct term for what is called a tidal wave.

A tsunami can be caused by any great event that causes a sudden large change in sea level such as an earthquake on the ocean floor, a submarine volcanic eruption, a landslide or even, very rarely, a meteorite impact. If the sea floor is uplifted five metres, a five-metre-high hump of excess water is created at the surface. Such a hump collapses to generate a succession of waves that radiate outward across the surface of the ocean and may cause damage and loss of life thousands of miles from their origin.

### What is a Wave?

Most waves are generated by the wind. Their depth is determined by their wavelengths, the distance from wave crest to crest. This length is short because only the sea surface makes contact with the wind. As the wave approaches the shore **it starts to 'feel'** the rising **bottom**, which slows it down. To carry the same amount of energy, the **height** of the wave **increases**. The wave then releases its energy by breaking. All ordinary waves, even the six-metre-high monsters that surfers love, release their energy relatively harmlessly by breaking on the shoreline, and this is repeated in the space of a few seconds.

But the **wavelengths** of tsunamis are **very long** — 100 to 500 kilometres. Therefore these waves set the **entire ocean** in motion, even in the 5-kilometre-deep ocean. Tsunamis cross an ocean of this depth at a **speed** of almost **800 to 1,000** kilometres per hour, just under the cruising speed of a jet plane.

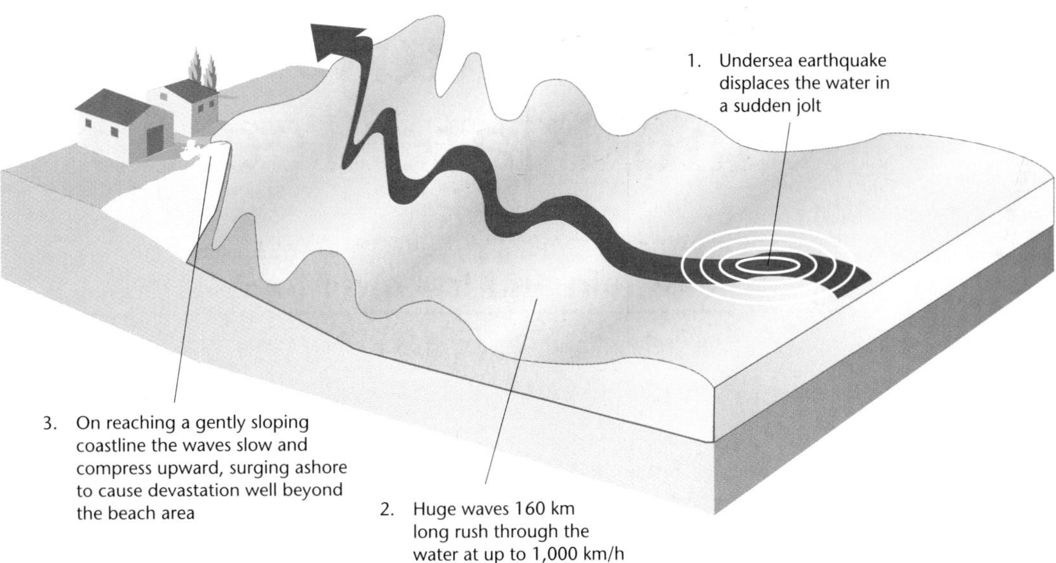
1. Undersea earthquake displaces the water in a sudden jolt
2. Huge waves 160 km long rush through the water at up to 1,000 km/h
3. On reaching a gently sloping coastline the waves slow and compress upward, surging ashore to cause devastation well beyond the beach area

Fig. 5.17
Devastating tidal waves, or tsunamis, occur when an earthquake at sea combines with shallow coastal waters

Consider a wave of 800 to 1,000 kilometres an hour, 1 metre high and 160 kilometres long rushing towards the shoreline. In the shallow water the front of the wave suddenly slows to about 30 kilometres per hour as 'it feels bottom', its wavelength shrinks to about 8 kilometres and the wave grows to about 5 metres in height. However, the back of the wave, still in deep ocean water, keeps rushing along at almost full speed. The wave races onto the beach without breaking and can devastate areas up to 1 kilometre inland.

## Damage Caused by Tsunamis

Tsunamis are thankfully rare in the Atlantic Ocean. Most originate along the 'Ring of Fire', a zone of volcanoes and earthquakes 40,000 kilometres long that encircles the Pacific Ocean.

The Japanese coast is especially vulnerable; history records at least fifteen major disasters there, some killing hundreds of thousands of people.

In 1933 an earthquake measuring 8.3 on the Richter scale caused a tsunami that killed 3,008 people in northern Japan.

The biggest tidal wave in modern history was the Sanriku tsunami, which hit northern Japan on 15 June 1886, killing an estimated 27,000 people following a jolt measuring 7.6 on the Richter scale.

Bangladesh also suffers from tsunami damage as it is a low-lying delta region along a narrowing continental coastline in the bay of Bengal in the Indian Ocean. This narrowing sea causes tsunamis to rise well above normal sea levels. Such waves regularly devastate coastal communities.

# Case Study: Ireland and Plate Tectonics

## Britain and Ireland Move North

The 'Ireland' and 'Britain' that we recognise today did not always exist, nor were they always lying between latitudes 50°N and 61°N.

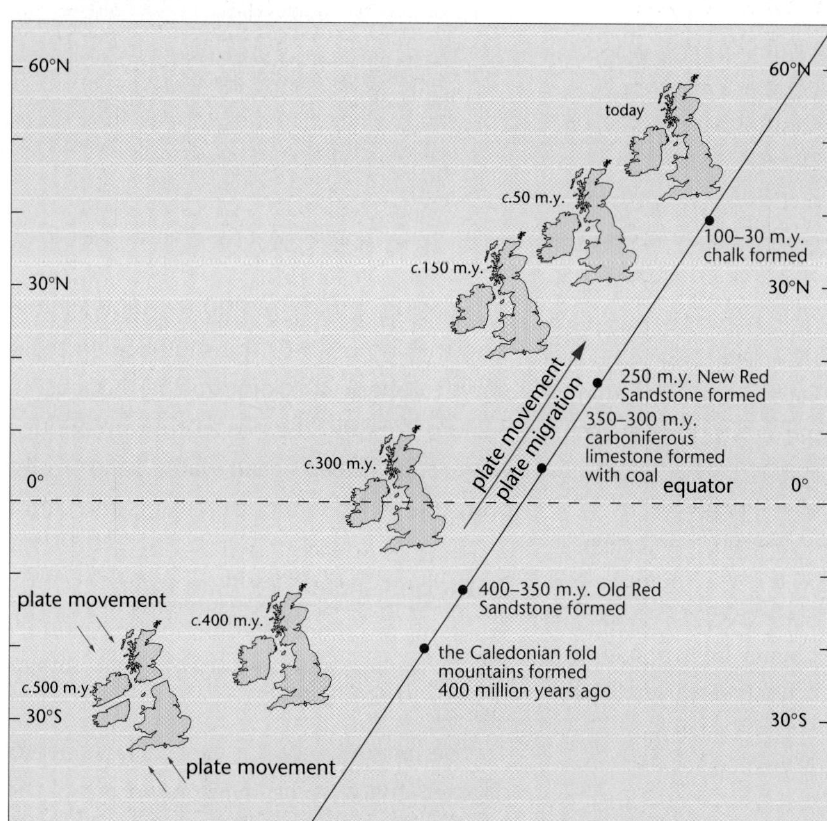


**Fig. 6.1**
Plate tectonics — Britain and Ireland — a summary

The Iapetus Ocean is often referred to as the Proto-Atlantic

American Plate North-western half of Ireland, Scotland

north-western Ireland

Iapetus Ocean

direction of plate movement

Wales and England

south-eastern Ireland

direction of plate movement

European Plate South-eastern half of Ireland, England, Wales

**Fig. 6.2**
The birth of Britain and Ireland

## Caledonian Fold Mountains, Fault-lines and Mineral Deposits

Ireland was formed about 400 million years ago at the end of the Silurian period. The American and European plates came together to form a great Euroamerican plate and the Caledonian fold mountains. (See p. 105.) As the plates approached each other, large pieces of continental crust, called terranes, each with its own distinctive geological character, were scraped from the surface of the ocean floor and pushed together. These terranes are divided by great north-east/south-west fractures and form Ireland's oldest rocks, which occur deep beneath the surface.

The fractures may have acted as routes for later hot-metal-rich fluids (hydrothermal solutions) that created the mineral deposits of the midlands such as Silvermines, Galmoy, Lisheen and Tara Mines near Navan in Co. Meath.

## Ireland's Buried Terranes

**What is a terrane?**
See p. 36.

**What are minerals?**
Minerals are the building blocks of the earth's crust. They are chemical elements or compounds which occur naturally within the crust.

**What are rocks?**
Rocks are aggregates or mixtures of minerals, the composition of which may vary greatly. Limestone, for example, is composed primarily of one mineral — calcite. Granite, on the other hand, typically contains three minerals — feldspar, mica and quartz.

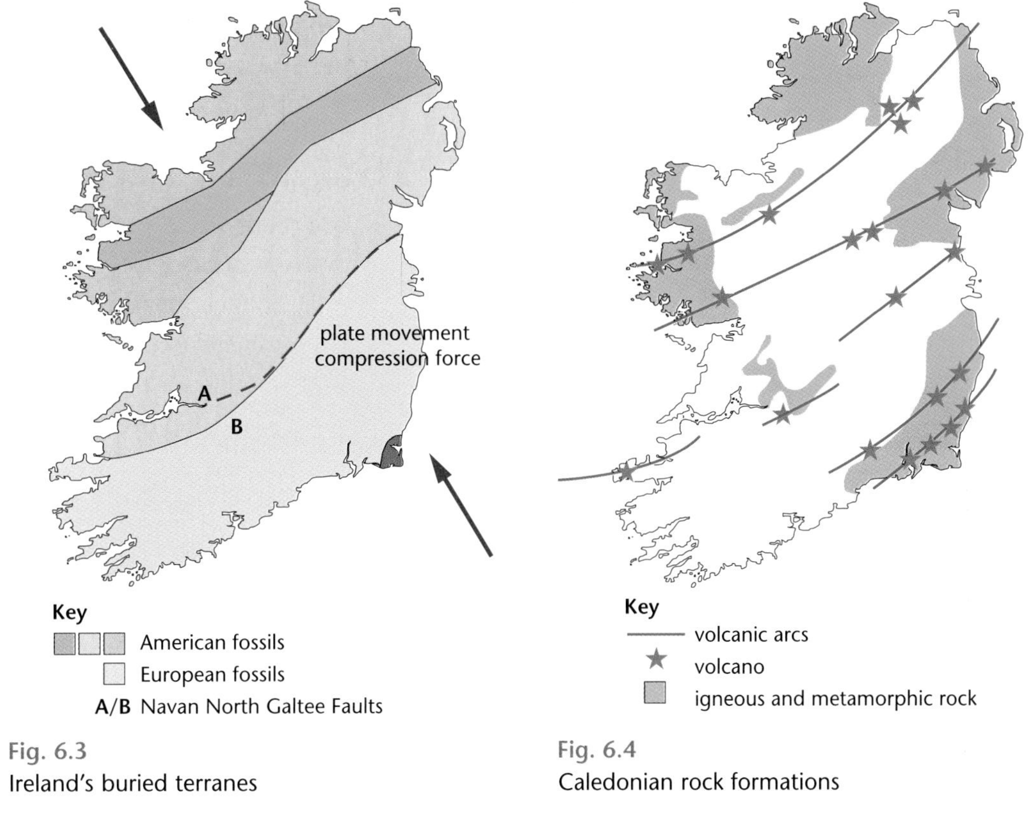


Fig. 6.3
Ireland's buried terranes

Fig. 6.4
Caledonian rock formations

Examine Fig. 6.3 showing Ireland's buried fault-lines and Fig. 6.4 showing the volcanic arcs associated with them. What similarity, if any, do you see in their distribution patterns?

## Ireland's Mineral Deposits Today

During the cooling process of the Caledonian mountain building era, liquids, with minerals in solution (hydrothermal fluids), accumulated near the top of the batholith magma deep within the mountains. Over millions of years, these hot fluids migrated great distances upward through fractures and joints of the overhead rocks, and created mineral deposits such as silver and gold veins.

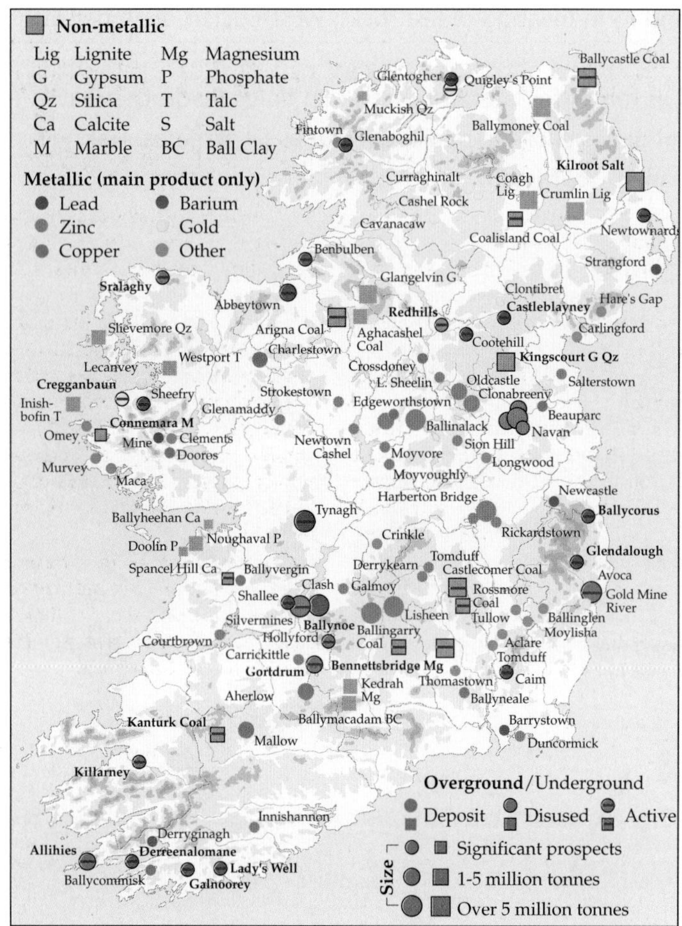


**Fig. 6.5**
Metalliferous mines, 1994. While the more important mineral deposits are shown, not all the workings are in continuous production. A significant trend of recent decades has been the discovery and exploitation of large deposits of heavy metals in the central lowlands, for example the copper, mercury and zinc mines at Tynagh (Co. Galway) and at Navan (Co. Meath), the latter of which is one of the greatest discoveries of zinc in the world. Major developments (lead, zinc) are now under way at Galmoy in Co. Kilkenny and Lisheen in Co. Tipperary. In the west of Ireland, gold is a current source of interest, located in rocks which underlie areas of great natural beauty and tourist appeal.

The mineral veins of Connemara, Mayo and Wicklow formed during or shortly after the Caledonian Foldings. These mineral deposits have an igneous origin (directly associated with magma) and are in igneous rock.

However, the ores of the central plain formed in the limestones some time after they were laid down in the Carboniferous period. Minerals such as zinc, lead and copper deposits formed as the hydrothermal fluids from deep down seeped into billions of hairline fractures and became part of the rock.

Today, Ireland is fast becoming a major mineral ore source/producer on a world scale. With the coming on stream of Lisheen in 1999, the country will be the world's seventh largest mineral producer in terms of tonnage of ore exported. The vast deposits of the central plain hold countless billions of Euros worth of zinc and other mineral deposits. Many major mining corporations are searching for more minerals. As a consequence of a 'new' extraction process, called the 'Froth Flotation Process', fine-grained ores, such as the zinc deposits of Galmoy and Lisheen, are now greatly sought after. Ireland's deposits are especially desired as they are close to the surface and have a high concentration of minerals in the ores.

**Class Activity**

1. Study Fig. 6.3, Ireland's buried terranes, Fig. 6.4, Caledonian rock formations, and Fig. 6.5, non-metallic and metallic mineral deposits. From your study of plate tectonics, so far, suggest some reasons for the pattern in the distribution of mineral deposits as shown in Fig. 6.5.

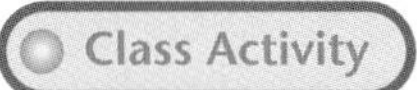


# Earthquakes in Britain and Ireland

What is the risk that Ireland would ever suffer a significant earthquake? Ireland is a stable land with little risk of a major earthquake. The early annals record earthquakes felt in Ireland but most did little or no damage. An exception was Sligo in 1490 when it was recorded that 100 people were killed along with many horses and cows, and a lake opened up.

There was a gentle tremor of 2.1 on the Richter scale, for example, on 21 November 1984, in Donegal, and a similar event in Midleton in Co. Cork in 1981. Some experts believe that some 5,000 years ago earthquakes hitting 6 or 7 were not unknown in Ireland, perhaps associated with an elastic resurgence of crustal layers relieved of the considerable weight of the recently retreated glacial ice.

Most earthquakes felt in Ireland are minor and many originate in Britain or the Irish Sea. There is a zone along the Irish coast from Cork to Dublin that is active and occasionally produces small earthquakes. The most active area is near Enniscorthy, Co. Wexford. In 1985 this centre produced an earthquake of magnitude 2 on the Richter scale. Activity was also reported in September 1988.

Irish earthquakes are minor events caused by the relaxation of stressed rocks. They are not connected with plate movement. The nearest plate boundary to Ireland is the Mid-Atlantic Ridge, about 2,500 km to the west. This is an active area and it is not impossible, although highly unlikely, that a major earthquake on the ridge could cause damage in Ireland.

In 1755 such an earthquake destroyed Lisbon, killing 60,000 people. The damage was completed by a tidal wave. These are produced at the sea bottom by the earthquake shock. They have great energy and wreak havoc when they reach land. Perhaps ancient tales of huge unexpected waves hitting the Irish coast record tidal waves produced by earthquakes in the mid-Atlantic.

On 1 June 1246, for instance, it is recorded that 'there happened so great an earthquake in England that the like had seldom been seen or heard. In Kent it was more violent than in other parts of the Kingdom, where it overturned several churches.'

Then in April 1580, in the same area, 'three distinct shocks were felt, so that at Dover, part of the "white cliffs" fell into the sea, carrying away a portion of the castle wall'. The same tremors caused masonry to fall from the eaves of St Paul's Cathedral in London and an apprentice was killed by stones falling from the nearby Christ Church. The greatest earthquake in Britain in recent years was centred near Colchester in Essex, and occurred at 9.18 a.m. on 22 April 1884; many chimney stacks were thrown down, and within a radius of seven miles of the town more than 1,200 buildings had to be repaired.

Article A

# Survey reveals new earthquake threat

A new survey has found large faults in rocks deep beneath Ireland and Britain — thus dispelling the belief that north-western Europe is earthquake-free.

The British Geological Survey revealed the presence of huge unsuspected faults which could generate tremors. About 300 small 'quakes shake Britain and Ireland every year with larger 'quakes occurring every decade.

Roger Musson, one of the authors of the report, said it vindicates the view of many geologists that both countries have been too lax in their earthquake precautions.

The report showed Cork and Donegal are the most earthquake-prone zones in Ireland.

Thousands of years ago, south Munster and north-west Ulster were hit by 'quakes which would have measured up to seven on the Richter scale — similar in intensity to 'quakes which wreaked havoc in Kobe in Japan in 1995 and Los Angeles in 1994.

Donegal was hit by a tremor in November 1984 which shook homes in Dunfanaghy. Three years earlier, a tremor hit Midleton, Co. Cork.

Article B

**Article A**

Study the article from the *Irish Times* newspaper and then answer the following questions:

1. What reason is given in the article for the earthquakes that affected Ireland some 500 years ago?
2. (a) Why, according to the article, is it unlikely that Ireland would be affected by earthquakes?
   (b) From your studies of terranes in Ireland, p. 75, and ancient faults in China, p. 63, is it wise to discount the possibility of earthquakes in Ireland?

**Article B**

1. What do you think is the origin of these deep faults under Ireland?
2. Which parts of Ireland are the most susceptible to earthquakes?

## Leaving Certificate Questions on Plate Tectonics and Associated Processes

Ordinary Level

**1998**

Major Earthquakes and Volcanoes occur in similar regions. Discuss.

**1994**

(i) Fold Mountains, Rift Valley, Volcano:
Select any **TWO** of the above landforms and for **EACH** one you select:
- Name one example of the landform.
- Describe and explain, with the aid of a diagram, how it was formed.
**(50 marks)**

(ii) 'Earthquakes can have major effects on human society and on the physical landscape.'
Explain this statement with reference to examples you have studied. **(30 marks)**

**1993**

(i) A study of patterns in the worldwide distribution of volcanoes and earthquake zones can help us to understand the causes of these events.
Examine this statement with reference to examples which you have studied.
**(50 marks)**

(ii) Describe some of the effects on human societies of the occurrence of a volcanic eruption or a major earthquake in populated areas. **(30 marks)**

**1991**

Examine this diagram, which shows some of the landforms produced by volcanic activity, and answer the questions which follow:

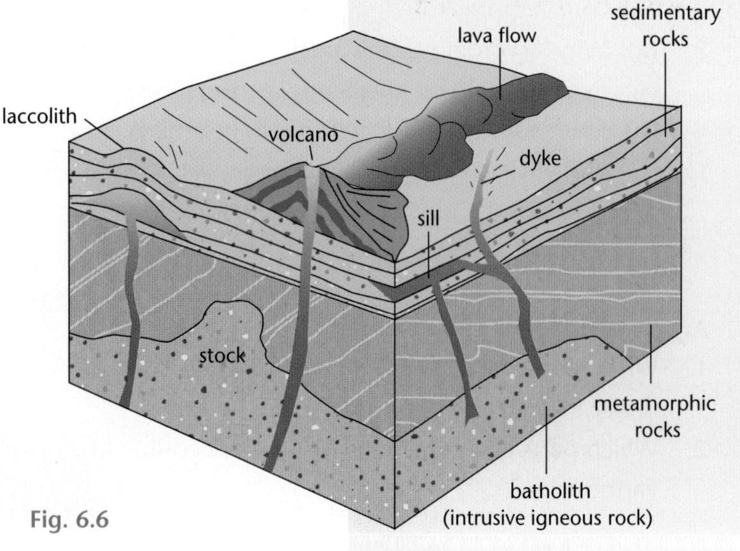


Fig. 6.6

(i) In the case of any **THREE** of these landforms, describe and explain the processes which shaped them. **(45 marks)**

(ii) Examine briefly ways in which volcanic activity can be of economic benefit to people. **(35 marks)**

**1990**

(i) Describe and explain the formation of mountains by Folding
**AND** by Faulting of the earth's crust. **(40 marks)**

(ii) Examine some of the consequences of a major earthquake occurring in a densely populated region. **(40 marks)**

**Higher Level**

**1997**

(i) Examine how Plate Tectonics has revolutionised our understanding of the geographical distribution and the causes of earthquakes. **(75 marks)**

(ii) 'The human cost of major earthquakes is influenced by socio-economic factors.' Analyse the accuracy of this statement. **(25 marks)**

**1996**

(i) Discuss how the Theory of Plate Tectonics helps us to explain the occurrence of volcanic activity and of earthquakes. **(60 marks)**

(ii) Examine some of the immediate **AND** the long-term effects of **EITHER** a volcanic eruption **OR** an earthquake which you have studied. **(40 marks)**

**1995**

The processes of folding, faulting and volcanic action have produced distinctive landforms worldwide.

(i) With reference to **ONE** landform in **EACH** case, examine how these processes have helped to shape the surface of the earth. **(60 marks)**

(ii) In the case of **ONE** of these landforms, examine **ONE** positive and **ONE** negative example of how human societies have interacted with it. **(40 marks)**

**1993**

(i) Earthquakes and volcanoes occur in quite predictable locations on the globe. Examine the theoretical basis for this statement. **(60 marks)**

(ii) The frequency of occurrence of earthquakes and volcanic eruptions is much more difficult to predict than their location. Assess the accuracy of this assertion with reference to examples you have studied. **(40 marks)**

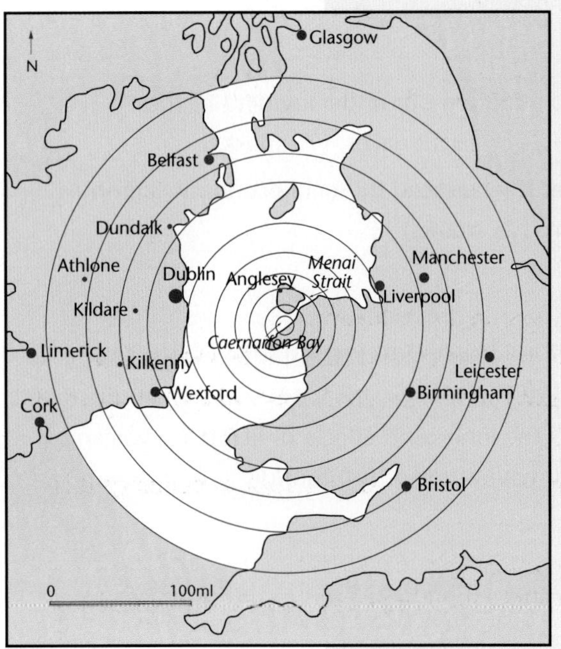


Fig. 6.7

**1992**

'The shudder which shook buildings in Dublin and most towns and villages of Leicester early yesterday morning marked the most severe earthquake registered in this country since scientific records began early in the present century. The earthquake — registering 5.5 on the Richter scale — demolished the myth that dangerous earthquakes cannot happen in this part of the world; in Britain, it has given rise to a new debate about the safety of nuclear power stations.' (*The Irish Times*, 20 July 1984)

(i) In the passage above, the writer refers to the 'myth' that dangerous earthquakes cannot happen in Ireland or Britain. Examine the current theories about the causes of earthquakes and explain why this part of the world has always been regarded as being safe from such events. **(50 marks)**

(ii) Explain some of the possible effects — including those referred to in the passage — of a severe earthquake happening in these islands and outline what steps governments could take in order to lessen these effects. **(50 marks)**

**1991**

(i) Describe and explain how the Theory of Plate Tectonics has added to our understanding of the forces which shape the surface of the Earth. **(75 marks)**

(ii) Examine how some of these forces can be harnessed to human advantage. **(25 marks)**

**1990**

Plate Tectonic Theory states that the crust of the Earth is made up of a number of rigid plates, which are in motion relative to each other, as they 'float' on the Mantle beneath.

(i) Explain how Plate Tectonic Theory has helped us to understand the world distribution of:

- Fold Mountains **AND**
- Volcanic Island-arcs **(60 marks)**

(ii) Another of the consequences of the movement of crustal plates is the occurrence of earthquakes. With reference to appropriate examples, describe the impact which a major earthquake can have on a human population and examine briefly attempts to lessen that impact. **(40 marks)**

# Understanding Maps and Aerial Photographs

## Reading the Photograph

Some important ways of examining a photograph are as follows:

1. A general glance at a photograph will give the reader a reasonable and generally sufficient amount of information to answer most questions, especially when dealing with topographical landforms.
2. For urban studies and the examination of individual features such as buildings it is recommended that the student do two things:
   (a) concentrate on a small area of the photograph at a time
   (b) use a magnifying glass (a seven-centimetre circular one is sufficient) to enlarge the area under observation.

   After much practice, a magnifying glass is unnecessary. However, one should practise such exercises regularly, otherwise much time may be lost during an examination.
3. A vertical photograph is best viewed with the shadows falling towards the observer. This effect creates a more natural landscape for the easy recognition of features.

## Types of Aerial Photographs

Studying photographs, especially aerial photographs, is an excellent way of understanding the various types of geographical landscapes. In order to use this system accurately one should study and practise the many skills of photograph interpretation that are necessary for the easy recognition and understanding of the formation of geographical landforms.

There are two main types of photographs:

1. vertical photographs and
2. oblique photographs.

LANDSCAPES OF THE WORLD

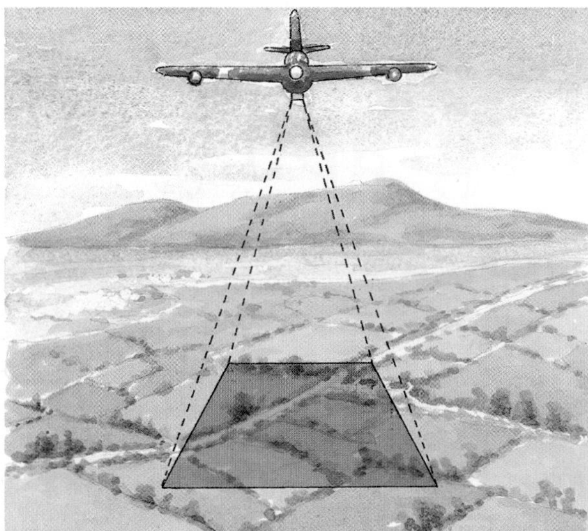

Fig. 7.1
Vertical aerial photograph
An aeroplane flies level and straight to take vertical photographs.
A vertical aerial photograph is true to scale.
The camera is facing vertically downward when taking such a photograph.

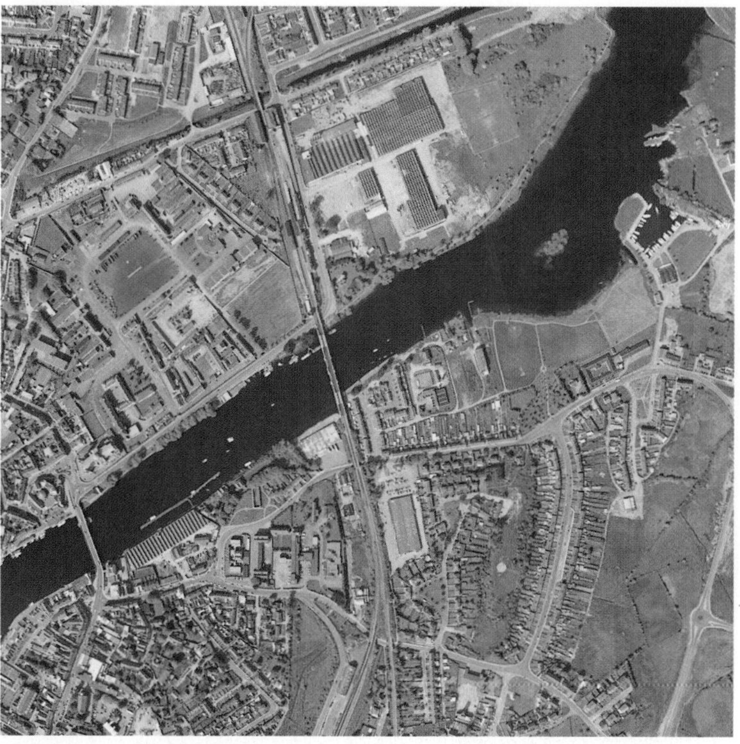

Fig. 7.2
Vertical aerial photograph of Athlone

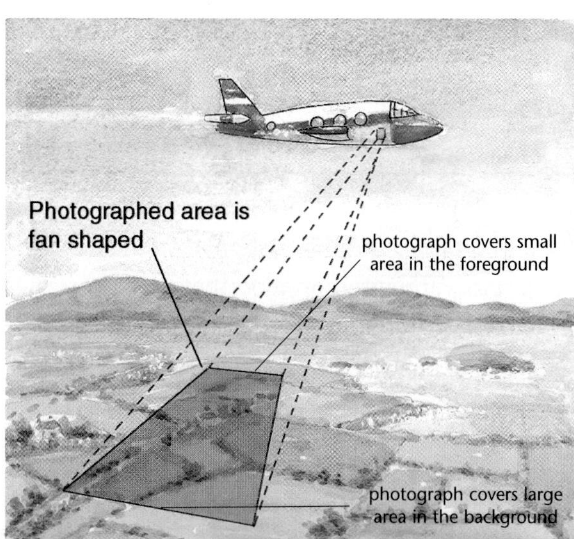
Photographed area is fan shaped
photograph covers small area in the foreground
photograph covers large area in the background

Fig. 7.3
Oblique aerial photograph
The photographed area is fan-shaped.
Although rectangular in shape, the plan of an oblique aerial photograph is similar to that of a truncated cone, the arms of which open out usually between 40° and 60°.

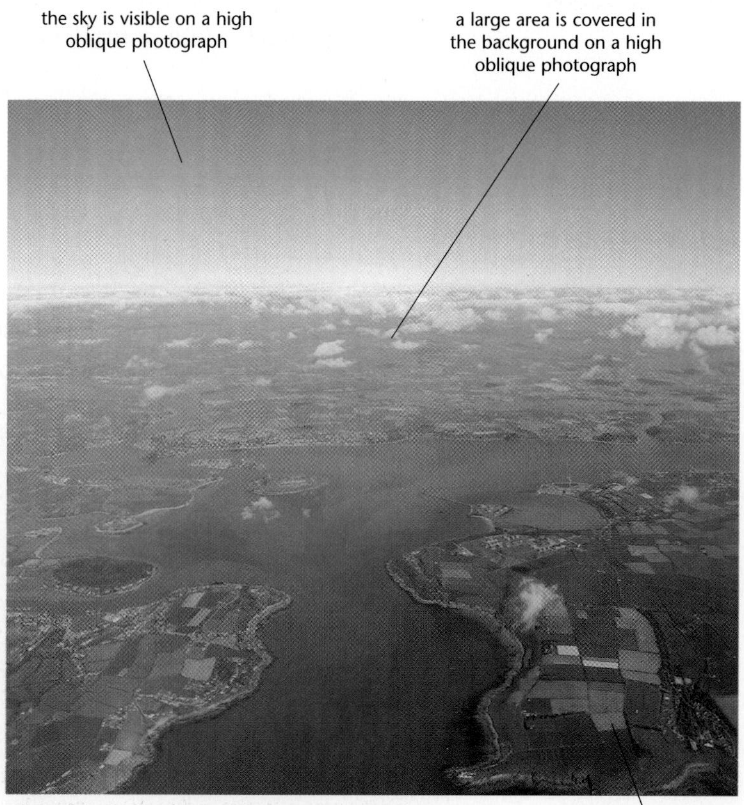
the sky is visible on a high oblique photograph
a large area is covered in the background on a high oblique photograph
a small area only is covered in the foreground of an oblique photograph

Fig. 7.4
A high oblique shows horizon (Cork Harbour)

no part of the horizon is visible

camera pointing steeply towards the ground

Fig. 7.5
A low oblique shows no part of horizon

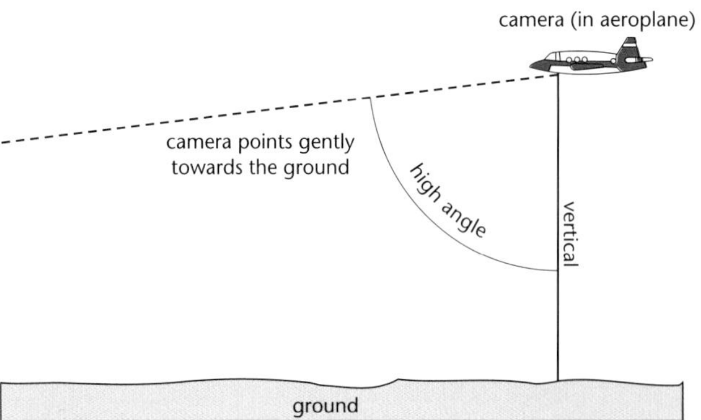


Fig. 7.6
High oblique aerial photograph

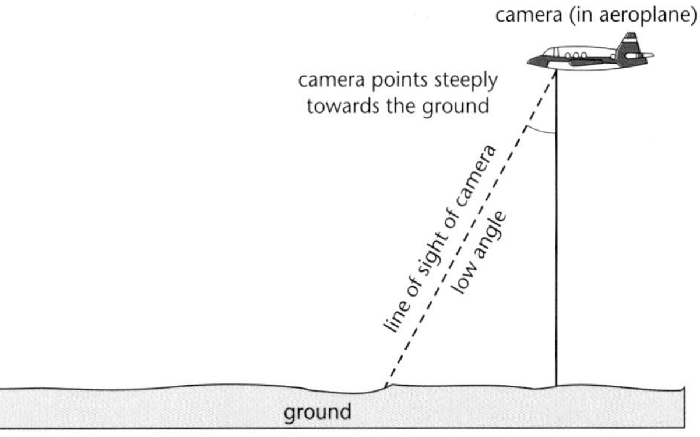


Fig. 7.7
Low oblique shows landscape only

## Locating Places or Features on an Aerial Photograph

For easy reference, a photograph may be divided into nine areas, as shown in Figs. 7.8 and 7.9.

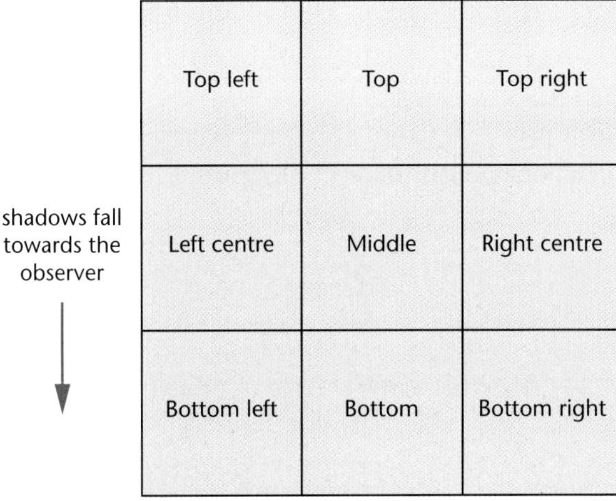


Fig. 7.8
Locating places or features on a **vertical** photograph when a direction arrow is not supplied

| | | |
|---|---|---|
| Left background | Centre background | Right background |
| Left centre | Middle centre | Right centre |
| Left foreground | Centre foreground | Right foreground |

Fig. 7.9
Locating places or features on an **oblique** photograph when a direction arrow is not supplied

Class Activity

Fig. 7.10

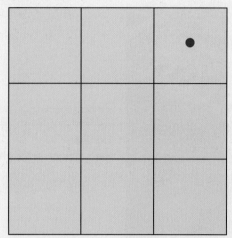

**Photograph A — Vertical**

1. Where is the circle located on this photograph? Is there a difficulty with this question/answer?

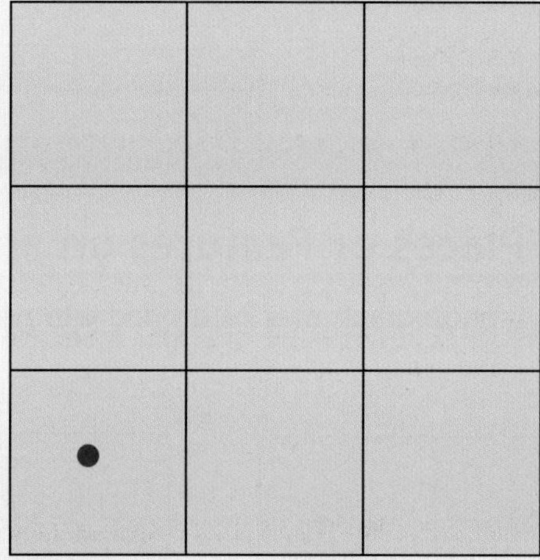

**Photograph B — Oblique**

2. Where is the circle located on this photograph?

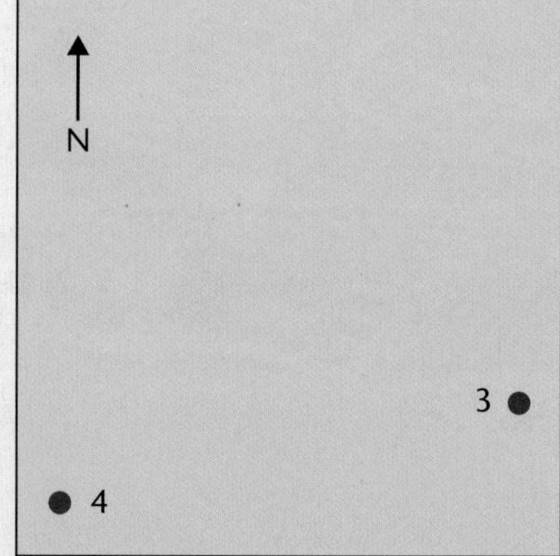


**Photograph C — Vertical**

Locating features when North arrow is supplied

3. Where is the circle 3 located?
4. Where is the circle 4 located?

Use the spaces provided to answer the questions.

Photograph A Q.1 ______________________

______________________________________

______________________________________

Photograph B Q.2 ______________________

______________________________________

______________________________________

Photograph C. Q.3 _____________________

______________________________________

______________________________________

& Q.4 _________________________________

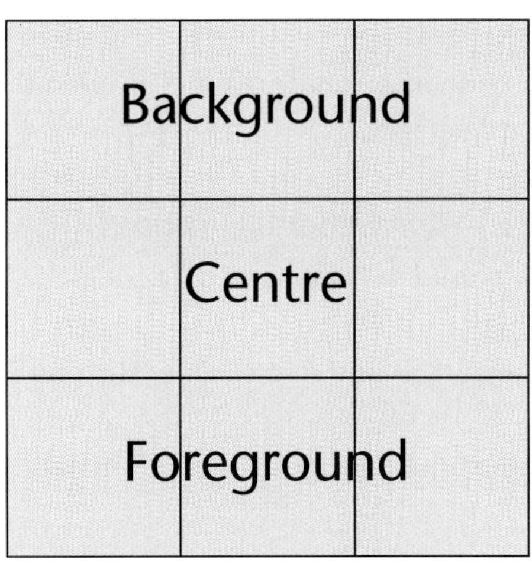


Fig. 7.11
Landform sizes on an oblique photograph

**Background**

- This area is furthest from the camera.
- Skyline or horizon may be shown.
- A large area is covered.
- Landforms/objects appear small as they are far away from the camera.

**Foreground**

- This area is nearest to the camera.
- A small area is shown.
- Landforms/objects appear large as they are near to the camera.

## Scale in Photographs

- **Foreground features**
  Scale is uniform on vertical photographs although there may be some slight distortion at the edges of the photograph because the camera lens is vertical only over the centre. On oblique photographs, objects in the foreground appear larger than objects of similar size in the background.
- **Area**
  On oblique photographs, the area covered by the photograph increases with distance from the camera position.
- **Distance**
  To measure distance on a photograph, note the features in the photograph between which one wants to measure. Then note the corresponding features on a map of the same area and measure accurately the distance between them on the map.

Fig. 7.12
The National Grid

## Locating Places on Ordnance Survey Maps of 1:50,000 Scale

### The National Grid

The National Grid is divided into lettered squares called **sub-zones**. Each Ordnance Survey map has printed on it, in blue, the letter(s) of the sub-zone(s) from which that map extract was taken. Each sub-zone is divided by a grid of lines called **co-ordinates**. Some of the lines are **vertical** and are called **eastings**. The other lines are **horizontal** and are called **northings**.

Both eastings and northings are numbered from 00 to 99. Eastings increase in value from left to right. Northings increase in value from base to top. (See Fig. 7.13.)

**Tip**
When giving a grid reference remember the word **ATLAS** —
**AT** represents **across the top**
**AS** represents **along the side.**

## Grid References

The **location** of places on Ordnance Survey maps is given in the form of grid references. There are two types of grid reference:

### 1. Four-grid Reference — for General Locations

Four-grid references are used to locate an entire square in which a specific feature is located. The grid reference for the bottom left-hand corner of that square is used for this purpose. For example, the grid reference for the ring-fort at Clonbouig is W 54 48.

For any four-grid reference, three pieces of information must be given in this order:

(a) give the **sub-zone** letter first

(b) give the two-digit **easting** number for the left side of the required square

(c) give the two-digit **northing** number for the base of the required square.

What is the four-grid reference for the holy well near Burren Bridge in the south of the map extract?

Fig. 7.13
Location by grid reference

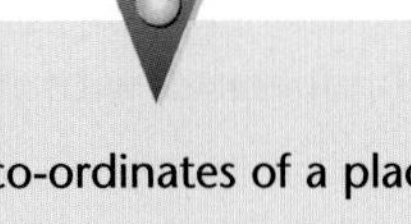

The co-ordinates of a place should be given as follows:
**sub-zone + easting + northing**
e.g. W 542 482

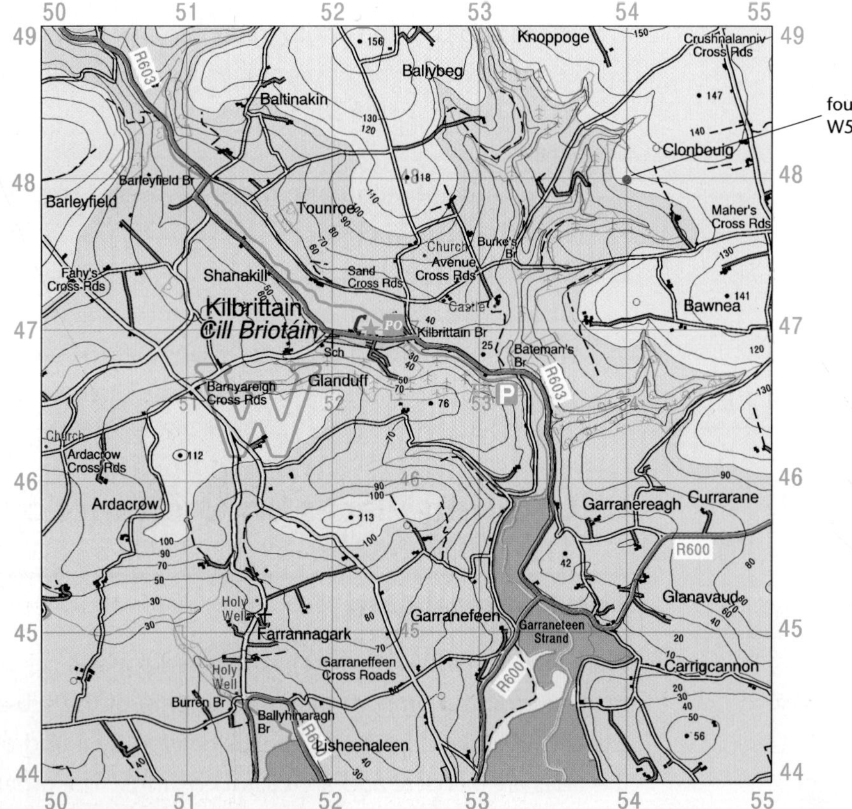


### 2. Six-grid Reference — for Exact Locations

Six-figure grid references are given to identify the **exact location** of a specific feature/place.

For this purpose one has to imagine that the space between any two easting numbers and any two northing numbers is divided into ten parts. For example,

MAPS AND AERIAL PHOTOGRAPHS

halfway between the eastings 54 and 55 is 54.5. Fractionally short of halfway is 54.4 and fractionally longer than halfway is 54.6.

The same applies to the northing co-ordinates. However, the **decimal point** is **not** included in the grid reference. For example, a six-figure grid reference for the ring-fort at Clonbouig is W 542 482.

What is the six-figure grid reference for the holy well near Burren Bridge in the south of the map extract?

Class Activity

Fig. 7.14

Use the spaces provided to identify the following features on the Kilbrittain map extract, Fig. 7.13.

| Name the features at the following grid references | |
|---|---|
| | **Feature** |
| W 531 472 | |
| W 526 475 | |
| W 515 452 | |
| W 546 472 | |
| W 532 466 | |

## Area Covered on an Oblique Photograph

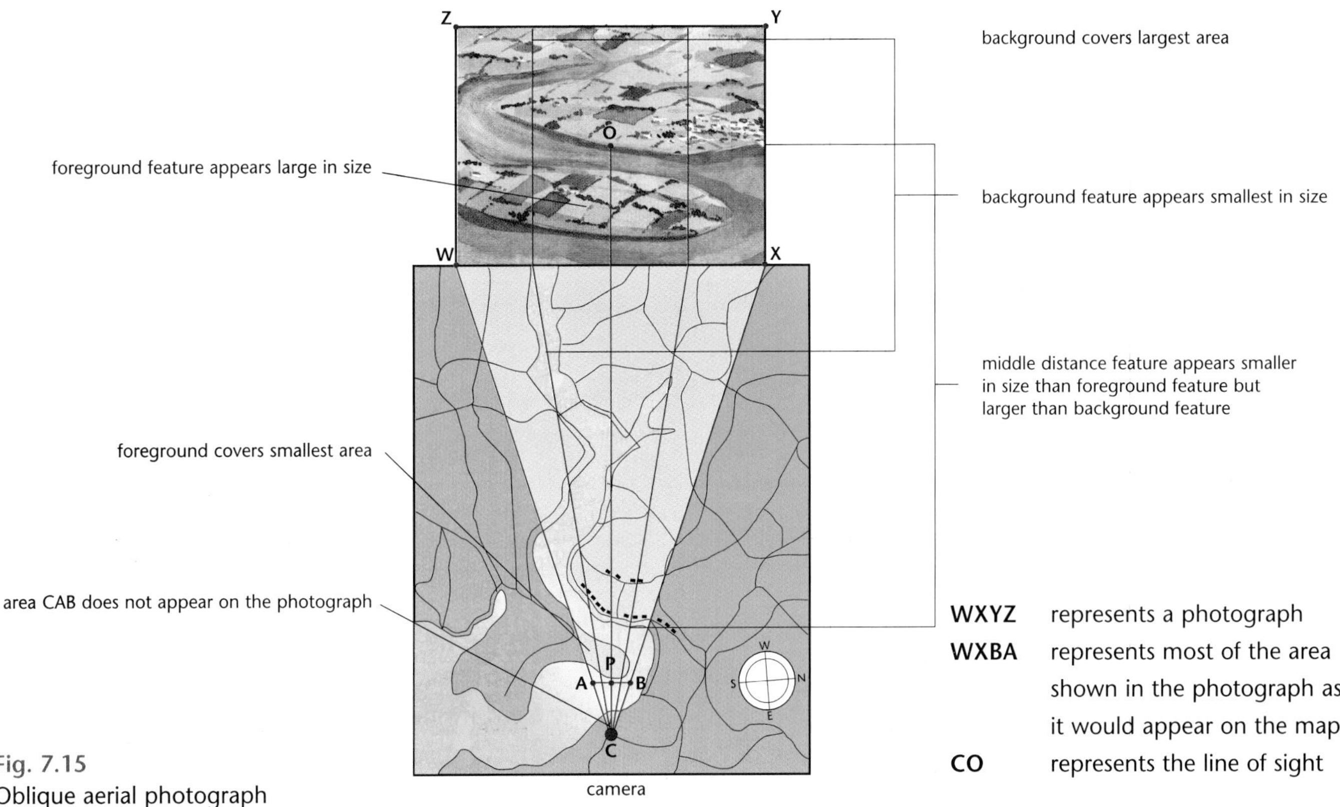
Z
Y
O
W
X
background covers largest area
background feature appears smallest in size
foreground feature appears large in size
middle distance feature appears smaller in size than foreground feature but larger than background feature
foreground covers smallest area
area CAB does not appear on the photograph
P
A
B
C
camera
W
S
N
E
WXYZ represents a photograph
WXBA represents most of the area shown in the photograph as it would appear on the map.
CO represents the line of sight

Fig. 7.15
Oblique aerial photograph

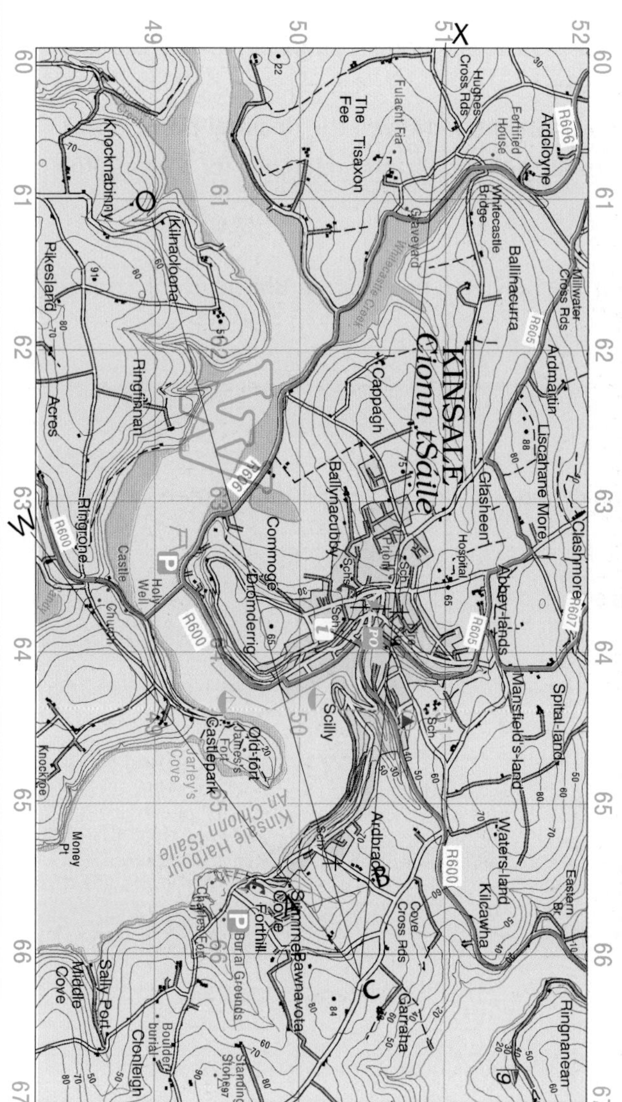


Fig. 7.16
Map extract showing the Bandon River estuary at Kinsale in Co. Cork

Fig. 7.17
Aerial view showing the Bandon estuary at Kinsale in Co. Cork

## Direction on Ordnance Survey Maps

Direction is usually given in the form of compass points. There are thirty-two points to the compass. However, one needs to use only sixteen, as represented in Fig. 7.18. The top of an Ordnance Survey map is always north. Therefore the bottom is south, the left is west and the right is east.

**Class Activity**

Use the spaces below, the compass points and the Ordnance Survey map extract, Fig. 7.19, to answer the following questions:

1. In which direction is Inishkeen from Louth? ______________
2. In which direction is Ballakelly Bridge from Castlering Bridge? ______________
3. In which direction does the Fane River run in the area shown on the map? ______

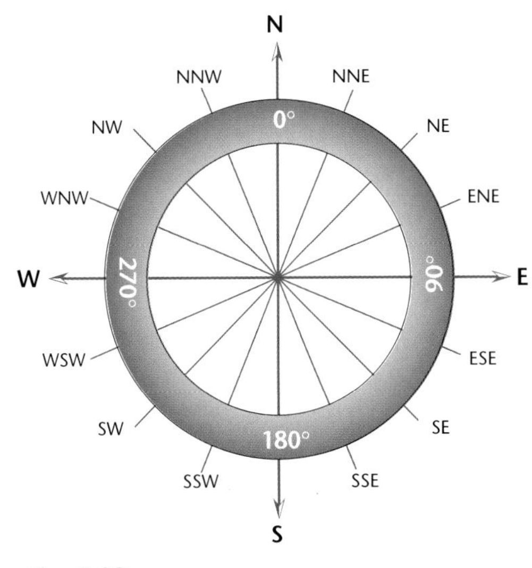


Fig. 7.18
Compass points

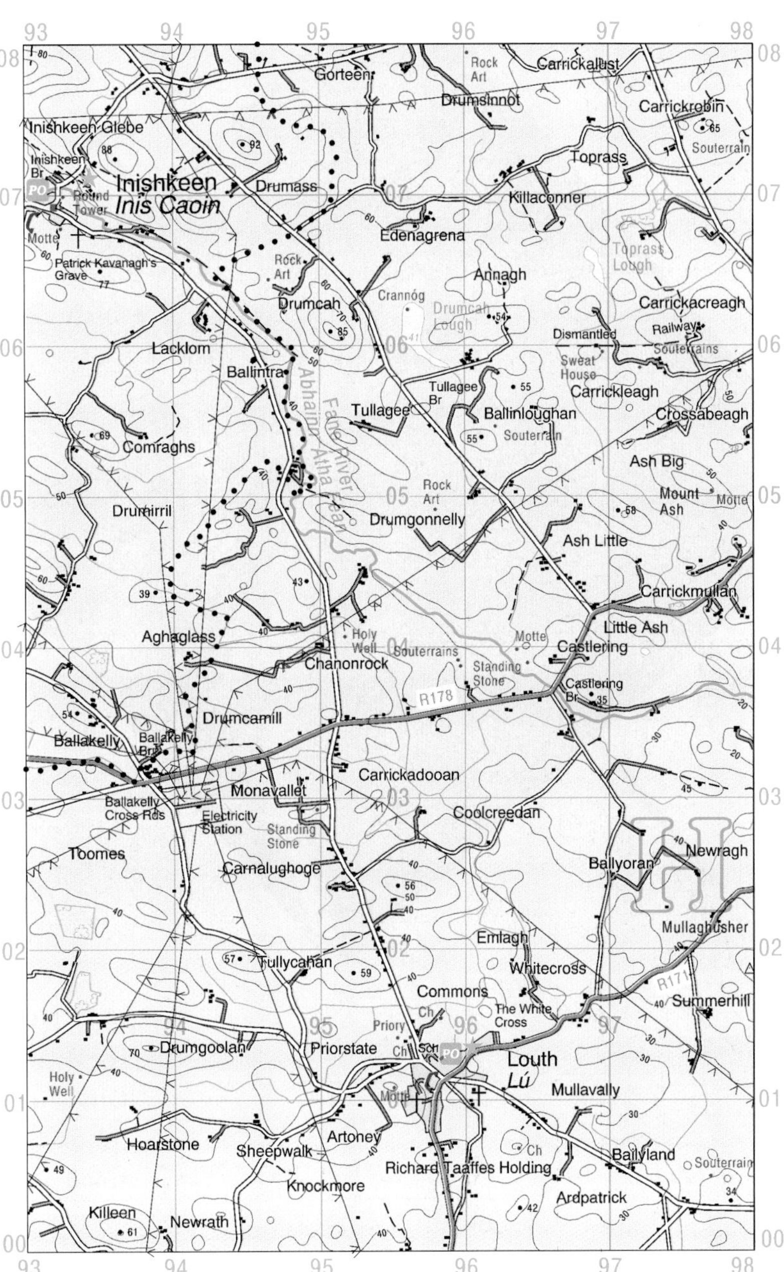


Fig. 7.19

## Finding Direction on Photographs Using Maps

The convention on Ordnance Survey maps that north is top does not always follow on aerial photographs. In order to find direction on any photograph, one must orientate the photograph in conjunction with landscape features such as buildings, bridges, stretches of roadway, lakes, woods, etc.

Fig. 7.20

## Finding Direction on a Photograph

1. Rotate the photograph until the canal and the road are parallel in both the map and the photograph.
2. The line of sight is then clear. In this case it is north-east. In other cases it may be south and may necessitate turning the photograph upside down.

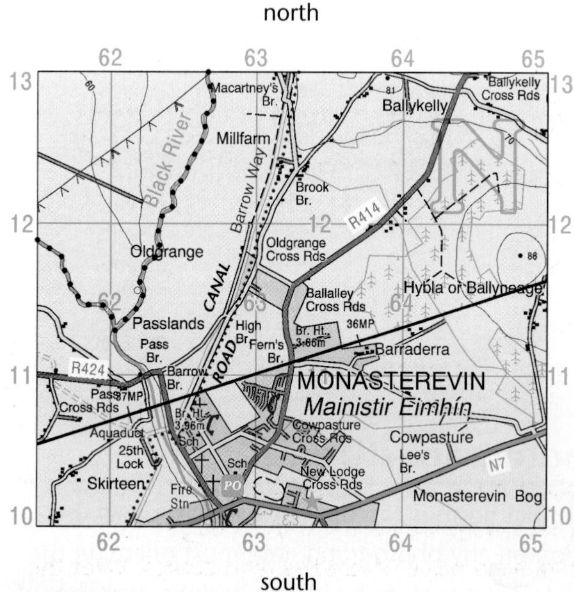


Fig. 7.21
The map remains fixed in its normal position. Rotate the photograph to match the common features.

### Steps to Help You Find Direction on a Photograph

1. Locate the area shown in the photograph on the Ordnance Survey map. Do this by identifying common features on both the photograph and the map, e.g. a straight stretch of roadway, a canal, a rail track, etc.
2. Make a point halfway across the top and the bottom of the photograph. Join these two points with a light pencil line. This pencil line is the line of sight of the camera lens.
3. Orientate your photograph until one or more of the common features in 1 above lie parallel to each other.
4. Remember the top of the map is north and the bottom is south. Compare the line of sight on the photograph with this north–south line on the map. From this comparison you should be able to determine the line of sight of the camera lens.

# Scale on Maps

## Representative Fraction (RF)

This particular example tells us that any single unit of measurement on the map corresponds to 50,000 similar units on the ground. So 1 centimetre on the map represents 50,000 centimetres (0.50 kilometres) on the ground.

Linear scale is a divided line which shows map distances in kilometres and miles. When measuring a distance on the map with a piece of string or a strip of paper, a direct reading can be gained by using this linear scale.

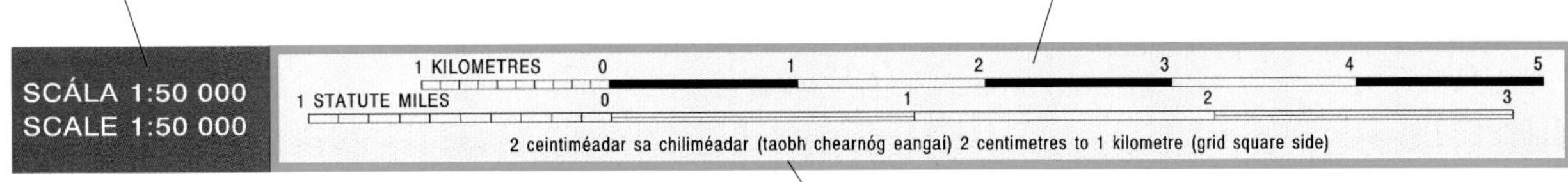


statement of scale

Fig. 7.22
Understanding scale

Fig. 7.23
Measuring a straight line

## Measuring Distance on a Map

### Straight Line Measurement

The shortest distance between two points on a map is often referred to as 'as the crow flies'.

**Method**

1. Place a straight edge of paper along a line which joins both places mentioned.
2. Mark exactly where each point touches the edge of the paper.
3. Place the paper's edge on the map's linear scale and measure carefully the distance between the two places mentioned. (See Figs. 7.23 and 7.24.)

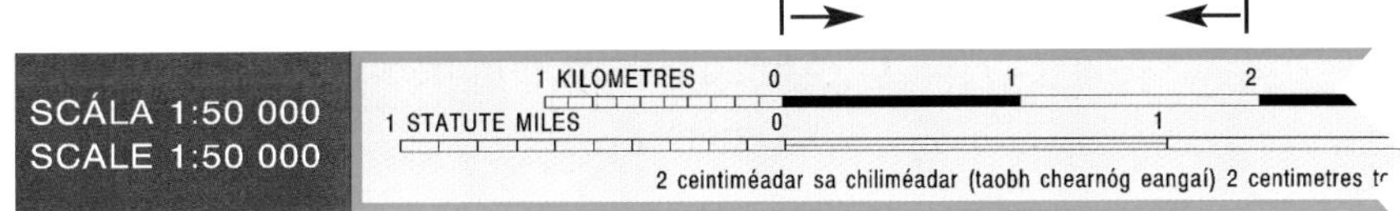


Fig. 7.24
Linear scale

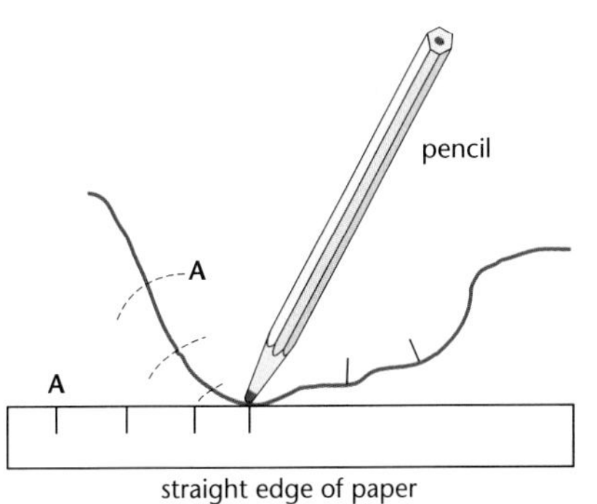


Fig. 7.25
Measuring a curved line

### Curved Line Measurement

Distances such as those along a road may be measured as follows.

**Method**

1. Lay a straight edge of paper along the centre of the road to be measured and mark the starting point on the paper's edge.
2. Use a pencil point to mark each place where the road curves. Pivot the edge of the paper along each section of roadway. (See Fig. 7.25.)
3. Mark the finishing point on the paper's edge.
4. Use the linear scale to measure the required distance.

# Calculating Map Area

## Method A

To calculate the area of all or part of a map:

1. Measure the length of the designated area in centimetres or with a strip of paper. Then translate this measurement to kilometres using the scale of the map (answer A).
2. Measure the breadth of the designated area in centimetres or with a strip of paper. Then translate this measurement to kilometres using the scale of the map (answer B).
3. Now multiply answer A by answer B. This result will give you the area of the designated area in square kilometres.

## Examples

Scale of map = 1:50,000 = 2 centimetres to 1 kilometre
Suppose length = 6 centimetres = 3 kilometres
Suppose breadth = 7 centimetres = 3.5 kilometres
Area = length x breadth = 3 x 3.5 = 10.5 square kilometres

**Or** by linear scale

Suppose length = 3 kilometres
Suppose breadth = 3.5 kilometres
Area = length x breadth = 3 x 3.5 = 10.5 square kilometres

## Method B

To calculate the area of all or part of a map:

1. Each blue grid box on a scale 1:50,000 is 1 kilometre x 1 kilometre in size.
2. Count the number of boxes along the length and breadth of the map extract.
3. Multiply these numbers together. The answer equals the area of the map in square kilometres.

## Example

Examine the map extract Fig. 7.23.
Scale of map = 1:50,000

1. Length of map = four boxes
2. Breadth of map = three boxes
3. Each box is 1 km x 1 km
4. Area = length x breadth = 4 x 3 = 12 square kilometres

# Understanding Symbols on Ordnance Survey Maps

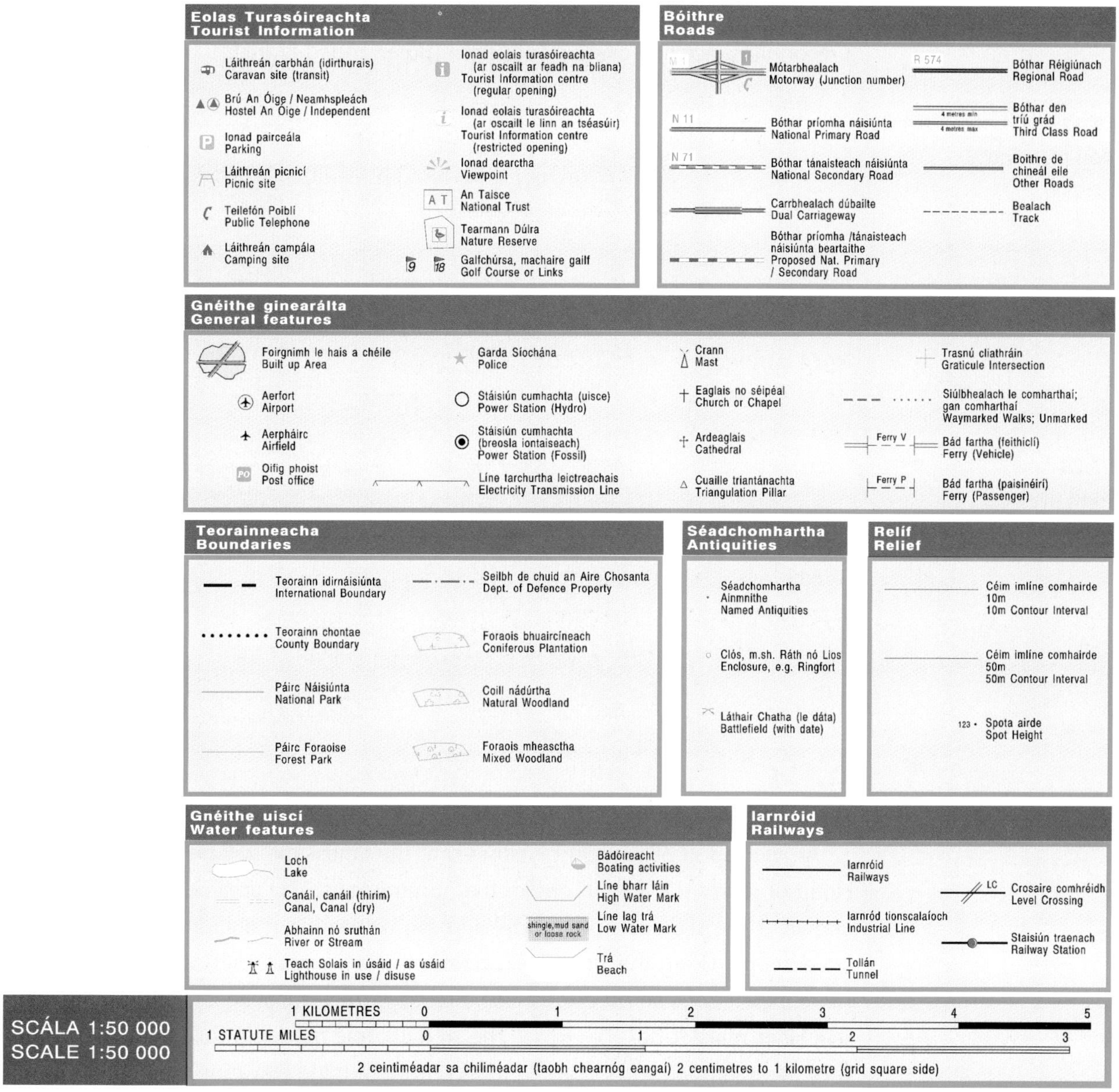

**Eolas Turasóireachta**
**Tourist Information**

- Láithreán carbhán (idirthurais) / Caravan site (transit)
- Brú An Óige / Neamhspleách / Hostel An Óige / Independent
- Ionad pairceála / Parking
- Láithreán picnicí / Picnic site
- Teilefón Poiblí / Public Telephone
- Láithreán campála / Camping site
- Ionad eolais turasóireachta (ar oscailt ar feadh na bliana) / Tourist Information centre (regular opening)
- Ionad eolais turasóireachta (ar oscailt le linn an tséasúir) / Tourist Information centre (restricted opening)
- Ionad dearctha / Viewpoint
- AT — An Taisce / National Trust
- Tearmann Dúlra / Nature Reserve
- 9 18 — Galfchúrsa, machaire gailf / Golf Course or Links

**Bóithre**
**Roads**

- M 1 — Mótarbhealach / Motorway (Junction number)
- N 11 — Bóthar príomha náisiúnta / National Primary Road
- N 71 — Bóthar tánaisteach náisiúnta / National Secondary Road
- Carrbhealach dúbailte / Dual Carriageway
- Bóthar príomha /tánaisteach náisiúnta beartaithe / Proposed Nat. Primary / Secondary Road
- R 574 — Bóthar Réigiúnach / Regional Road
- 4 metres min / 4 metres max — Bóthar den tríú grád / Third Class Road
- Bóithre de chineál eile / Other Roads
- Bealach / Track

**Gnéithe ginearálta**
**General features**

- Foirgnimh le hais a chéile / Built up Area
- Aerfort / Airport
- Aerpháirc / Airfield
- PO — Oifig phoist / Post office
- Garda Síochána / Police
- Stáisiún cumhachta (uisce) / Power Station (Hydro)
- Stáisiún cumhachta (breosla iontaiseach) / Power Station (Fossil)
- Líne tarchurtha leictreachais / Electricity Transmission Line
- Crann / Mast
- Eaglais no séipéal / Church or Chapel
- Ardeaglais / Cathedral
- Cuaille triantánachta / Triangulation Pillar
- Trasnú cliathráin / Graticule Intersection
- Siúlbhealach le comharthaí; gan comharthaí / Waymarked Walks; Unmarked
- Ferry V — Bád fartha (feithiclí) / Ferry (Vehicle)
- Ferry P — Bád fartha (paisinéirí) / Ferry (Passenger)

**Teorainneacha**
**Boundaries**

- Teorainn idirnáisiúnta / International Boundary
- Teorainn chontae / County Boundary
- Páirc Náisiúnta / National Park
- Páirc Foraoise / Forest Park
- Seilbh de chuid an Aire Chosanta / Dept. of Defence Property
- Foraois bhuaircíneach / Coniferous Plantation
- Coill nádúrtha / Natural Woodland
- Foraois mheasctha / Mixed Woodland

**Séadchomhartha**
**Antiquities**

- Séadchomhartha Ainmnithe / Named Antiquities
- Clós, m.sh. Ráth nó Lios / Enclosure, e.g. Ringfort
- Láthair Chatha (le dáta) / Battlefield (with date)

**Relíf**
**Relief**

- Céim imlíne comhairde 10m / 10m Contour Interval
- Céim imlíne comhairde 50m / 50m Contour Interval
- 123 · Spota airde / Spot Height

**Gnéithe uiscí**
**Water features**

- Loch / Lake
- Canáil, canáil (thirim) / Canal, Canal (dry)
- Abhainn nó sruthán / River or Stream
- Teach Solais in úsáid / as úsáid / Lighthouse in use / disuse
- Bádóireacht / Boating activities
- Líne bharr láin / High Water Mark
- shingle, mud sand or loose rock — Líne lag trá / Low Water Mark
- Trá / Beach

**Iarnróid**
**Railways**

- Iarnróid / Railways
- Iarnród tionscalaíoch / Industrial Line
- Tollán / Tunnel
- LC — Crosaire comhréidh / Level Crossing
- Stáisiún traenach / Railway Station

**SCÁLA 1:50 000**
**SCALE 1:50 000**

1 KILOMETRES 0 1 2 3 4 5

1 STATUTE MILES 0 1 2 3

2 ceintiméadar sa chiliméadar (taobh chearnóg eangaí) 2 centimetres to 1 kilometre (grid square side)

Fig. 7.26
Map legend

## Producing Ordnance Survey Maps (Photogrammetry)

In order to produce maps from aerial photographs, a set procedure must be followed. Two vertical views of a given area are provided by overlapping successive frames taken on a reconnaissance line overlap sequence (prearranged flight).

Fig. 7.27
Mission planning

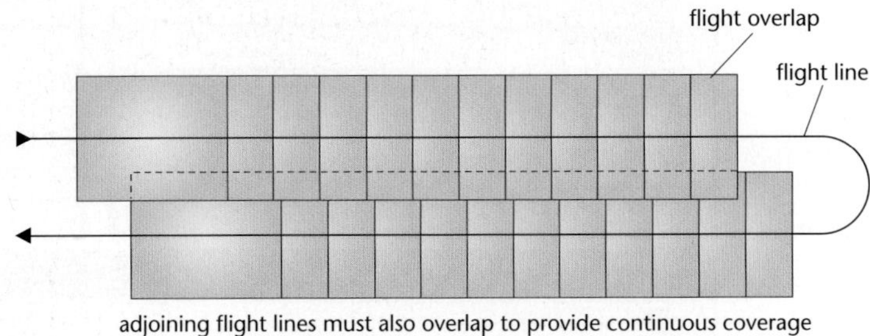


Fig. 7.28
Forward gain

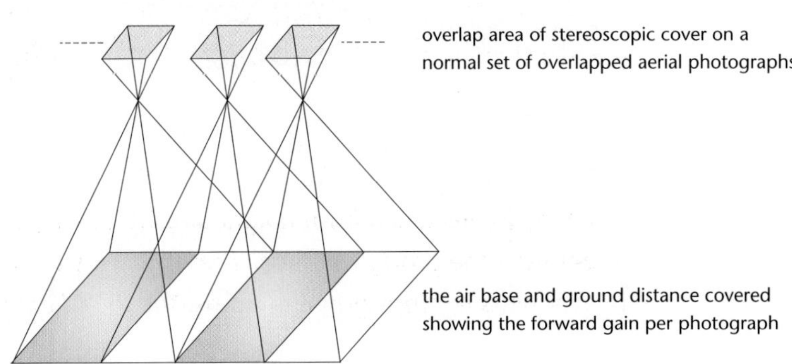


An overlap of 60 per cent ensures that the landscape will be covered even if broken cloud obscures the area in one of the photographs. The area common to each frame is known as the **stereoscopic model** and can be seen in three dimensions with the aid of a stereoscope. Using this principle, photogrammetric machines, which are much larger and more expensive, can produce maps directly from vertical aerial photographs. These machines make it possible to see the stereoscopic model in three dimensions. In addition, they are equipped with 'floating marks' to determine elevation differences of the terrain. The maps are produced from printers attached to the photogrammetric machines.

## Understanding the Hidden Landscape

The poorer and stony land of the west of Ireland, some upland areas and areas where grazing of animals is the local tradition may be regarded as those which preserved the historical landscape. In contrast, the arable lowland of the south and east of Ireland has levelled quite a lot of ancient habitation sites. Nevertheless, evidence of these destroyed sites may be reconstructed by the use of aerial photographs.

Fig. 7.29
'Visible' and 'invisible' characteristics of the landscape at Dunmanoge, in Co. Kildare

## How is this Done?

Photographs taken from a light aircraft can capture traces of early landscapes buried beneath the ground surface. These remains are revealed as subtle colour and height variations in crops growing above the foundations and pathways of ancient structures.

## Understanding Colour/Shades

Tone or shade of colour on aerial photographs is due to the amount of light which is reflected back to the camera lens. The amount of reflected light depends upon the nature and texture of the surface of the area being photographed.

### 1. Water

Calm, sunlit water may appear almost white. But calm water which lies in part of the camera field from which it reflects no light may appear almost black.

### 2. Arable Land

- Standing corn may be green during the growing season and light golden when ripened.
- Fields which have been harvested for grain may appear similar in shade to those harvested for silage. In such instances, the observer should note the uses of other fields on the farm in question.
- The character of the lowland terrain may help in identifying the type of farming practised in the area. For example, rolling or undulating land might be well drained and suitable for tillage, while flat, low-lying land near a river in its old stage of maturity would be more suitable for pasture and silage.

- Tilled ground will appear as dark brown patches. Farm buildings such as silage pits or grain silos are indicative of farming types.

## 3. Transport

Reflecting surfaces, which have been formed through pressure such as roads or tracks, show up clearly as dark green lines (see Fig. 7.29).

## 4. Historical

Prehistoric structures such as ring-forts and ancient field fences can often be identified only from aerial photographs. Varying depths of soil can produce varying growth rates in crops and pasture.

A deep soil will produce a rich growth, while a shallow adjoining soil will have a more restricted growth rate. These alternate growth rates may produce a recognisable pattern which can be seen only from the air. In addition, the taller, more luxuriant plants remain green for a longer period, while the remainder of the crop ripens. Old field boundary walls or buildings may lie immediately below the surface and can restrict the growth of plants due to their light soil cover. During long periods of drought in summer, such stone structures scorch plant roots, creating patterns which are easily identifiable from the air.

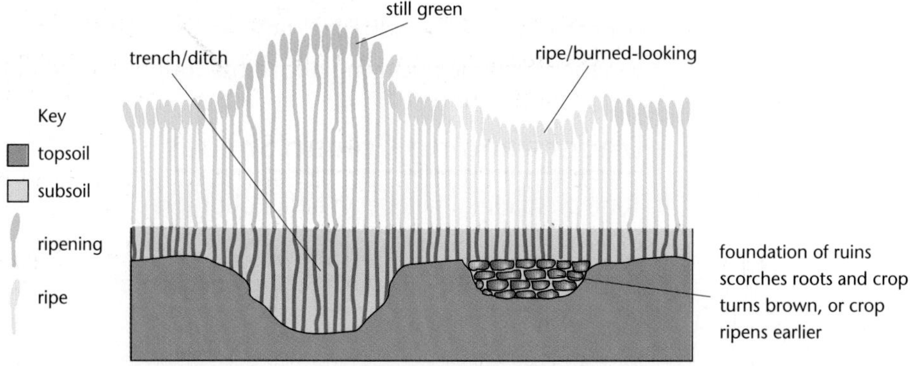


Fig. 7.30 Reconstructing our lost heritage from aerial photographs

# Time of Year and Photographs

## Spring

- Little foliage in trees. This represents the young leaves opening on the trees.
- Calves or lambs near their mothers in the fields.
- Some flowers in bloom.

## Summer

- Trees in full foliage.
- Hay and cereals appear ripe during July and August. Hay stacks, bales, numerous animals in the fields.
- Crowded beaches (depending on the weather).

### Autumn

- Foliage shows numerous shades of colour.
- Meadows cleared of hay, bales or stacks.

### Winter

- Absence of foliage on deciduous trees and hedges.
- Ploughed fields and fields without animals.
- Cattle enclosed in farmyards.
- Vehicle exhaust and chimney smoke clearly visible during cold spells.

## Time of Day

Long shadows are indicative of either morning or evening. Shadows around the base of trees indicate noon. If one knows the direction in which the photograph was taken, then a more precise time of day can be determined.

# Uses of Aerial Photographs

### 1. Military Uses

Aerial photographs are necessary in the preparation of maps for any military action. Satellites, which are constantly used to photograph land surfaces, are now used as foolproof navigation aids. This system (Global Positioning System) is based on a constellation of twenty-one satellites orbiting the earth at a very high altitude. These satellites use technology which is accurate enough to give pinpoint positions anywhere in the world, twenty-four hours a day. The Global Positioning System was developed by the US Department of Defense to simplify accurate navigation. It uses satellites and computers to triangulate positions which can be shown on high-resolution displays, giving one's position on a digitised chart of the area.

### 2. Road Construction

Local government bodies such as corporations and county councils use aerial photographs in the construction of ring roads and bypasses. Vertical photographs, when viewed through a photogrammetric machine, can determine the most desirable route in reference to cost and the least disturbance to people.

Amounts of trunking needed for hollows and depths of excavation needed in raised areas can be determined by the floating marks. From this information, the cost of construction can be estimated.

### 3. Industrial Uses

Large companies use photogrammetric machine personnel at the Ordnance Survey Office in the Phoenix Park. The ESB, for instance, uses this method to estimate the weight of coal in its stockpiles at Moneypoint. With this method, two photographs are used to determine the volume of the pile and thus the weight of the coal. This saves numerous working hours and so is cost-effective.

### 4. Sciences

Various sciences employ aerial photographs; for example they are used by geographers, geologists and meteorologists. Botanists, marine biologists, plantation and large farm owners and numerous other groups use various types of photographs for their studies, such as infra-red photographs which show different soil types. They may also be used to detect minerals, crop damage and so on.

### 5. Personal Use

Some aerial photographs are purchased by people who wish to display their house and gardens in a picture frame. Some may be just small, simple photographs; others may be elaborate prints on canvas.

### 6. Weather Prediction

High aerial and satellite photographs are used to track weather systems. These photographs provide detailed information that can often be used to predict the paths of hurricanes or tornadoes, which are capable of great destruction. These allow time for warnings to people who may then be able to secure homes and machinery before the weather system arrives.

## Understanding Slopes on Ordnance Survey Maps

### Slopes

Our landscape is not level. It rises and falls in hills, mountains and valleys. In other words, the landscape is made up of **sloping surfaces**. To be competent at map interpretation, it is necessary to be able to recognise different types of slope. It is then possible to give detailed descriptions of the hills, mountains, valleys and minor physical features which make up the landscape.

Recognition of slope types is easy, especially as there are only four basic slopes — uniform, concave, convex and stepped. These slopes are shown in Figs. 7.31 to 7.34.

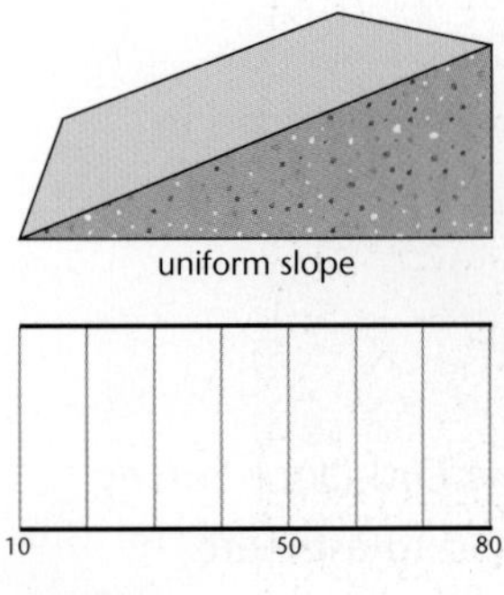


Fig. 7.31
Uniform slope

**Uniform slope:** Also referred to as an even or regular slope. It can be steep or gentle. Simple examples of this type of slope include the sloping top of a desk or a titled book. The contour pattern is easily recognised as the contours are regularly spaced.

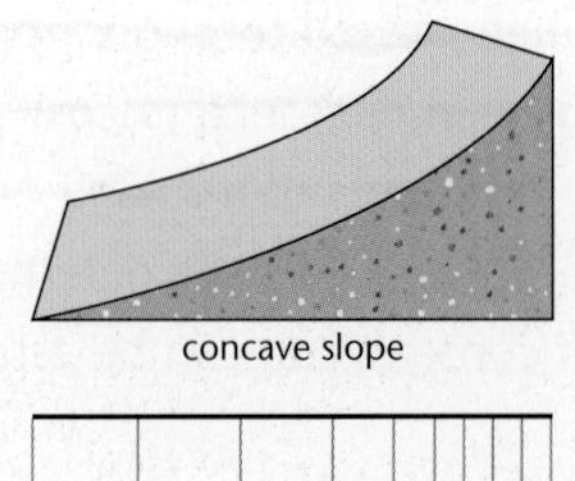


Fig. 7.32
Concave slope

**Concave slope:** Concave slopes begin as gentle slopes and gradually get steeper as one climbs uphill. The contour pattern is simple: widely spaced contours in the lower part, with more closely spaced contours in the upper part.

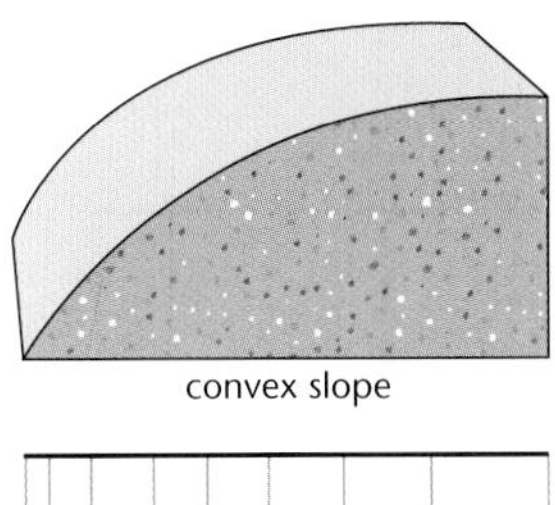


10 50 90

Fig. 7.33
Convex slope

**Convex slope:** Half an orange with its cut side down is a good example of a convex surface. A convex slope begins steeply. As one climbs the hill, the gradient becomes less steep. The contours are close together on the lower slope and further apart on the upper slope.

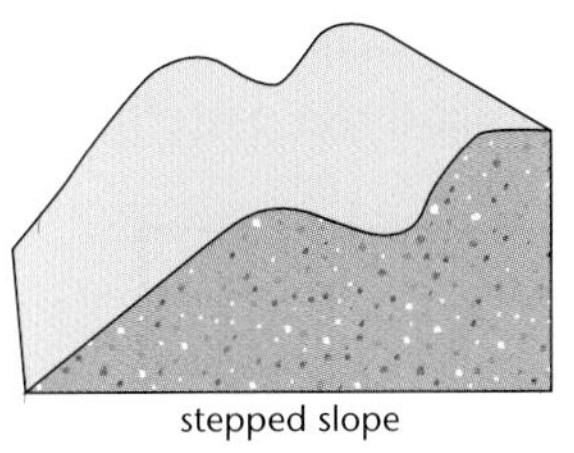


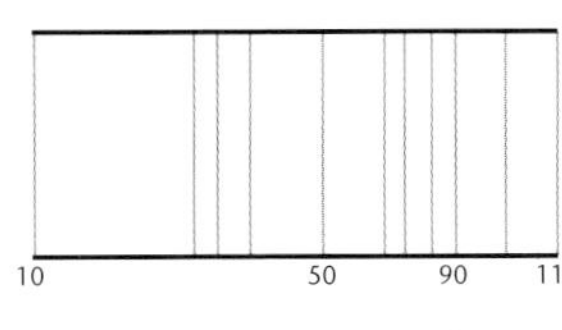


Fig. 7.34
Stepped slope

**Stepped slope:** A stepped slope is one which has alternating gentle and steep slopes. The lower slope can be gentle or steep. The contour pattern is easily recognised. Contours occur in groups, close together at first, then far apart, in a repeating pattern.

## Drawing Sketch Maps from Aerial Photographs

**Case Study**

## Aerial Photograph of Cashel in Co. Tipperary

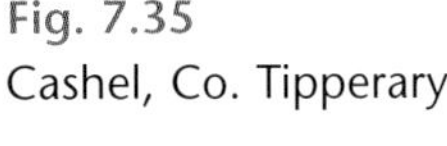

Fig. 7.35
Cashel, Co. Tipperary

On a sketch map of the urban area shown, mark the street pattern and identify six different land use patterns.

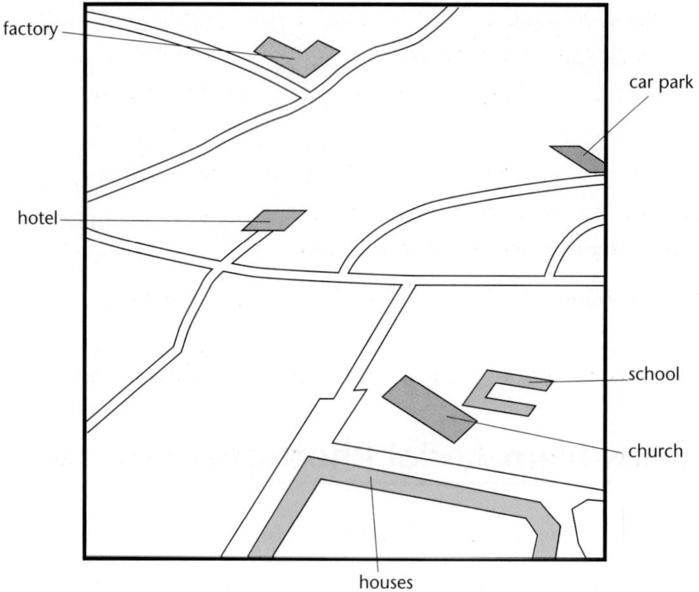


Fig. 7.36
Land use sketch map of Cashel

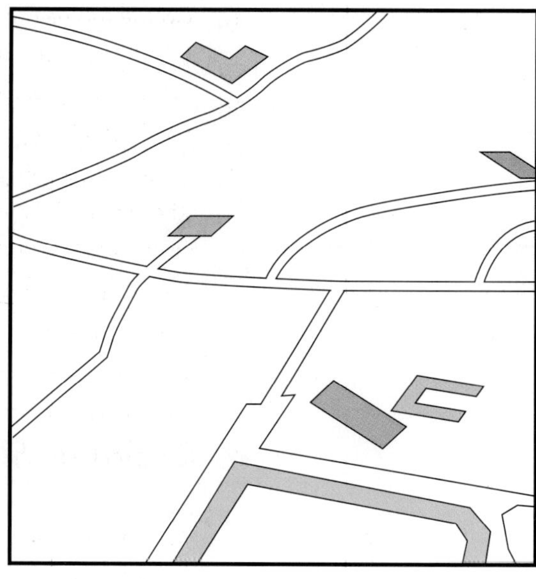

Fig. 7.37
Use this sketch map to mark and label any three other land uses in the town of Cashel

Class Activity

## Case Study: Sneem in Co. Kerry

Study the photographs of Sneem in Co. Kerry, Figs. 7.38 and 7.39. Then use the lines drawn on the photographs to draw a sketch map of the plan of the village. On it **mark** and **label**

1. the street pattern
2. the land uses
   - shop
   - church
   - filling station
   - dwelling
   - shed

Fig. 7.38
Guidelines for a sketch map

These lines should be lightly drawn and only barely visible. Divide the photograph into nine parts.

or

Fig. 7.39
Guidelines for a sketch map

Use diagonals and halving lines

## Tips

1. On the top half of your page, draw a rectangle with the sides in the same proportion to each other as they are on the photograph — but not necessarily the same size. Do not draw a sketch map greater than half the size of a foolscap page as it takes too long to draw and needs greater skill to produce.
2. Always use a soft pencil as it is easier to erase. Avoid using a biro. Divide your sketch and photograph into segments, or draw diagonals and halving lines, as shown in Figs 7.38 and 7.39. These will help you to locate coastline and landscape features.
3. Show and name only the features that you are specifically asked for. Do not include unnecessary detail.
4. When showing land use, define each land use area with a **heavy boundary mark**. Do not leave a land use area **undefined**.
5. Identify the features on your sketch by annotations (labelled arrows), or give a key to avoid overcrowding.
6. Colour is not essential. Use colour only if you have sufficient time.
7. Draw the sketch similar in shape to that of the photograph. For example, if the photograph is square, draw a square sketch. If the photograph is rectangular in shape then draw a rectangular sketch map.

## Drawing Sketch Maps from Ordnance Survey Maps

### Case Study 1

#### Barrow and Nore Basins

Study the Barrow and Nore map extract, Fig. 7.40. p. 102. On a sketch map (not a tracing) of this region, mark and label the following:

- the upland areas
- the highest point
- two named rivers
- three regional roads
- five third-class roads
- three settlements

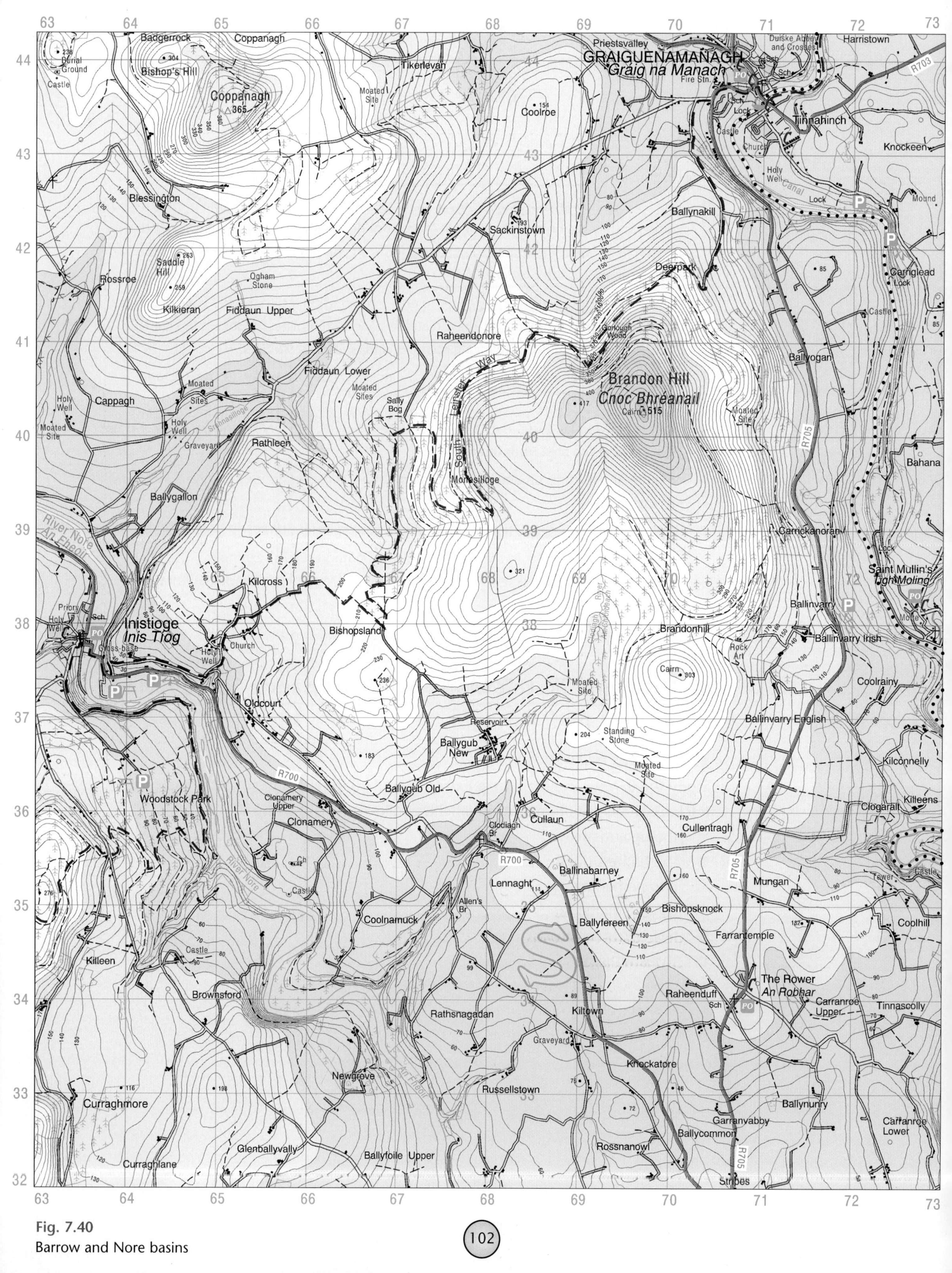


**Fig. 7.40**
Barrow and Nore basins

Fig. 7.41
Sketch map of the Barrow and Nore basins

Key
regional road
third-class road

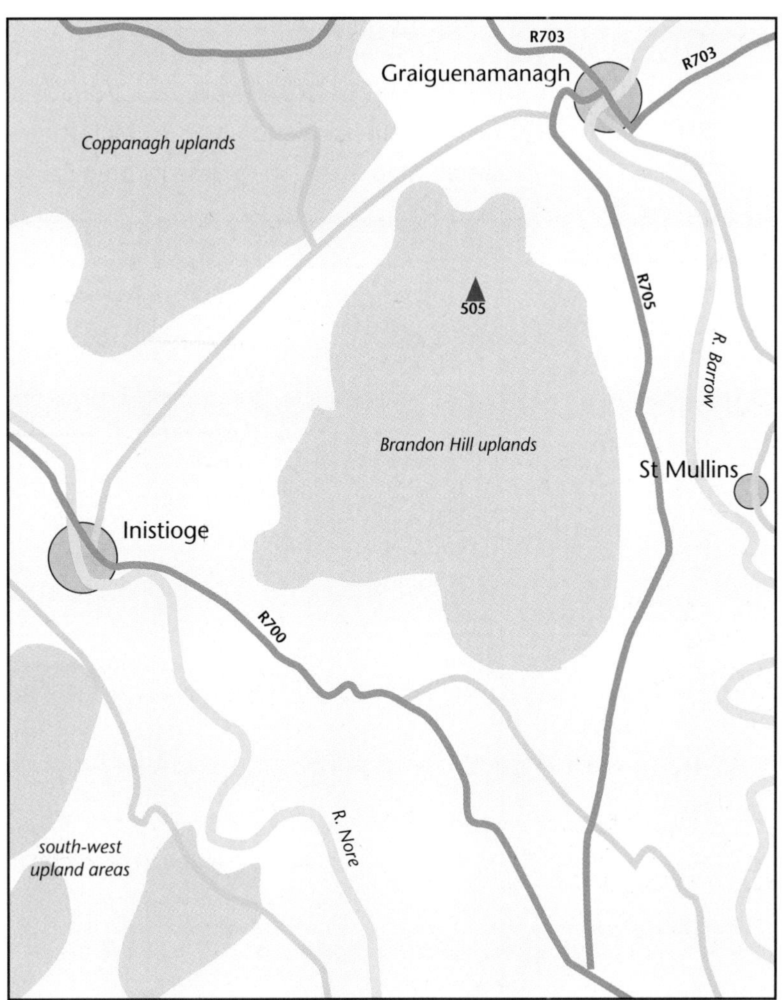


Class Activity

Study the Derryveagh map extract, Fig. 12.57, p. 228. On a sketch map of this region, mark and name the following:

- the regional roads
- three third-class roads
- two large lakes
- three important rivers
- one important mountain range
- the highest point on the map

## Case Study 2

### Inishowen Map Extract

Examine the Inishowen map extract on p. 107. Draw a sketch map of this area. On the sketch map, mark and name its physical regions.

Fig. 7.42
Sketch map of Inishowen map extract

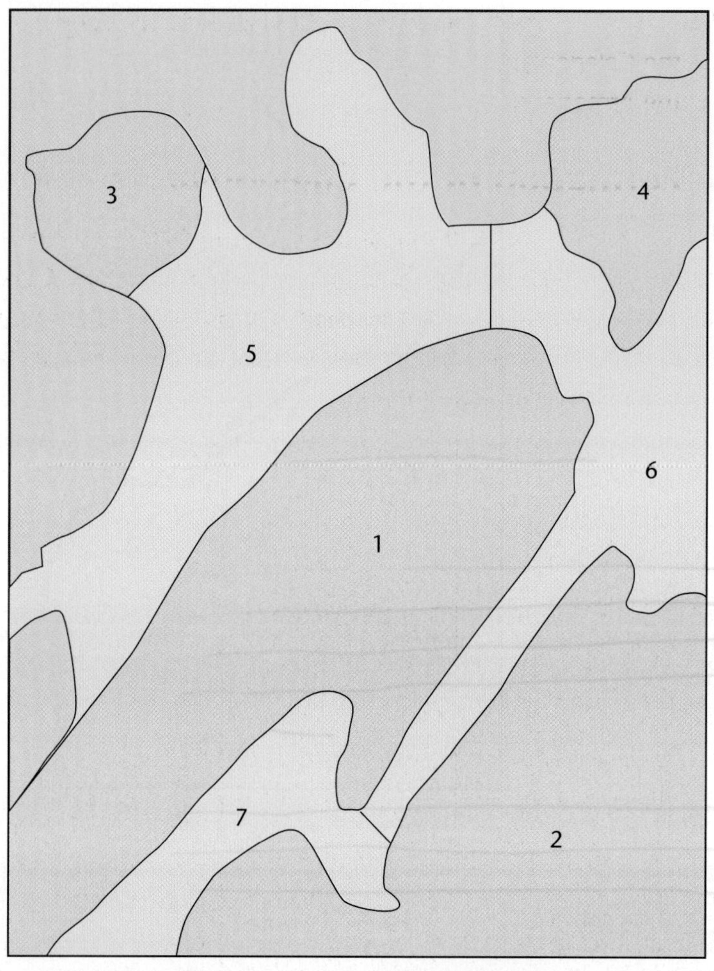


**Key — Physical Regions**

1. Raghtin More upland ridge
2. Bulbin uplands
3. Dunaff uplands
4. Binnion uplands
5. Dunaff coastal plain
6. Basin of the Ballyhallan-Clonmany rivers
7. Basin of the Owenerk River

(1–2: high uplands; 3–4: low uplands; 5–7: lowlands)

**Tips**

1. Give your sketch a title.
2. Draw a shape similar to that of the map.
3. Draw in the easting and northing lines lightly on your sketch as they appear on the map and use them as a guide only.
4. Draw a general impression of the elevated areas. Do not try to be over-exact.
5. Name each region according to a place name or its location on your sketch.
6. Identify features on your sketch by labelled arrows or give a key to avoid overcrowding.
7. Show only the most important routes, rivers and towns.
8. Colour is not essential. Use colour only if you have sufficient time.

**Class Activity**

Study the Barrow and Nore map extract, Fig. 7.40, p. 102. On a sketch map of this region, mark and name its physical regions.

### Definitions of Altitude — General Guide Only

Highland = Land with peaks over 700 m and large areas over 600 m
Upland = Land with peaks over 240 m and large areas between 180 m and 600 m
Lowland = Land of altitude between 0 m and 180 m

# Structural Trends in Ireland

There were two great mountain building periods in Ireland:

1. the Caledonian
2. the Armorican.

## Formation of the Caledonian Fold Mountains

The great north-east and south-west foldings which gave much of Ireland, Highland Scotland and Norway their backbones are generally called **Caledonian foldings**. Caledonia is an old name for Scotland and because of their common origin, geographers call all of these mountains the **Caledonides**, whether they occur in Newfoundland, Ireland, Scotland or Scandinavia.

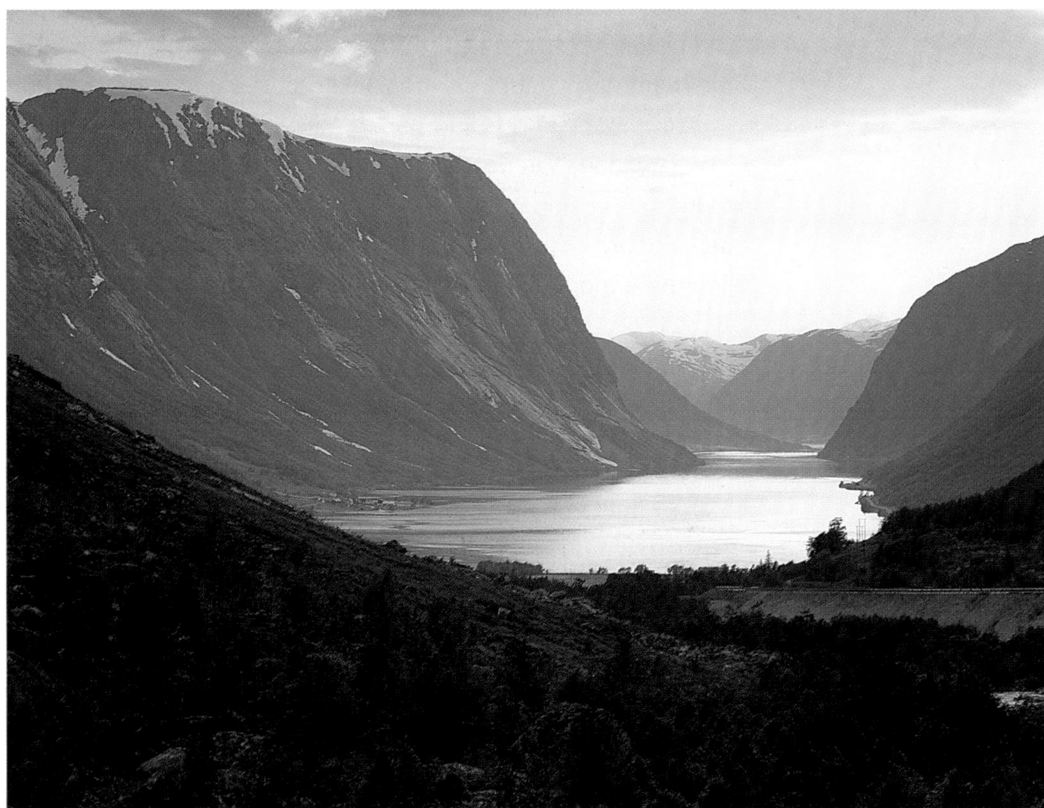

Fig. 8.1
Glaciated Caledonian fold mountains in Norway

From the study of plate tectonics, we know that approximately 400 million years ago, the North American and European plates moved towards one another. As the margins grew closer, immense forces caused great wrinkles to form on the earth's surface. Intense rock crumbling occurred (see p. 74). In our part of the globe, these wrinkles ran north-east and south-west. They forced up the areas which now form the Scandinavian mountains, the mountains of north-west Scotland, and in Ireland, the mountains of

Donegal, Mayo and Connemara, the Leinster Range, and the mountains running from Newry north-east into Co. Down and terminating in Slieve Croob. As these wrinkles ran north-east and south-west, so the trend of the mountains formed by them remains north-east and south-west to this day.

## Formation of the Armorican Fold Mountains

These foldings of Cork and Kerry are sometimes called **Armorican** because they are so well developed in Brittany, whose ancient name is Armorica. In Ireland, they are called the Cork and Kerry foldings as they are particularly marked in these counties.

Approximately 300 million years ago, there was a relative movement of the European and African plates. High mountains were formed in Europe such as the Harz Mountains in Germany. Disturbance was beginning to lose force as it approached Ireland. Its main effect was confined to the south of the country, where a great lateral thrust (force) from the south caused extensive rock deformation. The Old Red Sandstone and overlying Carboniferous deposits were folded into ridges and valleys which ran east–west.

Further north, the thrust progressively died away. Its weaker effects were influenced by the trend of the older Caledonian structures and followed a north-east/south-west trend.

This mountain building period gave Ireland all the mountains of the south-west, from Dungarvan in Waterford to the sea in Kerry. The foldings also extended northward. The Galtees, Silvermines, Slieve Bloom, the table-land of the Burren, Slieve Aughty and Slieve Bernagh are all products of this fold period. The foldings also thrust up the great limestone mountain mass of Ben Bulbin between Bundoran and Lough Gill in Co. Sligo.

**Class Activity**

Study the Ordnance Survey map extracts of the Inishowen peninsula in Co. Donegal (p. 107) and Cork Harbour (p. 108). Each region shown on these extracts displays an obvious structural trend. In each case, identify this trend and briefly account for its formation.

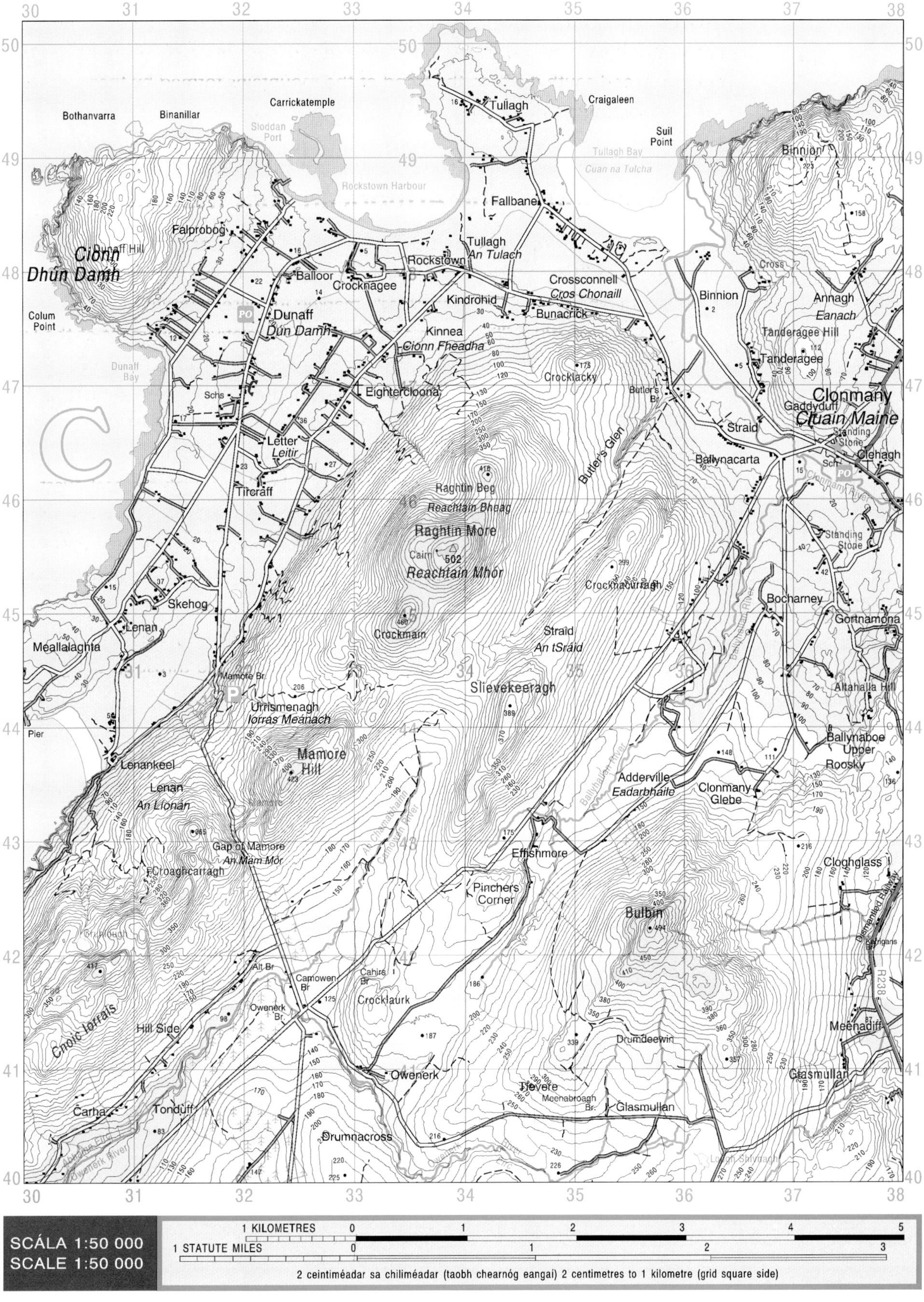

SCÁLA 1:50 000
SCALE 1:50 000

1 KILOMETRES 0 1 2 3 4 5
1 STATUTE MILES 0 1 2 3

2 ceintiméadar sa chiliméadar (taobh chearnóg eangai) 2 centimetres to 1 kilometre (grid square side)

Fig. 8.2
The Inishowen peninsula, Co. Donegal

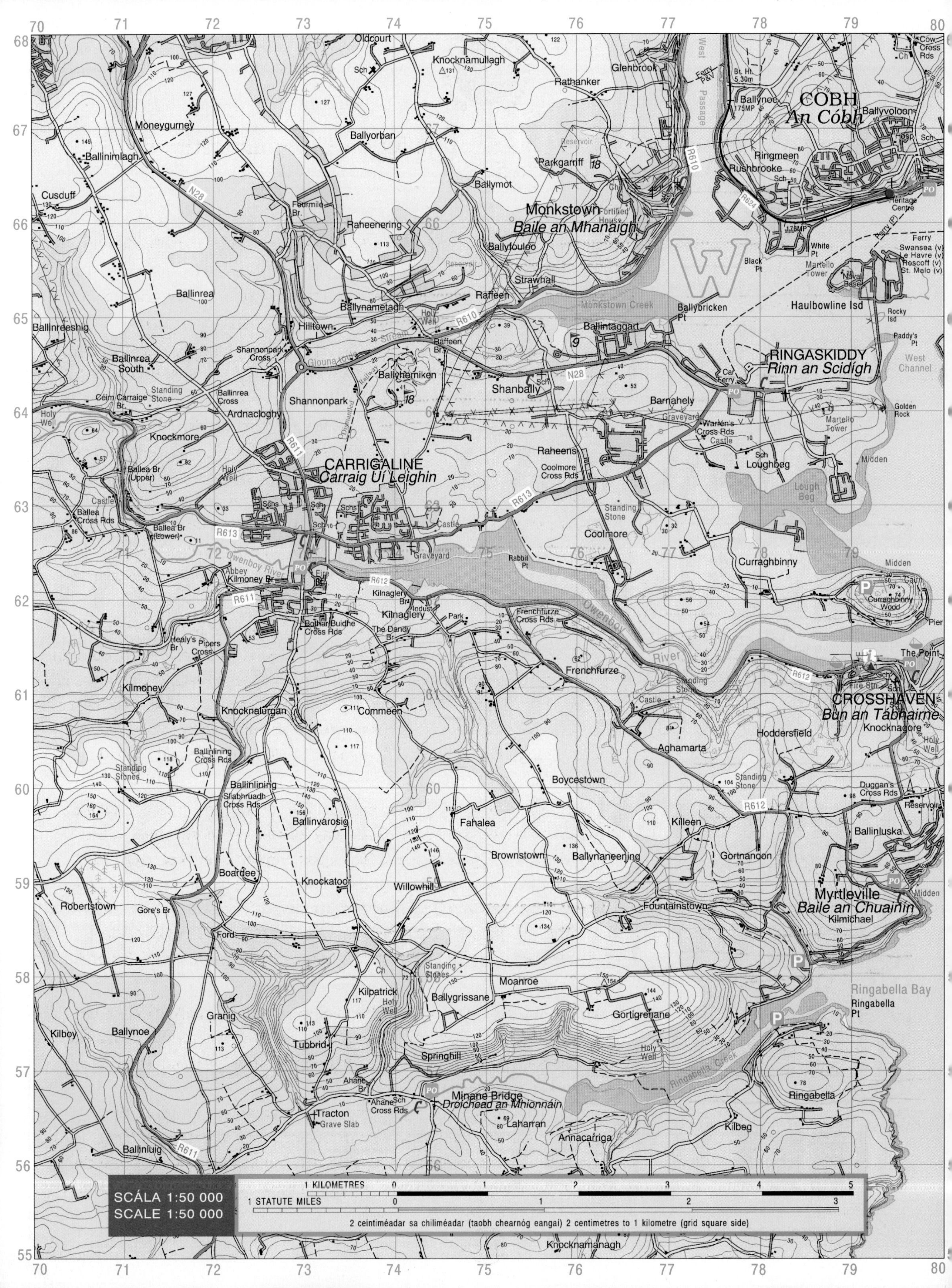

Oldcourt
Knocknamullagh
Glenbrook
Rathanker
Moneygurney
Ballinimlagh
Ballyorban
Parkgarriff
Ballymot
Cusduff
Fourmile Br.
Raheenering
Monkstown
Baile an Mhanaigh
Ballyfouloo
Strawhall
Raffeen
Ballinrea
Ballynametagh
Hilltown
Ballinreeshig
Monkstown Creek
Ballybricken Pt
Ballintaggart
COBH
An Cóbh
Ballynoe
Ballyvoloon
Ringmeen
Rushbrooke
Heritage Centre
White Pt
Black Pt
Martello Tower
Naval Base
Haulbowline Isd
Rocky Isd
Paddy's Pt
Ferry
Swansea (v)
Le Havre (v)
Roscoff (v)
St. Malo (v)
West Passage
West Channel
RINGASKIDDY
Rinn an Scidígh
Shannonpark Cross
Ballinrea South
Ballyhemiken
Shannonpark
Shanbally
Barnahely
Golden Rock
Céim Carraige Br.
Standing Stone
Ballinrea Cross
Ardnacloghy
Holy Well
Knockmore
Warren's Cross Rds
Castle
Loughbeg
Lough Beg
CARRIGALINE
Carraig Uí Leighin
Raheens
Coolmore Cross Rds
Ballea Br (Upper)
Ballea Cross Rds
Ballea Br (Lower)
Coolmore
Curraghbinny
Curraghbinny Wood
Graveyard
Rabbit Pt
Owenboy River
Abbey
Kilmoney Br
Fire Stn
Kilnaglery Br
Kilnaglery
Frenchfurze Cross Rds
Bothar Buidhe Cross Rds
The Dandy Br.
Healy's Br
Pipers Cross
Frenchfurze
The Point
Pier
Kilmoney
Knocknalurgan
Commeen
CROSSHAVEN
Bun an Tabhairne
Knocknagore
Hoddersfield
Aghamarta
Ballinlining Cross Rds
Boycestown
Duggan's Cross Rds
Ballinlining
Slabhruadh Cross Rds
Ballinvarosig
Fahalea
Killeen
Reservoir
Ballinluska
Brownstown
Ballynaneening
Gortnanoon
Boardee
Knockatoor
Willowhill
Myrtleville
Baile an Chuainín
Kilmichael
Midden
Robertstown
Gore's Br
Fountainstown
Ford
Standing Stones
Moanroe
Ringabella Bay
Ringabella Pt
Kilpatrick
Ballygrissane
Gortigrenane
Granig
Kilboy
Ballynoe
Tubbrid
Springhill
Ringabella Creek
Ringabella
Ahane Br
Ahane Cross Rds
Minane Bridge
Droichead an Mhionnáin
Tracton
Grave Slab
Laharran
Annacarriga
Kilbeg
Ballinluig
Knocknamanagh
SCÁLA 1:50 000
SCALE 1:50 000
1 KILOMETRES 0 1 2 3 4 5
1 STATUTE MILES 0 1 2 3
2 ceintiméadar sa chiliméadar (taobh chearnóg eangaí) 2 centimetres to 1 kilometre (grid square side)

◀ Fig. 8.3
Cork Harbour

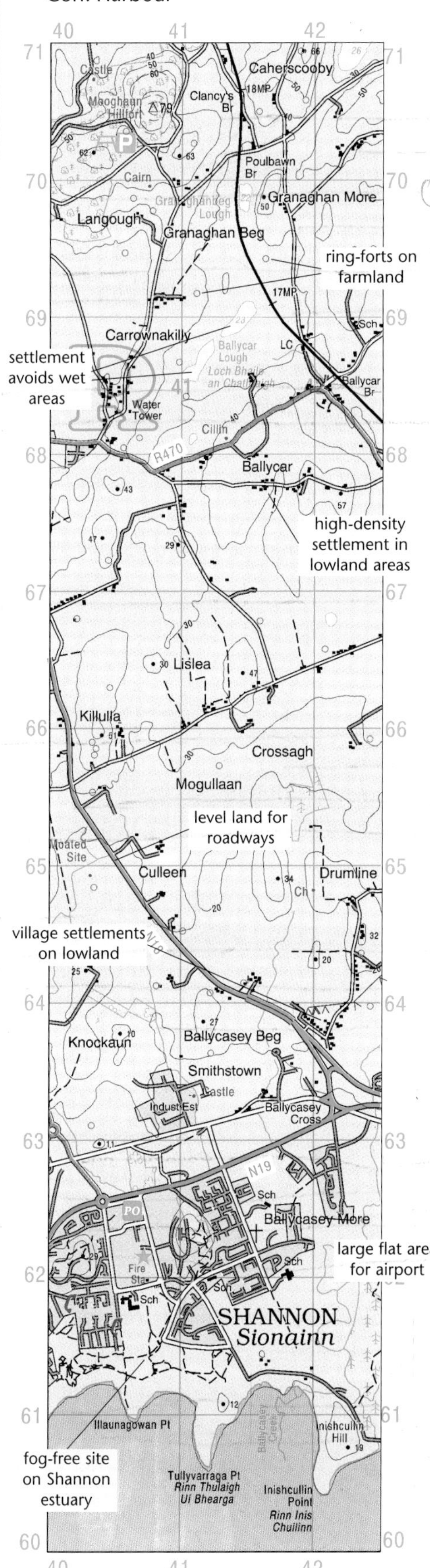


Fig. 8.4
Lowland area

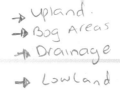

## The Effects of Relief on Communications in Ireland

### Lowland

Lowland areas generally have a high density of routeways. This occurs for a number of reasons.

1. Lowland areas are generally level or gently sloping. Route construction in such places is easy and inexpensive.
2. Settlement in Ireland is confined to lowland areas. Individual settlements such as farmhouses form dispersed patterns throughout lowland areas. All of these settlements are served by routeways so that regions with high-density settlement contain many routeways (such as the midlands and the east) while low-density areas have fewer routeways (such as the western part of Mayo and Galway).

   Villages and towns are generally situated in fertile lowland. They form **nodal centres** (focal points), so routes focus on these settlements. Thus areas with urban centres have a high density of routeways. Lowland areas therefore contain many national primary, national secondary, regional and third-class roads as well as railways.
3. Sometimes, however, routes avoid lowland. Large areas of flat land are not served by roadways in places such as the low-lying land along the flood plain of a river valley (e.g. near Clonmacnoise in Co. Offaly). Here, during times of heavy rainfall, a river may abandon its normal channel and spread across the level valley floor. Roads in such areas would be impassable for long periods of time throughout the year, while damage would be guaranteed to road surfaces and road foundations.
4. Level lowland between rivers is suited to the construction of canals. Where this exists, **watersheds** pose no difficulty for channel construction. In the Irish midlands, for instance, the land separating the Liffey basin from the Shannon basin is very low-lying, thus allowing the easy construction of the Grand Canal between both river systems.

   The land between the River Barrow and the Grand Canal is also low-lying, allowing barges on the Barrow to travel to either the River Shannon or the River Liffey.
5. Airports are best suited to large tracts of level land which allow for the easy construction of runways. Where possible, airports should also be sited away from mountains since aircraft need long stretches of level land when approaching a runway.

   Fog-free sites are particularly favoured for airport construction. The regular presence of fog means loss of landing fees and hazardous flying conditions. Shannon Airport is a particularly fog-free site since it is near the Shannon estuary and subject to the strong westerly winds from the Atlantic.

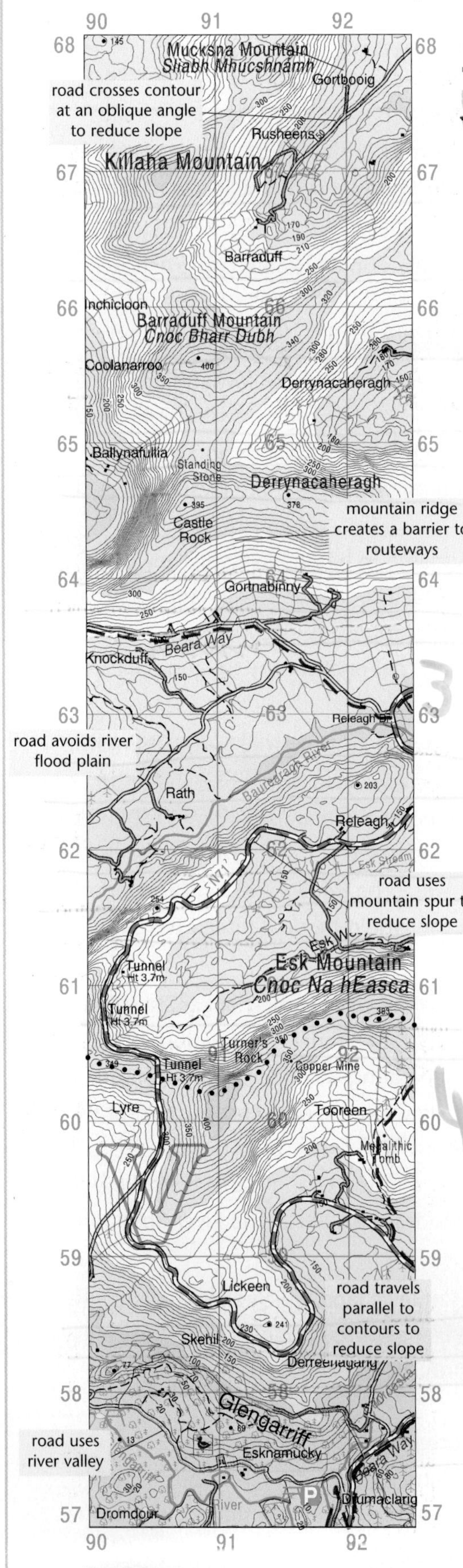


Fig. 8.5
Highland area

## Highland

Highland areas contain few routeways for a number of reasons.

1. Mountain areas are rugged with steep slopes, as in the elevated areas of Kerry, Galway and Mayo. Route construction in such areas is difficult and expensive. Routes avoid such areas where possible.
2. Intensive farming is absent from highland areas. Rainfall is high, and soils are thin and stony with large regions occupied by bog. Consequently, few people live in such areas.
3. Mountain areas experience severe weather conditions. In winter, heavy rainfall and subsequent run-off erode elevated routeways, often washing away parts of roadways.

   Frost action loosens surfaces such as tarmacadam which are then exposed to erosion by rainfall and passing traffic. Maintenance of mountain routeways is high because of the factors listed earlier, so the numbers of roadways in such areas are kept to a minimum. When routeways exist in highland areas, they are usually confined to valleys.

- **Valleys**

  Subsistence farming is often practised in these mountain valleys. Third-class routeways generally serve these isolated settlements.

  Valleys are also used as passageways through mountain areas. Route construction along valley floors is easy and relatively inexpensive as gradients (slopes) are low.

  Lowland areas are sometimes separated by a high mountain ridge. Where a gap exists in the ridge, routes will focus on it, taking advantage of the easier and shorter journey. In many instances, towns developed at the entrances to these passageways for both defensive and commercial purposes. Such towns are referred to as **gap towns**.

- **Saddles and mountain spurs**

  Saddles are wide troughs (low spaces) between upland peaks. Routeways take advantage of these lower altitude points for easier access through high ridges. These saddles allow routeways such as roads to travel through mountain terrain, thus shortening the route distance between lowland settlements.

  Roads use mountain spurs or valley sides to reduce the gradient when approaching mountain passes such as saddles. This is noticeable on Ordnance Survey maps when roads cross contours at an oblique (slanting) angle.

  Highland areas can often form barriers to communications. For example, long, high mountain ridges may prevent routeways from crossing over from one area to another. Where valleys and ridges run parallel to each other as in Co. Donegal, routeways must conform to the north-east/south-west trend of the Caledonian Mountains (see Fig. 12.57, p. 228).

STRUCTURAL TRENDS IN IRELAND

## 5 Drainage

Within valleys, such as those in Co. Donegal which run north-east/south-west and those in Co. Cork which run east–west, the routeways run parallel to the rivers for many kilometres, keeping up from the flood plains to avoid flooding. However, they take advantage of crossing points along river courses, causing the most important of these places to become nodal centres, such as at the lowest crossing points of rivers near the coast.

Lakes also affect the construction of communications. Routeways travel around lakes rather than across them. Due to their width and depth, the cost of bridges is high and often prohibitive. In such instances, routes generally focus on crossing points which are located immediately above and below the lakes, such as at Portumna and Killaloe on the River Shannon in Ireland.

## Case Study: The Influence of Relief on Communications

**Question:**

Examine the Inishowen map extract, Fig. 8.2, p. 107, then draw a sketch map to illustrate how the physical landscape has affected communications in the area.

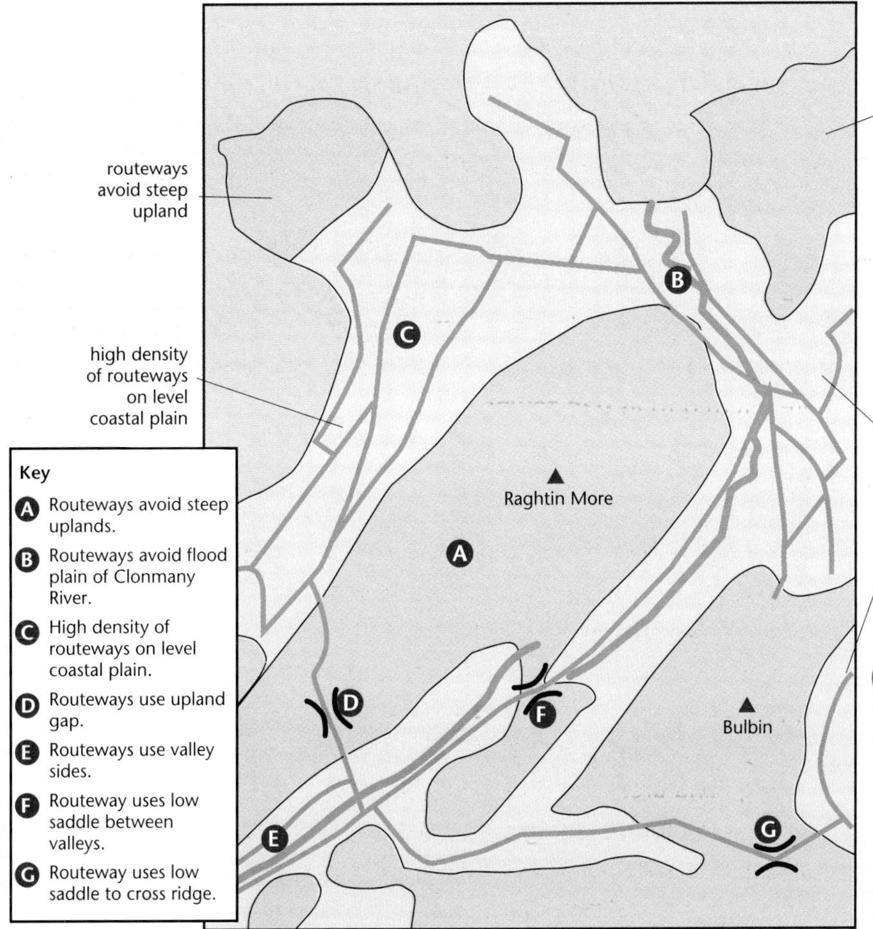
routeways avoid steep upland
routeways avoid steep upland
high density of routeways on level coastal plain
routeways meet where valleys join
routeway uses valley floor
Raghtin More
Bulbin
Key
A Routeways avoid steep uplands.
B Routeways avoid flood plain of Clonmany River.
C High density of routeways on level coastal plain.
D Routeways use upland gap.
E Routeways use valley sides.
F Routeway uses low saddle between valleys.
G Routeway uses low saddle to cross ridge.

**Fig. 8.6**
Sketch map of the Inishowen OS map extract

**Tips**

1. In an examination, mark *only four ways* in which the relief influences communications.
2. *Mark and name* the affected routeways.
3. *Mark and name* the associated landscape features which affect those routeways.
4. In your accompanying description, refer *only* to those examples marked on the sketch map.
5. Write a well-developed paragraph for *each* example. *Refer* regularly to the OS map extract to explain your answer.

**Class Activity**

Study the Ordnance Survey map extract of Brandon Mountain, p. 112. Then, using map evidence, describe how the physical landscape has affected communications in this area. Draw a sketch map to illustrate your answer.

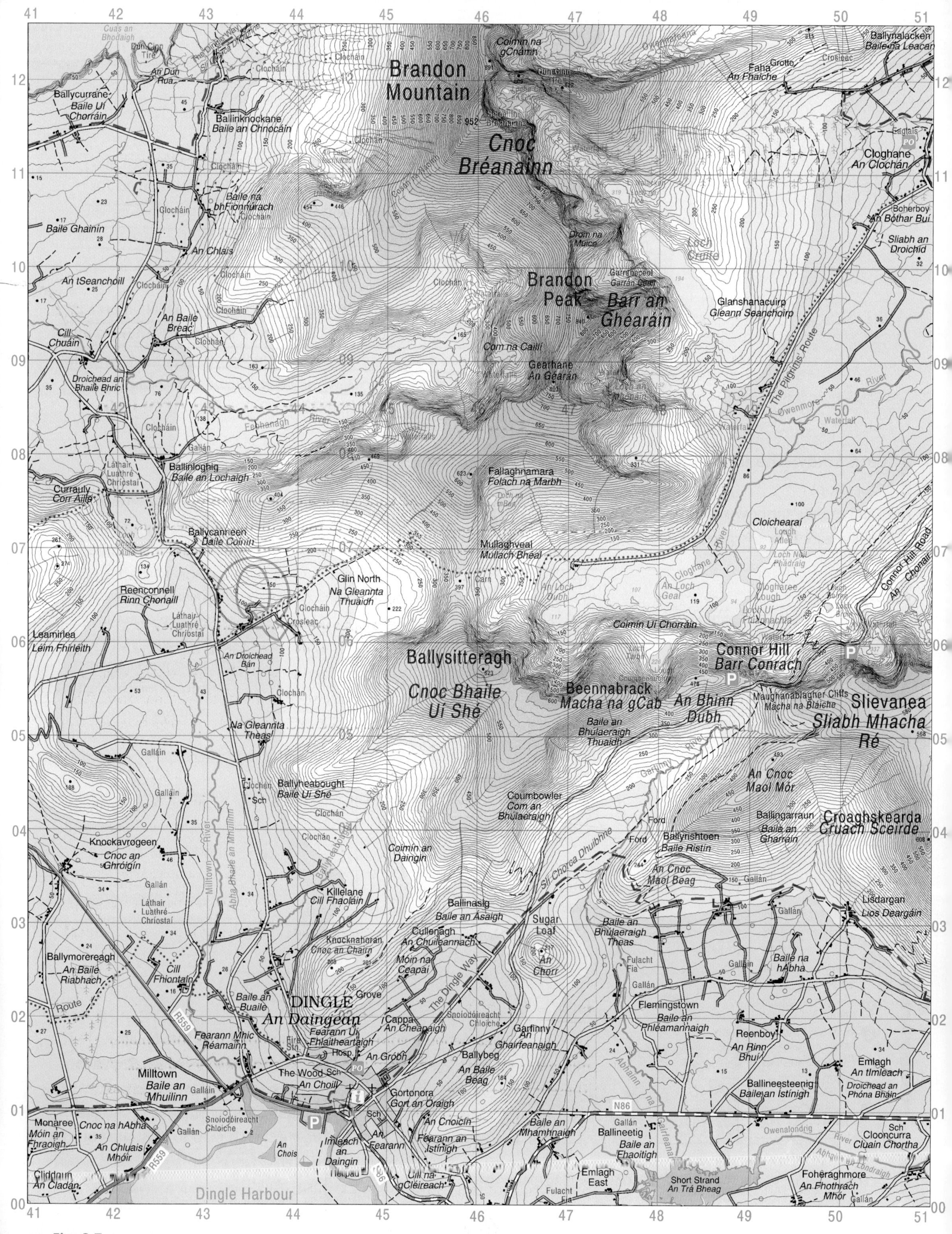


Fig. 8.7
Brandon Mountain map extract

# Weathering, Erosion, Mass Movement and Slopes

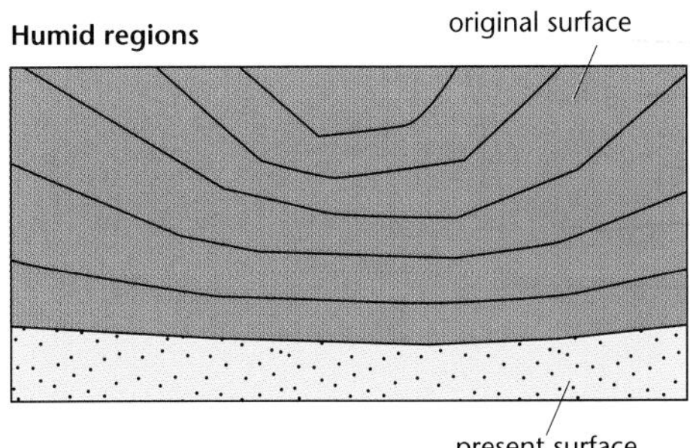


Fig. 9.1
Davis's concept of wearing down until a peneplain is formed

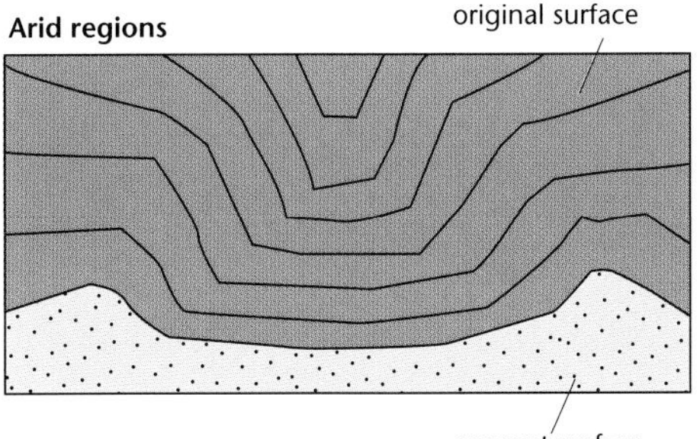


Fig. 9.2
Penck's and King's concept of parallel retreat until a peneplain is formed

**Exam Tip**
For describing the formation of a landform, remember the word
**FEED**
F = feature
E = example
E = explanation
D = diagram

Rocks at or near to the earth's surface are constantly being worn down by **denudation**. Denudation takes place due to the forces of **weathering** and **erosion**.

## Theories on the Weathering and Erosion of Slopes

### The Process of Slope Reduction

W.M. Davis introduced the concept of a **cycle of erosion**. From the instant a newly formed upland is exposed to the elements, it is destined to go through the stages of **youth**, **maturity** and **old age**. Valleys increase in width and depth and their sides or slopes reduce to become lower. Land between valleys (called **divides**) is lowered and rounded to form an undulating plain. When these slopes are reduced further, they form a **peneplain** (Fig. 9.1). (A peneplain is **almost** as level as a plain.)

Walter Penck and L.C. King disagreed with Davis. They stated that as slopes are actively weathered with constant removal of material from the scree slope, the whole slope will move back or retreat, parallel to itself, without getting any lower (Fig. 9.2).

The generally accepted theory now is that in humid regions, Davis's theory is applied while in arid regions, Penck's and King's theory is applied.

## Weathering

Weathering is the simple breaking down or decay of rocks which lie on or near the earth's surface. There are two types of weathering — **mechanical** and **chemical**.

### Mechanical Weathering

Mechanical weathering involves the breaking up of rock into smaller fragments ranging in size from large blocks and boulders to small grains. These fragments are generally coarse and angular in shape. Mechanical weathering is caused by:

1. frost action (freeze-thaw)
2. sudden temperature changes
3. plants and animals

4. crystallisation of salts
5. unloading (pressure release)

## Frost Action (Freeze-thaw)

Frost Action and Gravity

Process

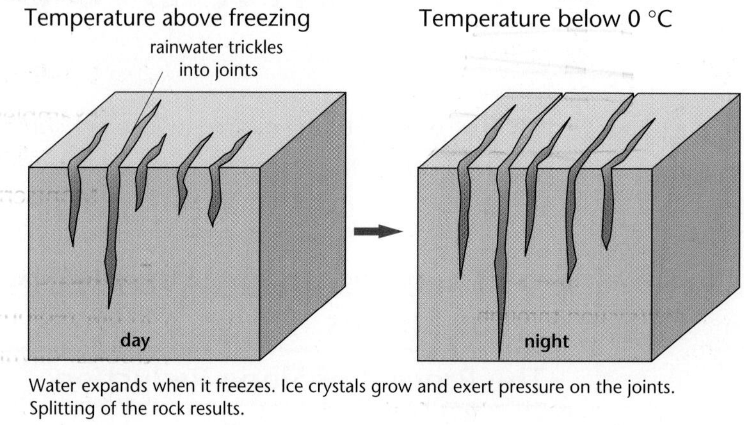

Temperature above freezing

rainwater trickles into joints

day

Temperature below 0 °C

night

Water expands when it freezes. Ice crystals grow and exert pressure on the joints. Splitting of the rock results.

Fig. 9.3
Freeze-thaw

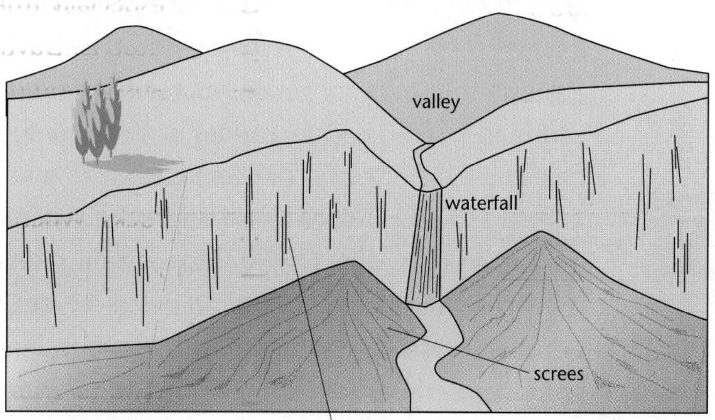


the steep cliff weathers to form rock particles which fall to the base of the cliff and form screes

water seeps into cracks and freezes at night, causing the rock to shatter

physical weathering on steep slopes often produces screes which collect at the bottom of the slopes

Fig. 9.4
Scree slopes

### Landform: Scree Slopes

Examples: Ben Nevis in Scotland
Snowdon in Wales
Croagh Patrick in Co. Mayo in Ireland

Process

### Formation

By day, water seeps into joints, pores and fractures in rock. At night, this water freezes, thus increasing its volume by 9 per cent (Fig. 9.3). The expansion exerts great pressure on the rock, causing it to shatter. This freeze-thaw disintegration of rock is common on the highland and upland areas of Ireland as indeed it is in all areas of high altitude throughout the world. The shattered fragments are pulled by gravity to the base of the rock, slope or mountain and accumulate to form less steep slopes called **scree** or **talus** (Fig. 9.4). Large scree accumulations in Ireland in areas such as Connemara, Donegal, Sligo and Kerry were mainly formed towards the end of the last ice age. Frost action still occurs today in Ireland, but on a reduced scale. Scree slopes are especially noticeable at the foot of rock cliffs in mountain areas.

Exfoliation: Sudden Temperature Changes

Process

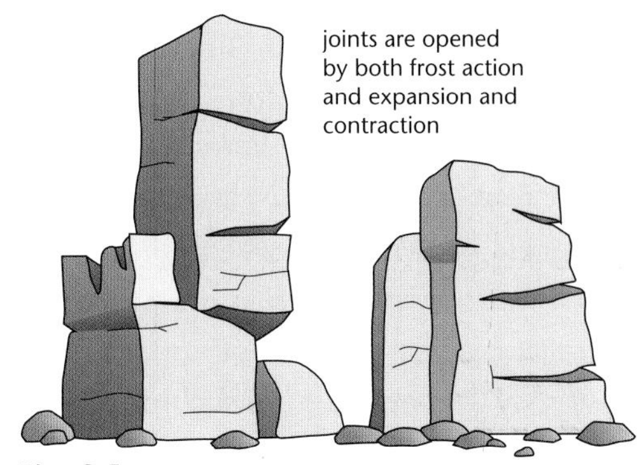
joints are opened
by both frost action
and expansion and
contraction

Fig. 9.5
Some rocks break up into large rectangular-shaped blocks under the action of mechanical weathering. This may be partly frost action and partly expansion and contraction through temperature changes. This is called block disintegration.

Process

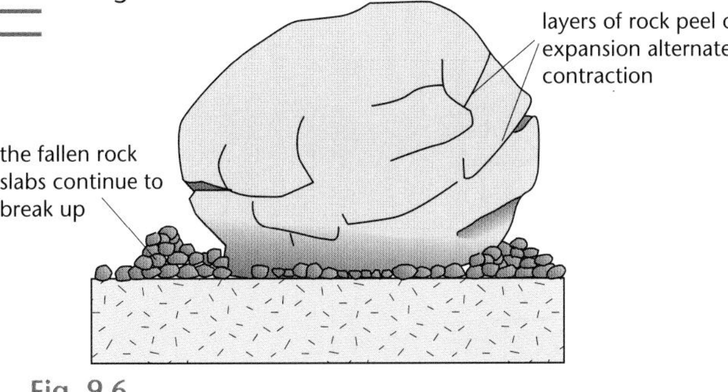
layers of rock peel off as
expansion alternates with
contraction
the fallen rock
slabs continue to
break up

Fig. 9.6
Large boulder showing break-up by exfoliation

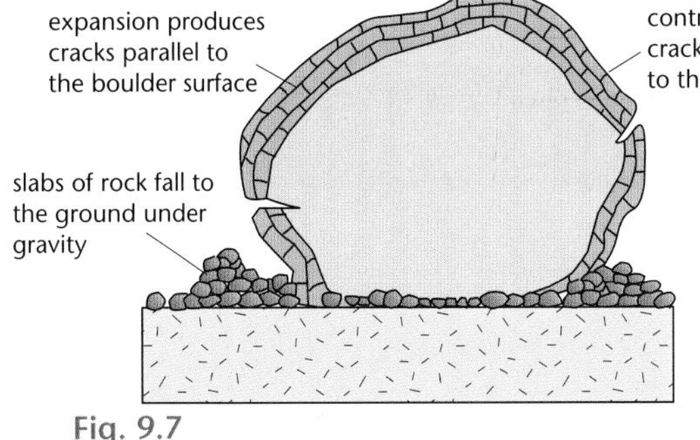
expansion produces
cracks parallel to
the boulder surface
contraction produces
cracks at right angles
to the boulder surface
slabs of rock fall to
the ground under
gravity

Fig. 9.7
Sectional view of the same boulder

Process

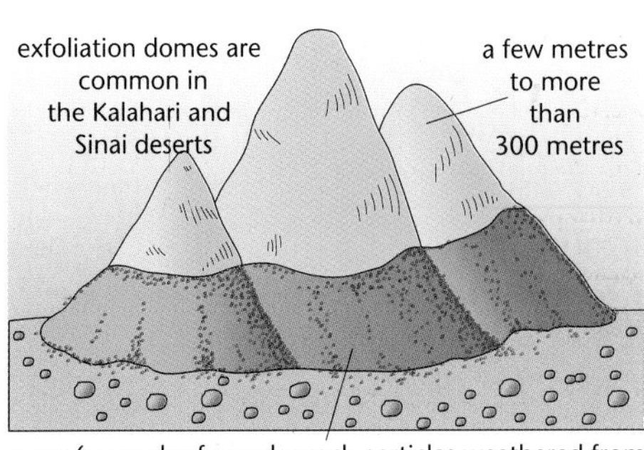
exfoliation domes are
common in
the Kalahari and
Sinai deserts
a few metres
to more
than
300 metres
scree (mounds of angular rock particles weathered from
the rocky masses and which collect around their bases)

Fig. 9.8
Exfoliation domes

Exfoliation, Sudden Temperature Changes, Block Disintegration

## Sudden Temperature Changes

### Landforms:

1. **Rectangular blocks**
   Example: Monument Valley, Arizona

2. **Exfoliation domes**
   Examples: Atacama Desert in Chile
   Australian Desert
   Montserrat mountain, near Barcelona

Processes

### Formation

In hot regions of the world such as the tropics, daytime and night-time temperatures often vary greatly. This is especially true in cloudless regions such as deserts. Daytime **insolation** (the absorption of the sun's energy by the earth's surface) and night-time radiation result in alternate heating and cooling of rocks. Where such a difference occurs, a place is said to have a **large diurnal range** in temperature. The rocks successively expand and contract and so tend to enlarge their joints. Their masses will ultimately break into smaller blocks, a process known as **block disintegration**.

In other rocks, their surfaces may lie exposed to the blazing sun, causing them to reach a higher temperature than the internal mass. In such instances, rocks heat at unequal rates. At night, these rocks cool at unequal rates, their outer 'shells' losing the greater amount of heat and in addition losing it faster than the core area.

When this process of alternate **heating** and **cooling** is associated with a small **moisture content**, it causes the surface 'shell' to 'peel off'.

Although rainfall in deserts may be limited, the rapid loss of temperature at night frequently produces dew. Also the mingling of warm and cold air on coasts

LANDSCAPES OF THE WORLD

Fig. 9.9
Exfoliation domes at Montserrat near Barcelona in north-east Spain

(coastlines of the Namib Desert in Africa and the Atacama Desert in South America) causes fog. There is sufficient moisture, therefore, to combine with certain minerals to cause the rock to swell (hydration) and the outer layers to peel off, a process known as exfoliation or onion peeling.

An isolated mass of rock, such as the 'Devil's Marbles' in central Australia, may be rounded off to form an exfoliation dome. These may also be found in such places as the Atacama Desert in Chile.

At present it would appear that exfoliation may be a consequence of either mechanical weathering or chemical weathering or both.

## Plants and Animals

Process

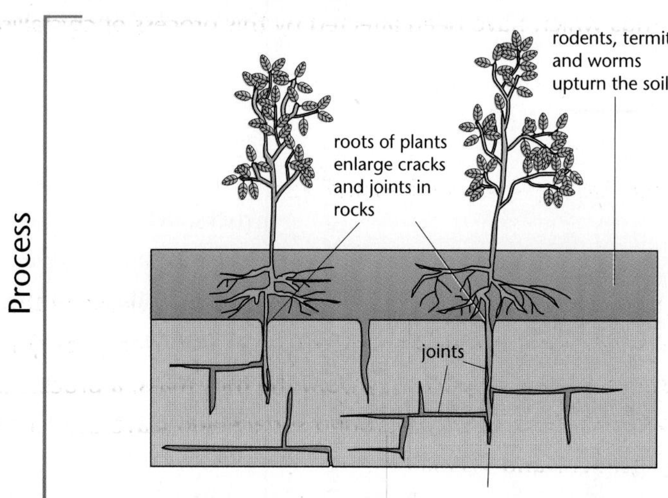
rodents, termites and worms upturn the soil
roots of plants enlarge cracks and joints in rocks
joints
as the root grows the joint is opened up

Fig. 9.10
Plants enlarge rock joints

### Landform: Local

Root Expansion Process

### Formation

**Plant Roots**

Seeds germinate in cracks in rocks to produce plants. The roots of plants, especially trees, penetrate into cracks and crevices in rocks, widening them as they grow larger and causing sections of rock to split off from the main body (Fig. 9.10). In a similar manner, trees are responsible for the destruction of man-made features such as walls when they are in close association with each other. Roots also disturb stone pavements and footpaths.

Creatures such as earthworms and burrowing animals are responsible for the continual upturning of the soil and thus expose fresh surfaces to the weathering process.

## Crystallisation of Salts

### Landform: Surface Flaking

Examples: National Library building and Christ Church Cathedral in Dublin

**Expansion, Mineral Formation, Crystallisation Processes**

#### Formation

Salts such as calcium sulphate and sodium carbonate may enter rocks in solution. As the water content reduces due to surface evaporation, salt crystals form within the rock. As this process continues, the crystals grow and exert pressure, causing granular disintegration and surface flaking of the rock.

In urban areas, some older buildings were built of porous rock such as limestone and sandstone. Sulphur emissions from chimneys and cars combine with moisture in the air to produce sulphuric acid. This acid reacts with such sedimentary rocks to form gypsum within the rock. Expansion then occurs which causes the surface of the rock to flake off, thus damaging the buildings. In Dublin, the National Library and Christ Church Cathedral are two buildings which have been affected by this process of crystallisation.

## Unloading (Pressure Release)

### Landform: Surface Sheeting

Example: Half Dome, Yosemite National Park

**Unloading, Sheeting Process**

#### Formation

As overlying rock layers are removed by denudation, the release of this weight-caused pressure allows the newly exposed rock to expand. This forms new joints in the rock, causing curved rock shells to pull away from the rock mass, a process known as **sheeting**. These new joints run parallel to the ground surface and leave the rock exposed to further weathering (Figs. 9.11 and 9.12).

Continued weathering eventually causes the slabs to separate and 'spall off', creating **exfoliation domes**.

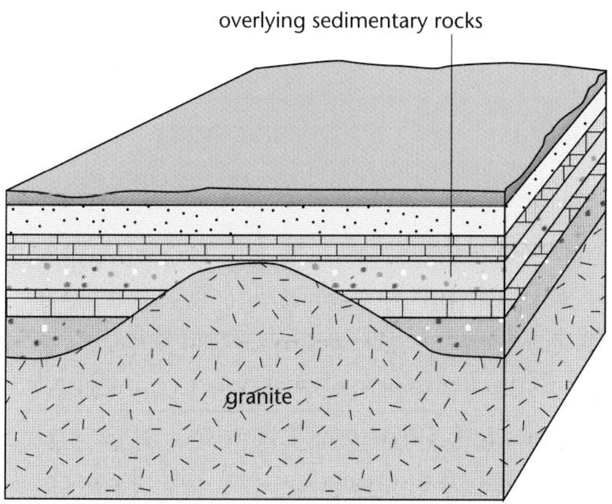


Fig. 9.11
Compressed granite

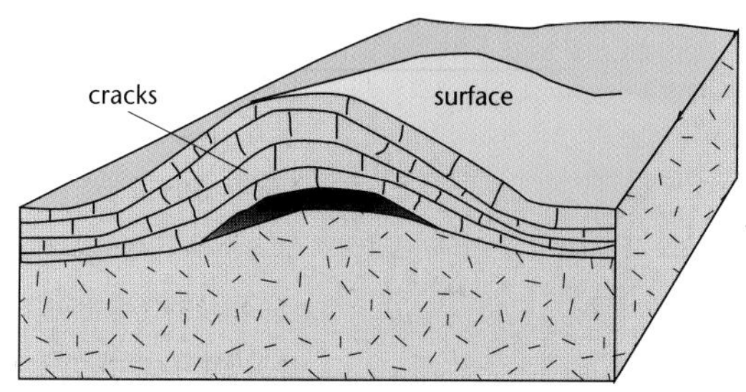


when overlying rocks have been eroded and removed, confining pressure is reduced from the granite and cracks appear

Fig. 9.12
Pressure removed from granite

**Fig. 9.13**
Chemical weathering of limestone in Bosnia

## Chemical Weathering

Most rocks are **aggregates**, or combinations of two or more minerals which are bonded together by cementing agents. Prolonged exposure to weathering agents weakens the bonding of these rock minerals, causing a disintegration of the rock, or, as in the case of rock salt, a removal of the rock altogether.

Mechanical weathering simply divides rock into smaller and more numerous particles, while chemical weathering produces new substances. Mechanical weathering is more widespread and is a more complex process. The first essential for chemical weathering is the presence of water. The processes involved in chemical weathering are as follows:

1. carbonation
2. oxidation
3. hydration
4. hydrolysis

## Carbonation

### Landform: Karst Landscape

Examples: Burren in Co. Clare
Ingleton in north Yorkshire

Chemical Reaction, Solution Processes

### Formation

Rainwater falling through the atmosphere combines with small amounts of carbon dioxide to form a weak carbonic acid. As it reaches and percolates through the ground, it alters carbonate compounds to soluble bicarbonate compounds. This process is particularly effective on chalk and limestone where carbonation converts calcium carbonate into calcium bicarbonate, which is soluble in water and may be removed in solution. Karst landscapes such as the Burren are greatly affected by this process.

Rain ($H_2O$) falling through the atmosphere joins with carbon dioxide ($CO_2$) and forms carbonic acid. This acid alters calcium carbonate in the limestone to soluble calcium bicarbonate, which is removed in solution to the sea.

*Equation:* $H_2O + CO_2 \rightarrow H_2CO_3$ (carbonic acid) $+ CaCO_3$ (limestone) $\rightarrow Ca(HCO_3)_2$ (calcium bicarbonate)

## Oxidation

The results of oxidation are best seen when rocks containing iron compounds are in contact with air. Water percolating through the ground or indeed moisture-laden air produces oxides of iron, resulting in a rusting of the rock particles. This chemical reaction gives a reddish-brown appearance to such rocks and soil. Deep clays are generally blue in colour due to an absence of air. If such clays are exposed to air, they will turn a reddish colour.

## Hydration and Hydrolysis

**Product:** kaolin or china clay, called after the hill in China (Kao-Ling) where it was mined for centuries

Example: Near Arklow in Co. Wicklow, Ireland

Chemical Reaction, Mineral Decay and Expansion Processes

### Formation

1. **Hydration**

   Certain minerals have the property of taking up water and thus expanding. This stimulates the disintegration of the rock containing these minerals by creating stresses and pressures within the rock. This process is called hydration. At this stage mechanical weathering aids the hydration process to shatter the rock.

2. **Hydrolysis**

   This is possibly the most significant chemical process in the decomposition of rocks and formation of clays. This process involves a chemical reaction between rock minerals and water. For example, granite is composed of feldspar, mica and quartz. Such feldspathic rocks crumble and form clays; in other words, the feldspars break down and the rock crumbles. When expanses of **granite** are exposed to the elements, they develop a graceful and **rounded appearance**. On the other hand, areas of **quartzite** have a **sharp and pointed appearance**. The Great Sugar Loaf Mountain in Wicklow, which is made of quartzite, has a pointed appearance, while the Dublin Mountains, which are made of granite, have a rounded appearance. The reason for this difference is the fact that in granite, the grains of quartz only bind the feldspars together, and are set free when it decomposes; in quartzite, the quartz grains themselves are held together by a strong quartz cement. So the two types of rock are acted upon differently by the weather.

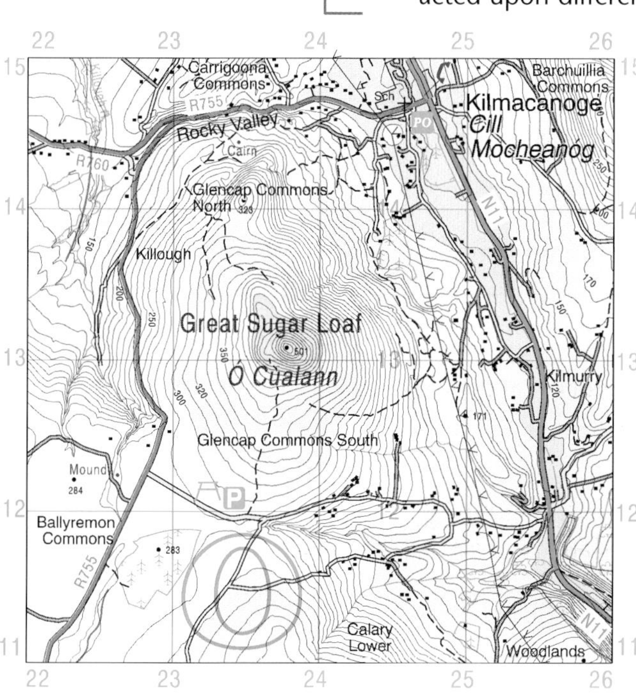

**Fig. 9.15** The Great Sugar Loaf Mountain, Co. Wicklow

**Class Activity**

Examine the Ordnance Survey map extract (Fig. 9.14) and the photograph (Fig. 9.15). Describe and account for the shape of the Great Sugar Loaf upland peak.

**Fig. 9.14** The quartzite peak of the Great Sugar Loaf Mountain, Co. Wicklow

## Chemical Weathering Creates Mineral Deposits

Leaching means the washing of minerals from surface layers to lower horizons in the soil.

Brick-red laterite (from the Latin 'later' meaning 'brick' or 'tile') soils develop in hot, humid climates that support rainforest or savannah vegetation. The drenching rains and high temperatures of these climates promote extensive soil leaching. The red colour comes from the oxidation of iron compounds in the soil (rust).

Weathering creates many important mineral deposits. It does so by concentrating small amounts of metals that are scattered throughout unweathered rock into economically valuable concentrations. Such a process is called **'secondary enrichment'**. This process takes place in association with percolating water and occurs in one of two ways.

1. A mineral deposit is concentrated near the surface when chemical weathering of the surface rock is coupled with percolating water. The percolating water removes undesired materials from decomposing rock or soil on the surface to lower horizons. This leaves the desired mineral concentrated in the upper zones. (See Fig. 9.16.)

### Example: Bauxite

Bauxite, the principal ore from which aluminium is made, is formed in this way. Even though 'aluminium' is the third most abundant mineral in the earth's crust, most is tied up in strong chemical compounds from which it is extremely difficult to extract the mineral.

Bauxite forms in rainy tropical climates. When aluminium-rich rocks are subjected to the intense and prolonged chemical weathering of the tropics, most of the common elements, including calcium, sodium and silicon, are removed by leaching. Because aluminium is extremely insoluble, it becomes concentrated at the surface as bauxite.

**Definition of a Mineral Resource**
A mineral resource may be defined as all discovered and undiscovered deposits of a useful mineral that can be viably extracted now or in the future. The processes of chemical weathering can form valuable mineral deposits.

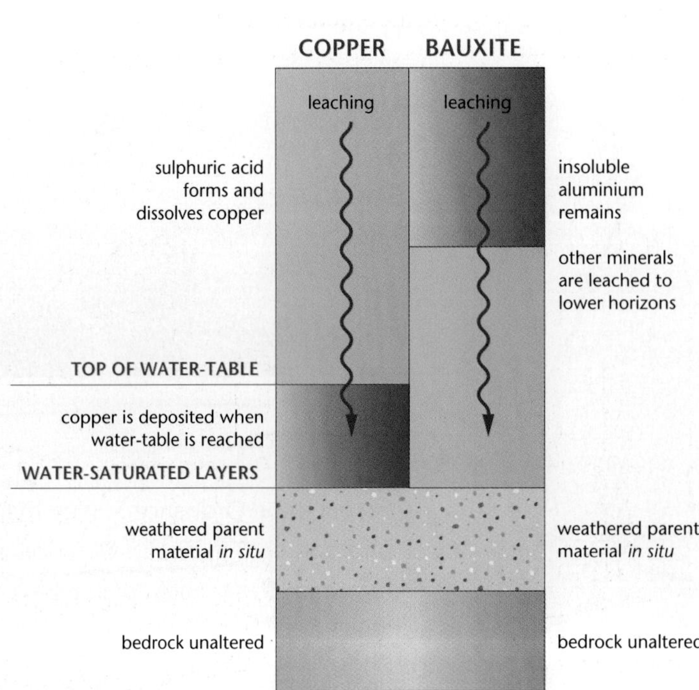


Fig. 9.16
Soil profile of a mineral-bearing laterite

2. The second way is basically the reverse of the first. The desired mineral that is found in low concentrations near the surface is removed and carried to lower zones by percolating water, where it is redeposited and becomes concentrated.

### Example: Copper

Some mineral deposits, such as some copper and silver ores, result when leaching removes the mineral from a low-grade surface deposit to lower horizons. Here it becomes concentrated to form a high-grade ore. Such enrichment occurs in deposits containing copper pyrites. Pyrite is important because when it chemically weathers, sulphuric acid forms. As this acid percolates down to lower horizons, it dissolves the ore metals until they reach the water-table (the zone beneath the surface where all the pore spaces are filled with water). Deposition occurs because of chemical reactions that occur in the solution when it reaches the water-table. In this way, the small percentage of metal that was scattered in the large volume of rock at the surface was redeposited to form a high-grade ore in a smaller volume of rock in lower horizons.

## Mass Movement

The movement of debris, the loose material derived from the weathering of bedrock, down a slope as a result of the pull of gravity alone, is known as mass movement. No actual transporting agency, such as running water, is involved. Its result is mass wasting.

Mass movement takes a variety of forms. Some movements are slow, almost imperceptible, and continue over a long period of time. Others, usually on a large scale, act suddenly, rapidly and sometimes catastrophically. Some are caused as a result of weathering, others as a result of erosion.

The more rapid movements may be 'triggered off' by some external influence. These may be natural or artificial.

1. Natural influences may be an intensive downpour, snow-melt, an earthquake, or erosion by a river, such as undermining of a river bank.
2. Artificial influences include quarrying, dam collapse, the clearance of trees from a hillside or the vibrations caused by a loud noise or passing train.

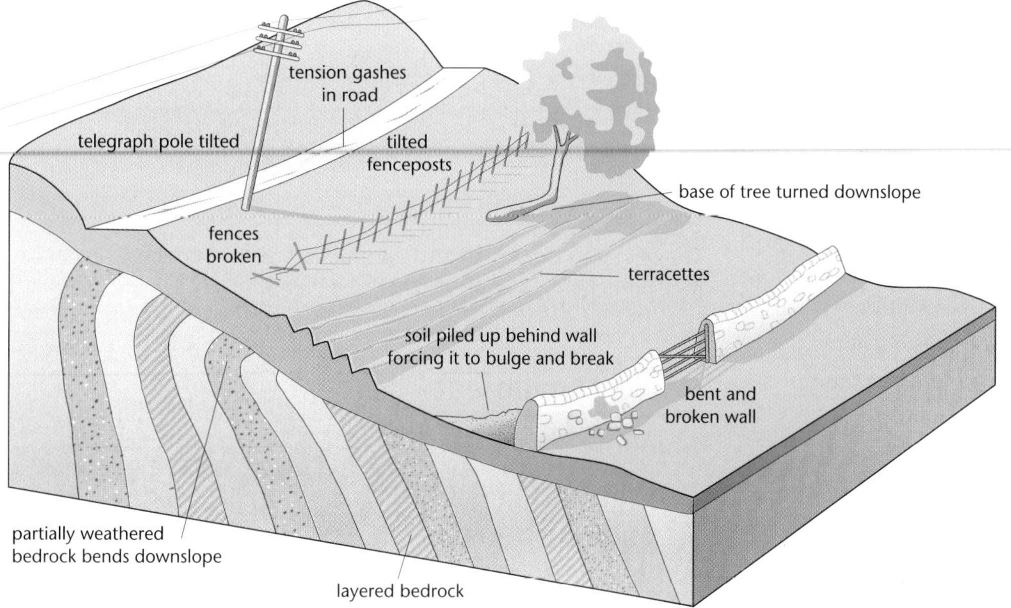


Fig. 9.17
Poles, fences and trees are tilted as a consequence of soil creep

## Slow Movement

### Landform: Terracettes

Examples: Local steep slopes in fields or on a mountain or hillside
Galtee Mountains in southern Ireland

### Formation

#### Soil Creep

**Soil creep** is the slowest type of mass movement. It involves a steady and almost indiscernible movement of weathered soil or rock particles downslope. It may only become apparent when posts, fences or trees are first tilted and then displaced downhill (Fig. 9.17). On short slopes found locally on farms or hillsides, a ribbed or stepped pattern may develop across the slope. These form **terracettes**. Also, soil tends to accumulate on the upslope side of fences, walls and hedges as a result of soil creep.

Vegetation reduces the process of soil creep as the roots of plants such as grasses and trees help bind soil particles together.

#### Solifluction

This process is most active in periglacial regions (areas of high latitude and high altitude). The greater the water content of a soil, the greater is the likelihood of soil movement. Besides adding to the weight of the material, water also causes some soil particles to swell. This swelling causes adjacent particles to move and it **lubricates** the soil, thus lessening resistance to gravity.

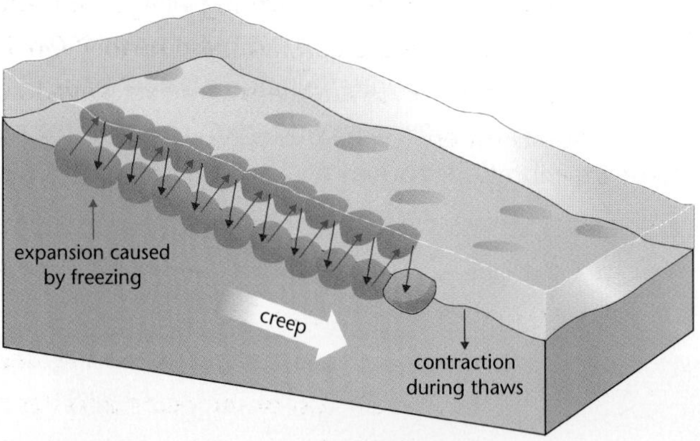


Fig. 9.18
The repeated expansion and contraction of the surface material causes a net downslope migration of rock particles — a process called soil creep

## Rapid Movement

### Mudflow

Mudflows are of two types:

1. Lahar
2. Mudflow

## 1. Lahar

### Landform: Lahar

Examples: Mt. St Helens in Washington State in the USA
Nevado del Ruiz in Colombia in South America

**Volcanic Eruption, Melting, Gravity**
Processes

### Formation

Mudflows occur when soils which contain a high percentage of clay particles become saturated with water. The steeper the slope, the higher the speed of the flow. Great destruction and death may result after such a soil movement. Soils on volcanic slopes have often become fluid as a result of an eruption, e.g. lahars.

As you may recall from chapter 4, snowcapped volcanic peaks are particularly prone to a type of mudflow called a **lahar**. Steep volcanic peaks frequently contain thick layers of loose pyroclastic debris (see p. 59) that was ejected from the volcano during an eruption or eruptions. Soils on such volcanic slopes have often become fluid.

In such instances, hot molten magma is ejected from the crater and may land on adjacent snow fields if the mountain is sufficiently high. This causes the snow, or ice, to melt, thus releasing vast quantities of water which saturate the ground. At the same time, the shaking of the mountain loosens debris, which mixes with the flowing water. Huge volumes of liquid mud, called lahars, may run down the volcanic slopes at high speeds, often covering villages and towns.

Such lahars are common and deadly. On 13 November 1985 a lahar descended the 5,400-metre slope of Nevado del Ruiz in Colombia in South America. Channelled within the valley of the Lagunilla River, it was 15 metres high and travelling at 70 kilometres per hour by the time it reached the town of Armero, 48 kilometres from the summit. The inhabitants of the town did not stand a chance. Approximately 25,000 people were buried alive in the mud.

The lahars from Mount St Helens, in Washington State in the USA, were the cause of most of the death and destruction as a consequence of the eruption in 1980.

Fig. 9.19
**Mudflow.** Carefully examine the mudflow trail in the centre of this photograph as it rushed downslope to Sarno in southern Italy in May 1998.

## 2. Mudflow

### Flooding Kills 100, Others Feared Buried in Rubble

Torrents of mud and water unleashed by forty-eight hours of torrential rain engulfed hundreds of homes in southern Italy on Thursday, 7 May 1998. At least 100 people died, over 1,000 people were left homeless and hundreds of others were listed as missing.

Fig. 9.20
A view of rescue workers in a muddy street in Sarno, near Naples. At least 100 people in Sarno were killed, 51 were feared buried under the mud, more than 1,000 people were left homeless, after torrential rain caused a massive landslide in southern Italy.

Rivers of mud burst into town centres, tearing apart houses and bridges, swallowing cars and sending panicked residents fleeing for their lives in the small towns — such as Sarno — and villages in the region between Salerno and Naples.

This region has experienced 631 landslides in the past seventy years. The government was accused of failing to predict the latest disaster. On Sunday, 10 May 1998, over 100 people were buried in a special plot amid scenes of grief-stricken relatives in the Sarno town cemetery.

In southern Italy, from Naples southwards, the landscape is one of steep mountain slopes, rock outcrops, coastal cliffs and terraced gardens and woodland. Dwellings and towns are perched on cliff-sides sometimes formed of soft limestones or volcanic tephra (volcanic ash, and pyroclastic materials). The landscape is steeply sloping causing a severe shortage of building land. With lax planning laws and uncontrolled construction, illegal and unplanned housing has been erected on these steep slopes over past decades. To allow for this development, woodland and other vegetation cover has been removed from surrounding slopes. Each of these factors has exposed mountainsides to erosion. This uncontrolled construction has been blamed on the corruption of government officials.

### Cause of Mudflow

When torrential rain seeps into deep soil on a steep slope, the soil becomes saturated and fluid and slips downslope as a mudflow under the pull of gravity. In this case, great depths of volcanic ash and lava cover much of this region. This ash became fluid where deforestation and illegal building occurred. With nothing to bind the soil particles together and no vegetation to absorb some groundwater, the full volume of rainfall percolated into the soil making it fluid.

## Landslide

### Landform: Landslides

Example: The Piave river valley in the Italian Alps

**Sliding Process**

### Formation

Landslides and rockfalls are very rapid slides of accumulated rock debris. They generally occur in glaciated mountain areas, in rock and gravel quarries, near large, steep rock outcrops or on steep slopes which have large accumulations of loose rock debris. Even though they occur quite suddenly, close inspection before the fact would reveal plenty of warning signs.

On 1 October 1963 a shepherd noticed that his flock of sheep refused to graze on the north slope of Mount Toc, in the Italian Alps. Eight days later, 300 million cubic metres of rock and soil slid down the north slope at more than 90 kilometres per hour into a reservoir at its base. Like an enormous plunger, it sent a huge blast of air, water and rock up the valley walls. It also sent a wall of water 100 metres high over the top of the reservoir dam. The surging waters spread upstream and downstream along the Piave River and flooded the populated valley on each side. Over 2,600 lives were lost. The event took about seven minutes from start to finish.

Sliding
Process

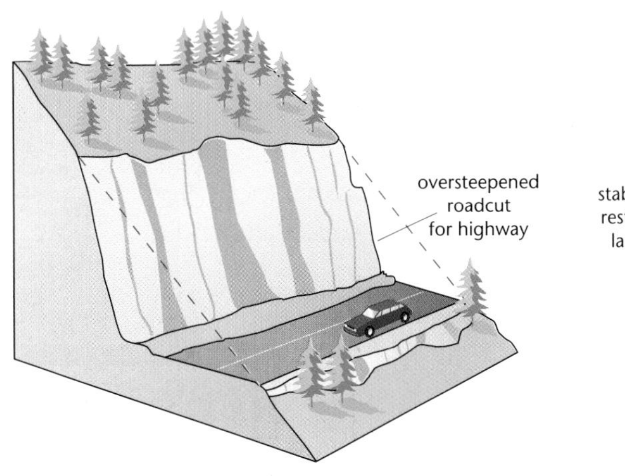


Fig. 9.21
Landslides
(a) Rotational slumping due to human activity
(b) Stable angle restored by landslide

Human Activity, Rotational Slumping,
Erosion by Waves or Rivers
Processes

## Rotational Slumping due to Erosion

Slumping involves both a downslope fall in material and a rotational movement of the falling mass. It is especially common in cliffs of clay which are under attack by waves or along the banks of rivers in their middle courses. This is particularly noticeable during times of flood.

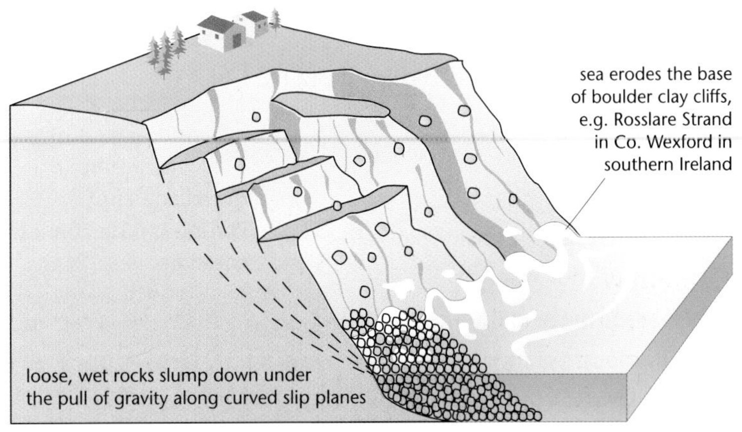


Fig. 9.22
Landslide caused by slumping

Landslides can sometimes be caused by people. Undercutting activities, such as road construction, and additional water from septic tanks and roofs, can make localised steep slopes unstable. Finally, they collapse creating a landslide.

## Landslides — Natural Causes

### Case Study

Fig. 9.23

Landslides may occur naturally. In August 1997 nineteen people were buried under tonnes of debris at Thredbo resort, 400 kilometres south-west of Sydney. The landslide occurred just before midnight at a time when Thredbo, at an altitude of 1,369 metres, was packed with thousands of Australian and foreign visitors. The resort sits on a ridge facing snow-covered mountain ranges in the Australian Alps.

Scientists blamed nature and not man's interference for the slide. Some environmentalists had claimed that overdevelopment of the ski resort might have been responsible. According to a Melbourne professor, soil condition and natural subsidence meant the slide was inevitable: 'Steep mountain areas with saturated thin soils are going to move.' Geologists believe that an underground spring may have destabilised the mountainside. It has been the scene of many landslides in the past.

## Landslides — Human Causes

### Case Study: A Mountain Moves

Fig. 9.24
An avalanche of rubbish buried a man alive and swamped the village of O Portiño on the Spanish coastline

A major ecological disaster is threatening a rugged stretch of Spain's north-western coastline, at La Coruña and neighbouring towns and villages. A million-ton mountain of rubbish, the area's accumulated household trash of more than two decades, is collapsing in a landslide that some locals say they had predicted for years. On 10 September 1996 the rubbish swamped one of the villages from which it emanated, killing a resident. Now it threatens to pollute the sea that provides the livelihood of many people in the area.

The avalanche began when deep fissures appeared in the Bens municipal dump, where the 400,000 people of La Coruña and sixteen smaller towns and villages have deposited the rubbish of everyday life for over twenty years. Heavy rains or perhaps a seismic disturbance are being blamed for the instability. However, the top has always leaked streams of foul-smelling liquid — evidence, some say, that it had poor drainage and began to collapse on its own liquid waste.

What officials now fear is that the rest of the mountain may come crashing down, burying everything in its path and stopping only when it reaches the Atlantic.

Source: adapted from *Time* magazine, October 1996.

Fig. 9.25
Cracks appear on the rubbish surface

## Rockfall

### Landform: Rockfall

Examples: Croagh Patrick in Co. Mayo in Ireland
Palisades Sill, Hudson Valley in New Jersey in the United States

Gravity Fall, Weathering, Freeze-thaw Processes

### Formation

(a)

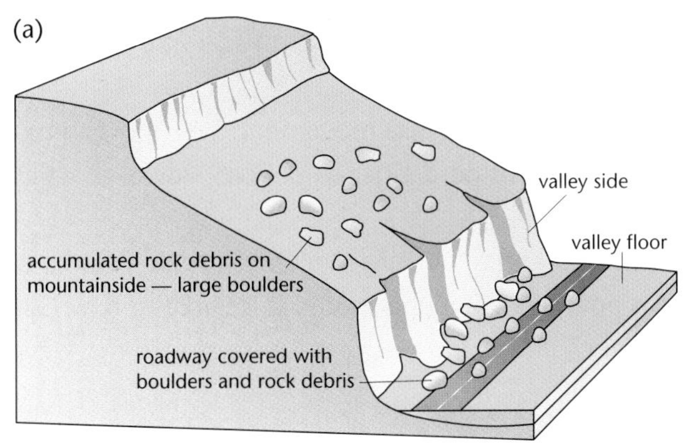


(b)

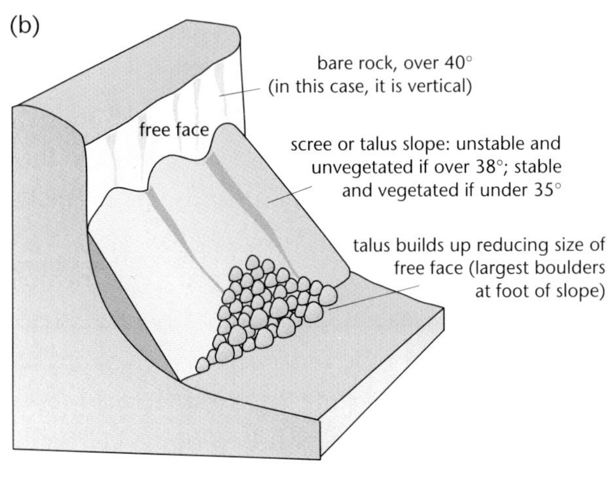

(c) Accumulated scree due to rockfalls near Lake Muskry in the Galtees, in Ireland. Frost action shatters the joints and sends large slabs of conglomerate rock crashing to the bottom of the slope.

Fig. 9.26
Rockfalls

Weathering Process

**Case Study: Rockfalls in the Palisades in New Jersey**

Weathering sometimes produces sudden dramatic rockfalls. Such a rockfall occurred along the Palisades Sill of basalt rock on the west bank of the Hudson River in New Jersey, in the United States. The sill of basalt rock is generally resistant to weathering and is divided by vertical joints into massive columns much like the columns of the 'Giant's Causeway' in Co. Antrim in Northern Ireland. However, the thin layer of softer rock at the base of the sill weathers more quickly. This leaves the huge vertical columns unsupported, which leads to their collapse to form a rockfall.

## Rapid or Slow Movement

### Landform: Earthflows

Example: Bog burst in east Limerick in southern Ireland

**External Influence — Rainfall**
Process — Gravity

### Formation

Earthflows occur frequently in areas of heavy rainfall where the rock is deeply weathered. Such deep soils become mobile when saturated with water and may suddenly slip downslope. This action leaves a curve-shaped (concave) scar at the origin of the slip and bulge (convex) at the base of the slope below.

Blanket bogs are susceptible to downslope movement. This movement may be slow or rapid. When it is slow there is little or no danger to people. However, when it is rapid, loss of life and property can result.

Peat has the ability to absorb and retain large volumes of rainwater. If a saturated bogland on a fairly steep slope gets additional water, it is liable to flow. Such was the case in east Limerick in Ireland in 1708.

**Bog burst**
**Bog bursts** are not uncommon in Ireland. After times of heavy rainfall, blanket bogs on medium to steep slopes, in upland areas, may become fluid and flow downslope. Such movement may be **rapid or slow-moving**.

**Case Study: Cooga Bog — A Bog Burst in 1708**

Castlegarde townland contains 200 hectares, of which one-fifth is bog and rough ground and the rest pasture and meadowland.

In the year 1708 the bog moved along a valley called Poulevard (which was another name for Castlegarde) in Doon in east Limerick and buried three houses containing twenty-one people. The flow of liquid turf was one mile long and a quarter of a mile broad and the mass of turf was twenty feet deep in some places. The flood of wet turf ran for several miles, crossed several roads, demolished several bridges and eventually flowed into a lake at Coolpish or Cool na Pisha. This account of the bog burst at Cooga is taken from the *Dublin Evening Telegraph* of 2 January 1897 and is discussed in the *Transactions of the Royal Irish Academy* by a Mr Ouseley. Between 1702 and 1896 there were fifteen similar bog bursts in Ireland; the last one was in Kerry at Owenaree, seven miles from Killarney, on 27 December 1896.

# Underground Water and Karst Landscape

## Sources of Underground Water

### Subterranean Water

A small amount of subterranean water may have remained in sedimentary rocks since their formation. This is known as **connate water**. During tectonic activity, some of this water is released, usually having been heated and mineralised. It reaches the surface as hot springs, geysers or pools and is called **juvenile water**.

### Meteoric Water

This type of water percolates into the ground from the earth's surface and is directly derived from rainfall or snow-melt. Certain factors will determine the amount of water which will eventually percolate to the water-table.

1. The **frequency of rainfall** in an area.
2. The **rate of precipitation**. The faster the rate of fall, the less able the ground is to absorb it, so much of it flows into streams and gullies as run-off.
3. The **slope (gradient)** of the ground. The steeper the slope, the greater the amount of run-off. On some extremely steep slopes, run-off will represent 100 per cent of precipitation.
4. **Vegetation** holds water long enough to enable the ground to absorb a larger quantity. Bare rock surfaces such as karst landscapes (see p. 139) encourage run-off.

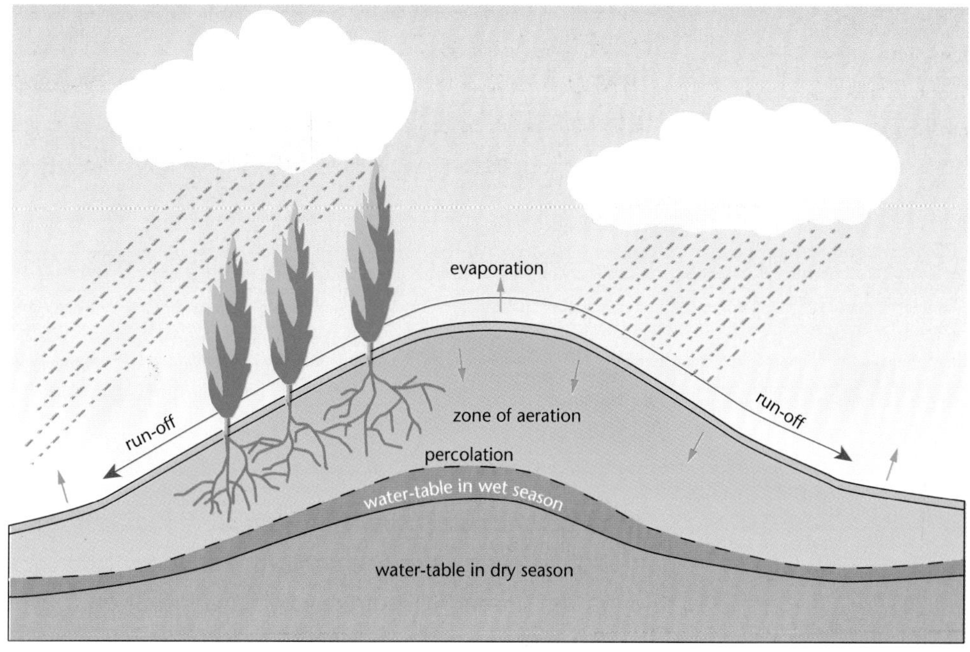


Fig. 10.1
Meteoric water

## Water-table

### Definitions

#### Impermeable rocks

Rocks which do not allow water to pass through them.

#### Permeable rocks

Rocks which will allow water to pass through them. Permeable rocks may be of two types:

1. **porous rock** — rock containing pores or spaces which allow the passage of water, e.g. chalk
2. **pervious rock** — rocks which allow the passage of water through their joints, bedding planes and cracks, e.g. limestone

Even though a rock may be porous, it may also be impermeable. Clay, for instance, is porous (it absorbs water), but once it is saturated, it will not allow water to pass through it. This is the reason for pools of water occurring at gaps in fields during winter.

## The Hydrologic Cycle and Groundwater

Less than 1 per cent of the water on the earth is groundwater. Although this volume sounds little, it is four times larger than the volume of all the water in freshwater lakes, or flowing in streams, and nearly a third as large as the water contained in all of the world's glaciers and polar ice.

Fig. 10.2
The hydrologic cycle

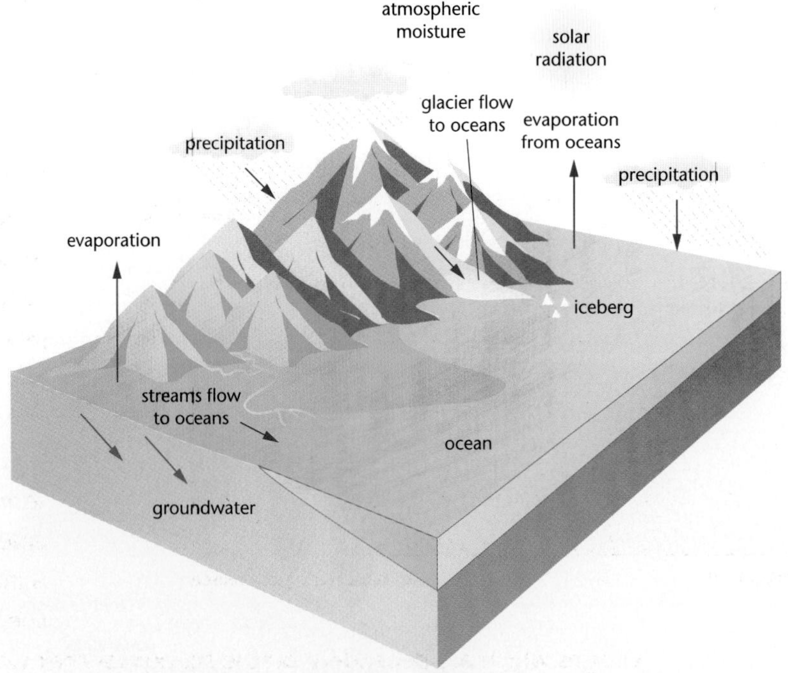


Most groundwater originates from rainfall, which soaks into the ground and becomes part of the groundwater system. It moves slowly towards the ocean, either directly through the ground or indirectly by flowing out on to the surface and joining streams.

Fig. 10.3

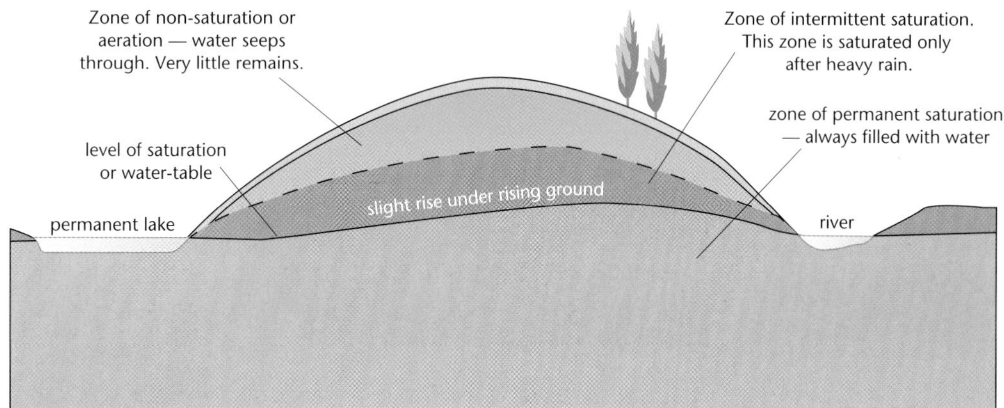


Water entering surface rocks will eventually move downward until it reaches a layer of impermeable rock. At this stage, it can no longer travel downward and so saturates the overlying rock layers, filling all pores and crevices. The portion of the rock where water is permanently stored is called an **aquifer**. The upper level of this water supply is called the **water-table**. Its level generally runs parallel to the ground surface, rising slightly under high ground and dipping down under low ground. However, the water-table may be permanently exposed in hollows, creating permanent water pools in fields or in deserts (**oases**).

Some pools may be of a seasonal nature. In winter, for instance, slight hollows may fill with water while in summer they dry up and disappear. These occur in parts of counties Clare and Galway in Ireland and are called turloughs (see pp. 144–5).

## Springs and Wells

### Springs

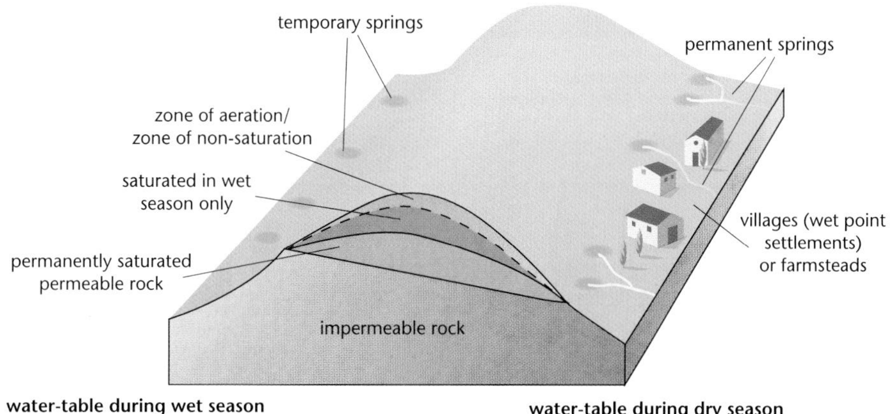


Fig. 10.4
Permeable rock overlying impermeable rock on a hillside

Natural outflows of underground water are called springs. Such outflows of water may only be small seepages or they may be strong streams. The following are some of the more common types of springs.

Where a line of springs appears, the term **spring line** is used. All these springs share the characteristic of emerging at a point where the aquifer is intersected by a slope. This spring line is often indicated by a string of villages which are dependent on the springs for their water supply. Such settlements are common in the Cotswolds of southern England. In Ireland, villages or farm settlements such as clachāns in Co. Kerry may appear in a line as a result of springs issuing from the foot of a mountain.

Fig. 10.5
Well-jointed rock in sloping terrain

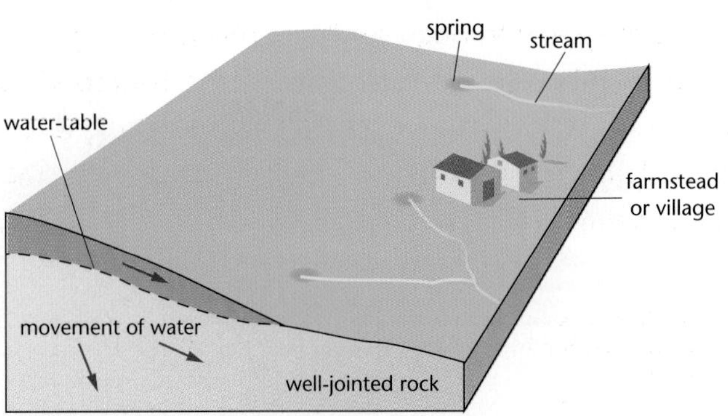


Fig. 10.6

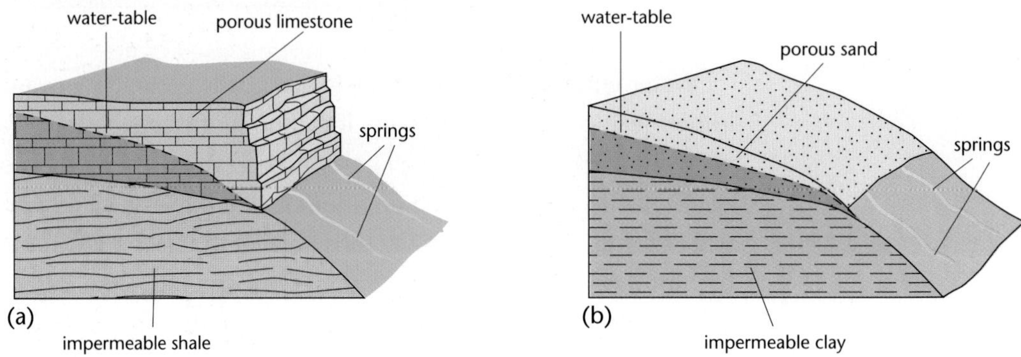


(a) A spring discharges water at the contact between a porous limestone and underlying impermeable shale.

(b) Springs lie at the contact between a porous sandy soil cover and an underlying impermeable clay.

Fig. 10.7

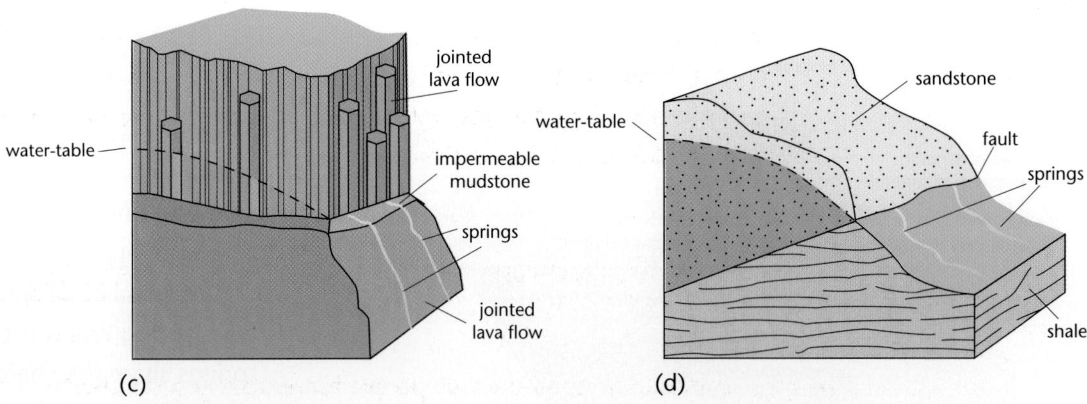


(c) Springs emerge along the contact between a highly jointed lava flow and an underlying impermeable mudstone.

(d) Springs emerge as a consequence of impounding of water by a dyke across a layer of permeable rock, or along a fault-line where the impermeable rock (in this case shale) meets the land surface. The water-table here rises on the upslope side of the dyke or fault-line, resulting in springs.

Rivers that disappear through swallow-holes (see p. 141) may reappear further downslope as a spring. The point at which an underground river emerges from the earth is termed a **resurgence**. Such waters are not true springs and their waters have not been filtered by bedrock. In Ireland the Aille River in the Burren and the Gort River in Co. Galway are examples of such features.

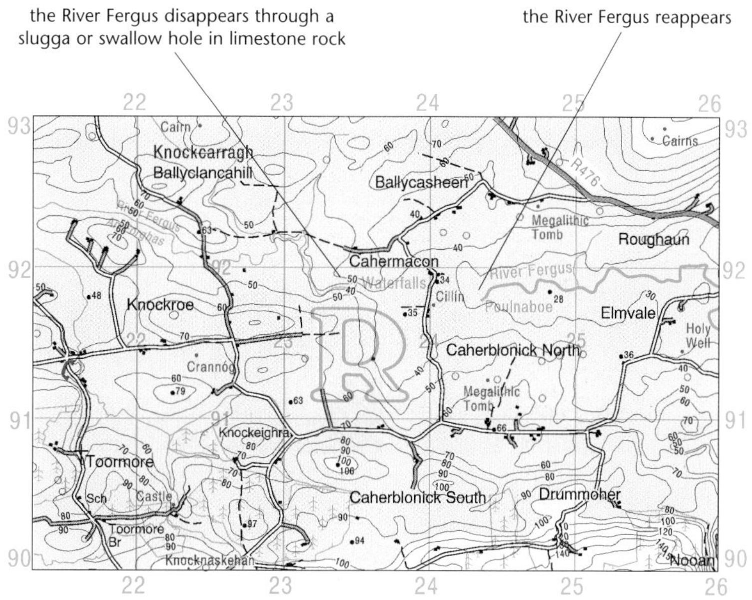


Fig. 10.8 An example of a resurgence

Many springs contain minerals dissolved from the rock by the moving water. They are known as **mineral springs**. The belief that these springs relieve ailments has popularised them as health resorts. Lisdoonvarna in Co. Clare in Ireland hosts such a mineral spring. Montecatini, near Florence in Italy, is another.

The Belgian town of Spa, near Liège, became so famous for its mineral springs that health resorts throughout the world have become known as spas.

Bath, in southern England, has warm springs and mineral waters. The town was founded by the Romans, who used the springs.

## People, Processes and Underground Water

### Wells

The most common method used by people for removing groundwater is a well. A well is an opening (like a pipe) bored into the ground until it reaches some distance below the water-table. Wells serve as reservoirs into which groundwater seeps from the adjoining saturated rock or aquifer, and from which it can be pumped to the surface.

When water is pumped from a well, the rate of withdrawal exceeds the rate of local groundwater flow. The imbalance creates a drop in the water-table in the vicinity of the well. This drop in water level is called 'a cone of depression' since it occurs in the shape of an inverted cone as water is drawn into the well from all directions.

Fig. 10.9
Wells on level land

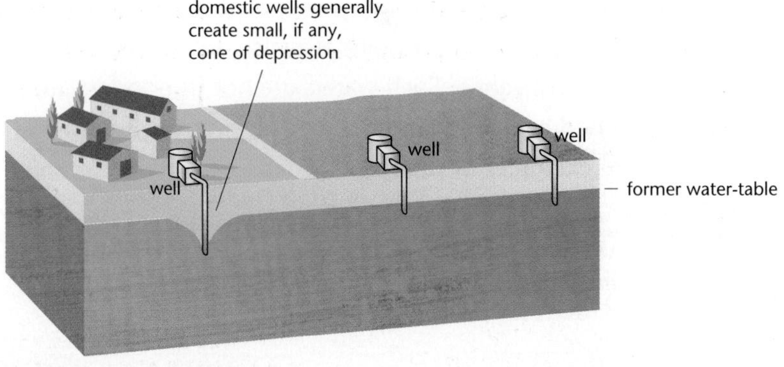


**Before heavy pumping**

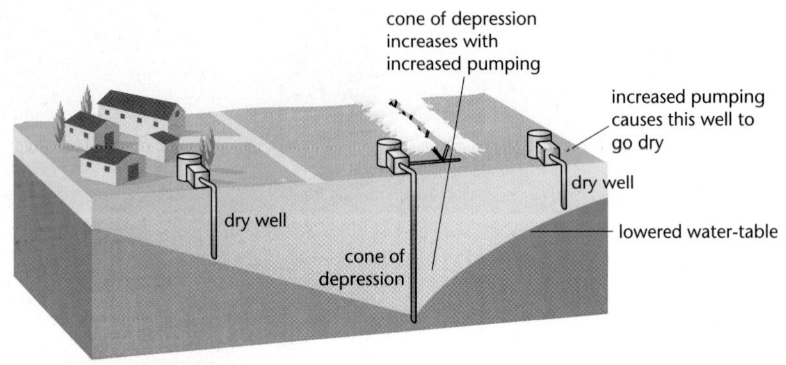


**After heavy pumping**

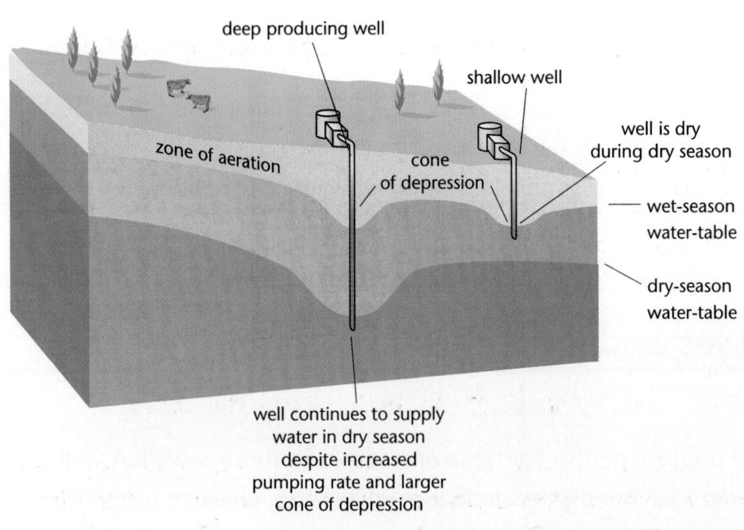


Fig. 10.10
Wells on sloping land

Extraction sometimes greatly exceeds the rate at which the groundwater of the area is recharged (replaced) by water percolating from rain or river-beds. This happens especially in areas of dry climate where large expanses of land are irrigated for agricultural purposes or where urban areas have expanded even into desert environments, or both, e.g. in California.

Digging for water dates back to nomadic times, long before the beginning of Western civilisation, and continues to be an important method of obtaining water. The Bushpeople of the Kalahari Desert, today, show us how water can be sourced even in desert environments. They search for tubers of certain plants which store liquid. Hollows in trees also contain some water.

The level of the water-table may rise or fall considerably in the course of a year, dropping during times of drought and rising during periods of rainfall. In order to ensure a continuous supply of water, a well should be bored many metres below the lowest estimated level of the water-table.

Generally, for a domestic well, the cone of depression is hardly noticeable. However, for irrigation or industrial purposes, the withdrawal of water can be great enough to create a very large cone of depression. This may subsequently lower the water-table in an area and cause nearby shallow wells to become dry.

## Artesian Wells

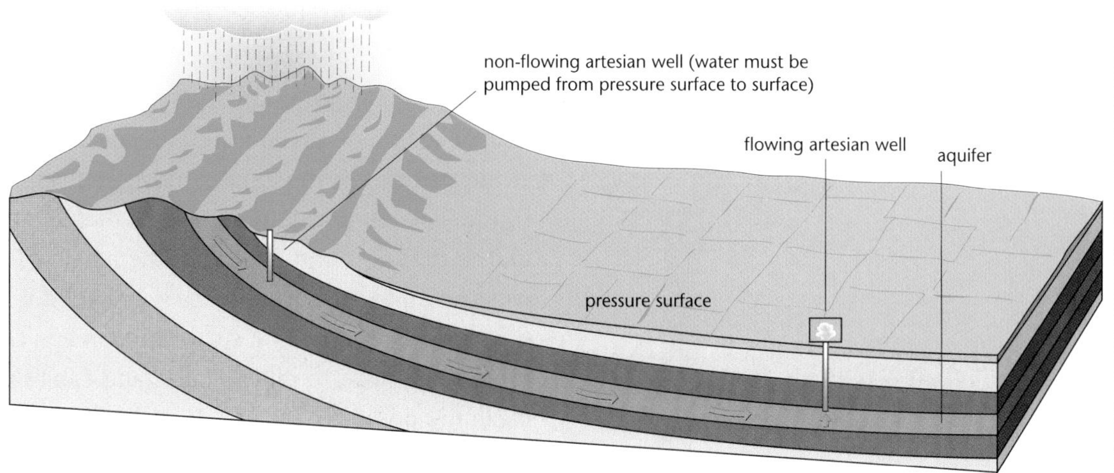


**Fig. 10.11**
Artesian systems occur when an inclined aquifer is surrounded by impermeable beds

**Remember!**
The portion of bedrock where water is permanently stored is called an aquifer.

Aquifers occur when a syncline formed of permeable rock, with one or both ends exposed, is enclosed above and below by impermeable rock. Rainwater enters the aquifer at the exposed end(s). It seeps towards the centre through gravity, producing sufficient hydraulic pressure to flow up a well-shaft (Fig. 10.11).

Flowing wells such as these are called **artesian wells**, their name being derived from Artois in France, where they were first used. The greatest artesian basin in the world is in eastern Australia, and stretches from the Gulf of Carpentaria in the north to the Darling River in the south, and from the Great Dividing Range in the east to Lake Eyre in the west.

In the London Basin, wells bored into the aquifer have reduced the level of the water-table by seventeen metres. The water pressure is so low in places that water must be pumped to the surface.

## Groundwater — A Renewable or Non-renewable Resource?

Until now, geography books have treated groundwater as a renewable resource. But is it? For most people, supplies of groundwater appear to be endless, being continually replaced by rain and snow. But are they? People need clean water supplies. Some water may be available but it may contain harmful bacteria.

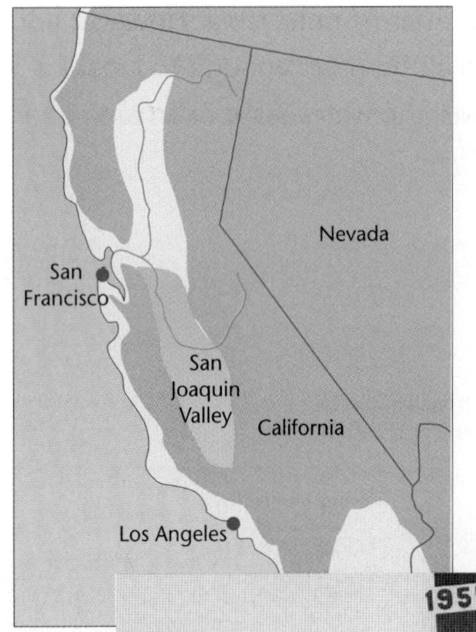


### Farming and Domestic Needs

In the past, the small well needed to supply the home and livestock needs of a family farm was dug by hand. This took the form of a large cylindrical or square hole which was lined with masonry. Today, modern wells are drilled by powerful machinery which can, if required, bore a hole 40 cm (16 inches) or more in diameter to depths of 300 m (1,000 ft) or more.

### Groundwater and Subsidence

Surface subsidence can result from natural processes related to groundwater.

Water pressure in the pores of an aquifer helps to support the weight of the overlying rocks or sediments. When groundwater is withdrawn, the pressure is reduced and the particles of the aquifer move and settle slightly. The amount of subsidence depends on how much the water pressure is reduced and on the thickness and texture of the aquifer. Such land subsidence is widespread in the south-western United States. Here, withdrawal of groundwater has caused structural damage to buildings, roads and bridges, damage to pipes, drains and cables and an increase in areas liable to flooding.

Structural damage can be especially severe where water is pumped from beneath cities. Mexico City displays some of the most severe effects of land subsidence. This city, once the Aztec capital, rests on soft lake-bed sands, silts and clays. Excessive pumping of water from these aquifers has caused homes to sink as much as nine metres. This has turned some intervening streets into elevated highways and once ground floor entrance halls are now below street level.

Land subsidence can be seen in the San Joachin Valley of California. Groundwater extraction to feed this rich agricultural region has caused a lowering of the land surface by nine metres in fifty years (Fig. 10.12).

Fig. 10.12
The shaded area on the map shows California's San Joaquin Valley. The marks on the utility pole in the photo indicate the level of the surrounding land in preceding years. Between 1925 and 1975 this part of the San Joaquin Valley subsided almost nine metres because of the withdrawal of groundwater and the resulting compaction of sediments.

## Groundwater and Pollution

### Toxic Wastes and Agricultural Poisons

Vast quantities of domestic refuse and industrial wastes are deposited each year in natural hollows, quarries, mountain bogland or excavations at the earth's surface. When such dumping areas are full to capacity, they are generally **covered with clay**, and then **revegetated** with **grass seed**.

When rainwater seeps down through the waste matter, the same processes take place as those associated with chemical weathering. Percolating rainwater dissolves and carries away soluble substances. In this way, harmful toxic chemicals leach into the aquifers (groundwater reservoirs) and contaminate them.

Pesticides and herbicides, such as DDT, can also find their way into groundwater by leaching. Fertilisers, such as nitrate, are harmful in even small quantities in drinking water.

Liquid and solid wastes from septic tanks, sewage plants, farmyards and animal feedlots and slaughterhouses may contain bacteria, viruses and parasites that can contaminate groundwater. The amount of pollution this will do will then depend on the nature of the aquifer rock.

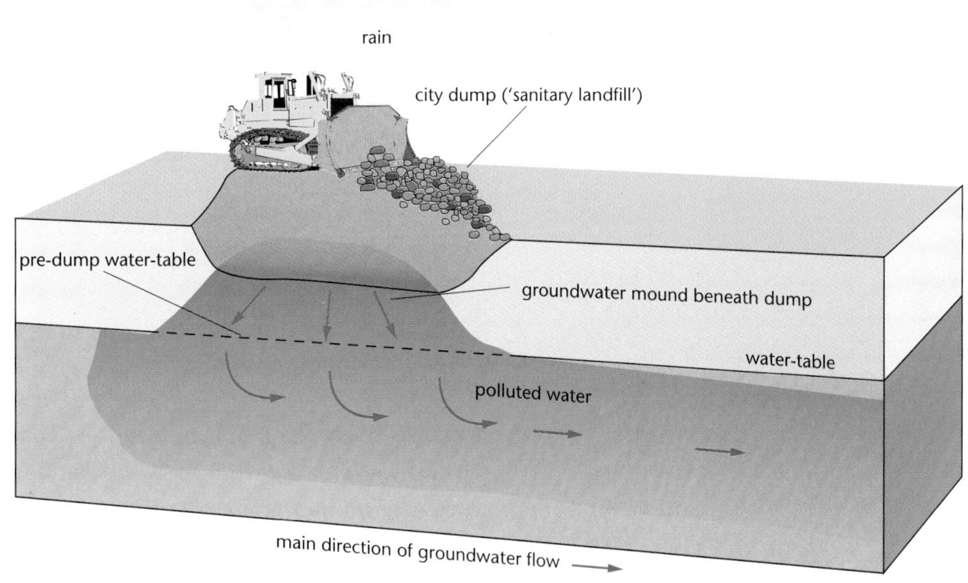


Fig. 10.13
Seepage from urban and rural dump site

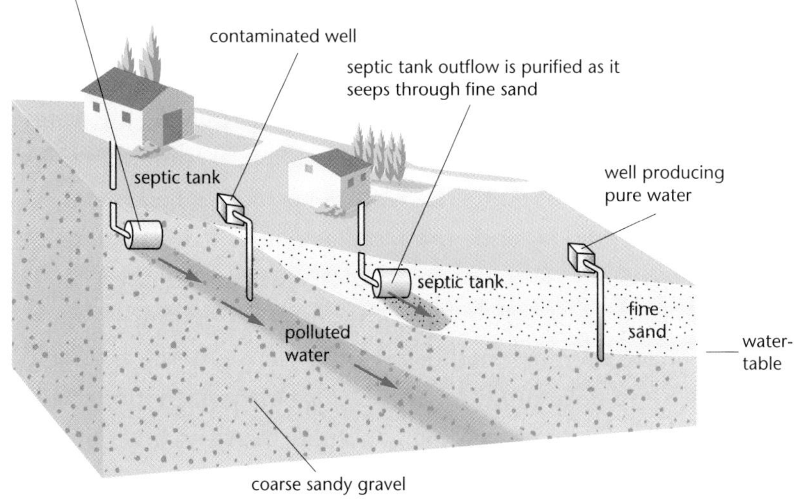


Fig. 10.14
Purification of groundwater contaminated by sewage. Pollutants percolating through a highly permeable sandy gravel contaminate the groundwater and enter a well downslope from the source of contamination. Similar pollutants moving through permeable fine sand higher in the stratigraphic section are removed after travelling a relatively short distance and do not reach a well downslope.

For instance, if the contaminated liquid, such as domestic sewage outflow from a septic tank, seeps through permeable sandstone, it may well be purified over a short distance — as little as fifty metres. However, if toxic liquid enters a more permeable rock, such as a well-jointed limestone bedrock, rather than seeping through the rock it may travel along the joints and bedding planes and may contaminate groundwater some kilometres away. In this case, the water flows too rapidly and is not in contact with the surrounding material long enough for purification to occur.

Not all groundwater pollutants disperse like the above.

Petrol (gasoline), which leaks from filling stations at hundreds of thousands of locations, is less dense than water and floats upon the water-table. Some liquids are

heavier than water and sink to the bottom of the water-table. These types of liquids may travel in unpredicted directions and determining the extent and flow direction of such pollution is a lengthy and expensive process.

Fig. 10.15

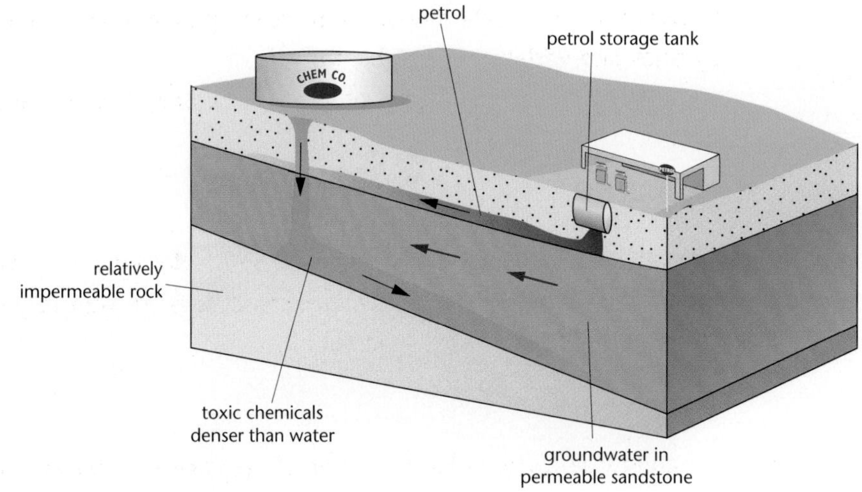


It must be emphasised at this stage that not all natural groundwater supplies are pure and fit for human consumption. Naturally occurring minerals within rock and soil may contain harmful substances such as arsenic and mercury. Other sources may have a high salt content and are undrinkable. Like a 'bad waterhole' depicted in a cowboy film, some springs contain such high levels of toxic elements that the water can sicken or kill humans and animals that drink it.

## Underground Water and Limestone/Karst Landscapes

Karst landscape is unique, in that surface water flows down to the water-table without first seeping through soil and rock pores. Thus polluted surface water directly enters the water-table without first being purified in any way. Surface water may disappear through a sink-hole and emerge many kilometres away as a resurgence, without being purified.

Because karst landscape is characterised by underground drainage, the potential for pollution is enormous. In the United States, for example, about 20 per cent of the freshwater supply passes through karst landscape. Numerous tonnes of domestic, agricultural and industrial wastes have been pumped down into cavern systems, changing pure freshwater supplies to sewers and polluting their waters. *Time* magazine reported that on several occasions during the 1980s, in Bowling Green, Kentucky, benzene and other chemical fumes rose up from the caverns, endangering schools and homes. Luckily, however, geology specialists from Western Kentucky University used their knowledge of groundwater movement to identify the sources of the pollution. (See 'Karst Landscape' below.)

**sink-hole/swallow-hole**
A circular hollow or depression that forms when the roof of an underground cavern or channel collapses.

## Karst Landscape

Karst landscape is the term given to areas which display a pattern of denudation similar to that of the Karst region of Slovenia along the eastern coast of the Adriatic Sea. These landscapes are best developed in moist temperate to tropical regions which are underlain by thick and widespread soluble rocks.

There are three main types of karst landscapes, classified according to their stage of maturity.

### 1. Sink-hole Karst

The landscape of sink-hole karst is level and is dotted with sink-holes of various sizes and shapes. Examples of such landscapes can be seen in Kentucky and Indiana in the United States and Jamaica in the Caribbean Sea.

### 2. Cone Karst

The landscape of cone karst consists of many closely spaced conical or pinnacle-shaped hills which are separated by deep sink-holes. Examples of such landscapes can be seen in Mexico, Cuba and Puerto Rico as well as some islands in the South Pacific.

### 3. Tower Karst

Tower karst landscapes consist of tower-like limestone hills separated by flat plains of alluvium. These vertical-sided hills rise up to 200 metres high. If riverborne alluvium gathers in the bottom of sink-holes, the depressions slowly fill with sediment. This allows rivers to overflow and meander across the land between the towers. In this way alluvial plains are created between the steep-sided towers. One of the most famous tower karst landscapes is near Guilin in south-eastern China.

Fig. 10.16
A karst region in Croatia

The limestones of Britain and Ireland contain many fossils, including corals. These fossils indicate that the rock was formed during Carboniferous times (350–300 million years ago) on the bed of a warm sea that covered much of these islands when they lay in warmer latitudes at the equator.

Limestone rock creates a distinct type of scenery, known as **karst**, when the rock is exposed at the surface. Limestone is found in thick beds in Britain and Ireland; these formations are separated by almost horizontal bedding planes with vertical joints.

Karst landscape is clearly developed in the Peak District and Yorkshire Dales National Parks, in Britain, and in the Burren in Co. Clare, in Ireland. Karst landscapes also occur in many parts of the United States, such as parts of Kentucky, Tennessee, Alabama, southern Indiana and central and northern Florida.

The rare combination of karst landforms, Arctic, Alpine and Mediterranean plants, Atlantic coastline, prehistoric dwellings and megalithic tombs makes the Burren a special place.

## Surface Features

### Landform: Limestone Pavement

Examples: The Burren in Co. Clare, in Ireland
The Karst region in Slovenia
The Kentucky Plateau in the United States

**Solution Process**

### Formation

Limestone pavement forms over a very long time in areas where limestone rock is exposed at the surface. Its distinctive landscape forms for three main reasons:

1. Limestone is a rock which offers varied resistance to rainfall and surface run-off.
2. Limestone is pervious, which means that water is able to pass freely through its vertical joints and horizontal bedding planes.
3. Limestone is dissolved by a weak acid called carbonic acid. Rainwater ($H_2O$) falling through the atmosphere joins with carbon dioxide ($CO_2$) and forms carbonic acid. This acid alters calcium carbonate in the limestone to soluble bicarbonate, which is removed to the sea. Simply, when the carbonic acid falls onto the exposed limestone surface, it dissolves the rock.

*Equation:*

$H_2O + CO_2 \rightarrow H_2CO_3$ (carbonic acid)

$\rightarrow H_2CO_3$ (carbonic acid) + $CaCO_3$ (limestone)

$\rightarrow Ca(HCO_3)_2$ (calcium bicarbonate)

Fig. 10.17
Limestone pavement showing a grike through the centre and clints on each side

Because limestone is well jointed, it encourages water to follow the path of least resistance through these joints, rather than percolating into the rock itself. Rainfall or carbonic acid acts upon these joints, enlarging them into long parallel grooves called grikes. Between these grikes are narrow ridges of rock called clints. Such a surface resembles a paved/slabbed area and is called limestone pavement (Fig. 10.17).

The Burren in Co. Clare has the largest single area of limestone pavement in Europe.

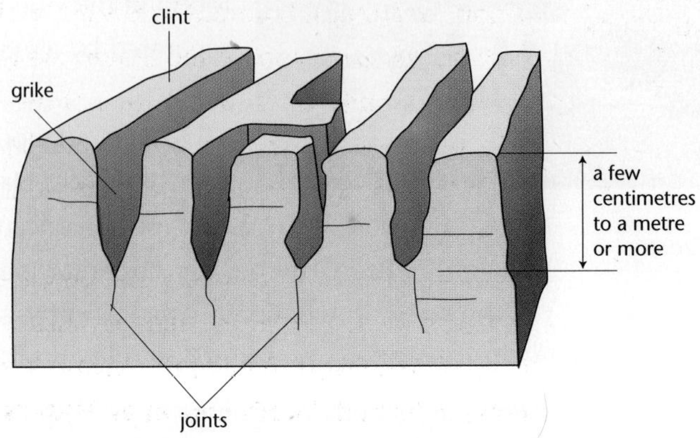


Fig. 10.18
Weathering of limestone rocks

## Landform: Sink-holes/Swallow-holes

Examples: Poll na gColm in the Burren in Co. Clare, in Ireland
Gaping Gill on the slopes of Ingleborough, in Britain



**Did You Know?**
Gateway to the underworld of ancient Maya belief, the ancient *cenote* (sink-hole), or well, was a crucial source of water for the Maya of the northern Yucatan peninsula, where no surface rivers flow. Humans sometimes were ritually sacrificed in such wells.

### Formation

Sink-holes are sometimes known as swallow-holes, sluggas or dolines (although the term doline may refer to another feature which is dealt with later). Sink-holes are openings in the beds of rivers which flow over limestone rock. These holes allow the river water to disappear underground into solution channels which wind their way under the surface.

Rainwater which lands on a limestone rock surface quickly disappears underground, flowing downward through the narrow crevasses, called grikes, on the surface. This rainwater follows underground bedding planes and joints which it enlarges into solution channels and where it erodes and dissolves caverns of various sizes.

When parts of the roof of some of these channels and caverns collapse, openings, called sink-holes, are formed at the surface, in the shape of inverted cones. If one or more of these sink-holes occurs on the bed of a river on the surface, the river water will disappear underground into a solution channel or cavern.

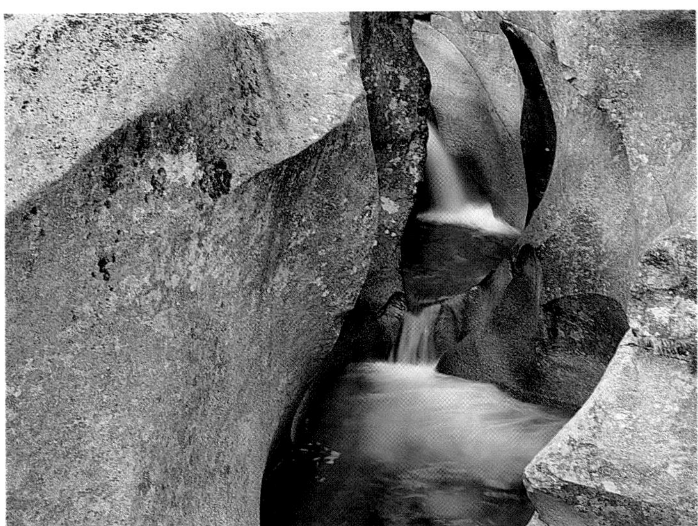

Fig. 10.19
A solution channel in limestone rock in the Swiss Alps

These sink-holes may be several metres in diameter as well as over a hundred metres in depth. In the Burren, however, they are not very deep; examples there include Poll na gColm, Poll Binn and Poll Eilbhe. (The word 'Poll' is the Gaelic word for a hole.)

Rivers that disappear through swallow-holes may reappear further downslope as a resurgence (see p. 133).

Sink-holes generally form over many years without any noticeable change on the surface. However, they can also form suddenly and without warning when a cavern roof collapses under its own weight; those created in this manner are steep-sided and deep. Sink-holes may occur in densely populated areas and may create a serious geological hazard. On 8 May 1981 a crater-like sink-hole began forming in Winter Park, in Florida in the United States. Newspaper accounts of its formation were front-page news and made this sink-hole one of the most publicised ever. In one small area of about 25 km$^2$, in Florida, more than 1,000 collapses have occurred in recent years.

In Florida, Kentucky and southern Indiana, there are tens of thousands of these sink-holes.

The names of some rivers that disappear through holes often give a clue to their fate. In Kentucky, for example, there is Sinking Creek, Little Sinking Creek and Sinking Branch. In Hertfordshire, in Britain, the Mymmshall Brook disappears through cone-shaped sink-holes in its bed, locally known as 'Water's End'.

## Landform: Dry Valley

Examples: Section of the Glenaruin river valley in the Burren in Co. Clare, in Ireland
The Devil's Dyke, on the South Downs in England

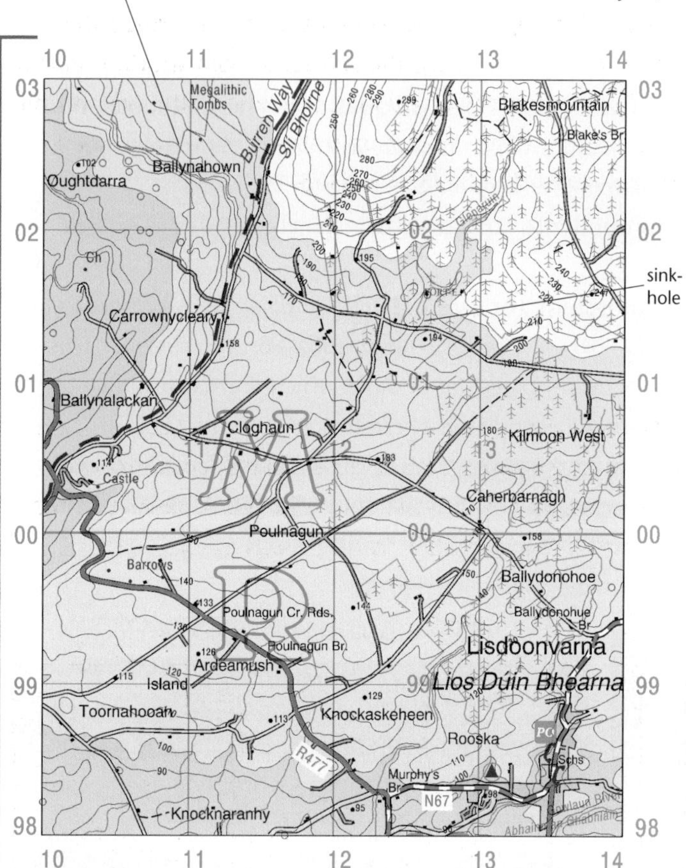


**Fig. 10.20**
Closely examine this Ordnance Survey map extract. The Glenaruin River disappears down a sink-hole as labelled. Can you find any others?

The word 'Ballynahown' suggests 'the place of the river'. What does this tell you about the river which flowed through this area at one time? What does it also tell you about the type of underground rock?

### Formation

Dry valleys are extinct or partially used river valleys which are without running water for all or most of the year.

Normal river valleys are formed by the processes of hydraulic action and abrasion. With hydraulic action the force of flowing water erodes the landscape and the eroded material is washed away by the river's ability to carry material, an ability called its 'competence'.

Abrasion is the ability to erode the bed and sides of the river, by the river's water and by the eroded material which it carries.

In the case of limestone areas, the river continues to erode a valley as described above, until such time as the water meets a sink-hole on its bed. At this stage the waters flow down into the sink-hole and travel underground, possibly resurfacing elsewhere as a resurgence (see p. 133). The sink-hole and underground channels were formed by solution of the limestone causing collapse of a channel roof.

However, not all dry valleys were formed in this way. They were generally formed in one of the following ways:

1. They were carved by glacial meltwater when the ground was frozen and so impermeable. These were left dry as the ice disappeared.

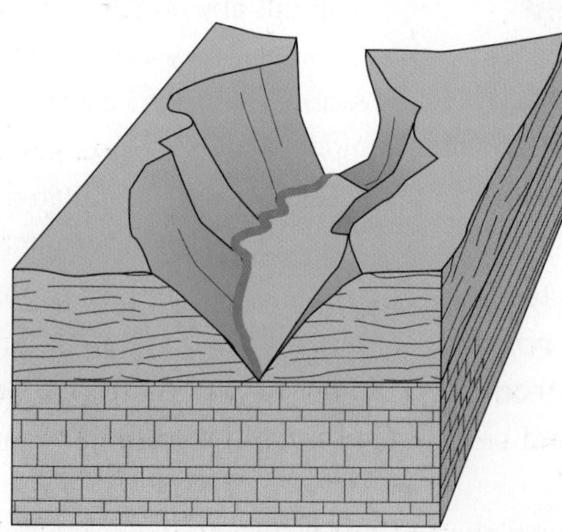

**Fig. 10.21**
A river cuts through impermeable shale until it reaches limestone rock

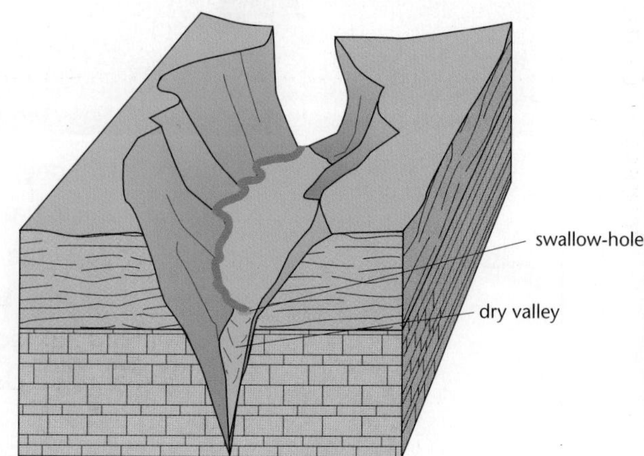


**Fig. 10.22**
The river disappears through a sink-hole in limestone rock

2. They were formed when large rivers cut vertically down faster than other neighbouring streams, thereby causing a fall in the water-table. This leaves the smaller streams without a water supply, so their valleys are dry.
3. When a surface stream enters a sink-hole, the valley downstream is left dry. Later other sink-holes are formed upstream. In this way, the surface stream shortens and the dry valley increases in length. During spells of heavy rain, the level of the water-table may rise above the level of the sink-hole and a temporary stream may occupy the valley. This occurs near Lisdoonvarna in Co. Clare, in Ireland.

## Landform: Cuesta or Escarpment

Example: North Downs and South Downs in England

### Formation

The most characteristic feature of chalk landscape is the 'escarpment' or 'cuesta'. These chalk landforms were gently tilted when the African and Eurasian plates collided. They have a **steep 'scarp' slope on one side** and a **gently sloping 'dip slope' on the other**. Such landforms are found in the North Downs and the South Downs in England.

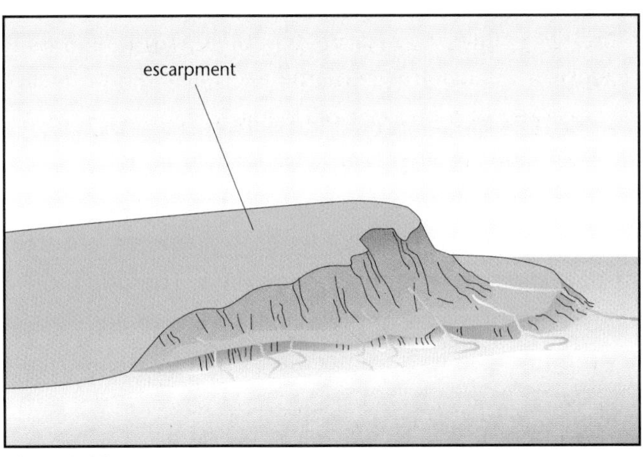


Fig. 10.23

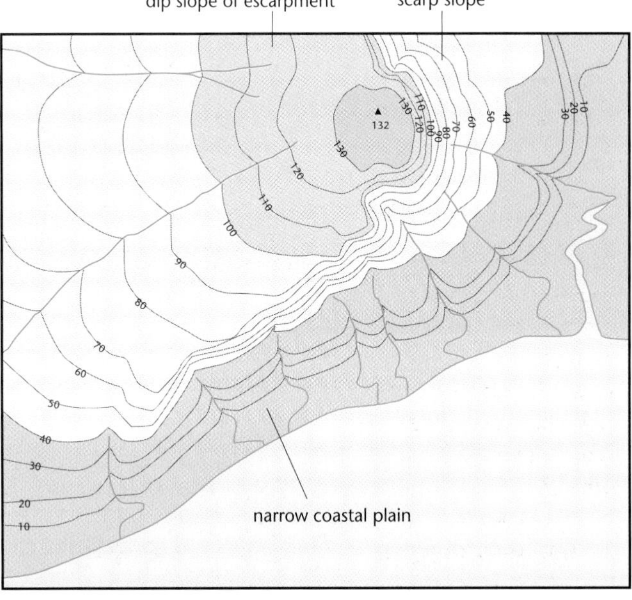


Fig. 10.24

Chalk is a soft, white pure limestone that was compressed into rock. In contrast to limestone landforms and scenery, chalk landscape consists of rounded gently rolling hills. The chalk bedrock is seldom exposed on the surface and generally supports only a thin soil cover.

Chalk was laid down during the Cretaceous period, between 100 and 65 million years ago, in warm, calm sea water which was rich in algae and free from sediment. These algae secreted small amounts of calcium carbonate which formed a chalk-mud on the sea floor. The sediment-free sea water allowed its white colour to be retained.

Fig. 10.25
Chalk landscapes in southern England support sheep farming

## Other Surface Features

### Karren

When rainwater weathers limestone, it may create a landscape of tiny hollows which are separated by narrow ridges (Fig. 10.26).

Fig. 10.26
Karren landscape

### Uvalas

When two or more sink-holes coalesce, they form a much larger depression called an uvala, which may be in excess of 200 metres in diameter. The origin of uvalas is sometimes attributed to the joining together of a number of dolines.

### Poljes

These huge depressions in the landscape (up to several kilometres in diameter) are steep-sided with flat floors. They are best developed in the former Yugoslavia. The origin of poljes may be attributed to the coalescence of uvalas or downfaulting (tectonic movement). During glaciation, they are believed to have been enlarged by the movement of ice.

### Dolines

Sink-holes are sometimes called dolines. Generally, however, the term doline refers to a closed depression or hollow in a limestone landscape. Dolines vary in size from a few metres to many kilometres in diameter; they may be anything from a few metres to 150 metres in depth. Such depressions occur as a result of the joints in underlying limestone growing wider. This causes the overlying land to collapse or slump, forming saucer-shaped hollows. The floors of such hollows often have a coating or blanket of clay upon which water may stay during spells of wet weather.

### Turloughs

Turloughs are seasonal ponds which form from time to time in depressions such as dolines. Depending on the season and/or the level of precipitation, the water-table may rise or fall. In winter, for instance, heavy rains will cause these hollows to fill with water while in summer, due to a shortage of water, they will dry up. Soil gathers in such hollows and produces a rich grass which is grazed in summer if the water-table is sufficiently low.

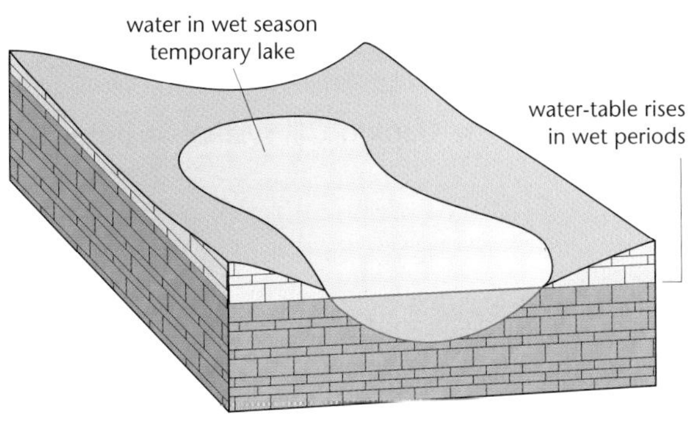


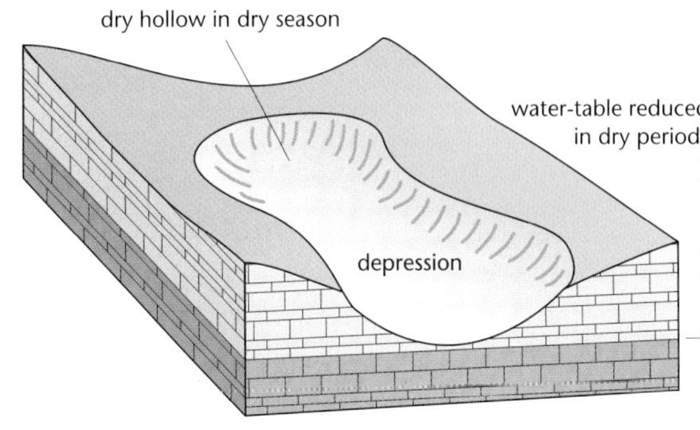


Fig. 10.27
Turlough

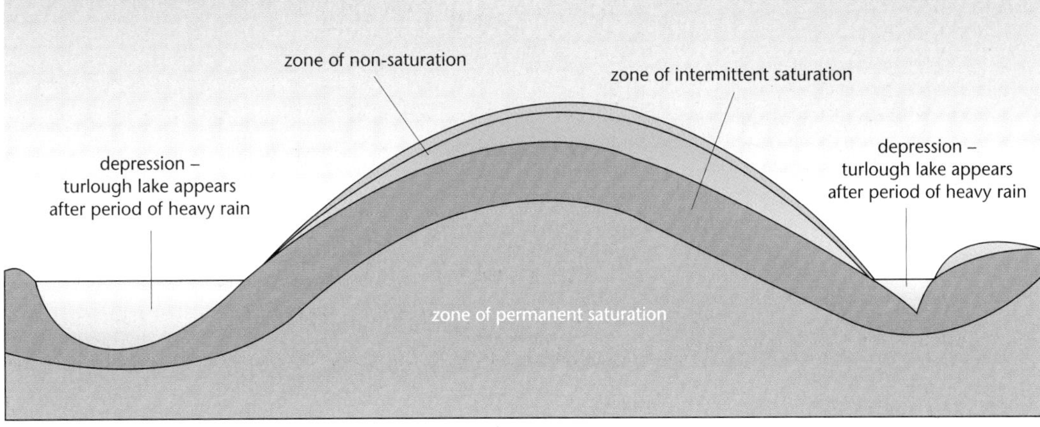


Fig. 10.28
Formation of a turlough

**Class Activity**

Examine the Ordnance Survey map of the Burren in Co. Clare, in Ireland, Fig. 10.38, p. 152. Then do the following:

1. Locate Carran depression at grid reference R 285 985. Use a pencil to follow the 120-metre contour nearest to this grid reference. Then examine the surrounding contours.
   (a) What does this examination suggest to you about the land enclosed by the 120-metre contour?
   (b) From your study of limestone landscapes, what surface landforms can you identify within this enclosed area?
   (c) There is a depression at grid reference R 250 980. Into which category of depressions, uvala, polje, doline or turlough, would you place this landform? Explain your answer.
2. From your studies so far on limestone landscapes, what other surface landforms can you identify on the map extract?

# Underground Features

## Landform: Cavern

Examples: Crag Caves in Co. Kerry in Ireland
Mammoth Cave in Kentucky in the United States
Jenolan Caves, west of Sydney in Australia
Stalactite and stalagmite chambers in the Gaping Gill system in Britain

Processes: Solution, Hydraulic Action

### Formation

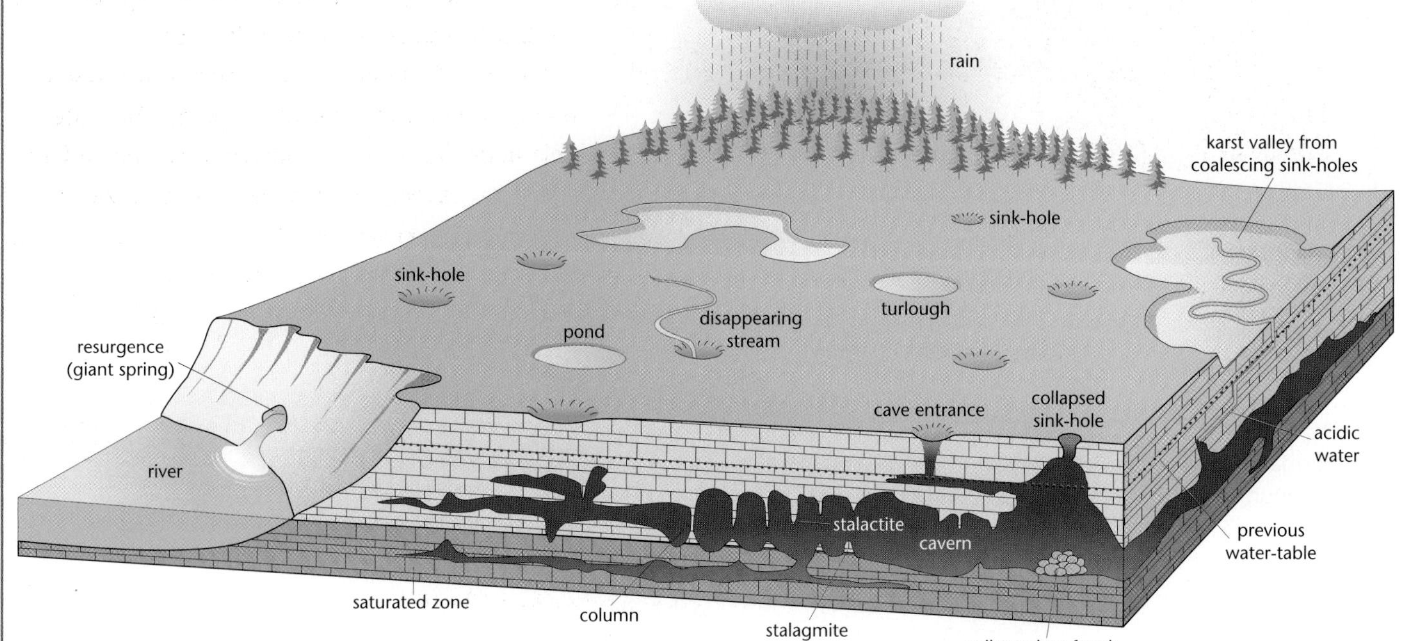


Fig. 10.29
The development of a limestone cave network begins at the top of the zone of saturation and follows the lowering of the regional water-table

The formation of most caverns takes place in the zone of saturation, **at or below the water-table**. Weak (dilute) carbonic acid in groundwater reacts with calcite in limestone rocks to form calcium bicarbonate, a soluble material that is then carried away in solution.

Groundwater follows lines of weakness and acts upon bedding planes and joints in the limestone bedrock. As time passes, the acidic groundwater dissolves the rock to create cavities and gradually enlarges them into caves. The dissolved rock is carried by streams to the sea.

Finally, huge underground passages with caves and an interconnected system of cave chambers is formed. These chambers are called 'caverns'. The Carlsbad caverns in south-eastern New Mexico, in the United States, formed in this case by sulphuric acid, include one chamber 1,200 metres long, 190 metres wide and 100 metres high. This particular cavern has an area equal to fourteen football pitches and enough height to accommodate the US Capitol building in Washington DC.

Some cavern systems are completely flooded or partially flooded. Howe caverns in upper New York state are partially flooded and tourists can avail of a boat trip through

Fig. 10.30
Lechuguilla in the USA. Identify the stalactites, stalagmites and pillars.

some of the system. Marble Arch caverns in Co. Fermanagh in Ireland also provide boat trips in underground caverns.

Underground streams in limestone landscapes form a maze of channels that run, in some cases, for many kilometres underground. Their waters chiefly come from streams pouring through shallow holes on the surface of the ground. Groundwater also seeps into these passages from the surrounding bedrock. In some interglacial and immediately postglacial periods, these underground streams carried much more water than at present. In limestone regions, the water-table fluctuates regularly and because the surface run-off is so rapid, the underground channels and caverns often quickly fill to capacity. A sudden downpour may cause such a rapid rise that cave explorers (spelaeologists) may be trapped or may have to make a speedy retreat to the surface. It is advised that an individual should never enter such an underground system unless accompanied by an experienced guide.

Processes: Solution, Evaporation

## Landform: Dripstone (Flowstone)

Examples: Crag Caves in Co. Kerry in Ireland
Mammoth Cave in Kentucky in the United States
Jenolan Caves, west of Sydney in Australia

### Formation

Evaporation takes place as water seeps from limestone joints in cavern roofs. When this happens, some carbon dioxide is released from the solution and the 'water' is unable to hold all the calcium carbonate. Calcium carbonate deposits are left on the ceiling or wall of the chamber or on the floor as the water falls due to gravity. All these calcium carbonate deposits are called dripstone.

## Dripstone Deposits

### 1. Stalactites

Continuous seepage of water through cavern ceilings produces constant dripping and evaporation at specific locations. As drops of water fall from the ceiling of the cavern, they leave behind a deposit of calcium carbonate. Deposition of the calcium carbonate occurs fastest at the circumference of the drop. A hard ring of calcite develops and grows down to form a tube which eventually fills up to form a solid stalactite. Calcite is a mineral formed from calcium carbonate. In its purest form it is white in colour. However, as seeping water will have impurities in solution or

suspension, they will discolour the calcite, causing brownish or other discolorations. These discolorations are especially noticeable on stalagmites.

2. **Stalagmites**

   If seeping water does not entirely evaporate on the cavern ceiling, it falls to the floor or to sloping sides along the width of the cavern. Wider and shorter stumps or domes of calcite deposits build up to form stalagmites. Their shape results from dropping water 'splashing' onto the floor, spreading out and thus forming a larger base than that of the stalactite.

3. **Columns**

   As stalactites grow downward they join with stalagmites growing upward. When they meet they form columns or pillars.

4. **Curtains**

   As water seeps out of a continuous narrow fissure or cavern or cave roofs, a curtain-like feature of dripstone is formed.

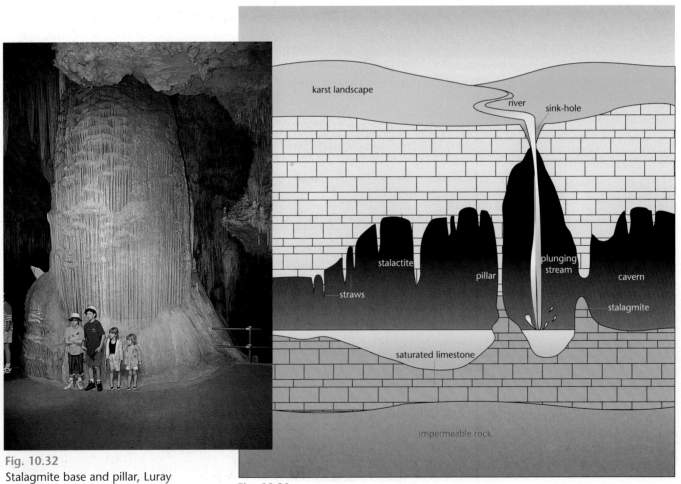

Fig. 10.32
Stalagmite base and pillar, Luray Caverns, Virginia, USA

Fig. 10.31

# The Cycle of Erosion in a Karst Region

Fig. 10.33
Karst topography, Guangxi province, China

## Stage of Youth

Denudation wears away overlying bedrock until underlying limestone is exposed. Surface streams continue to erode into the limestone, opening up joints and bedding planes. Soil is thin and scarce as weathered material is easily either washed away by rivers or blown away by the wind (Fig. 10.34a).

## Stage of Maturity

At this stage, all surface drainage has disappeared through swallow-holes (sluggas) into underground systems and huge tunnels and caverns are formed below the surface. The roofs of tunnels and caverns collapse while joints in underlying limestone grow wider to form dolines. Dry valleys and turloughs also form on the surface. The surface level between these depressions gradually reduces and the level of the general area is lowered (Figs. 10.34b and 10.34c).

## Stage of Old Age

Poljes and other depressions coalesce to lower the landscape so that only small hills called **hums** remain. Underground cavern systems disappear and surface drainage resumes (Fig. 10.34d). During these various stages of erosion the processes of river erosion and wind erosion also operate and help to reduce and erode the landscape.

**Class Activity**

Examine the cycle of erosion in a karst region above. Then refer to the three main types of karst, p. 139. From this information, classify the type/stage of maturity of the karst region shown in the photograph, Fig. 10.33.

(a) Normal surface drainage

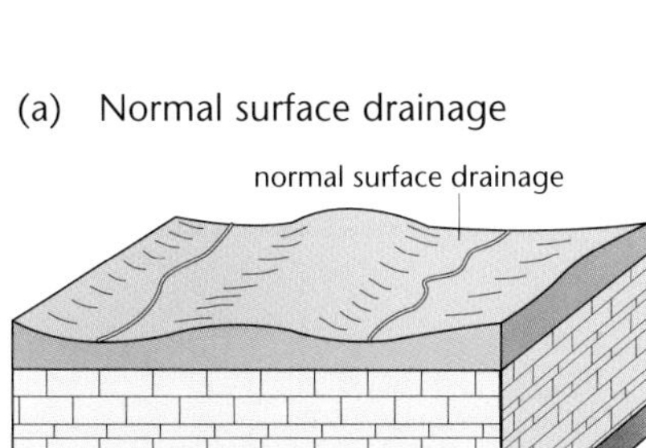


youth: streams flowing normally but beginning to work down into the limestone

(b) Dry valley

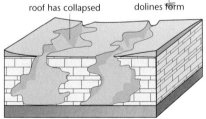

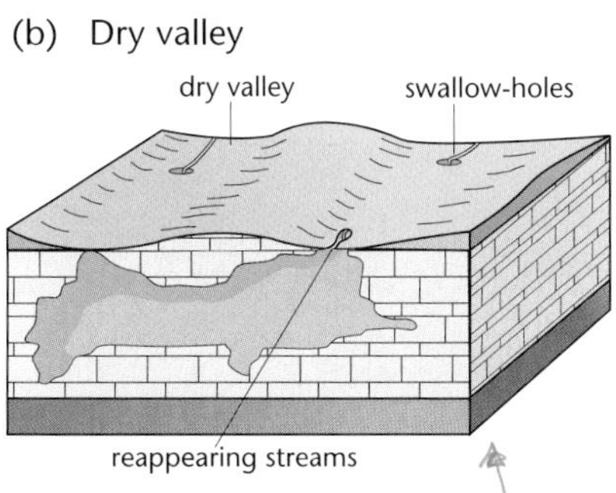


(c) Roof has collapsed

roof has collapsed
dolines form

maturity: surface drainage non-existent

(d) Hums

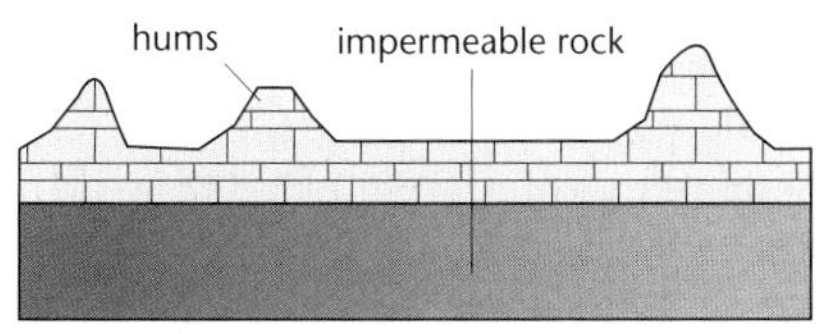


old age

Fig. 10.34
The cycle of erosion in karst areas

## Case Study: The Karst Region of the Burren in Co. Clare, in Ireland

Study the sketch map of the Burren, Fig. 10.35 and the Ordnance Survey map of the Burren, Fig. 10.38, p. 152. Then complete the following activities:

1. On the sketch map, Fig. 10.35, mark in, as accurately as you can, the outline of the area represented by the Ordnance Survey map, Fig. 10.38.

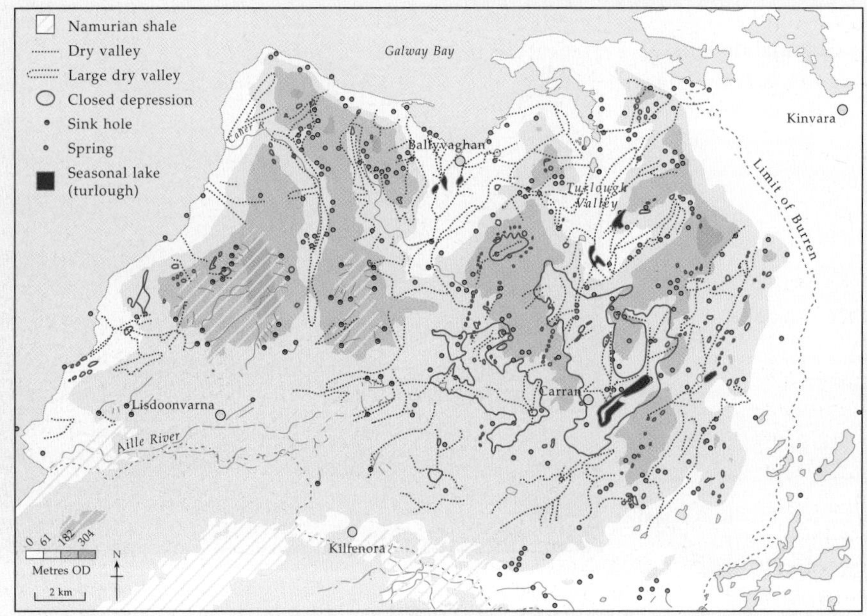


**Fig. 10.35**
Sketch map of the geology and landforms of the Burren in Co. Clare in Ireland

2. On the Ordnance Survey map, Fig. 10.38, locate two examples of each of the landforms outlined on the sketch map, Fig. 10.35.
3. For each of the landforms (a) sink-hole, (b) turlough, (c) dry valley, complete the following exercises:
   (i) With the aid of a labelled diagram, describe the formation of the landform, including the processes involved in its formation.
   (ii) Name two examples of each landform, one local and one foreign.
4. The photograph, Fig. 10.36, shows portion of the limestone surface with its thin soil cover. At present, this area has a very low population density. Suggest why it was able to support a large population and why it was suited to farming activities during prehistoric and early Christian times.
   Use the following headings to develop your answer:
   **High** pre-Christian population
   (a) Tillage and well-drained soils
   (b) Simple tools and light soils
   (c) Shellfish
   **Low** present population
   (a) Light soils, coastal area, strong winds and karst
   (b) Upland area and machinery
5. Carefully examine the photograph of Ballykinvarga near Kilfenora, Fig. 10.36. Then identify the topographical features A–F.

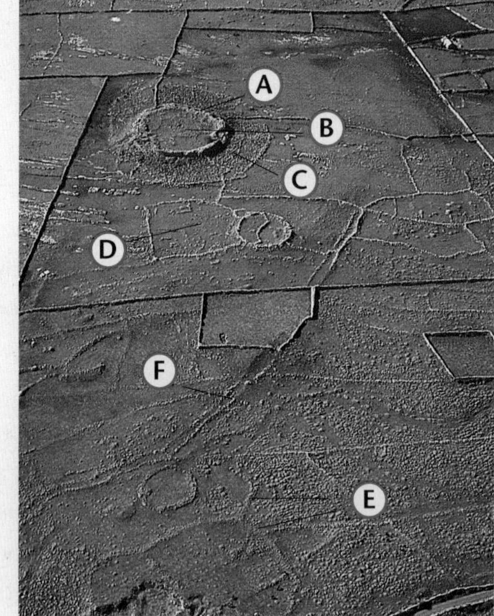


**Fig. 10.36**
The stone fort of Ballykinvarga near Kilfenora, in the Burren in Co. Clare, in Ireland. The large fort of prehistoric or early historic period age dominates the photograph but also preserved are myriad other enclosures large and small and smaller cahers (stone forts in Gaelic), all of unknown age.

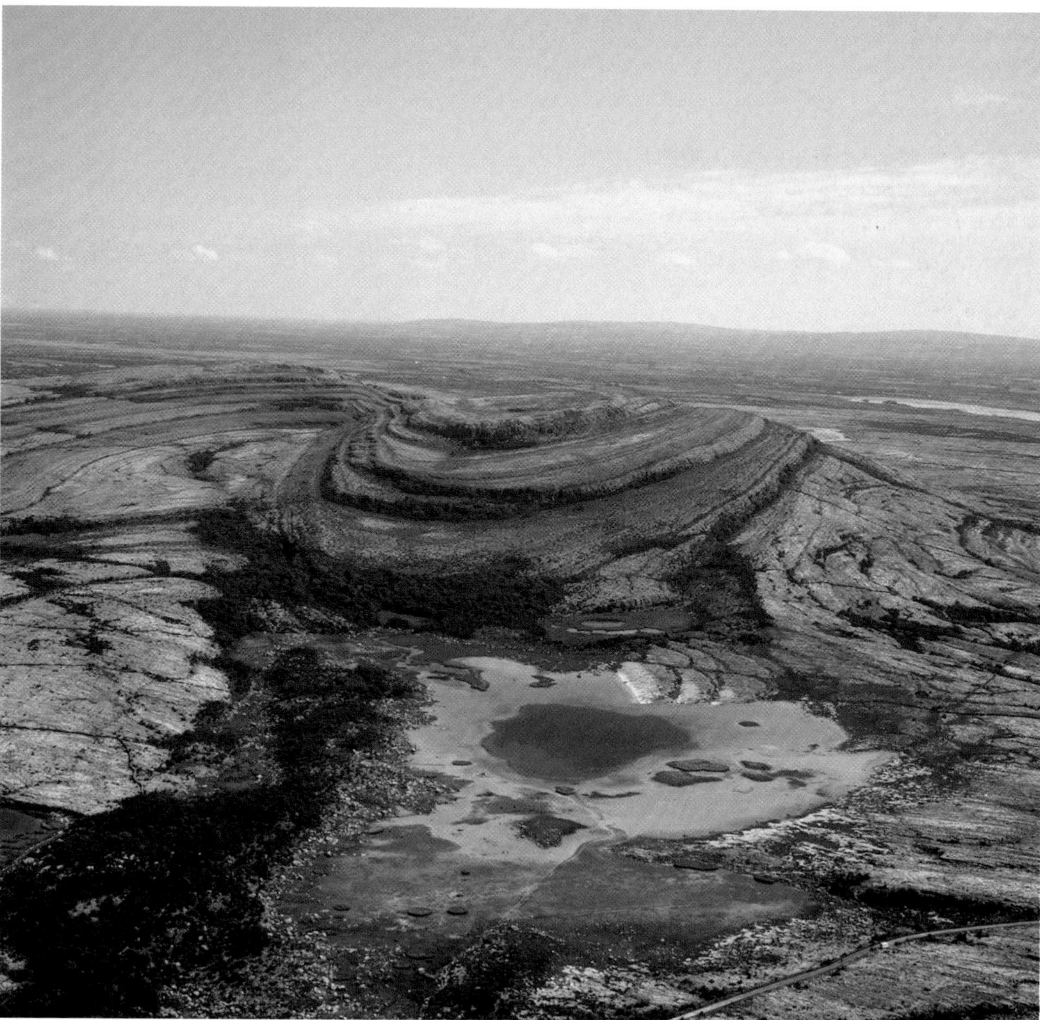

Fig. 10.37
Mullaghmore mountain in the extreme south-east of the Burren in Co. Clare. Here, the limestones have been folded by earth movements and the topography faithfully reflects the underlying rock structures. Mullaghmore is developed in a downfold (syncline). At the foot of the mountain is Lough Gealáin — part turlough and part permanent lake on the left middle of the photograph. In dry weather the lake contracts to a small water-filled hollow some 14 m deep. In wet conditions, however, this lake and other turloughs are filled by water from springs and extend over the surrounding area as in the photograph.

Study the photograph of Mullaghmore in Co. Clare (Fig. 10.37) and answer the following questions:

1. Classify the area shown, justifying your answer from evidence in the photograph.
2. Explain how limestone is affected by the weathering process.
3. Choose three landscape features found in an area such as this and: (a) name the features; (b) draw a sketch of the features, noting their characteristics with labelled arrows; and (c) explain how each of these features was formed.
4. Outline the cycle of erosion which occurs in karst areas.
5. Conflicts of interest arise, from time to time, over plans to develop natural regions, such as the Burren, as tourist destinations. Develop two positive and two negative arguments as to why associated structures should or should not be built to accommodate such developments.
   In your answer, develop at least one point on the Burren region. Two other points should refer to natural regions of your choice, in other parts of the world.
6. Describe the type of aerial photograph shown. Be specific in your choice of terminology.

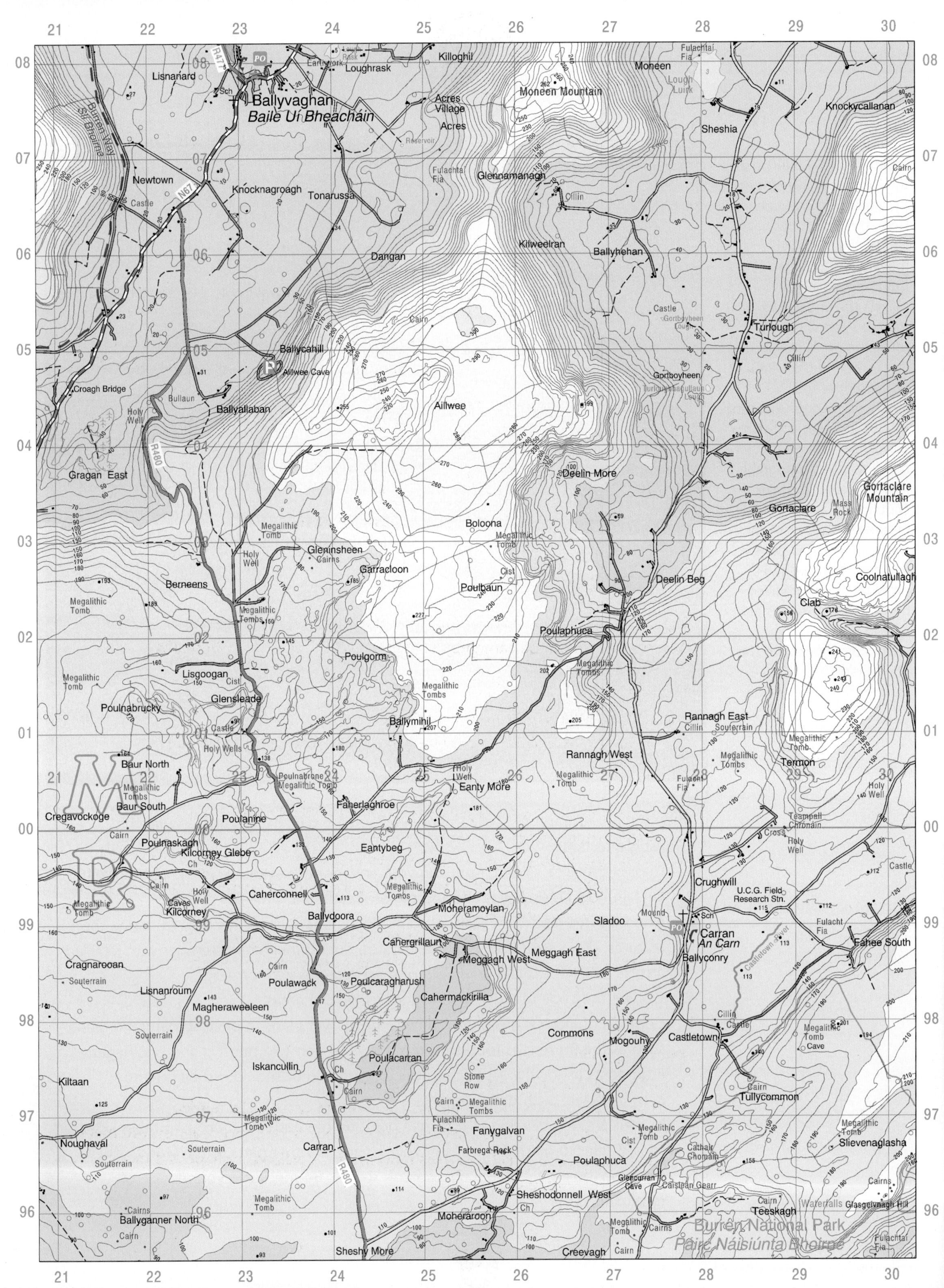

Ballyvaghan
Baile Uí Bheacháin
Lisnanard
Loughrask
Killoghil
Moneen
Moneen Mountain
Acres Village
Acres
Sheshia
Knockycallanan
Newtown
Knocknagroagh
Tonarussa
Glennamanagh
Kilweelran
Ballyhehan
Dangan
Turlough
Ballycahill
Aillwee Cave
Gortboyheen
Croagh Bridge
Ballyallaban
Aillwee
Gragan East
Deelin More
Gortaclare Mountain
Gortaclare
Boloona
Gleninsheen
Garracloon
Poulbaun
Deelin Beg
Coolnatullagh
Berneens
Clab
Poulaphuca
Poulgorm
Lisgoogan
Glensleade
Poulnabrucky
Ballymihil
Rannagh East
Rannagh West
Baur North
Termon
Eanty More
Baur South
Faherlaghroe
Cregavockoge
Poulanine
Poulnaskagh
Kilcorney Glebe
Eantybeg
Caherconnell
Crughwill
U.C.G. Field Research Stn.
Moheramoylan
Kilcorney
Ballydoora
Sladoo
Carran
An Carn
Fahee South
Cahergrillaun
Meggagh West
Meggagh East
Ballyconry
Cragnarooan
Poulawack
Poulcaragharush
Lisnanroum
Cahermackirilla
Magheraweeleen
Commons
Mogouhy
Castletown
Iskancullin
Poulacarran
Kiltaan
Tullycommon
Fanygalvan
Noughaval
Slievenaglasha
Carran
Poulaphuca
Sheshodonnell West
Ballyganner North
Teeskagh
Moheraroon
Burren National Park
Páirc Náisiúnta Bhoirne
Sheshy More
Creevagh

◀ Fig. 10.38
The Burren in Co. Clare

## Leaving Certificate Questions on Weathering, Mass Movement and Slopes, Karst and Underground Water

### Ordinary Level

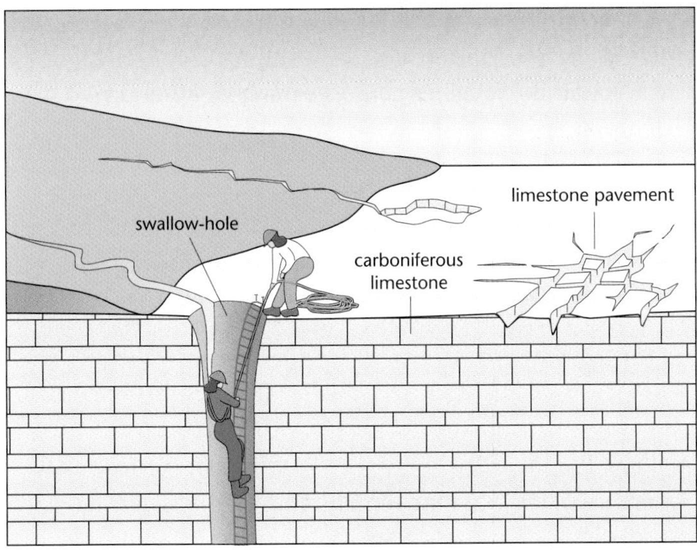


Fig. 10.39

**1998**

**Karst Landscapes**

Limestone Pavements, Swallow-Holes, Limestone Caves, Dry Valleys.

(i) Select **TWO** of the above Karst features and using a diagram describe how they were formed. **(40 marks)**

(ii) 'The Burren is the best-known Karst region in Ireland.'
Briefly explain its importance and in your answer refer to:

- Heritage aspects
- Environmental issues. **(40 marks)**

**1998**

**The Physical World** — Explain the following:

(iii) Weathering can be caused by physical **or** chemical action. **(40 marks)**

**1996**

**The Physical World** — Explain the following:

(iii) Limestone caves have interesting geographical features. **(20 marks)**

**1995**

Soluble Limestone or Karst regions, such as the Burren, contain a great variety of landscape features, both over and under the surface.

(i) Name any **THREE** of these features. **(15 marks)**

(ii) For **EACH** feature named above, describe and explain, with the aid of a diagram, how it was formed. **(45 marks)**

(iii) The Burren is Ireland's best-known soluble limestone region.
Briefly explain its importance, referring to some of the following:

- Heritage and Tourism
- Flora and Fauna **(20 marks)**

**1994 and 1993**

**The Physical World** — Explain the following:

(iii) Weathering can be due to physical **or** chemical action. **(20 marks)**

(iv) Limestone regions have distinctive underground and surface features. **(40 marks)**

(i) The influence of gravity on mass movement has helped to shape and modify many landforms. **(40 marks)**

**1989**

(i) With reference to **ONE** example in **EACH** case, explain how the rocks of the earth's surface are affected by the natural processes of mechanical **AND** chemical weathering. **(40 marks)**

(ii) Describe and explain **ONE** way in which natural weathering processes can affect human-made objects. **(20 marks)**

(iii) Explain briefly **ONE** way in which human activities can accelerate weathering processes. **(20 marks)**

**1997**

**The Physical World** — Explain the following:

(iii) The Burren is Ireland's best-known limestone region. (40 marks).

**Higher Level**

**1998**

Examine the illustration below which classifies mass movement and then answer the questions which follow:

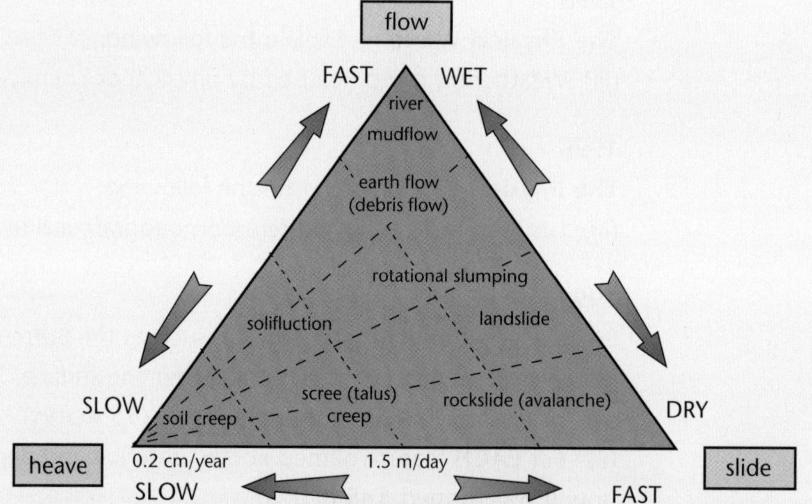


(i) With reference to examples which you have studied, explain fully **one** example of slow and **one** example of fast mass movements. **(50 marks)**

(ii) Examine **two** ways in which human activities can accelerate mass movements. **(50 marks)**

**1997**

(i) The Karst regions of the Burren and the Aran Islands contain many characteristic landforms both above and beneath the surface.
Examine, with reference to specific examples, the formation of any **three** such landforms. **(60 marks)**

(ii) As a result of tourism pressure, these regions have vulnerable environments. Examine **one** example of this vulnerability and suggest how a balance may be found between economic and conservation needs. **(40 marks)**

**1995**

The processes of weathering, together with gravity, are important factors in shaping landscapes.

(i) Explain this statement, referring to **THREE** weathering processes. **(75 marks)**

(ii) Examine how human activities can accelerate or intensify any **ONE** of the weathering processes referred to above. **(25 marks)**

**1993**

(i) 'The Karst topography of the Burren region exhibits many karstic landforms both above and below the surface.'
Select **two** features above and **two** features below the surface and explain with the aid of diagrams how each of these features was formed. **(80 marks)**

(ii) Give your views on recent plans for an interpretative centre in the Burren region. **(20 marks)**

# Rivers, River Processes and Landforms

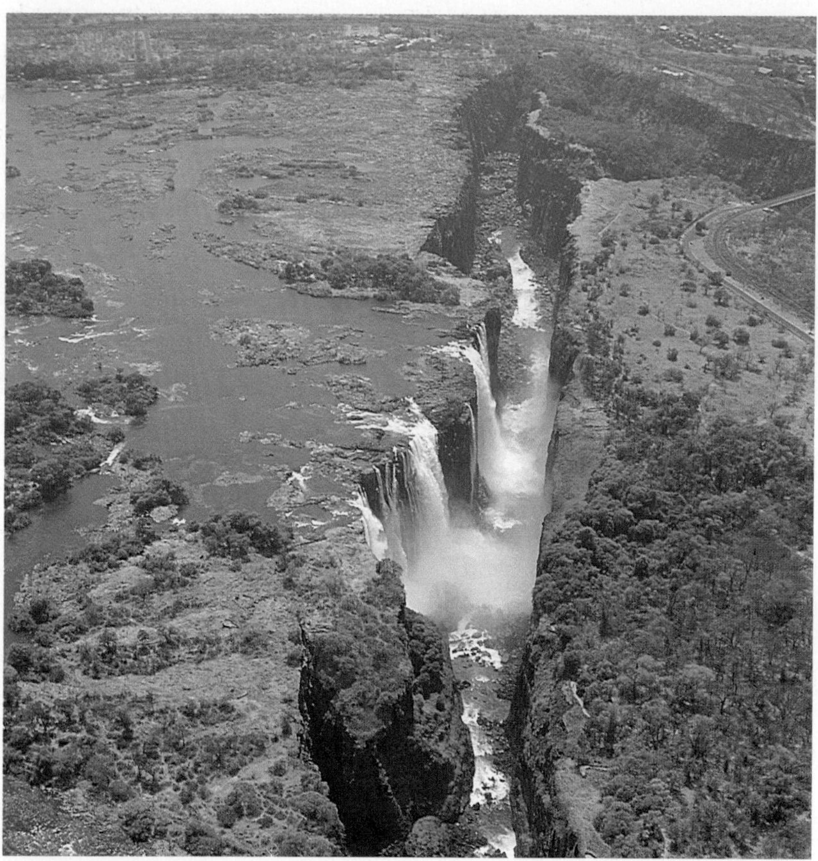

Fig. 11.1
Victoria Falls, Zimbabwe

**river basin**
A river basin is the area drained by a river and all its tributaries.

**watershed**
A watershed is a ridge of higher ground, which separates one river basin from another.

Rivers perform three basic functions. They **erode**, **transport** and **deposit** material, so they are constantly changing the surface of their basins. The energy of a river depends upon:

1. its **volume**
2. its **speed** or **velocity**

Much of this energy is used up by the river in transporting its **load**. This is the material carried by a river, which is derived from weathering and erosion. The ability of a river to carry its load is known as its **competence**. This competence increases with the river's velocity (speed). The weight of the largest fragment that can be carried in a load increases with the sixth power of the river's speed. Therefore, if the speed becomes twice as great, the maximum particle size increases sixty-four times, i.e. two to the power of six. During times of flood, a river's volume is increased and even large boulders can be moved along its bed. The water turns brown due to huge amounts of suspended matter. When a flood subsides and normal levels are again attained, the brown colour disappears and only tiny particles can again be moved.

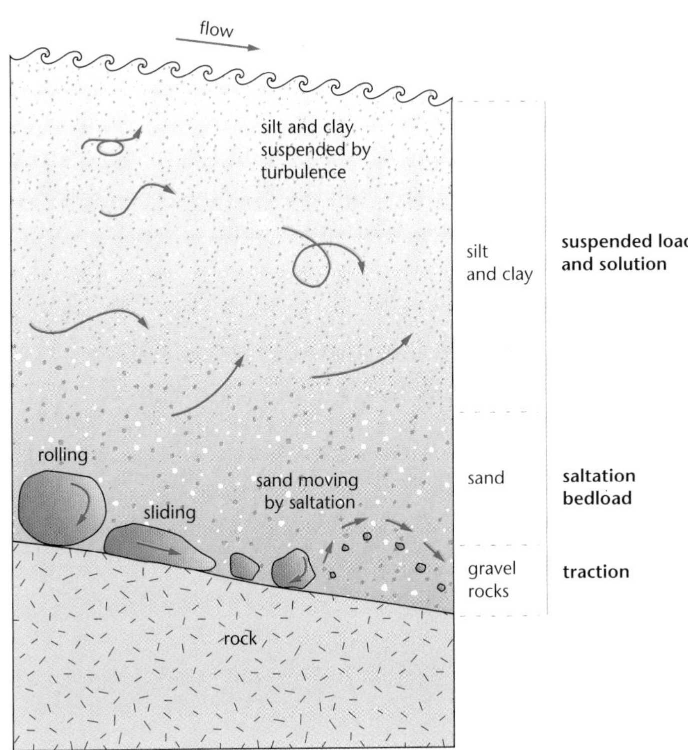


Fig. 11.2

The river's load is carried in the following ways (Fig. 11.2).

1. **Suspended load**. Most particles are carried in suspension by a river, including fine clay and silt. Hydraulic action may initially cause finer particles to be lifted from the river-bed, but once in suspension, the turbulence of the water keeps them up and the particles are transported downstream.
2. **Solution**. Rivers which flow over soluble rock such as limestone will carry some matter in solution. Chalk streams, for instance, may appear to be carrying no load at all, whereas they may have large amounts of soluble minerals dissolved in their waters.
3. **Saltation**. Some particles are light enough to be bounced along the river-bed. They are lifted from the river-bed by hydraulic action. Because they are too heavy to form part of the suspended load, they fall back onto the river-bed to be picked up once more. This process is repeated and so the pattern of bouncing stone is achieved.
4. **Traction (bedload)**. The volume and speed of a river is greatly increased during times of flood. Pebbles, large stones and sometimes huge boulders are rolled along the river-bed during these periods of high discharge. This process is often referred to as **bedload drag**.

## Processes of River Erosion

### Hydraulic Action

Hydraulic action is caused by the force of moving water. By rushing into cracks, the force of moving water can sweep out loose material or help to break solid rock. Turbulent water and eddying may undermine banks on a bend of a river, a process known as **bank-caving**.

Erosion also occurs because of **cavitation**. Cavitation takes place when bubbles of air collapse and form shock waves against the banks. Loosely consolidated clays, sands and gravels are particularly vulnerable to this type of erosion.

### Abrasion or Corrasion

This is a wearing away of the banks and the bed of a river by its load. The greater the volume and speed of a river, the greater its load. Thus a river attains its greatest erosive power during times of flood. This type of erosion is seen most effectively where rivers flow over flat layers of rock. Pebbles are whirled round by eddies in hollows in the bed, so cutting potholes. In mountain streams large deep pools are similarly worn.

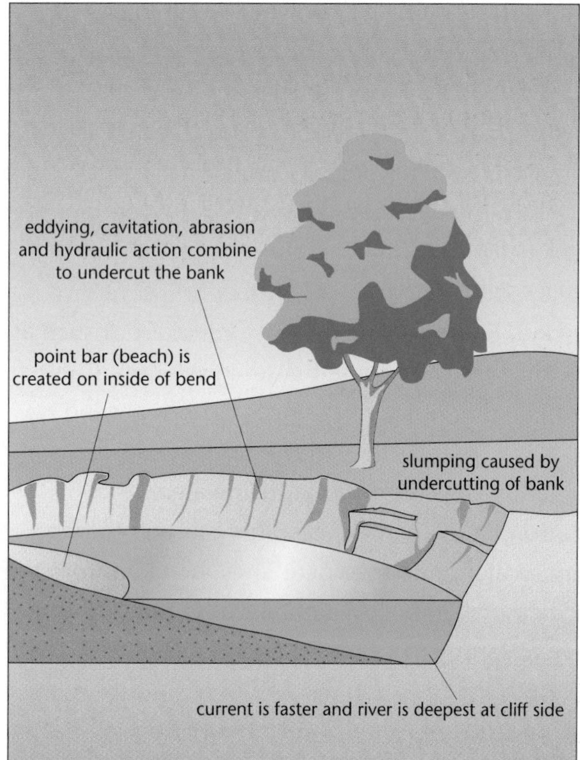


**Fig. 11.3**
Processes involved in bank-caving on a river bend

### Attrition

As a river carries its load, the particles are constantly in collision with each other and with the bed of the river. These particles are therefore getting progressively smaller in size as they move downstream. Because of this, boulders and pebbles in a river are always rounded and smooth in appearance.

### Solution

This is a chemical erosion process whereby a river dissolves rocks such as limestone and chalk as it flows across their surface. Rainwater is a weak carbonic acid. It reacts with limestone or chalk, carrying some of it away in solution.

## Process of Deposition

A river deposits material due to a reduction in energy. This generally occurs because of:

1. **decreasing velocity** due to
   - a change of slope
   - reduction in volume
   - entering a lake or sea
2. a **reduction in volume** due to
   - ending of a wet period
   - flowing through a desert
   - a period of drought
   - flowing over porous rock
3. an increase in the **size of its load** due to
   - a fast-flowing tributary adding extra material
   - heavy rainfall

**Velocity and load both increase during times of flood.**

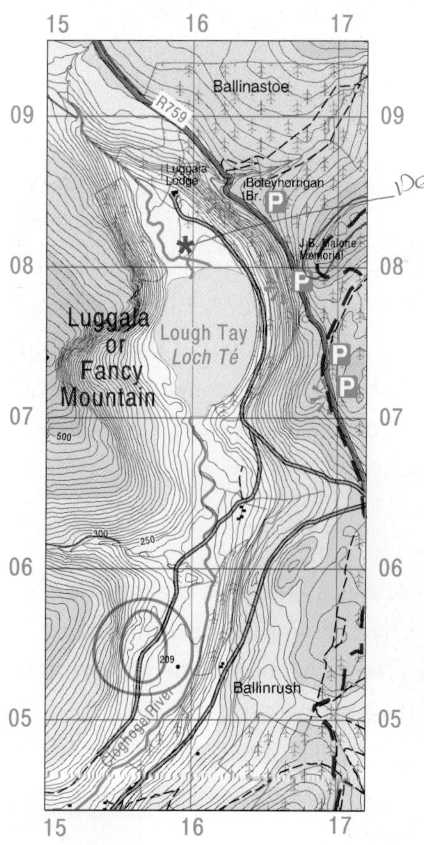


**Fig. 11.4**
Study this Ordnance Survey map extract and 'Process of Deposition' above. Then answer the following:

1. In which direction is the Cloghoge River flowing?
2. A lacustrine delta has formed at * on the map extract. Why has a delta formed at this location?

## The Long Profile of a River

A river's activity concentrates on the creation of a slope from source to mouth, a slope which will result in such a velocity (speed) that erosion and deposition are exactly in balance. At this stage the river is said to be **graded** and to have achieved a **profile of equilibrium**. Such a profile is rarely if ever achieved, however. Variations in volume, changes in base level or unequal resistance of underlying rocks all prevent a river from ever achieving a graded profile.

Fig. 11.5

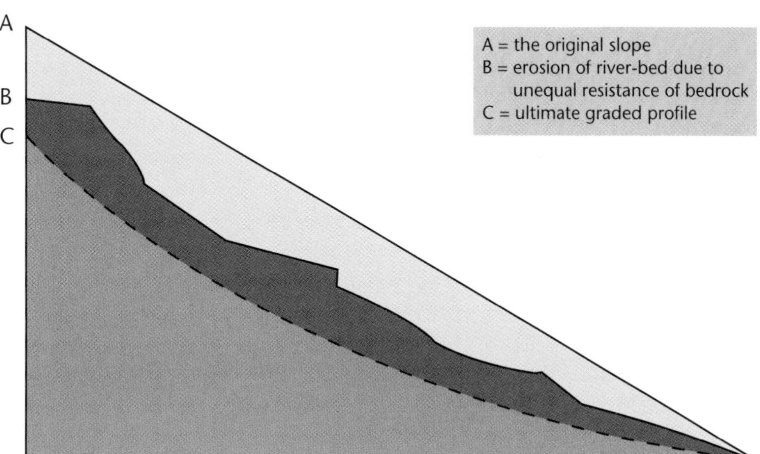


### Typical River Profile

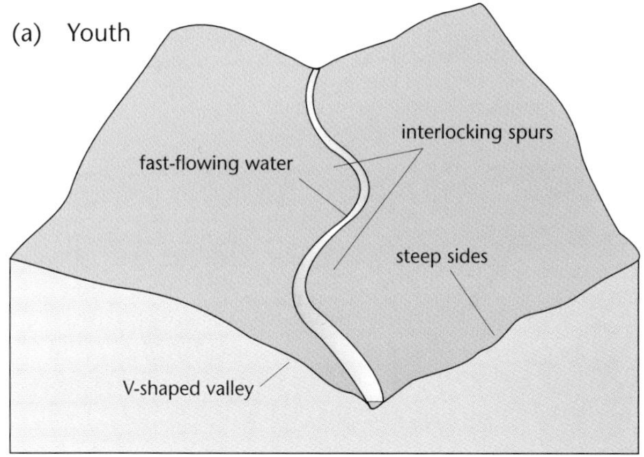


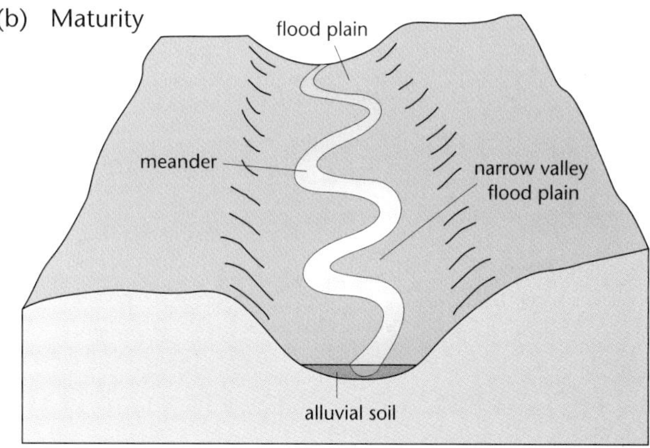


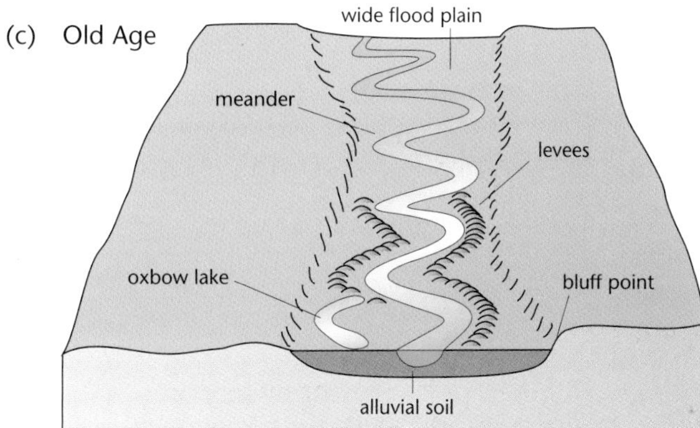


Fig. 11.6
The three stages in the development of a river valley

## Upper Course: Young River Valley

### Landform: V-valley

Examples: The Avoca River in Co. Wicklow, in Ireland
The headstreams of the Ganges in the Himalayas and the Yangtze Kiang in China
The headstreams of the Murray-Darling River in New South Wales, in Australia

### Young River Valleys in a Glaciated Highland Area

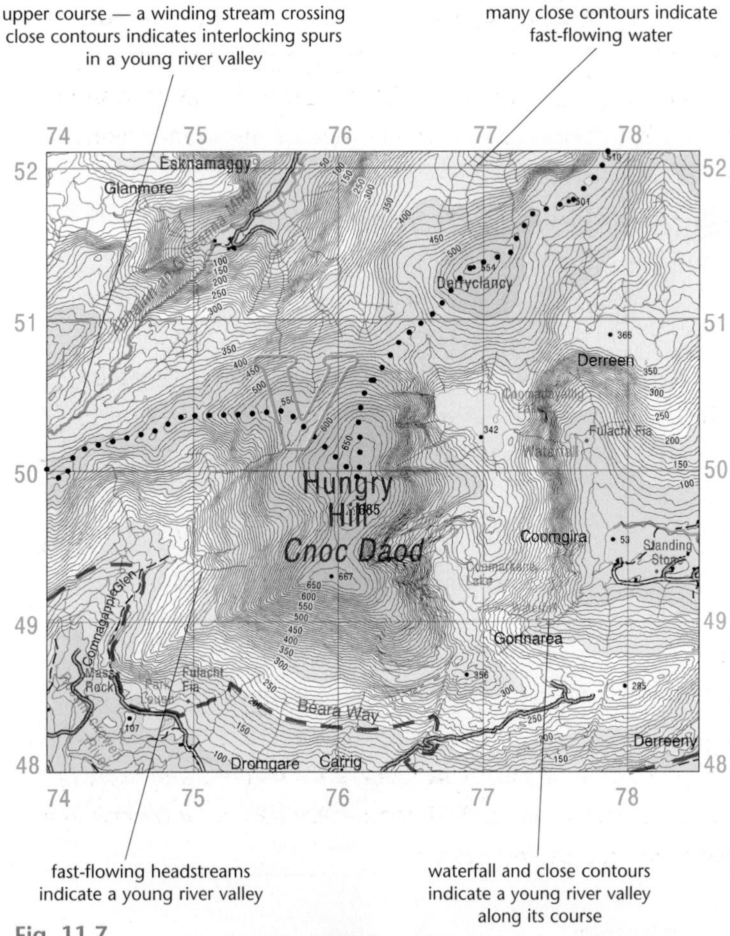

upper course — a winding stream crossing close contours indicates interlocking spurs in a young river valley

many close contours indicate fast-flowing water

fast-flowing headstreams indicate a young river valley

waterfall and close contours indicate a young river valley along its course

Fig. 11.7

### Young River Valley in a Lowland Area

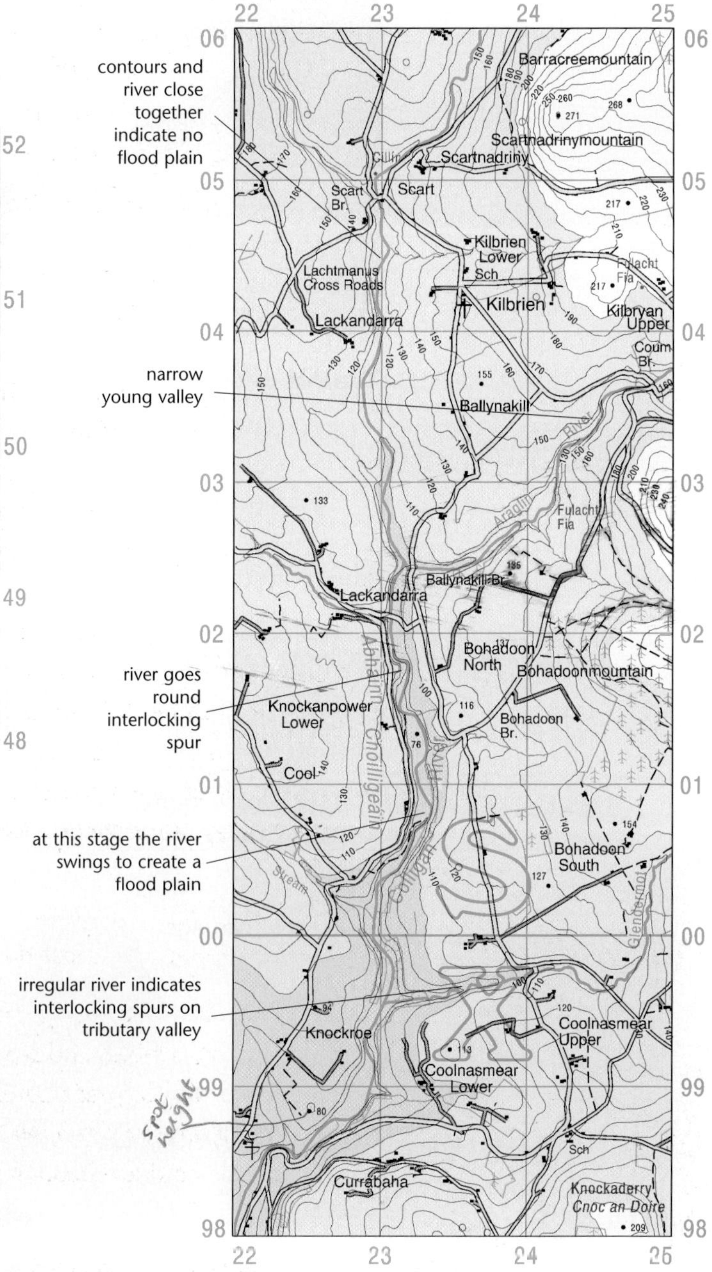

contours and river close together indicate no flood plain

narrow young valley

river goes round interlocking spur

at this stage the river swings to create a flood plain

irregular river indicates interlocking spurs on tributary valley

Fig. 11.8

> **Student Note**
> In order for students to become familiar with processes in river valleys, the author has repeated the relevant processes involved with each landform explanation.

Processes: Hydraulic Action, Corrasion, Cavitation, Slumping Erosion, Suspension, Saltation, Traction

## Formation

- In its upper course, a river is primarily concerned with vertical erosion. The steepness of the valley sides and the depth of the valley therefore depends on:
  — the volume and speed of the river
  — the type of bedrock over which the river flows and
  — the speed of weathering and mass movement on the valley sides

In this youthful stage, level land on either side of a river does not exist, so the stream occupies the entire but limited valley floor.

- Stream channels in the upper course of a river change from following a relatively straight course initially to following a winding course. This may occur for a number of reasons. Irregularities in the stream's bed such as patches of hard resistant rock may deflect the stream from side to side. In other places, banks consisting of soft material allow bank erosion. The current of the stream tends to be strongest on the outside of a bend. As a result, bends become more pronounced. Spurs project from both sides of the valley and interlock, forming **interlocking spurs**.

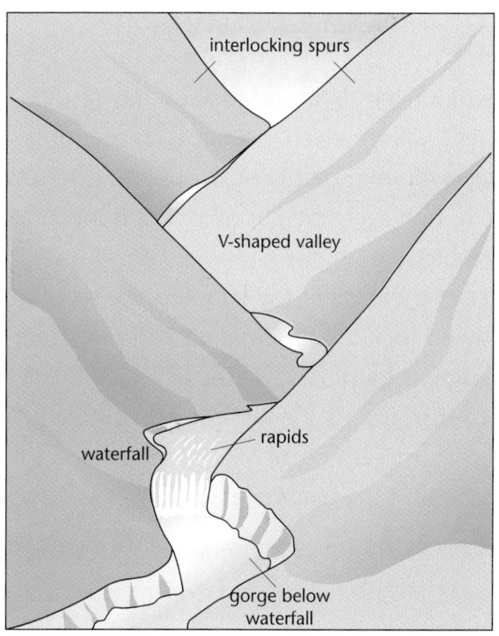



Fig. 11.9

Erosion processes are most effective when a youthful river rushes down a mountainside. Hydraulic action, caused by fast water which rushes into cracks, sweeps out loose material and helps to break up solid rock. This fast-moving water also tries to find the easiest route to lowlands. As a result it winds and twists its way around obstacles, such as hard, resistant rock.

This winding, fast-flowing water carries eroded material and, by corrasion, uses it to undermine land on the outside of bends, a process known as bank-caving. Another process, that of cavitation, occurs as bubbles of air collapse and form shock waves against the banks. As land is undermined, it slumps into the river channel. The

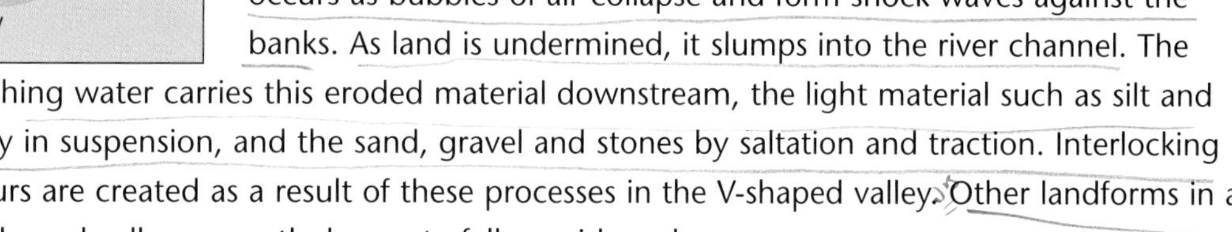

rushing water carries this eroded material downstream, the light material such as silt and clay in suspension, and the sand, gravel and stones by saltation and traction. Interlocking spurs are created as a result of these processes in the V-shaped valley. Other landforms in a V-shaped valley are potholes, waterfalls, rapids and gorges.

### Landform: Potholes

Examples: Most rivers in their youthful stage

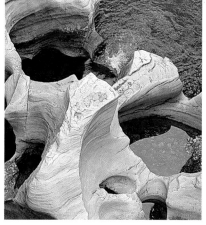

If the bed of a young, fast-flowing river is uneven or rough, the water swirls, moving pebbles and stones in a circular pattern in some places. These rotating stones cut circular depressions into the bed of the river, forming potholes. Once the depressions deepen, stones become trapped and the erosion continues until deep cylindrical hollows form.

Fig. 11.10
'Bourke's Luck' potholes in South Africa created by pebble action in a young river valley

LANDSCAPES OF THE WORLD

## Landform: Waterfall

Examples: Ashleagh Falls on the Erriff River in Co. Mayo, in Ireland
Niagara Falls on the border between the USA and Canada

### Formation

Fig. 11.11

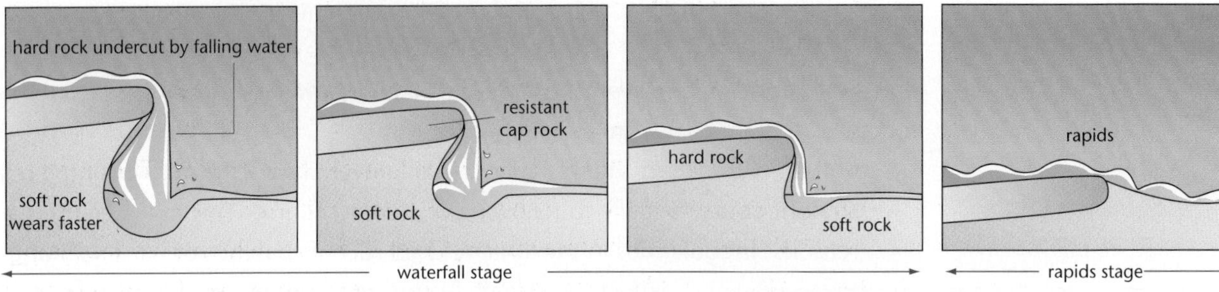
hard rock undercut by falling water
soft rock wears faster
resistant cap rock
soft rock
hard rock
soft rock
rapids
waterfall stage
rapids stage
undercutting of the hard rock causes waterfall to recede upstream, as it moves upstream its height decreases

Processes: Hydraulic Action, Corrasion, Undermining, Rejuvenation, Tectonic Movement

1. When waterfalls occur in the upper course of a river, their presence usually results from a bar of hard rock lying across the river's valley, which interrupts its attempts towards a graded profile. If this slab of rock is dipping gently downstream, it results in a series of rapids with much broken water. If, however, the bar of rock is horizontal or slightly inclined, a vertical fall in the river results.

   The hydraulic action of the falling water at the base of the falls and the scouring action of its load undercuts into the underlying rock usually creating a plunge pool.

   Undermining causes an overhanging ledge of hard rock, pieces of which break off from time to time and collect at the base of the falls. As the falls retreat upstream, a steep-sided channel with rapids is created downstream of them. This feature is called a gorge. Niagara Falls in the USA and Canada and the waterfall on the Yellowstone River in Wyoming formed in this way.

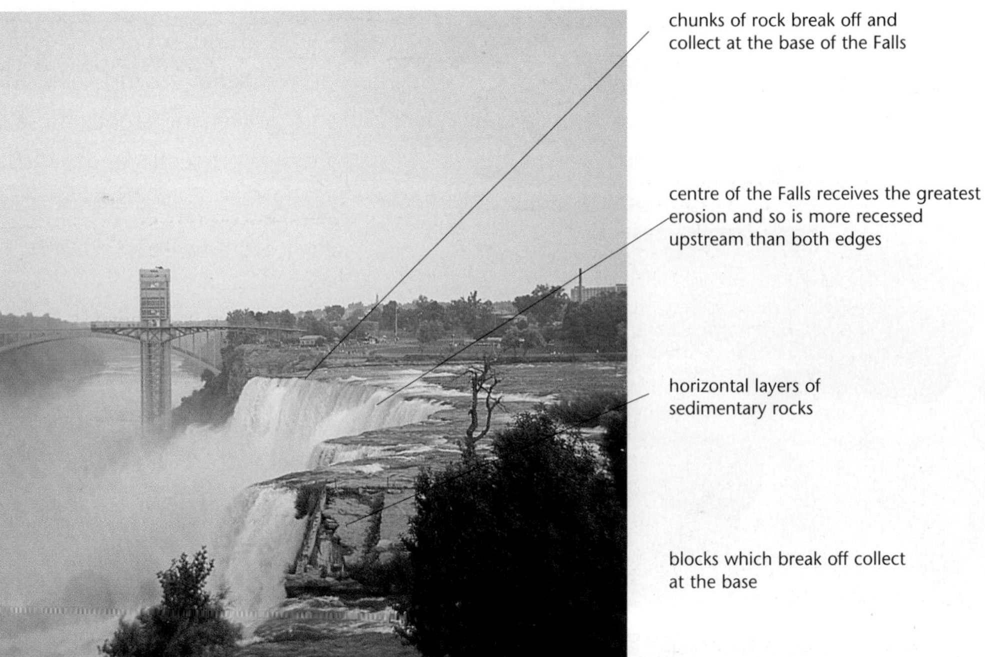
chunks of rock break off and collect at the base of the Falls
centre of the Falls receives the greatest erosion and so is more recessed upstream than both edges
horizontal layers of sedimentary rocks
blocks which break off collect at the base

Fig. 11.12
The American Falls at Niagara. The centre of the falls recedes faster upstream than both edges due to severity of erosion in the centre of the river.

2. If a waterfall appears in the middle course of a river, it may be the result of **rejuvenation** (*rejuvenate* means 'make young again'). This may be caused by a fall in sea level, by a local uplift of land or by the presence of glacial debris. This causes a steeper slope and a greater river velocity which renews downcutting or vertical erosion. Where a change of slope occurs, it is known as the **knickpoint**, often noted by the presence of rapids. Downstream of the knickpoint, the river cuts into its former flood plain, leaving terraces on either side of the valley.
3. Some waterfalls are due to faulting. Tectonic movement causes some parts of the earth's surface to subside, thereby interrupting the course of some rivers. As a result, a vertical drop may occur on the river-bed, creating a waterfall. The famous Victoria Falls on the Zambezi River in Africa formed in this way (see Fig. 11.1, p. 156).
4. Waterfalls are commonly found in glaciated districts where over-deepening of the main valley leaves hanging valleys and cirques high above the main floor. Rivers flowing from these glaciated valleys plunge to the larger and deeper valley floors below. In some cases, the falling water forms a cascade which plunges over many layers of rock until it finally reaches level ground. Powerscourt Falls in Co. Wicklow and Torc Waterfall in Co. Kerry, in Ireland, and the falls at Lauterbrunnen, near Interlaken in Switzerland, are examples of cascades.

## Landform: Gorge/Canyon

Examples: The Grand Canyon on the Colorado River in Arizona, in the USA
The 'Strid' on the River Wharfe above Bolton Abbey in Yorkshire, in Britain
The Glencullen river gorge near Enniskerry in Co. Wicklow, in Ireland

### Formation

The term gorge is used to refer to a long steep-sided trench valley which is deep in proportion to its width. A river gorge is found where river erosion cuts down more rapidly than the forces of weathering can wear back and 'open up' the sides. The Wicklow Mountains in Ireland and the Maritime Alps in Provence in France have part of their upper courses contained in a gorge.

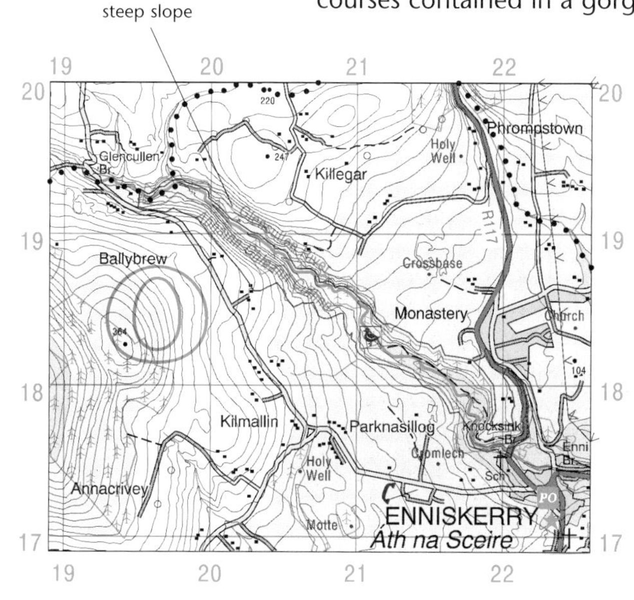


Fig. 11.13 The Glencullen river gorge near Enniskerry in Co. Wicklow

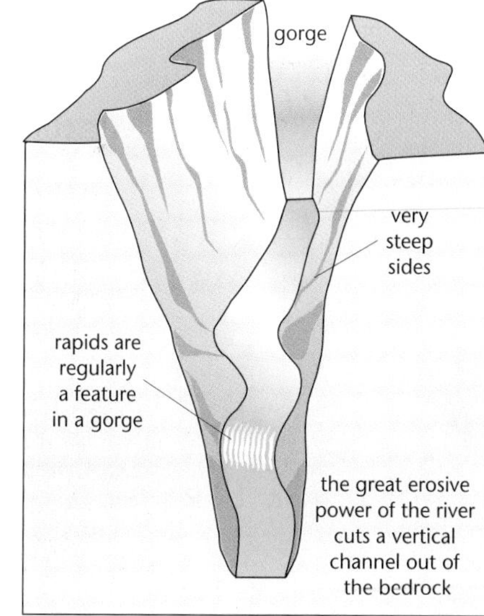


Fig. 11.14

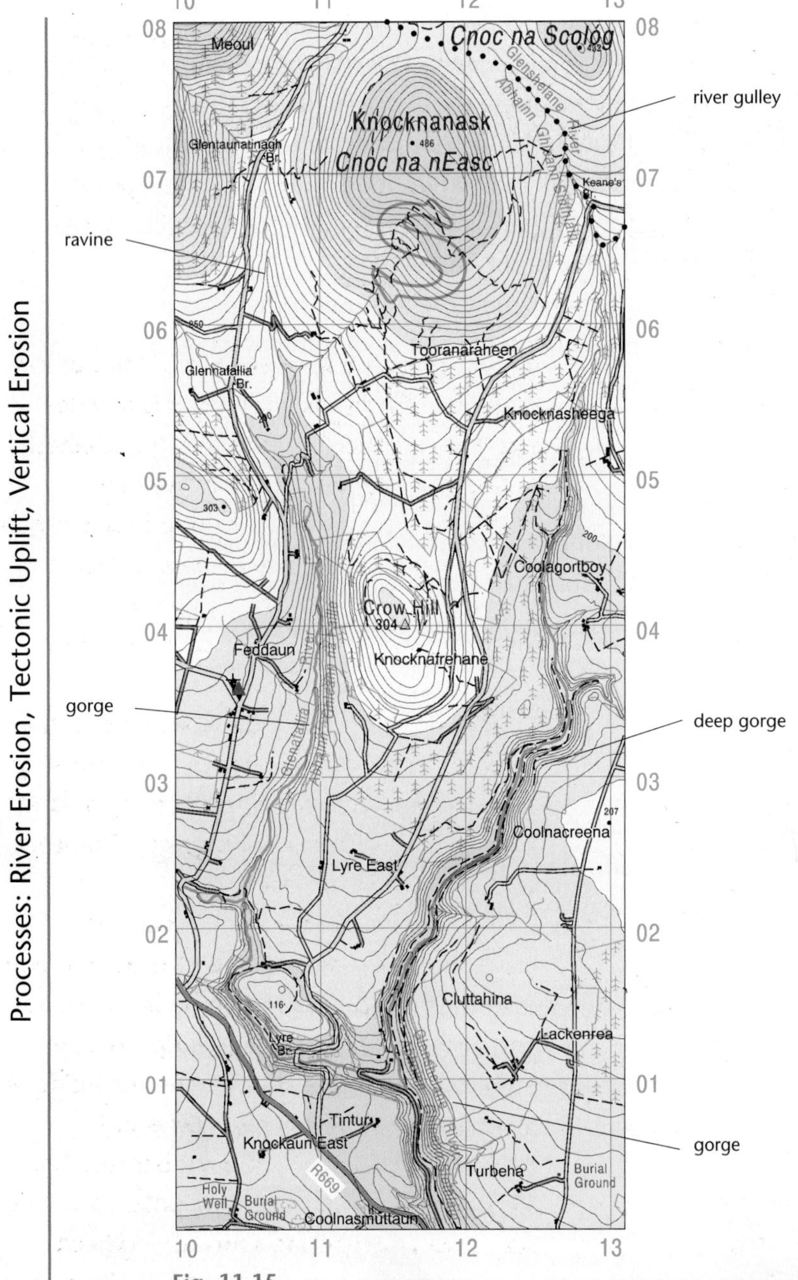


Fig. 11.15

1. Gorges form when a waterfall recedes upriver, leaving a gorge downstream of the falls, such as on the Niagara River in the USA.
2. A second type is generally associated with arid climates. In such cases a river deriving its water from snow-melt on mountains upstream beyond the desert is able to maintain its flow and its erosive power, e.g. the Nile Gorge.
3. A third type of gorge may be found where a river retains its course as uplift of the surrounding land continues due to earth movement. The result is a river system where the main river and its tributaries divide a plateau into a series of blocks, for example the Nore River at Inistioge in Co. Kilkenny in Ireland.

The Ganges and some of its headstreams have gorges which exceed five kilometres in depth. Other examples are the Rhine gorge between Bingen and Bonn in Germany.

In Ireland, the Glencullen River has cut a narrow steep-sided gorge as it flows down from the Wicklow Mountains towards the coast, see Fig. 11.13.

## Middle Course: Mature River Valley

**Lateral erosion** is dominant at this stage of development. This results in a small flood plain, the size of which varies from one side of a river to the other. The valley sides are less steep due to weathering and erosion and the valley profile is one of gently sloping sides with a flat floor (Fig. 11.16).

### Landform: Flood Plain

Examples: Blackwater Valley near Fermoy in Co. Cork, in Ireland
Mississippi Valley near St Louis, in the United States
The Rhine river valley in the Netherlands

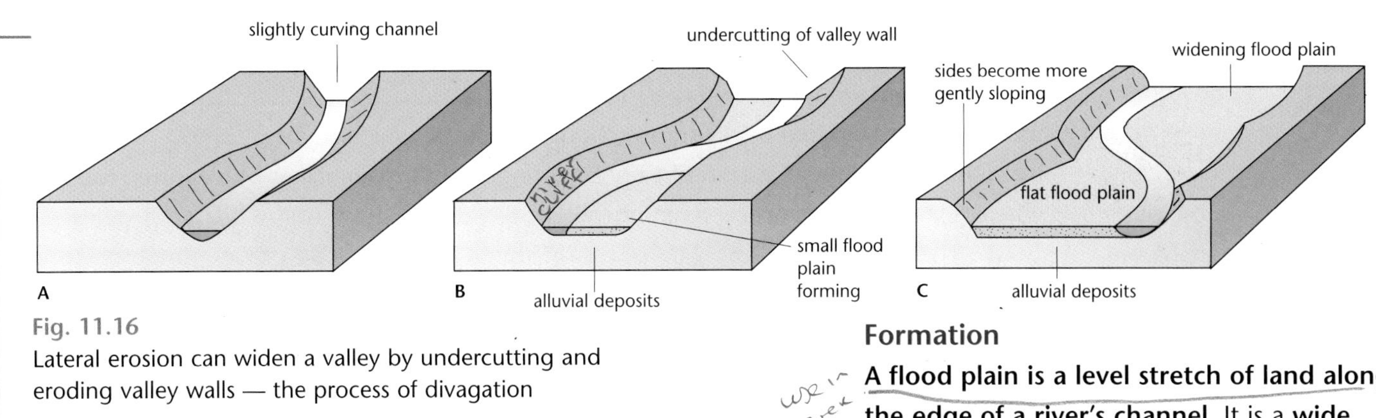


**Fig. 11.16**
Lateral erosion can widen a valley by undercutting and eroding valley walls — the process of divagation

**Fig. 11.17**
Some contour patterns which represent a mature river valley

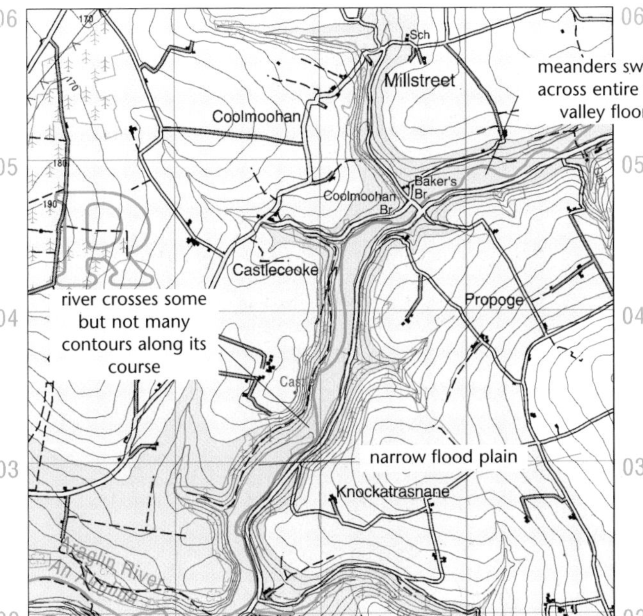


**Fig. 11.18**
Some contour patterns which represent a mature river valley

## Formation

**A flood plain is a level stretch of land along the edge of a river's channel.** It is a **wide** and **flat** valley floor which is often **subjected to flooding** during times of heavy rain.

When a river swings from side to side it is said to **meander**. Each river loop is called a meander. Flood plains are created by meanders and they are pronounced in the middle course of a river.

When meanders migrate downstream, they swing to and fro across the valley. Interlocking spurs are removed, and a level stretch of land is created on both sides of the river. This level land is called a flood plain.

As the water flows around a bend, it **erodes** most strongly **on the outside**, forming a river cliff. **Undercutting** of the bank occurs and **slumping** takes place (see Fig. 11.3, p. 158). Little erosion takes place on the inside of a bend, but deposition often takes place, causing a gravel beach or point bar (see Fig. 11.3). The valley has been straightened at this stage, with interlocking spurs having been removed by the lateral (sideways) erosion of the meanders. This lateral erosion is called the process of '**divagation**'.

When in flood, the river spreads across the flat flood plain and deposits a thin layer of alluvium with each successive overflow on the flood plain and on the river's channel. This combined depositional process is called '**aggradation**'. Over thousands of years, aggradation builds up a thick sediment blanket, 'the flood plain' through which the river flows.

**alluvium**

Alluvium is fine material consisting of silt and clay particles. It is rich in mineral matter which is transported by a river in times of flood and it is deposited on the flood plain and in the river's channel in the middle and lower courses of rivers.

## Alluvial Fan

When a mountain torrent enters a main valley, it often deposits a fan-shaped mass of material. This is caused by a **sudden change of slope** which reduces the river's velocity — from the steep slope of a torrent to the flat floor of the main valley. This is often found where a stream pours over a hanging valley onto the flat floor of the glaciated valley below. The river deposits build up to form either a fan- or a cone-shaped mound of material against the valley side.

In the photograph, Fig. 11.19, an alluvial fan developed as a cascade of water poured onto the glaciated valley below. The flat valley floor reduced the speed of the falling water and so it deposited part of its load to form an alluvial fan which accumulated over a period of time.

When a river is overladen with material, it drops some of its load on the stream's bed. This creates 'islands' (or bars) of gravel which later gather silt and sand, forming soil in sufficient quantities to support vegetation.

Braiding occurs in rivers which are heavily laden with material. It happens where a river's channel widens out and shallows, thus losing the ability to carry its load, which it drops in midstream. This gravel deposit may either split up a stream into numerous small channels, or it may appear as an island in a river. Braiding occurs most often where the river's banks are composed of easily erodible sands and gravels, or where a mountain torrent enters a river in its mature or old stage (Figs. 11.20 and 11.21).

**Fig. 11.19**
The Lauterbrunnen valley in the Swiss Alps

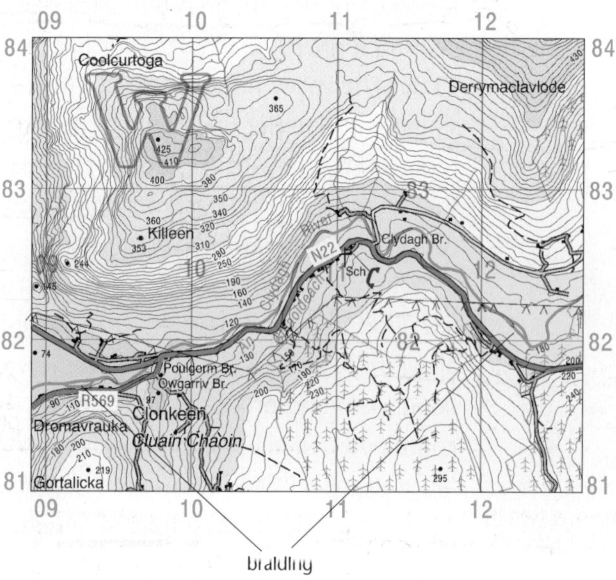


**Fig. 11.20**

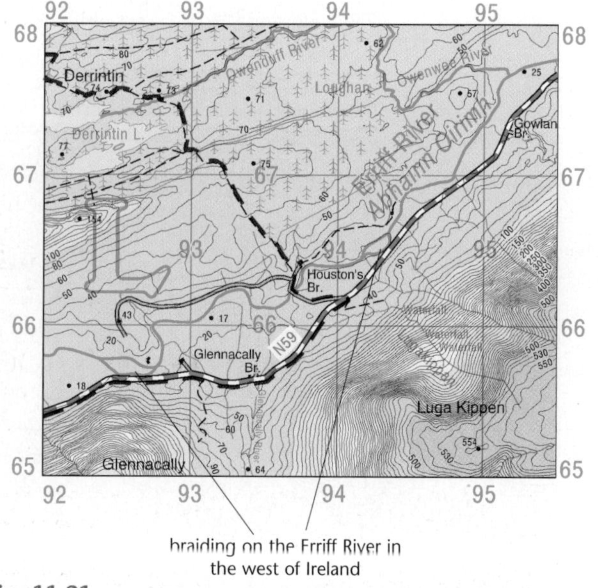


**Fig. 11.21**

1997 OS.

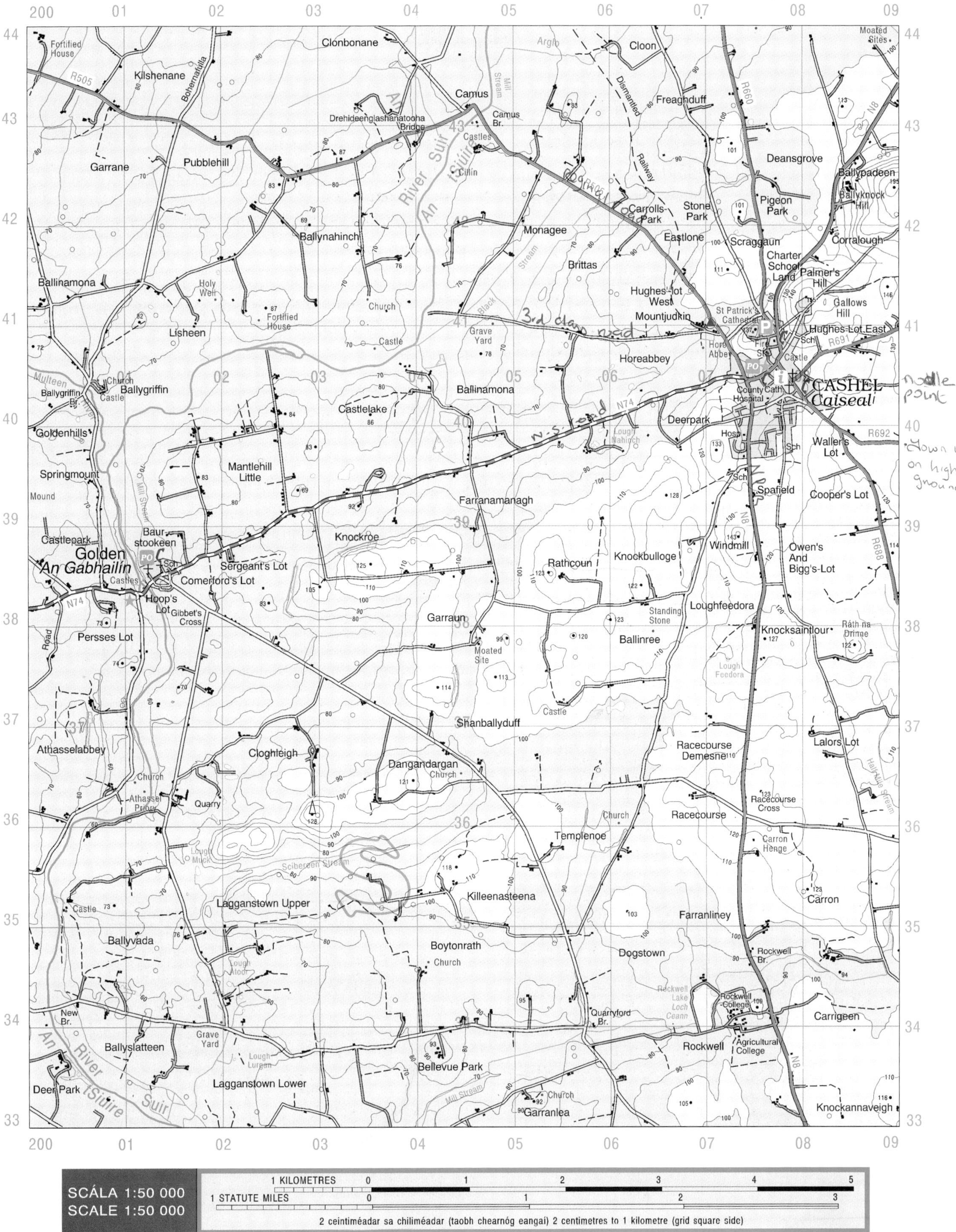


Fig. 11.22

**Class Activity**

Study the Ordnance Survey map extract of Cashel/Golden, Fig. 11.22, p. 167. Then complete the following:

1. Identify the river landform at S 027 407. With the aid of a sketch map, explain the processes involved in its formation.
2. Identify the stage of maturity which the River Suir displays along the stretch from Golden, grid reference S 013 385 to New Bridge, grid reference S 003 342. Then, with the aid of a sketch map, justify your choice of stage of maturity.

## Lower Course: Old River Valley

In its lower or old age, a river wanders over its extensive and flat flood plain in a series of sweeping meanders (Fig. 11.6c, p. 159). The edges of the valley are bounded by low bluffs which have been reduced by weathering. The valley of an old river may be distinguished from that of a mature river by the relationship between meanders and valley sides. In a **mature river valley**, the **meanders fill** the width of the valley floor; in an **old river valley**, the flood plain is **far wider** than the meanders.

### Landform: Oxbow Lake (or Cut-off)

Examples: Moy River at Clongee near Foxford in Co. Mayo, in Ireland
Arkansas River in the United States
The Shannon River at Roosky in the midlands in Ireland
The Amazon River in Brazil

Fig. 11.23

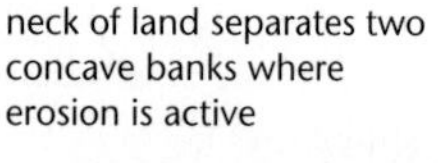


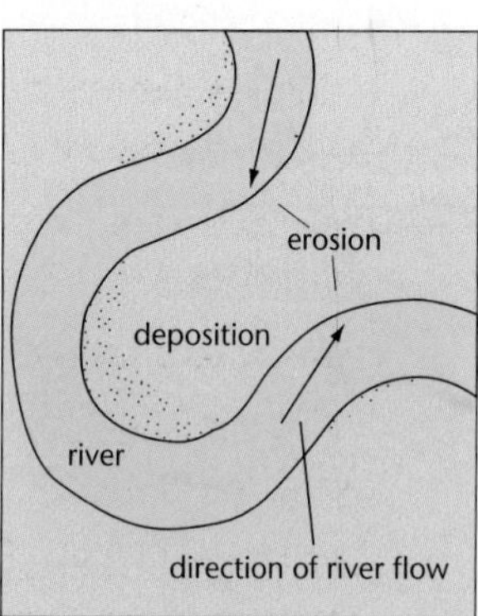


A looping meander occurs where a narrow neck of land separates two concave banks which are being undercut.

The neck is ultimately cut through. This may be accelerated by river flooding.

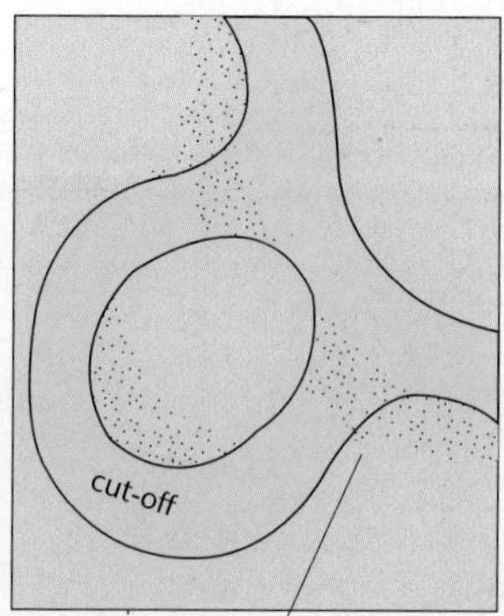


deposition begins to seal up the ends of the cut-off

Erosion has broken through the neck of land. This generally happens when the river is in flood. The meander has been cut off.

deposition seals the cut-off which becomes an oxbow lake

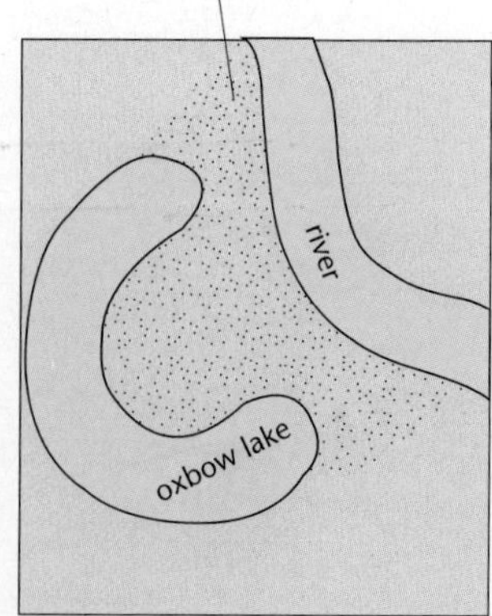


Deposition takes place along the two ends of the cut-off and it is eventually sealed off to form an oxbow lake.

Oxbow lakes are relics of former meanders and are often called **cut-offs**. They may occasionally occur in a mature valley but are common on the lower courses or old valley floors of rivers (Fig. 11.6c, p. 159). As meanders move downstream, erosion of the outside bank leads to the formation of a loop in the river's course, enclosing a 'peninsula' of land with a narrow neck.

As water sweeps around this looping meander bend, the zone of highest speed swings towards the outer stream bank. Strong water turbulence in association with the river's load causes undercutting and slumping of sediment where this fast-flowing water meets the steep bank.

Abrasion, cavitation and hydraulic action combine to cause this undercutting.

Meanwhile, along the inner (near) side of each meander loop, where water is shallow and slow-moving, coarse sediment, such as gravel and sand, accumulates to form a point bar.

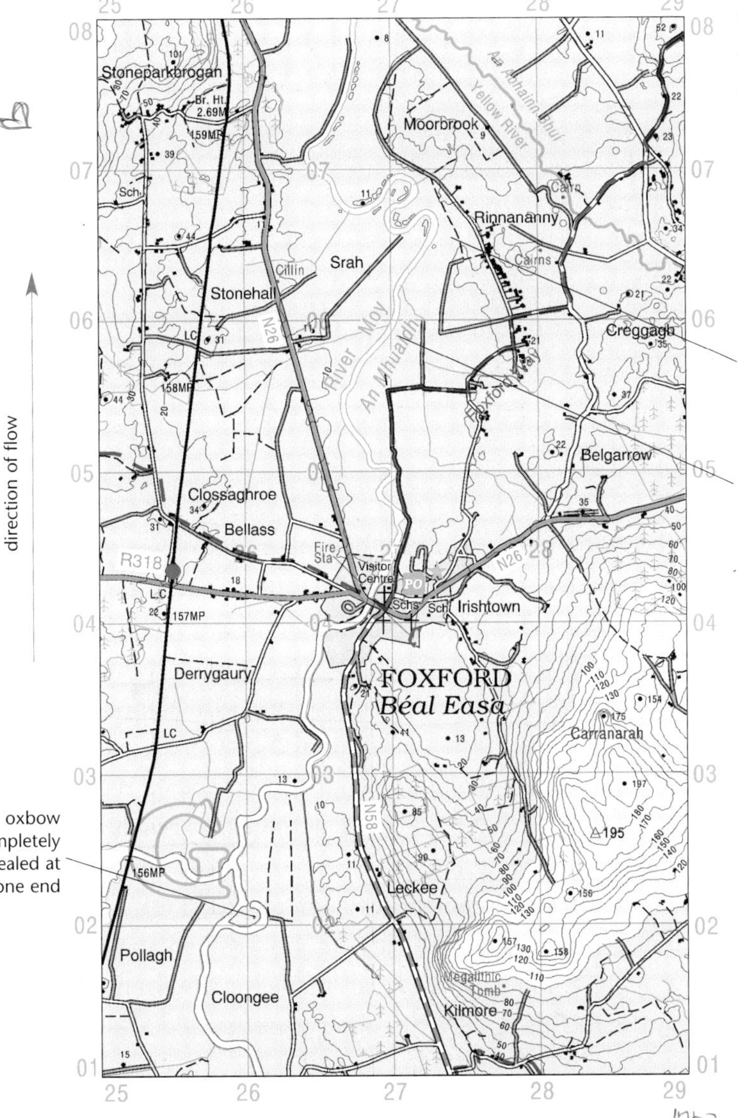


**Fig. 11.24**
The lower course of the River Moy in Co. Mayo in western Ireland showing an oxbow lake at a very advanced stage of formation

In this instance the 10-metre contours indicate a wide and flat flood plain. The looping meanders do not 'fill' the flood plain.

Finally, during a period of flood, the river cuts through this neck and continues on a straighter and easier route, leaving the cut-off to one side. Deposition occurs at both ends of this cut-off to form an oxbow lake (Fig. 11.23, p. 168).

After a long period of time, these oxbow lakes are filled with silt from flood water and they finally dry up. At this stage, they are called **meander scars** or **mort lakes** and are clearly visible from aerial photographs on some old valley floors. Drainage schemes may create 'man-made' cut-offs, which are sometimes used as nesting areas for wildlife such as ducks, coots and moorhens.

## Landform: Levees

Examples: The Mississippi River in Illinois, in the United States
The Yangtze Kiang and the Hwang-Ho in China

**Fig. 11.25**
The absence of levees creates widespread flooding on extensive flood plains

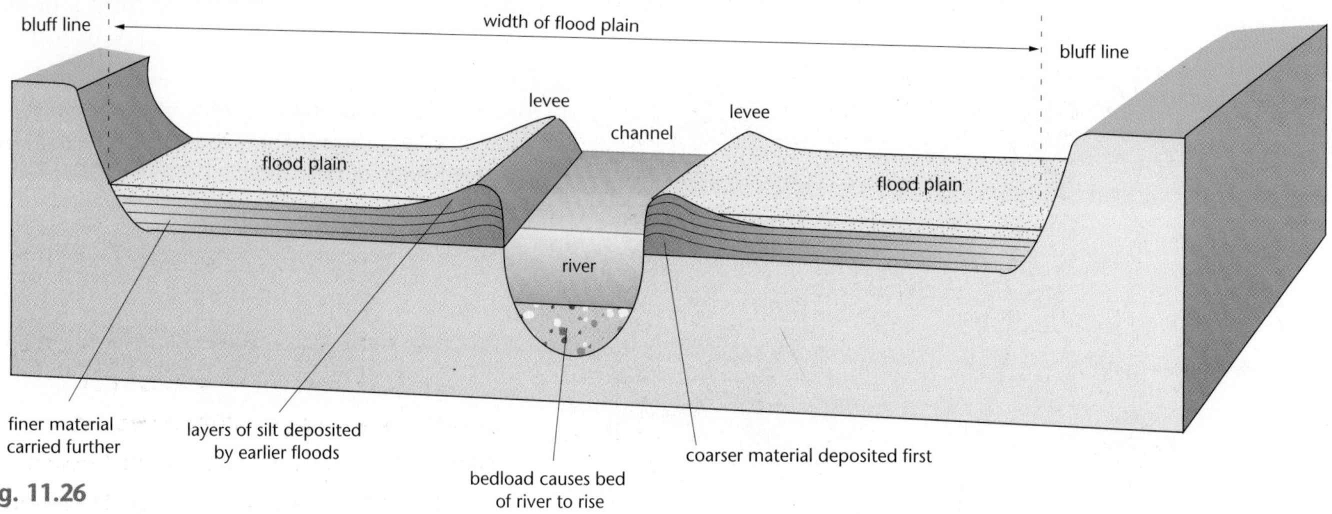


**Fig. 11.26**
Cross-section of a flood plain showing levees and bluffs

Processes: Flooding, Aggradation, Deposition

### Formation

A natural levee is a broad, low ridge of fine alluvium built along the side of a channel by debris-laden flood water. As the sediment-laden flood water flows out of the completely submerged channel during a flood, the depth, force and turbulence of the water decreases sharply at the channel margins. This sharp/fast decrease results in sudden/rapid deposition of the coarser materials (usually fine sand and coarse silt), along the edges of the channel, building up a natural levee. Further away from the channel, finer silt and clay settle out in the quiet waters which cover the flood plain. This combined depositional process is called 'aggradation'. Over thousands of years, aggradation builds up a thick sediment blanket creating the levees and flood plain.

Where natural levees do not exist and in an effort to contain flood waters within a defined channel, people often dredge gravel and silt from a river's bed and drop it along the channel edge to form a raised embankment similar to a levee. Thousands of miles of levees have been built along the Mississippi River in order to control flood water.

### Disadvantages of Levees

See Fig. 11.59, p. 190.

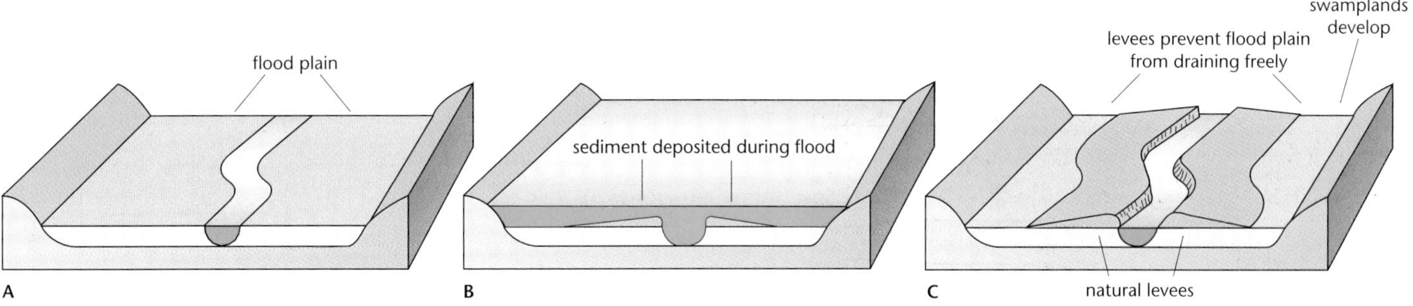


Fig. 11.27
Natural levee deposition during a flood. Levees are thickest and coarsest next to the river channel and build up from many floods, not just one. (Relief of levees is exaggerated.)
(A) Normal flow. (B) Flood. (C) After flood.

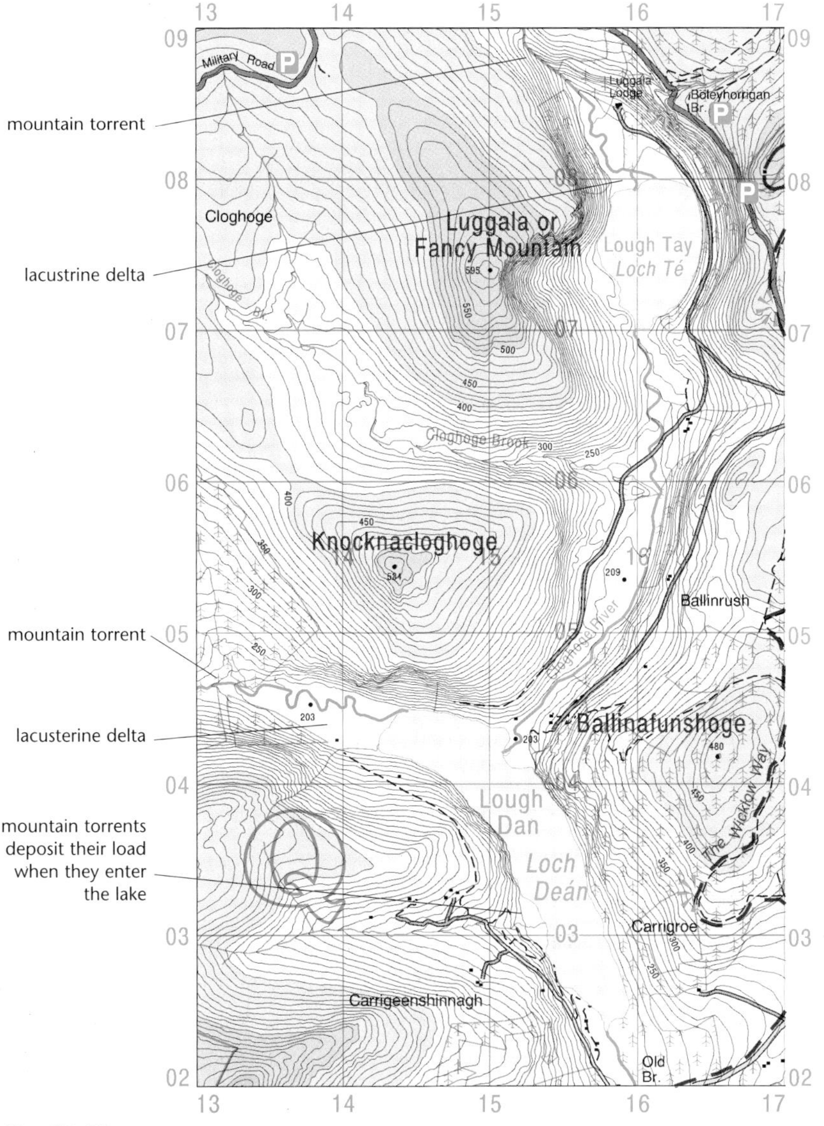


Fig. 11.28
Lacustrine deltas in the Wicklow Mountains in Ireland. The area shown on this Ordnance Survey map extract has been glaciated.

## Landform: Delta

Examples: Mississippi in the Gulf of Mexico
Po estuary in Italy
Roughty River estuary in Kenmare, in Ireland

When a river carries a heavy load into an area of calm water such as an enclosed or sheltered sea area or a lake, it deposits material at its mouth. This material builds up in layers called **beds** (Fig. 11.31). These form islands which grow and eventually cause the estuary to split up into smaller streams called **distributaries**. Should this occur in a lake, it is called a **lacustrine delta** (Fig. 11.28). If it occurs at a coast, it is called a **marine delta** (Fig. 11.29). The material which builds up to form the delta is composed of alternate layers of coarse and fine deposits, reflecting times of high and low water levels respectively in the river. Mountain streams flowing into glaciated valleys often build deltas within ribbon lakes. This causes a filling in of the lake, thereby reducing its length over time or dividing ribbon lakes into smaller ones. Such deltas may occur at any stage of a river's course. This has occurred at the western end of the Upper Lake in Glendalough in Co. Wicklow, in Ireland, where large amounts of sediments have been deposited by a mountain torrent. A lacustrine delta now occupies a large area at the upper end of this glacial lake.

The upper and lower lakes at Glendalough once formed a single lake. The north-flowing Pollanass River deposited material in the lake, forming a delta. This delta grew across the lake, dividing it into the Upper Lake and the Lower Lake.

In the case of marine deltas, stretches of sea are surrounded by deposited sediments. These sediments are derived mostly from the river, but are also added to by material transported along the coast by longshore drift. Bars, spits and lagoons may therefore form at the seaward edges of growing deltas. The lagoons gradually silt up to form swamps and marshes (Fig. 11.29).

**Fig. 11.29**
Some features of an old river valley

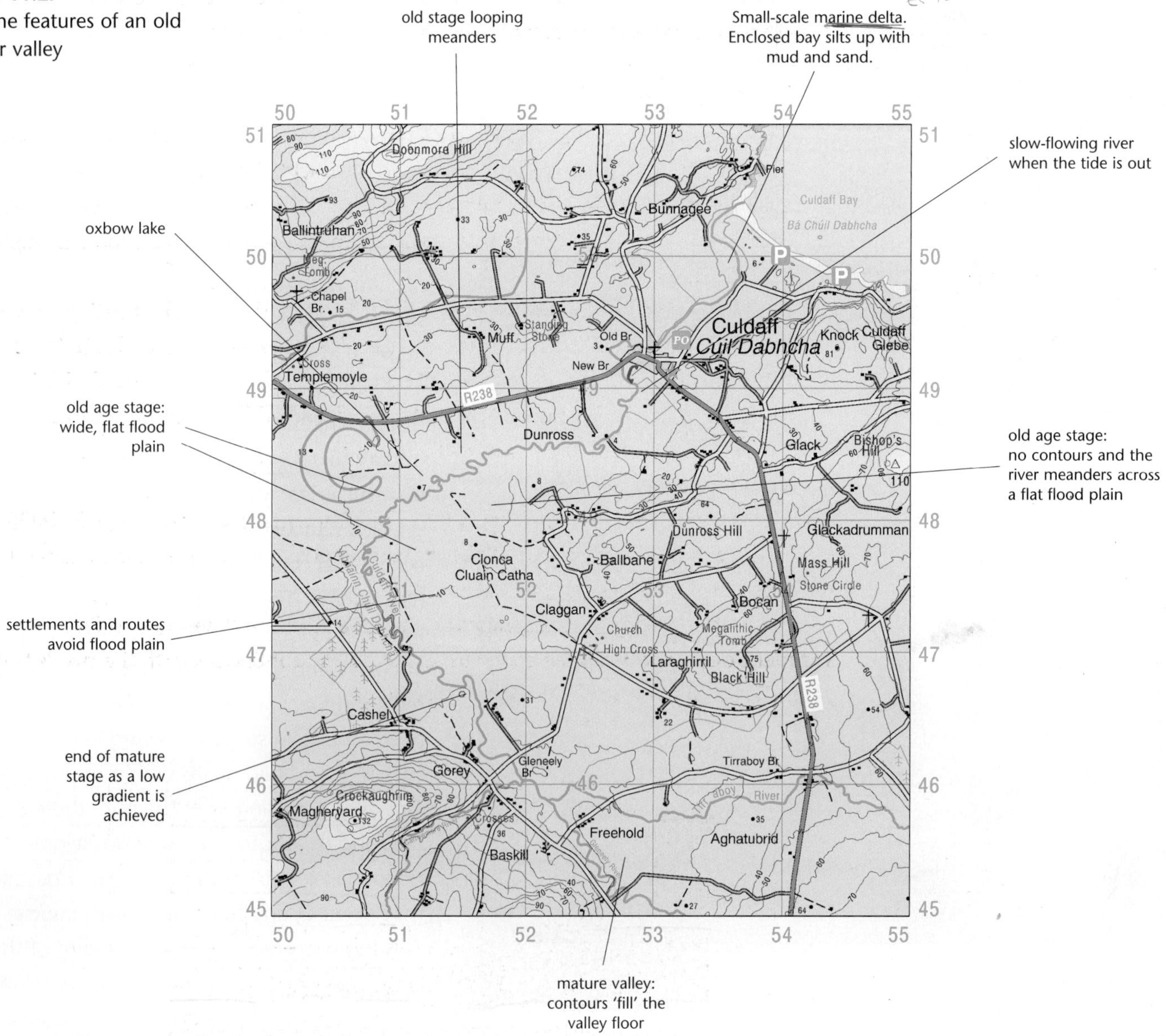

old stage looping meanders

Small-scale marine delta. Enclosed bay silts up with mud and sand.

slow-flowing river when the tide is out

oxbow lake

old age stage: wide, flat flood plain

old age stage: no contours and the river meanders across a flat flood plain

settlements and routes avoid flood plain

end of mature stage as a low gradient is achieved

mature valley: contours 'fill' the valley floor

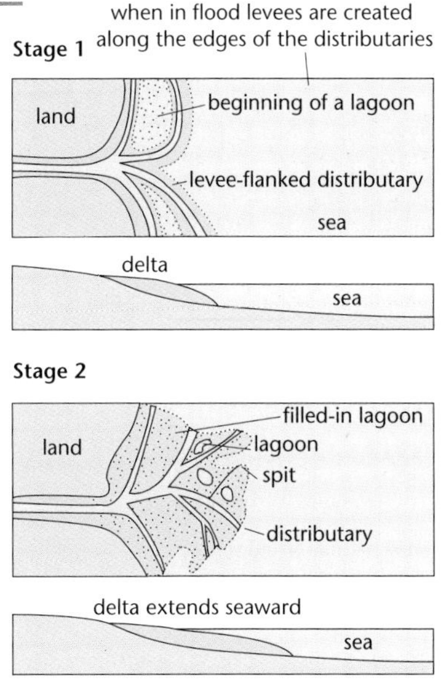


Fig. 11.30
Stages in the development of a delta on a large scale

There are three main types of marine deltas.

### 1. Arcuate

This type is triangular in shape, like the Greek letter 'delta'. The apex of the triangle points upstream. Arcuate deltas are composed of coarse sands and gravels. They are found where sea currents are relatively strong, which limits delta formation beyond the original estuary. This type of delta is constructed from porous deposits. Thus distributaries are numerous with very few communicating channels. *Examples*: the Nile in Egypt; the Po in Italy; the Irrawaddy in Burma; and the Hwang-Ho in China.

### 2. Estuarine

Such deltas form at the mouths of submerged rivers. The estuarine deposits form long, narrow fillings along both sides of the estuary. *Examples*: the Elbe in Germany; the Seine in France; and the Shannon estuary in Ireland.

### 3. Bird's Foot

These deltas form when rivers carry large quantities of fine material to the coast. Such impermeable deposits cause the river to divide into only a few large distributaries. Levees develop along these distributaries, so long projecting fingers extend out into the sea to form a delta similar in shape to a bird's foot. *Example*: the Mississippi.

When rivers like the Mississippi or the Nile reach the sea, the meeting of fresh and salt water produces an electric charge which causes clay particles to 'clot' and to settle on the seabed to form a delta, in a process known as '**flocculation**'.

**The materials deposited in a delta are classified into three categories.**

1. Fine particles are carried out to sea and deposited in advance of the main delta. These are the **bottomset beds**.
2. Coarser materials form inclined layers over the bottomset beds and gradually build outward, each one in front of and above the previous ones, causing the delta to advance seaward. These are the **foreset beds**.
3. Sediment of various grain sizes, ranging from coarse channel deposits to finer sediment deposited between channels, is laid down, continuous with the river's flood plain. These deposits are called the topset beds.

Fig. 11.31

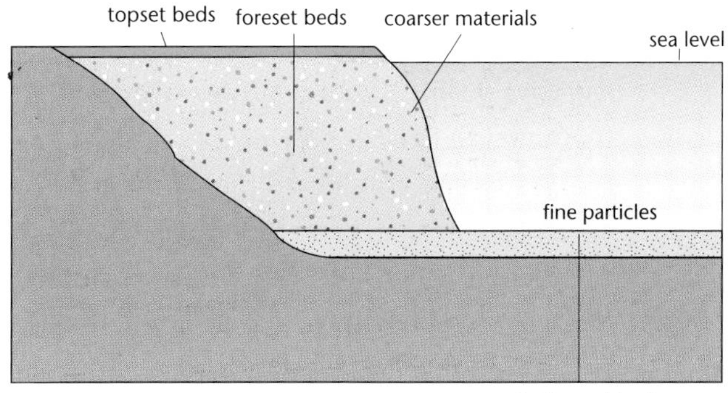

Class Activity

Study the Ordnance Survey map of Avoca, in Co. Wicklow, Fig. 11.32, p. 175, then do the following:

1. (a) Trace the outline of the complete map extract (the edge where the colour ends). Then trace the course of the Avoca River and ALL its tributaries ONLY. Use a fine-tipped pen and very clear tracing paper, otherwise this exercise will not be of any use.

   When you have completed this exercise, then observe how much of the map these streams make up. What is the geographical term for the area drained by these streams?

   (b) With a different-coloured fine-tipped pen, trace in all of the other streams. Draw a line (from north to south) dividing the streams of the Avoca River from the other streams.

   When you have completed this second exercise, then observe if this line coincides with any landform/s on the map. What is the geographic term for this land which divides river basins?

2. Draw a sketch map of the area shown on the map of Avoca. On it mark and name the following:

   (a)
   - the Avoca River and its tributaries
   - the low upland areas and their highest points
   - one national primary road
   - five third-class roads

   (b) Then observe from the map and sketch map if the heights of the low upland areas suggests anything about the original land surface before the streams eroded their valleys.

3. With the aid of a sketch map explain how the physical landscape has affected communications in this area.
4. With the aid of diagrams, describe the patterns of drainage in the area shown.
5. Discuss the patterns in the distribution of woodland shown on the map.

## Stream Rejuvenation

A serious interruption to the development of a profile of equilibrium (see p. 159) is a change in **base level.** This may cause a steeper slope and greater velocity, and therefore renewed downcutting in a river valley.

There is a limit to how deeply a river can erode and that limit is the stream's base level.

- Sea level is considered the ultimate base level, because it is the lowest level to which stream erosion can erode the land.
- Temporary or 'local' base levels include lakes and resistant rock layers which interrupt a stream from achieving its ultimate base level. For example, when a stream enters a

**Fig. 11.32** ▶
Avoca region

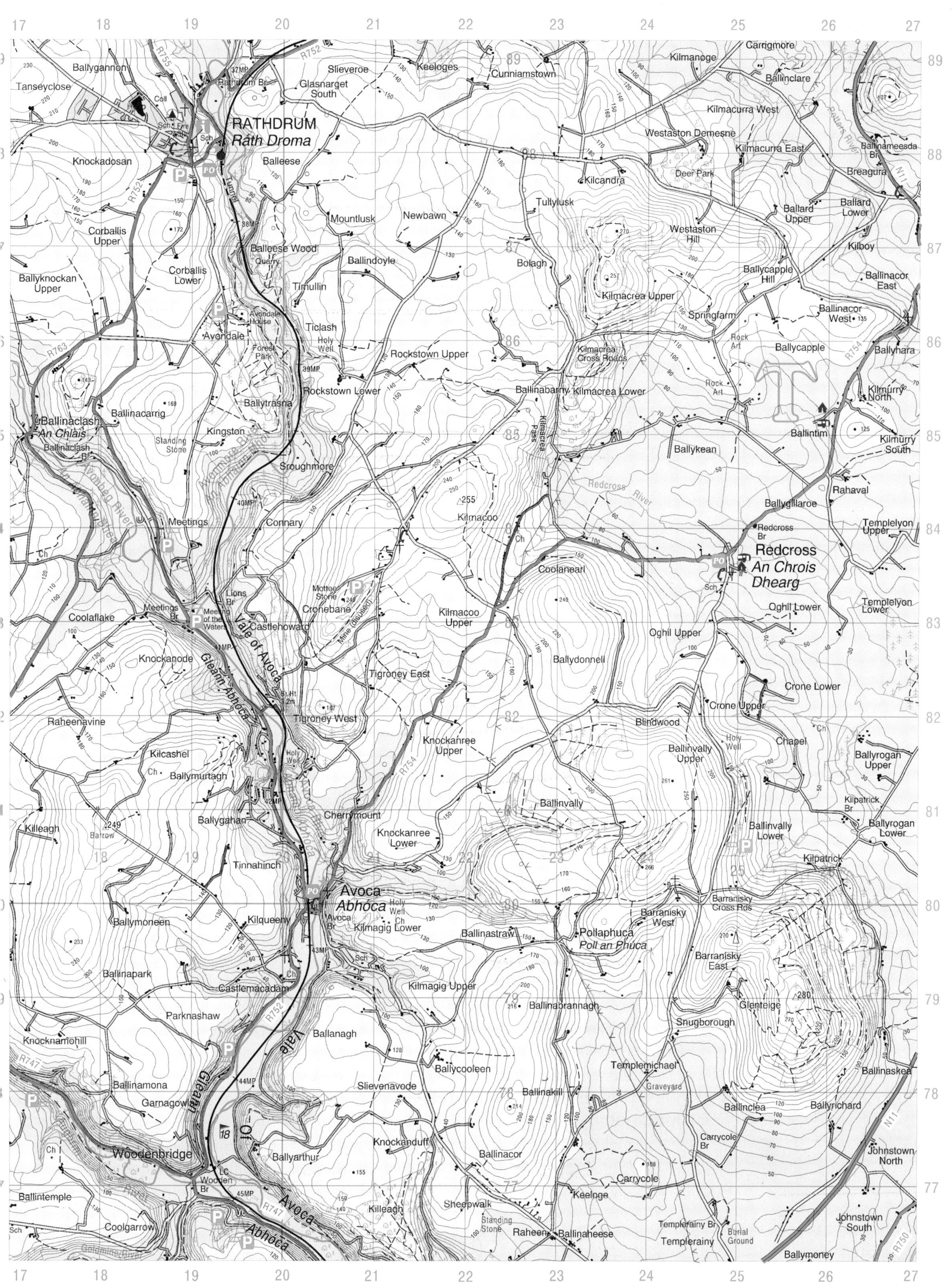

RATHDRUM
Ráth Droma
Ballygannon
Tanseyclose
Knockadosan
Corballis Upper
Corballis Lower
Ballyknockan Upper
Slieveroe
Glasnarget South
Keeloges
Cunniamstown
Kilmanoge
Carrigmore
Ballinclare
Kilmacurra West
Westaston Demesne
Kilmacurra East
Ballinameesda Br
Breagura
Deer Park
Kilcandra
Balleese
Tunnel
Mountlusk
Newbawn
Tullylusk
Westaston Hill
Ballard Upper
Ballard Lower
Kilboy
Balleese Wood
Quarry
Ballindoyle
Bolagh
Ballycapple Hill
Ballinacor East
Timullin
Kilmacrea Upper
Springfarm
Ballinacor West
Avondale House
Avondale
Ticlash
Holy Well
Forest Park
Rockstown Upper
Kilmacrea Cross Roads
Rock Art
Ballycapple
Ballyhara
Rockstown Lower
Ballinabarny
Kilmacrea Lower
Kilmacrea Pass
Kilmurry North
Ballinaclash
An Chlais
Ballinacarrig
Ballytrasna
Kingston
Standing Stone
Ballinaclash Br
Ballintim
Kilmurry South
Ballykean
Sroughmore
Redcross River
Rahaval
Avonbeg River
Ballygillaroe
Meetings
Connary
Kilmacoo
Redcross Br
Templelyon Upper
Coolanearl
Redcross
An Chrois Dhearg
Lions Br
Mottee Stone
Cronebane
Mine (disused)
Meetings Br
Meeting of the Waters
Coolaflake
Castlehoward
Vale of Avoca
Kilmacoo Upper
Ogphil Upper
Oghil Lower
Templelyon Lower
Knockanode
Gleann Abhóca
Ballydonnell
Tigroney East
Crone Lower
Crone Upper
Tigroney West
Raheenavine
Blindwood
Knockanree Upper
Chapel
Kilcashel
Ballymurtagh
Ballinvally Upper
Ballyrogan Upper
R754
Ballinvally
Kilpatrick Br
Killeagh
Barrow
Ballygahan
Cherrymount
Knockanree Lower
Ballinvally Lower
Ballyrogan Lower
Tinnahinch
Kilpatrick
Avoca
Abhóca
Avoca Br
Kilmagig Lower
Barranisky Cross Rds
Barranisky West
Ballymoneen
Kilqueeny
Ballinastraw
Pollaphuca
Poll an Phúca
Barranisky East
Ballinapark
Castlemacadam
Kilmagig Upper
Glenteige
Ballinabrannagh
Parknashaw
Snugborough
Knocknamohill
Ballanagh
Ballycooleen
Templemichael
Ballinaskea
Ballinamona
Slievenavode
Graveyard
Ballinakill
Garnagowlan
Ballinclea
Ballyrichard
Knockanduff
Carrycole Br
Woodenbridge
Ballyarthur
Ballinacor
Johnstown North
LC
Wooden Br
Carrycole
Ballintemple
Keelogue
Killeagh
Sheepwalk
Templerainy Br
Coolgarrow
Standing Stone
Raheen
Ballinaheese
Templerainy
Burial Ground
Johnstown South
Goldmine River
Ballymoney
R752
R755
R763
R747
R750
N11

lake its force is immediately checked and its ability to erode is reduced. In this way, the lake prevents the stream from eroding below its level at any point upstream from the lake. However, the lake outlet stream can reduce the lakewater's level or indeed drain the lake completely by eroding downward. Thus the lake is only a temporary hindrance to a stream trying to achieve its 'profile of equilibrium'.

When a fall in sea level occurs, new land is added to the lower course of the river which lies between the original estuary and the new river mouth. Because the gradient of this new extension is greater than the slope of the river's lower course, the river at this point acquires a renewed capacity for vertical erosion. Features of a 'youthful' river valley quickly begin to appear.

Streams/rivers that receive renewed energy from local land movements of uplift, fall in sea level, or increased volume due to long-term climatic changes are said to be rejuvenated. Rejuvenation, in simplest terms, means that the river has been made to flow faster for various reasons. For a meandering stream that flows slowly on a flood plain, the method of achieving this task is to cut deeply into the thick flood plain sediment.

Fig. 11.33

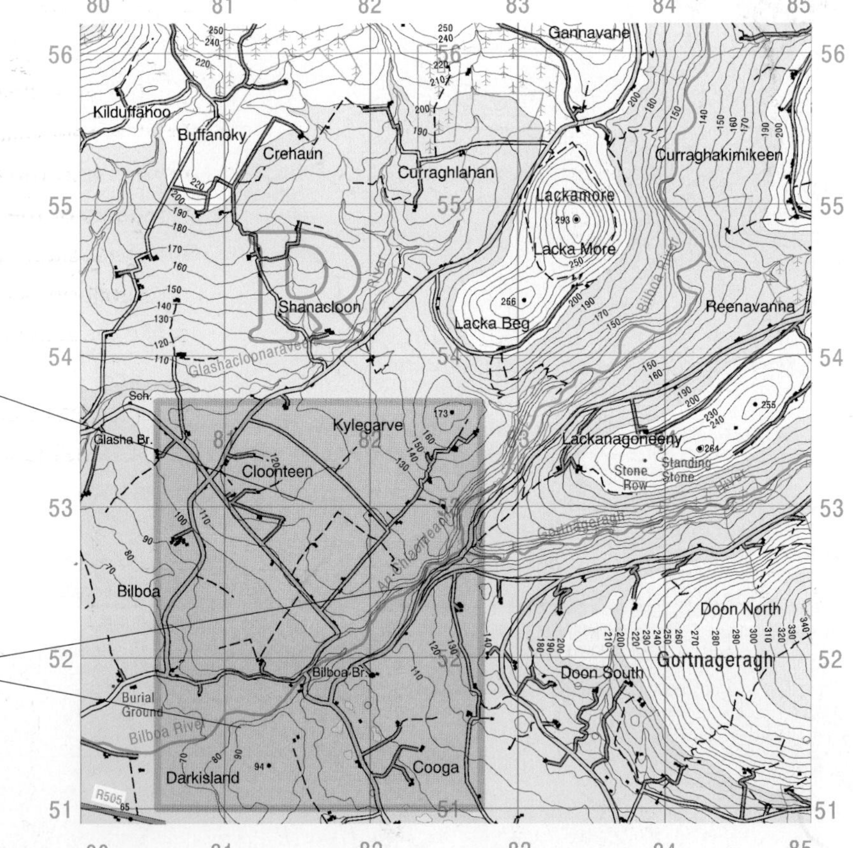

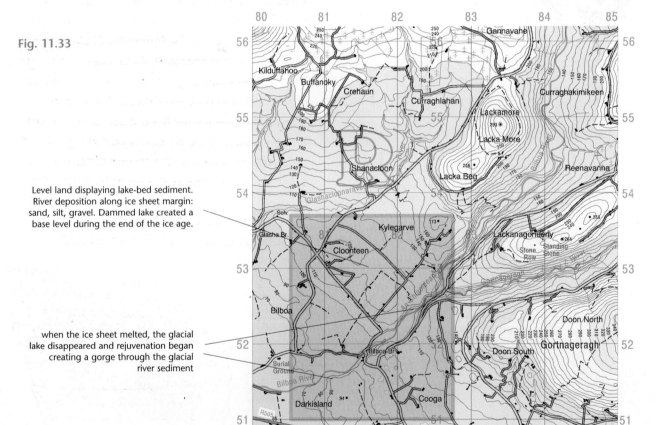

Level land displaying lake-bed sediment. River deposition along ice sheet margin: sand, silt, gravel. Dammed lake created a base level during the end of the ice age.

when the ice sheet melted, the glacial lake disappeared and rejuvenation began creating a gorge through the glacial river sediment

## The Effects of Rejuvenation on the Long Profile

### 1. Knickpoint

When a fall in base level occurs, the river begins to cut upstream from its mouth. This produces a new curve or profile of erosion which intersects with the old curve at the knickpoint. The knickpoint is distinguished by a marked break of slope at the junction of

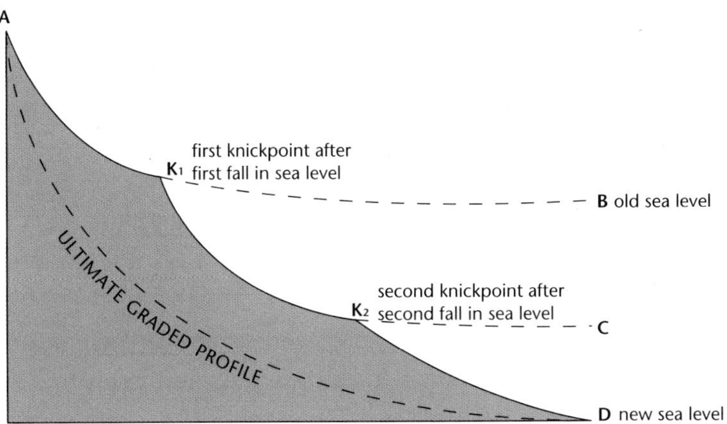


Fig. 11.34

AB original profile, graded to sea level at B.

$AK_1C$ profile attained after fall of sea level to C.

$AK_1K_2D$ profile attained after second fall of sea level.

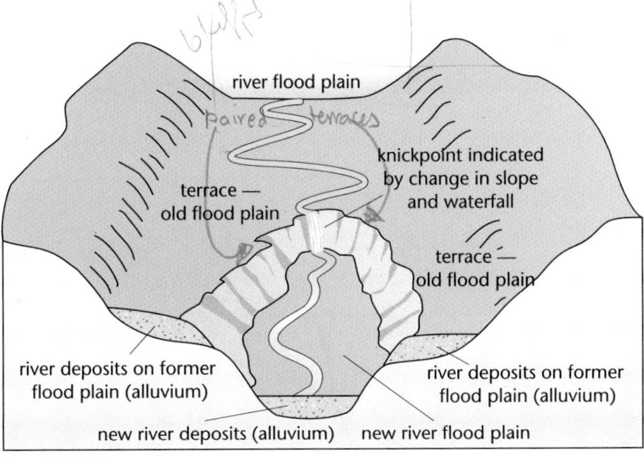


Fig. 11.35

the old river profile and the new profile; it may be marked by rapids or a waterfall. Such features indicate the presence of a hard outcrop of rock. The knickpoint may linger here for some time until the feature disappears altogether. Thus the knickpoint recedes upstream at a rate which depends upon the resistance of the rocks.

Some Irish rivers experienced several stages of rejuvenation and display several knickpoints, each separated by a gently sloping section of the long profile graded to the corresponding sea level. The Barrow, Nore, Suir, Erne and Shannon all display polycyclic (many cycles) profiles.

## 2. Terraces

Because of rejuvenation, a river's energy is renewed, and downcutting occurs. The river sinks its new channel into the former flood plain, leaving the former valley floor well above the present river level. A new valley is gradually widened by lateral erosion and the process of divagation so that remnants of the former flood plain form **terraces** on either side. These are called **paired terraces**. If rejuvenation occurs for the second time, a second set of terraces will form at a lower level than the first. These are called **stepped terraces**. Each set of terraces may frequently be matched with corresponding knickpoints in the long profile.

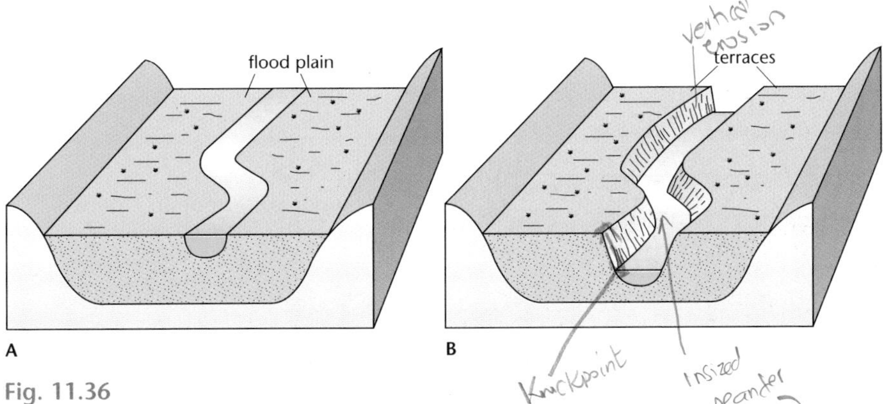


Fig. 11.36

Terraces formed by a river cutting downward into its own flood plain deposits.

(A) River deposits thick flood plain deposits.

(B) River erodes its flood plain by downcutting. Old flood plain surface forms terraces.

(C) Lateral erosion forms new flood plain below terraces.

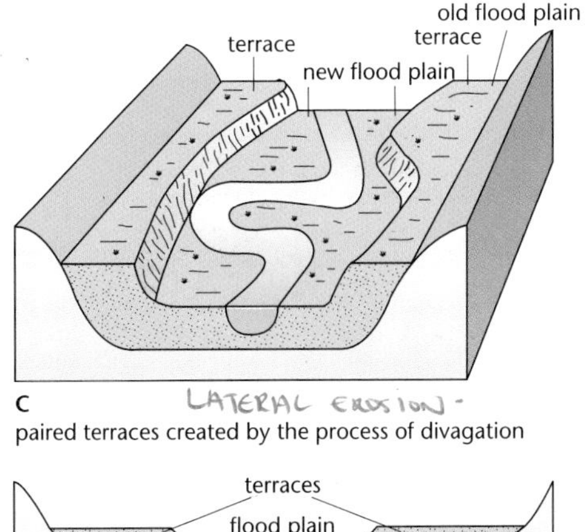


paired terraces created by the process of divagation

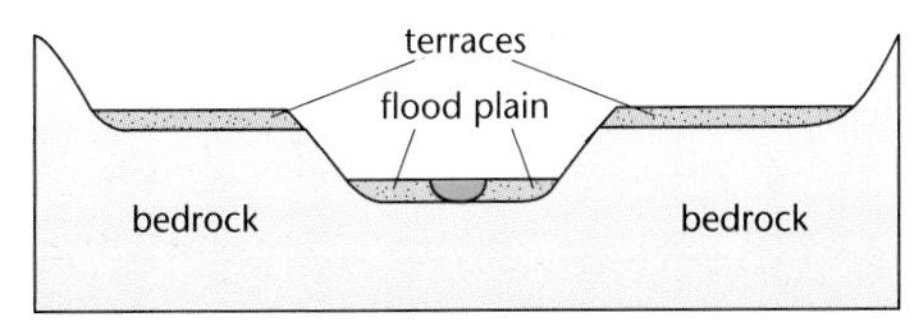


terraces may be erosional benches cut in rock

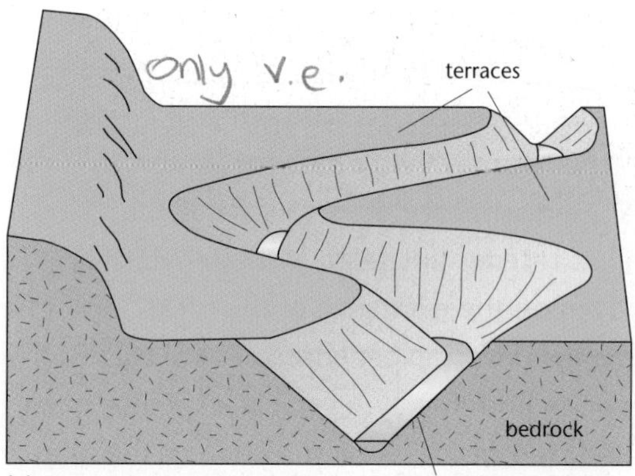


(a) symmetrical V-shape

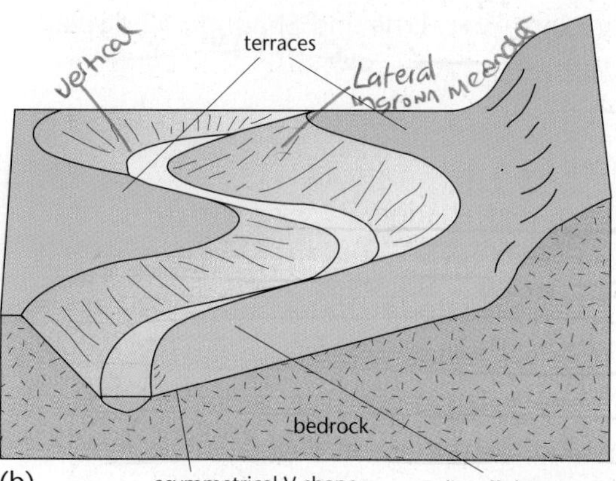


(b) asymmetrical V-shape slip-off slope

Fig. 11.37

### 3. Incised Meanders

Because of rejuvenation, downcutting is sometimes severe, causing a deep erosion of the river's channel into the alluvium and the bedrock while it maintains a winding course.

Incised meanders may be classified into two types.

1. **Entrenched meanders** — In this case, the valley sides are steep and symmetrical in cross-profile due to vertical erosion alone (Fig. 11.37a).
2. **Ingrown meanders** — Here, both vertical and lateral erosion continue after rejuvenation. Thus meanders develop with one valley side steep and the other more gentle, producing a more open valley with slip-off slopes (Fig. 11.37b).

In Ireland, as the granite block of southern Leinster was slowly raised, the rivers Barrow, Nore and Slaney eroded across it as fast as it rose. By maintaining their original courses, these rivers now find themselves leaving open country, meandering through a steep-sided gorge and emerging into open country once more.

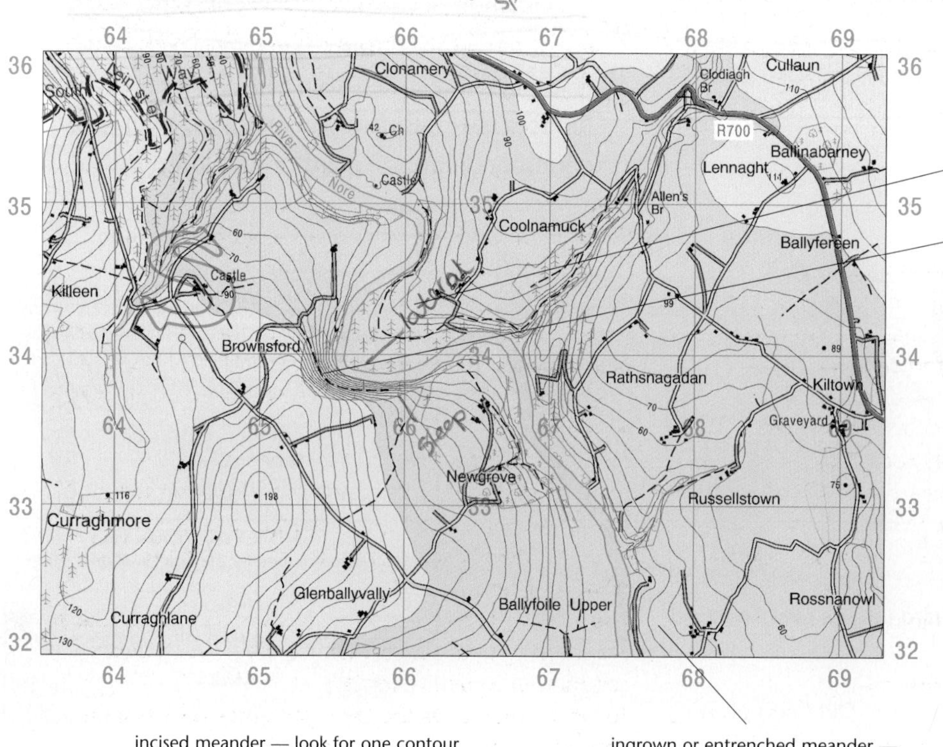


gentler slope of slip-off spur

steep slope shows lateral erosion

these features identify an asymmetrical V-shape and so indicate an ingrown meander

**Fig. 11.38**
Contour pattern of an ingrown meander. A section of the gorge on the River Nore in Co. Kilkenny in Ireland. The meandering course indicates that at one time this river flowed through an old or mature valley. As the surrounding land was uplifted, the river retained its meandering course. It now flows through a gorge with relatively steep sides.

Fig. 11.39

Study the photograph of Carrick-on-Shannon in the Shannon river basin in Co. Leitrim, Fig. 11.39. Then answer the following:

1. Locate Carrick-on-Shannon in Ireland in your atlas. Then examine the photograph and identify the stage of maturity of the river. (Remember the Shannon River flows very quickly between Castleconnell and Limerick city — 230 km to the south.)

   Now let us examine the river together and state the information presented to us in the photograph in a logical manner. Delete the incorrect words.

   Then rewrite the article in your copy.

   The river is winding/flowing 'snake-like' over a flat/sloping region.

   These huge twists in the river are called oxbow lakes/meanders.

   The brown patches along the river edge are levees/callows.

   This suggests that the river overflows/builds up its banks after a period of light/heavy rain. The brown colours are caused because the water floods/is dirty in these areas from pollution/regularly.

   Such are the characteristics of an/a old/mature river valley.

   So the River Shannon is in its stage at this point in its course. Does this make sense? Remember, downstream near Limerick, the Shannon flows quickly. This type of rough and fast water is found in the course of a relatively young river.

   Now concentrate again on the question.

   For the river to be at this stage of maturity something must have happened to alter the speed of the water channel. Identify and explain the interfering process. (Hint — refer to p. 177.)

2. The brown patches along the river's edge are precious to water-loving bird life. To what extent are such areas common in Europe? Explain.

3. To what extent does the town of Carrick-on-Shannon depend on the river for its survival? Use evidence from the photograph to support your answer.

Fig. 11.40
Carrick-on-Shannon in Co. Leitrim

Examine the photograph of the Shannon River, Fig. 11.40 and answer the following:
NB Do NOT use tracing paper when answering this question.

1. (a) The river has had an impact on the morphology (shape/layout) of this settlement. Examine one way in which the photograph provides evidence of this.
   (b) Residential and transport functions are key elements well illustrated on this photograph. With the aid of a sketch map describe in detail the main types of (i) residential land use, (ii) transport land use in evidence on the photograph.
2. Rivers are regularly used for commercial as well as recreational purposes. With reference to the photograph, discuss this statement.
3. Give reasons why the bridge which crosses the Shannon River was constructed on this site to the right of the photograph and not constructed in the foreground of the photograph.

Fig. 11.41

Study the photograph of the River Shannon at Roosky, in Co. Leitrim, Fig. 11.41. Then do the following:

1. At what stage of maturity is the river?
   Explain your answer fully, using evidence from the photograph.
2. The river seems to have a major influence on the economic and social life of this area. Discuss.
3. The fields vary in colour. Suggest a reason/reasons for this.
4. The local industrial activities have taken some precautions to reduce pollution of the area. Discuss this fully.
5. Suggest what kind of economic activity the large factory is involved in. Discuss this fully in class.
6. Suggest why a village settlement developed at this location.
7. In which direction on the photograph is the river flowing?

## Cross-sections

Doing cross-section exercises regularly, while studying Ordnance Survey maps, has the following advantages:

- It gives a student some concept of particular contour spacings relative to their corresponding slopes on the ground.
- It provides a side-view (profile) of the relief (shape) of a particular stretch of land.
- It develops and nurtures the following skills:

— pencil work skills
— recognition of detail
— neatness
— accuracy and speed in a given time-frame

a cross-section gives us a side-view of the relief or shape of a landscape

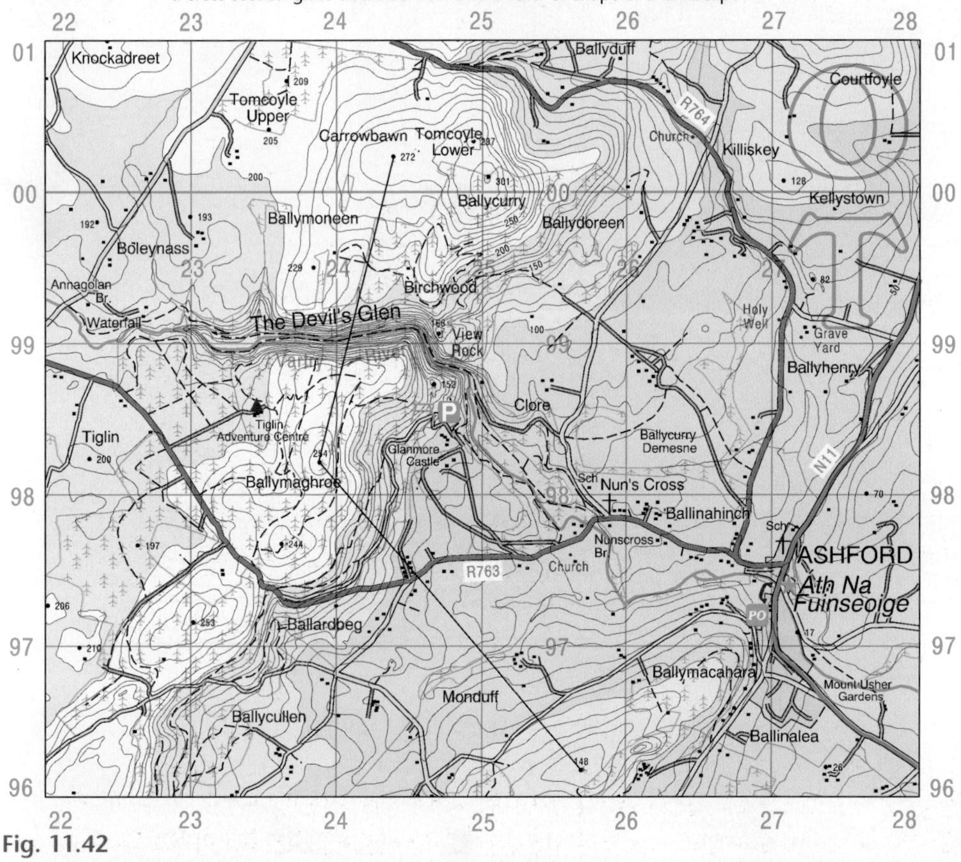


Fig. 11.42
Cross-section from Carrowbawn 272 m to Ballymaghroe 254 m to spot height 148 m at grid reference T 257 962 looking east

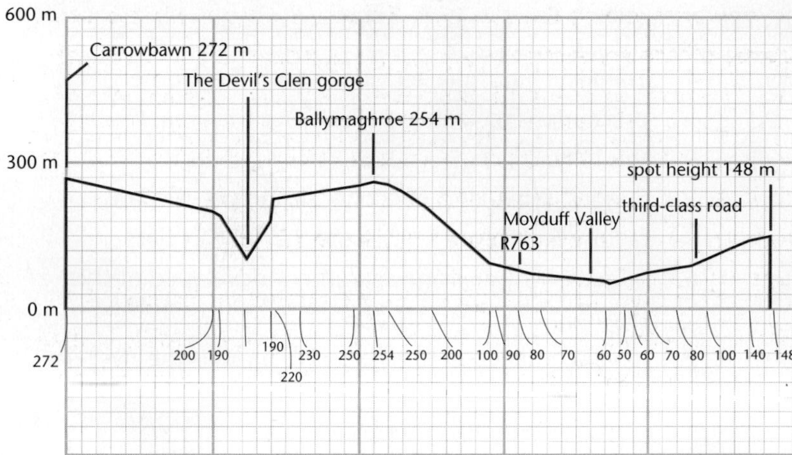


Fig. 11.43
Cross-section from Carrowbawn 272 m to Ballymaghroe 254 m to spot height 148 m at grid reference T 257 962 looking east

Procedure to be followed in completing a cross-section:

1. Draw a light line joining the places mentioned.
2. Place a long strip of white paper along the bottom edge of this line.
3. Place two fingers on the paper strip to avoid movement.
4. Use a sharp pencil to mark off contours, upslopes (hilltops), downslopes (valleys), rivers and other relevant features.
5. Transfer the marking from the paper strip to squared paper and finish the cross-section.
6. Give your section a title.
7. Label the cross-section. Name the important landforms (features) on the section.
8. Give both vertical and horizontal scales.

1. Draw a labelled cross-section on the Avoca region map, p. 175, from spot height 169 m grid reference T 186 854 to the Church at grid reference T 213 838.
2. Carefully examine the contour pattern of the Devil's Glen grid reference T 240 990, Fig. 11.42, p. 182.
   Suggest one reason why this valley is exceptionally steep-sided (gorge) (hint — glacial lake). (See 'Landform: Glacial Spillway' on p. 208.)

## Patterns of Drainage

### Dendritic Pattern

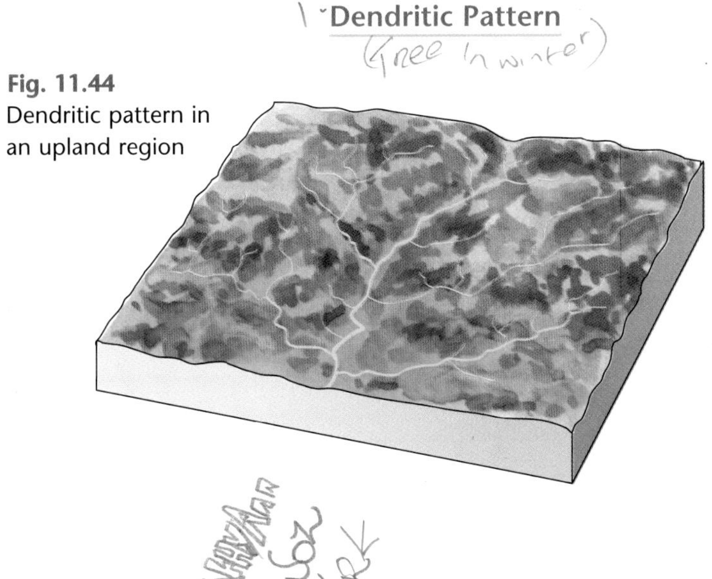

Fig. 11.44
Dendritic pattern in an upland region

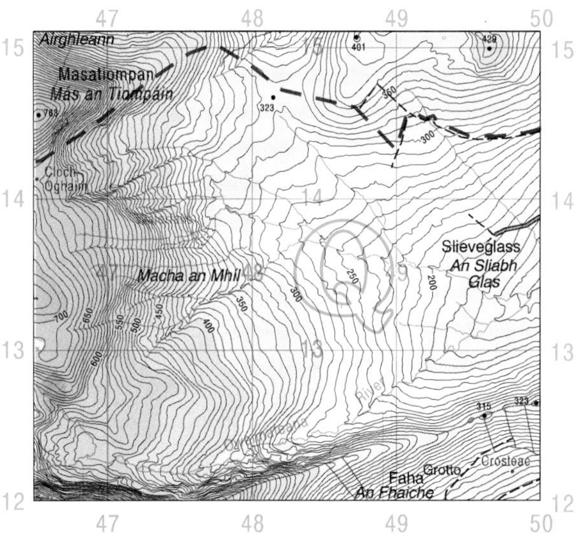


Fig. 11.45
Dendritic pattern in a highland area

*Dendros* is the Greek word for a tree. Thus a dendritic pattern is **tree-shaped**. On a newly formed landscape, the first streams and rivers will flow according to the fall of the land. Their direction is thus consequent upon that slope, so these rivers are called **consequent streams**. As they develop, tributaries flow towards these main valleys, joining the parent river **obliquely**, with minor tributaries joining them in turn. If the rocks in the river's basin have **equal resistance** to erosion, each consequent stream will then become the centre of a converging stream pattern. This is called **dendritic drainage**. Every river appears to consist of a main trunk, fed from a variety of branches, each running into a valley proportional to the river's size.

### Trellised Pattern

When tributaries flow into the main river at **right angles** a trellised pattern is formed. If the land surface consists of rocks of **varying degrees** of resistance — in other words, if the land is composed of bands of hard and soft rocks at right angles to the consequent stream

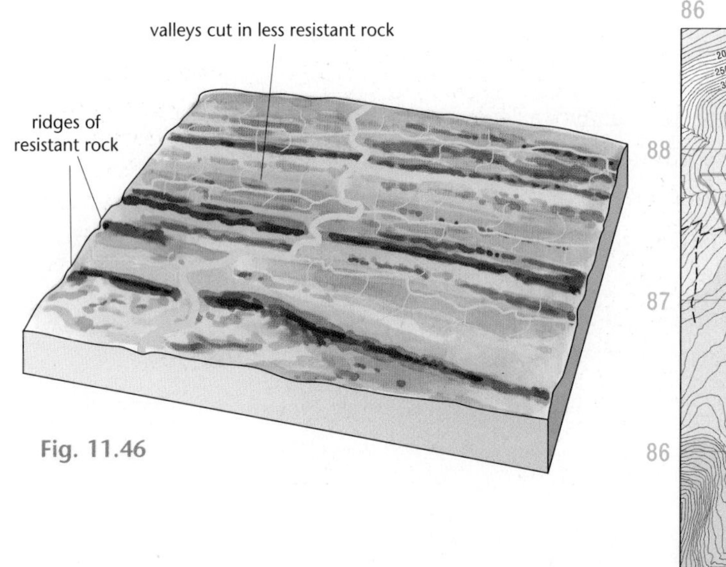


Fig. 11.46

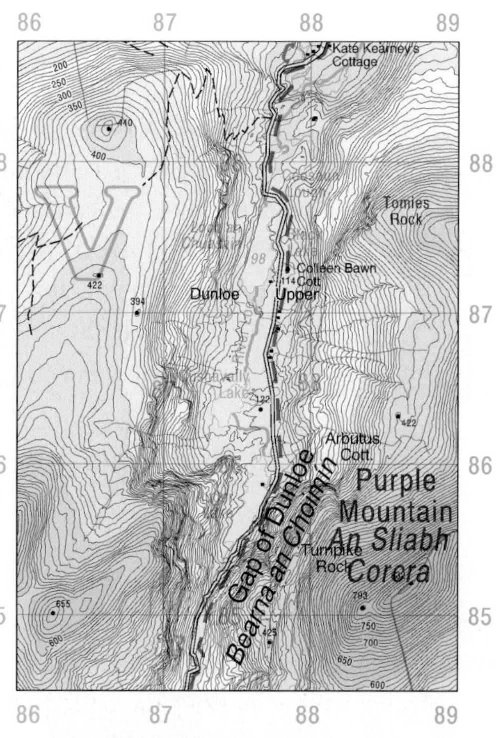


**Fig. 11.47**
Trellised drainage pattern in a glaciated valley. When tributaries flow into the main river at right angles, a trellised pattern is formed. This pattern occurs when valleys and ridges run parallel to each other.

— streams called **subsequent streams** will develop along the softer bands of rock. These subsequent streams will form broad valleys through the process of headward erosion and will be flanked on either side by parallel ridges. Tributaries will develop and flow from these ridges to join the subsequent streams or streams at right angles (Fig. 11.46). Tributaries to the subsequent streams are called **secondary consequents** and **obsequents** or **anti-consequents**. River capture frequently occurs in areas of trellised drainage, such as on the Blackwater River at Cappoquin in Munster.

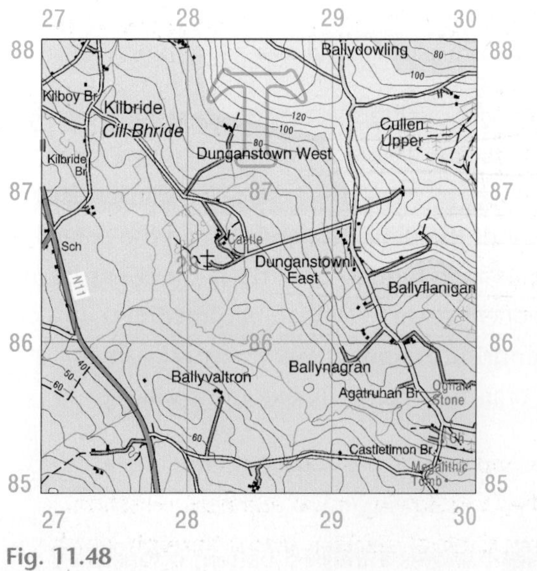


**Fig. 11.48**
Rectangular patterns are similar to trellised patterns, except that both the main streams and the tributaries follow courses with right-angled bends. Lines of weakness such as faults and joints may be responsible for such patterns.

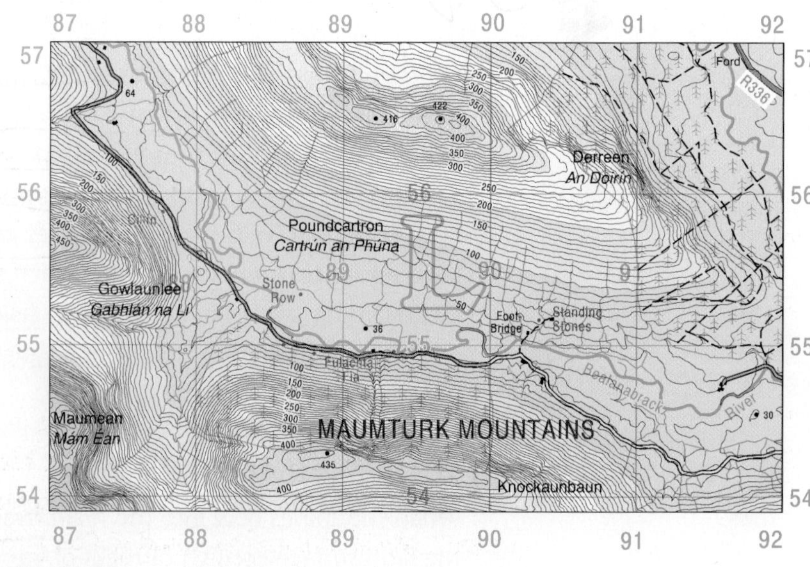


**Fig. 11.49**
Trellised pattern in a river valley

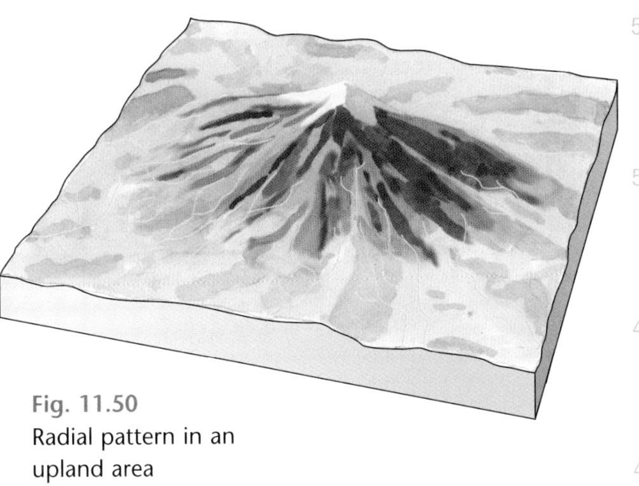

Fig. 11.50
Radial pattern in an upland area

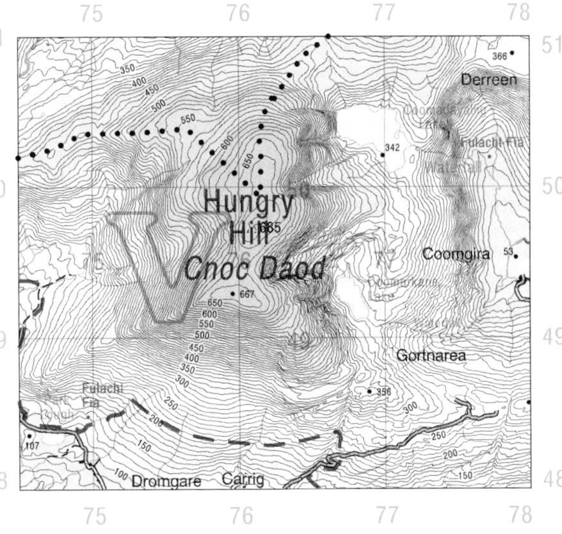


Fig. 11.51
Radial pattern in an upland area. When several streams flow outward (radiate) in all directions from a mountain or hill, they form a radial pattern of drainage.

## Radial Pattern

Rivers which radiate outward from a mountain form a radial pattern. This is best displayed in well-defined circular or oval-shaped upland areas. Some of these rivers may in fact display a different drainage pattern from another, but together they may radiate outward (north, south, east or west) from a central elevated area. They all share a common watershed at their highest source streams.

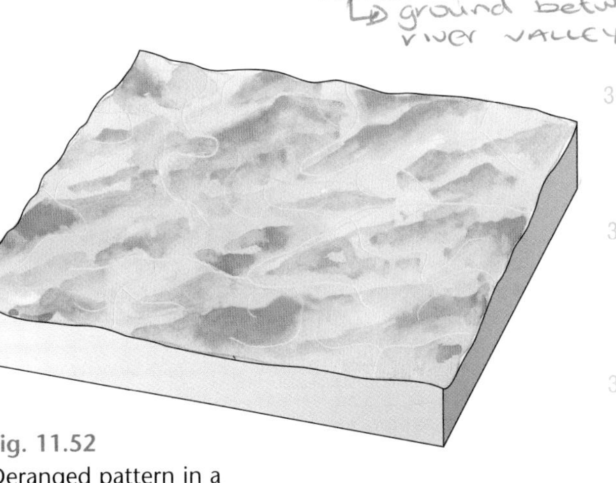

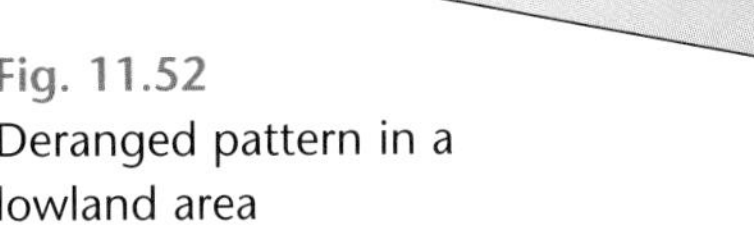

Fig. 11.52
Deranged pattern in a lowland area

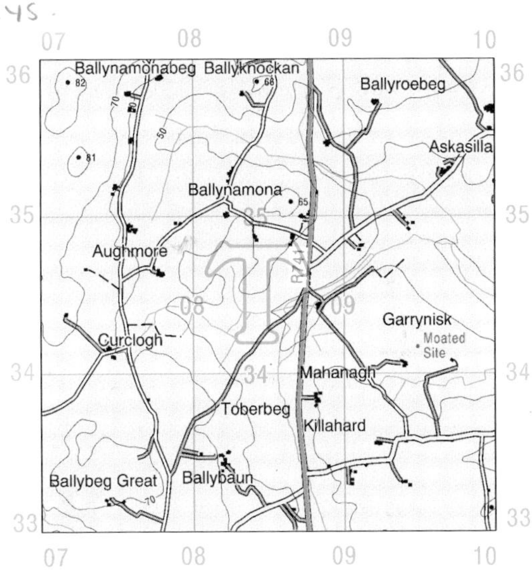


Fig. 11.53
Deranged pattern in a lowland area. This is a river pattern which generally develops in a lowland area where glacial drift exists. Rivers have a chaotic appearance.

## Deranged Pattern

Deranged drainage generally develops in a lowland area. Rivers have a chaotic appearance, with streams intersecting each other and flowing in no apparent direction. It usually develops as a result of widespread deposition of glacial material through which post-glacial streams have had to find a route. An example of deranged drainage can be found on the coastal plain west of Cahore Point in Co. Wexford, in Ireland.

## Case Study: Munster Rivers — Bandon, Lee, Blackwater in Southern Ireland

On a newly formed landscape, the first streams and rivers will flow according to the fall of the land. Their direction is thus consequent upon that slope, so these rivers are called **consequent streams**. If the **rocks do not vary** in resistance over such a landscape, tributary streams will join the consequent stream and a **dendritic drainage pattern** will form.

If, however, the land surface consists of **bands of hard and soft rock**, other patterns will develop. If the bands of more and less resistant rock occur at right angles to the consequent stream, subsequent streams will develop along the softer bands of rock and broad valleys will be formed. The consequent stream will continue to erode vertically and steep-sided gaps will be cut into the harder bands of rock. A **trellised pattern** of drainage then develops (Figs. 11.48 and 11.49, p. 184).

In Munster, consequents developed on a north–south sloping landscape. Having eroded their south-flowing channels, they encountered east–west valleys and ridges. As explained above, they initially cut steep gaps in the sandstone ridges and then developed east-flowing tributaries along the softer limestone valleys.

By the process of headward erosion, the tributaries captured the headwaters of the south-flowing consequents and caused the noticeable right-angled bends on such rivers as the Suir and Blackwater. This process is known as **river capture**. In such an instance, the steep-sided gap becomes a **wind gap** and the tiny stream which occupies it is known as a **misfit stream** (Fig. 11.55).

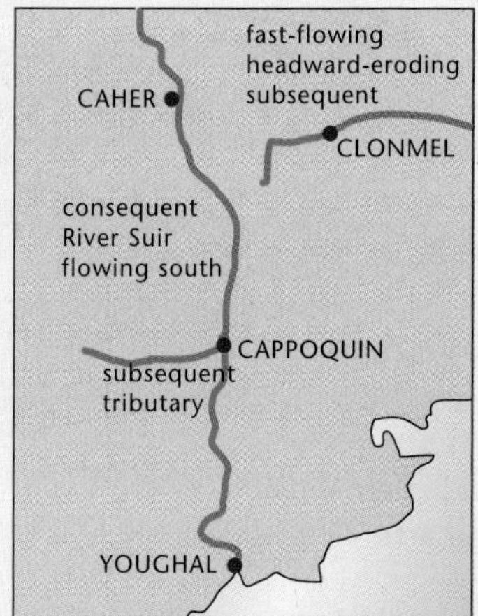


**Fig. 11.54**
Process of river capture

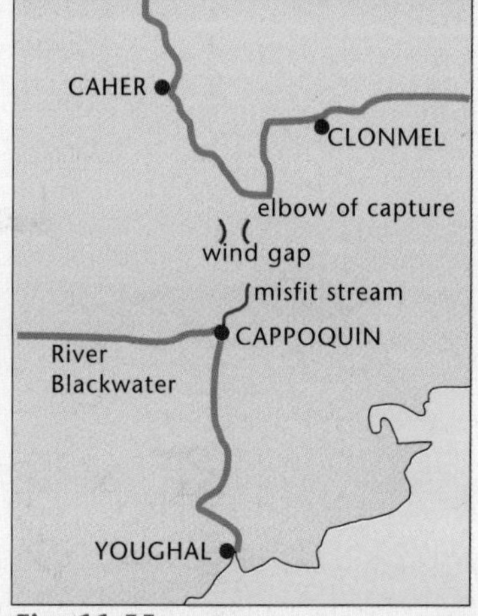


**Fig. 11.55**
Right-angle bend of river capture

**How to Describe the Drainage of an Area**

1. Are there many or few rivers? Name some of the largest rivers.
2. What patterns of drainage do they form?
3. In which direction(s) do they flow?
4. Are they fast or gently flowing?
5. Is the land low-lying and liable to flooding?
6. Is the land upland with fast-flowing streams?
7. Are there many lakes in the area? Name some.

**Each River**

1. In which direction does it flow?
2. Has it many or few tributaries? Name some.
3. What pattern of drainage do the tributaries form?
4. At what stage of maturity are the tributaries? Explain.
5. At what stage of maturity is the main river? Explain.
6. Are there lakes on the river? Name some.
7. Is it a wide or narrow valley or is it on a plain?
8. Is it a wide river?

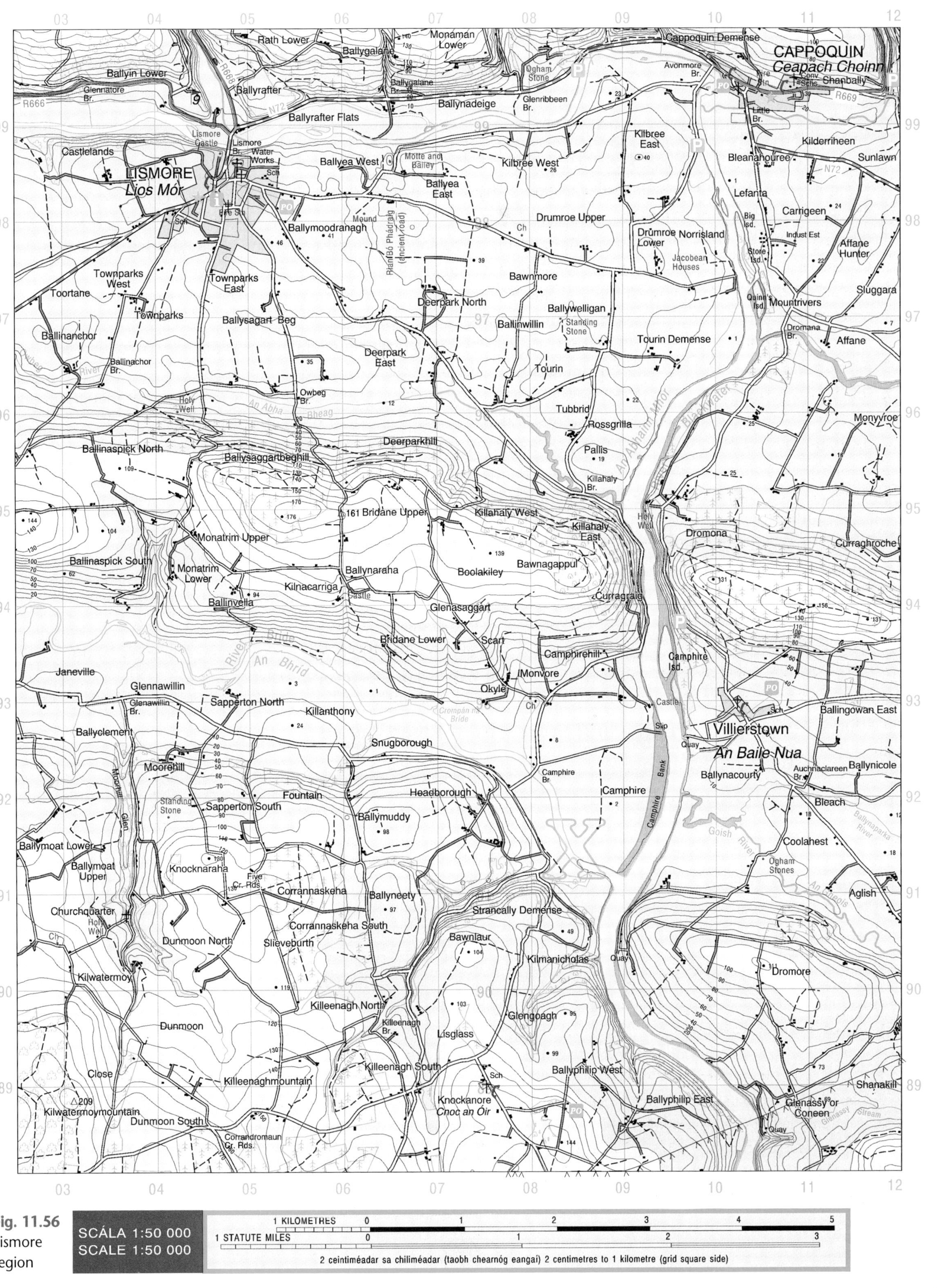


Fig. 11.56 Lismore region

LANDSCAPES OF THE WORLD

Class Activity

Carefully examine the Ordnance Survey map of the Lismore region, Fig. 11.56, p. 187. Then do the following:

1. Describe the patterns of drainage which exist on the map. Use diagrams to explain your answer. Give one example, from the map, of each pattern.
2. Explain fully how the dominant pattern of drainage on the map evolved over time (see 'Case Study: Munster Rivers', p. 186).
3. With the aid of a sketch map, explain how the physical landscape has influenced the communications network of the area shown on the map.
4. In this section of its course, the River Blackwater displays some features that are typical and others that are untypical of that stage of development.
   (a) Comment on the stage of maturity of the river in the map extract. Use evidence from the map extract to support your answer.
   (b) (i) Identify one feature that is typical of the stage of development shown. With reference to the map extract, locate the feature and give a detailed account of its formation.
   (ii) Identify one feature that is untypical of that stage of development. With reference to the map extract, locate the feature and give a detailed account of its formation.
5. On the Inishowen map extract (Fig. 8.2, p. 107) describe the course of: (a) the Owenerk River; (b) the Ballyhallan–Clonmany rivers.

## People and Processes in River Valleys

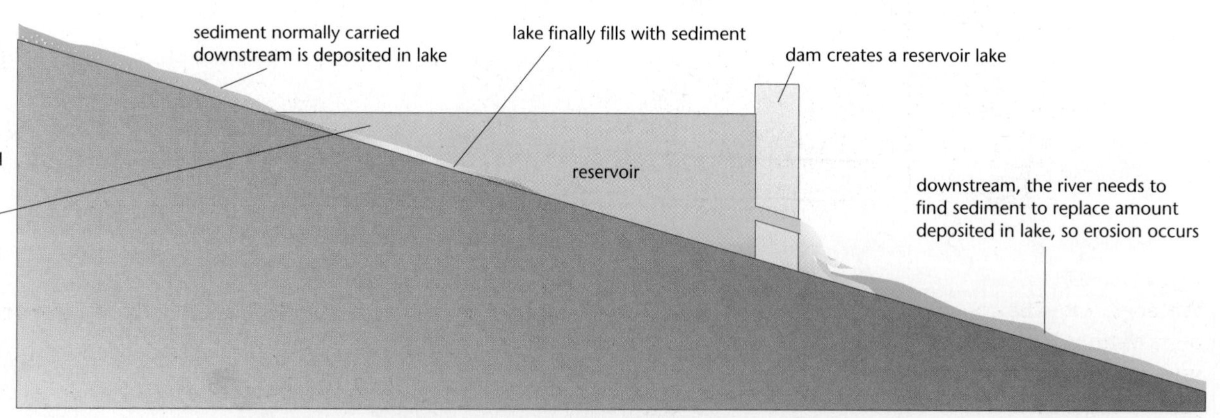
sediment normally carried downstream is deposited in lake
lake finally fills with sediment
dam creates a reservoir lake
upstream — natural flow of water downslope is prevented by constructing a dam across the valley
reservoir
downstream, the river needs to find sediment to replace amount deposited in lake, so erosion occurs

Fig. 11.57

### Dam Construction — Positive

Dams are constructed across a river's channel for the purpose of generating hydroelectricity and creating reservoirs for irrigation and urban water supplies. Such dams interrupt the natural flow of rivers and reduce the ability of rivers to carry sediment downstream from their sources to the sea or flood plains. However, these activities are generally regarded as positive influences. For example, let us look at a case study of hydroelectric power generation in a river valley.

RIVER PROCESSES & LANDFORMS

## Hydroelectric Power Generation

Hydroelectric power is generated from the potential energy of water in rivers as they flow downslope to the sea. Water-power has been used in small ways for thousands of years, but only in the twentieth century has it been widely used for generating electricity. Unlike coal and oil, hydropower cannot be used up; it is a renewable resource.

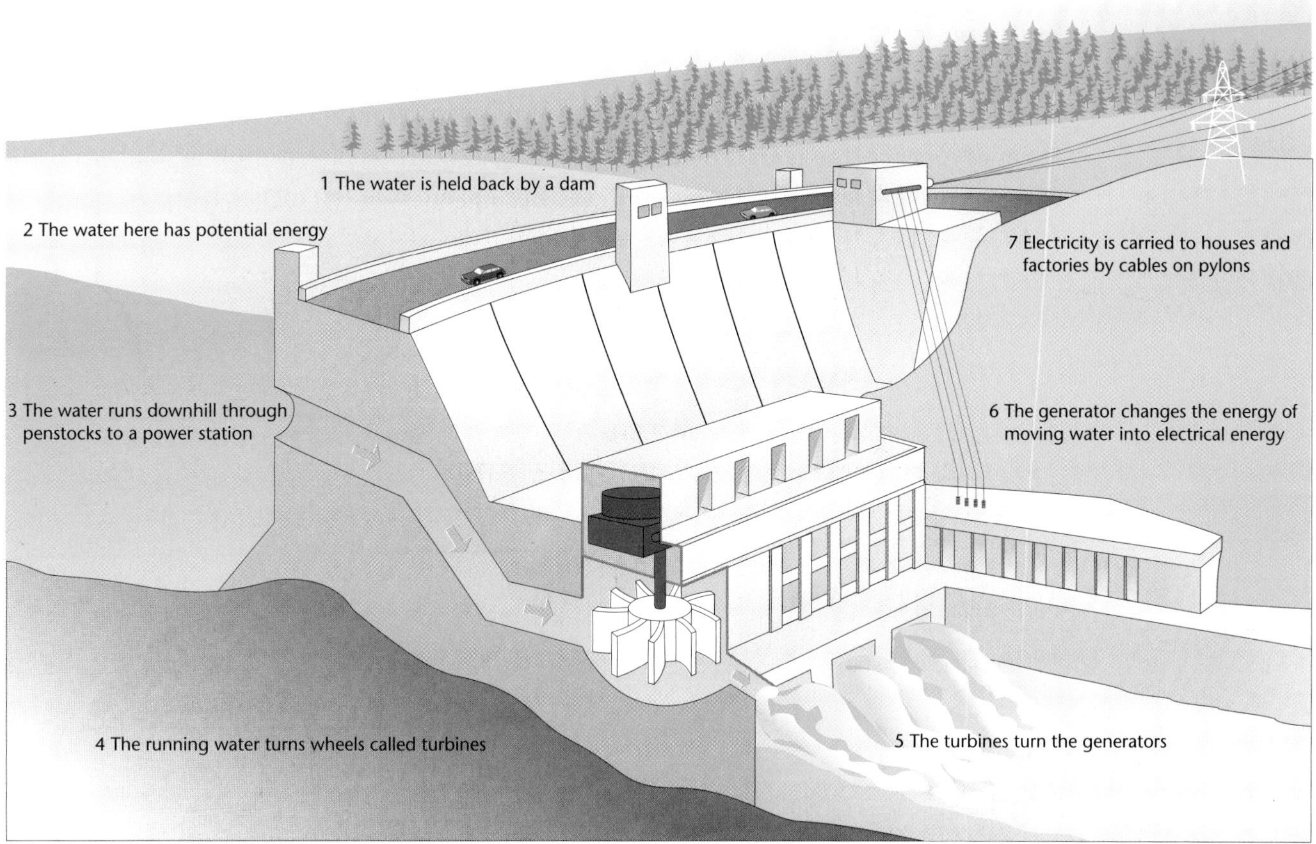
1 The water is held back by a dam
2 The water here has potential energy
3 The water runs downhill through penstocks to a power station
4 The running water turns wheels called turbines
5 The turbines turn the generators
6 The generator changes the energy of moving water into electrical energy
7 Electricity is carried to houses and factories by cables on pylons

Fig. 11.58
**Water-power.** The energy of moving water has been used for thousands of years. People built waterwheels along rivers as long as 2,000 years ago.

The energy of moving water is now used to produce electricity in **hydroelectric power stations**. Hydroelectricity provides over 5 per cent of the energy used in the world today. Because the water comes from rain or melting ice, it never runs out. Only countries that have lots of water can produce electricity this way. Scandinavia, North America and Russia are able to produce large amounts of their electricity from hydroelectric power.

The fast-flowing streams from the Alps, such as the Adige, have been harnessed to develop hydroelectric power. Huge pipes carry water from the mountain benches down to the hydroelectric stations at their base. These high heads of water are ideal for electricity generation. Factories concerned with electrometallurgical, chemical and textile industries are built near the power stations.

## Dam Construction — Negative

### Dambursts

A major fear of dam construction is that of damburst.

Some dams have been shattered on purpose. During the Second World War, some dams in industrial areas were 'blown up' by Allied bomber planes to reduce the efficiency of German factories. However, some dams have collapsed due to natural causes.

In 1852 Johnstown in Pennsylvania, in the USA, was devastated by water from a reservoir when a dam on the Conemaugh River burst. Over 2,000 citizens died.

## Construction of Levees

### Man-made Levees

The term **'levee'** comes from the French word *lever*, which means 'to raise'. In the United States of America, the term is used to describe walls or dykes built along the southern part of the Mississippi River. The levees on the Mississippi are over ten metres high.

Fig. 11.59
How a dyke or levee fails

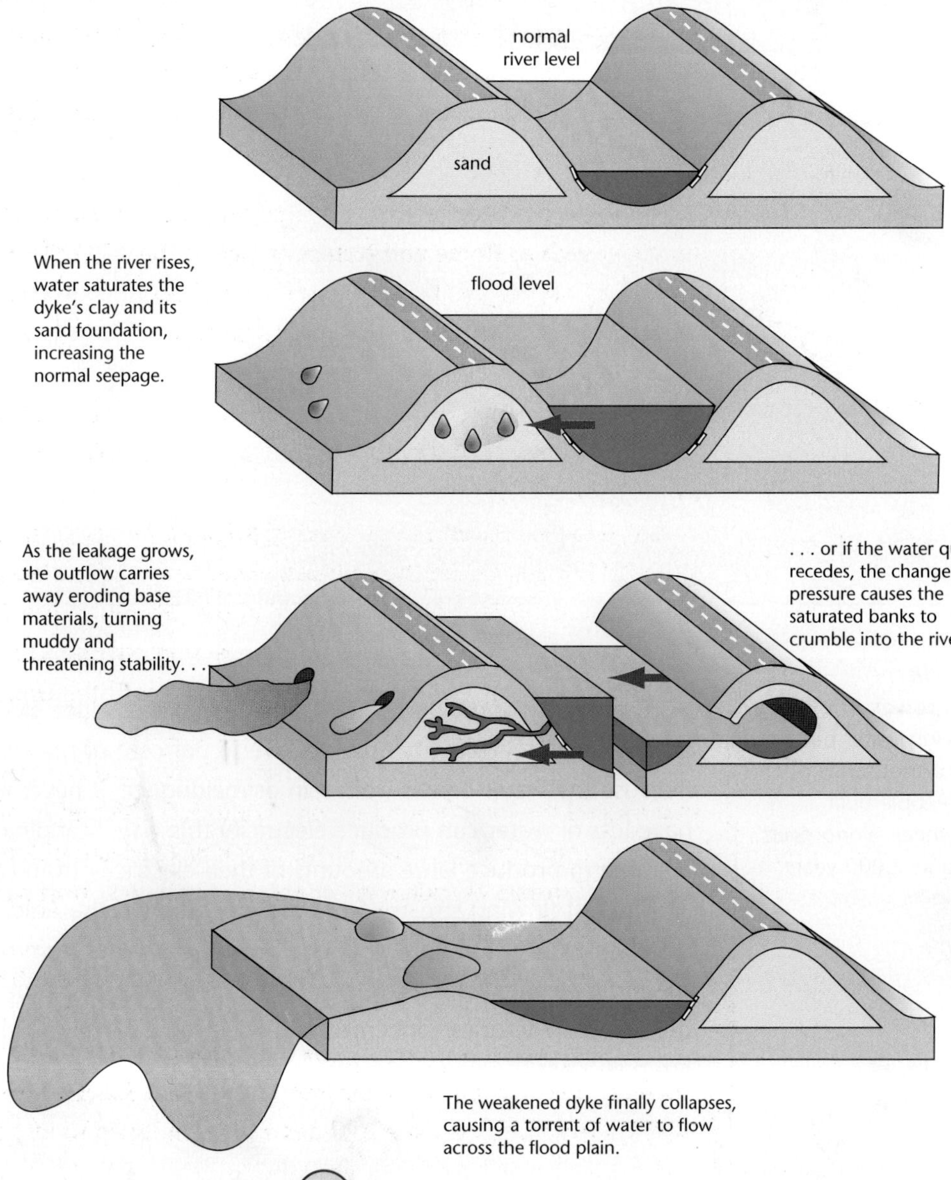

normal river level

sand

When the river rises, water saturates the dyke's clay and its sand foundation, increasing the normal seepage.

flood level

As the leakage grows, the outflow carries away eroding base materials, turning muddy and threatening stability. . .

. . . or if the water quickly recedes, the change in pressure causes the saturated banks to crumble into the river

The weakened dyke finally collapses, causing a torrent of water to flow across the flood plain.

## Natural Levees

Some levees are created naturally. In such cases flood waters quickly drop their coarse sediment on the banks of a river as the flood waters overflow onto the surrounding flood plain. In time, wide, gently sloped raised banks form, which help to retain further flood water. Most levees, however, are man-made.

Some authorities object to the building of levees that enclose a river so much that the water is high above the surrounding countryside. This makes floods even more dangerous when they occur. If a levee were to break under these circumstances, a 'wall' of water would rush across the countryside, smashing everything in its way. Experts who oppose high-level levees believe that regulating floods by headwater control (building dams upstream) is better than attempting to regulate them with levees.

Over the past few centuries and especially over the past century, people have encroached onto flood-prone areas. Instead of avoiding such areas, people now try to tame nature by building dykes and levees to retain flood waters. Vast areas of existing farmland, towns and cities now lie below water when rivers are in flood.

Much of the lower Rhine flood plain, especially in the Netherlands, has been reclaimed from the Rhine and the sea. This landscape is below both river level and sea level. In 1995 hundreds of thousands of people were evacuated as their water defences were severely tested.

## Irrigation and Water Transfers

Throughout the Mediterranean, irrigation practices support the market gardening industry for cities such as Rome and Naples as well as the 150 million tourists who visit the region annually. The long, hot and dry summers of countries such as Greece, Italy, southern France and Spain in Europe, as well as parts of north Africa, mean that irrigation plays a vital role in their way of life.

This practice is not new, it has continued in various areas for thousands of years. Even the Roman Empire, based on urban control centres (towns), transferred water from far away by aqueducts.

Los Angeles city and large areas of irrigated farmland in southern California are supplied by water from the Colorado River below Lake Mead; the Sacramento River in northern California and Owens Valley in the foothills of the Sierra Nevada also provide a water supply. Ninety per cent of this water transfer is used to irrigate the world's richest cash crop farmland, which produces one-quarter of the United States' fruit and vegetables, as well as cotton, rice, soya beans and sugar beet.

What happens when the government or the local population demand water supplies that exceed the local ground supply? Water is transferred from elsewhere to meet this need.

But it comes with a price. These water transfers are bitterly opposed by residents and especially farmers of source areas who fear that their own rich farms and wetlands will dry up as a consequence of such practices.

Water transferred from the rivers Amu Darya and Sry Darya is used to irrigate vast desert areas that surround the Aral Sea in central Asia (see 'A Sea Turns to Dust', p. 424). As a consequence, this once waste land now produces cotton, rice and vegetables. However, because of the water transfer, the Aral Sea itself is shrinking. Some places that were on the seashore some decades ago are now up to fifty kilometres from the sea.

**Fig. 11.60**
Cork city

Study the photograph of Cork city and then answer the following:

1. The River Lee which flows through the city appears to have a number of channels. The main channel in the left foreground splits into two channels while another waterway appears from the centre foreground. All of these waterways enter the Lee estuary in the centre background. Does this suggest anything about the site of the city?
2. The island, its land and buildings in the centre of the photograph appear to have a distinct character of their own. Account for this using evidence from the photograph.
3. The waterway entering the photograph in the centre foreground differs from the other water channels. Identify this difference and suggest a reason for it.
4. The city in the photograph appears to have retained its 'old character' even though modern development has taken place over the past few decades. Can you suggest some reasons for this, using evidence from the photograph to support your answer?

Study the photograph of Carlow town. Then do the following:

1. Riverside towns and cities are often subjected to flooding after spells of torrential rain. Explain fully, with the aid of diagrams, why this process occurs.
2. Carlow, in the photograph, is flooded by the River Barrow. Explain
   (a) the immediate effects and
   (b) the long-term effects of
   such flooding on the town.
3. The Normans, generally, were excellent developers of fortified settlements. In the case of Carlow, did the Normans choose a poor or good site for their settlement? Discuss, using evidence from the map to support your answer.
4. Refer to the other photograph of Carlow on p. 432. Then state, using evidence from both photographs, if recent developers have taken measures to cope with such flooding in the future.

**Fig. 11.61**
An aerial view of Carlow town after the River Barrow burst its banks

## Leaving Certificate Questions on Rivers

### Ordinary Level

**1998**

'Rivers have helped to shape the Irish landscape.'

(i) Name **THREE** landforms found along the course of a river and, with the aid of a diagram, describe how each was formed.
Name a specific location where the feature may be found. **(60 marks)**

(ii) Explain briefly **TWO** ways in which rivers are of value to man. **(20 marks)**

**1996**

V-shaped valley, Oxbow Lake, Delta, Levees, Interlocking spurs, Waterfall.

(i) In the case of **EACH** of the above features, found along the course of a river, state whether it is formed by erosion or deposition. **(18 marks)**

(ii) Select any **THREE** of these features and, with the aid of a diagram, describe how each was formed. **(48 marks)**

(iii) 'Large-scale flooding by rivers has caused enormous problems for local communities.'
Discuss this statement. **(14 marks)**

**1995**

**The Physical World** — Explain the following:

(i) Erosion by rivers has resulted in the formation of distinctive landscape features. **(40 marks)**

**1994**

(i) Erosion and deposition by rivers have helped shape the landscape. Name **THREE** landforms which result from river action. **(15 marks)**

(ii) For **EACH** landform selected, describe and explain, with the aid of a diagram, how it was formed. **(45 marks)**

(iii) 'Rivers are of great use to people and yet are often abused by people.'
Explain this statement. **(20 marks)**

**1993** — Explain the following:

(iii) The point where a river enters the sea can be marked by a number of distinctive landforms. **(40 marks)**

**1991**

(i) Explain, with reference to **THREE** typical landforms, how rivers help to shape the landscape. **(60 marks)**

(ii) Examine briefly, using examples which you have studied, how human activities change the operation of natural processes in river valleys. **(20 marks)**

**1989**

(i) Describe and explain the formation of any **THREE** landforms which can be found in a typical river valley. **(60 marks)**

(ii) A change in the relative levels of land and sea may cause a river valley to be rejuvenated.
Explain briefly **TWO** ways in which the landforms of the valley may be changed in such an event. **(20 marks)**

**Higher Level**

**1998**

(i) With reference to processes of erosion **and** to processes of deposition, examine **three** ways in which rivers shape the Irish landscape. **(75 marks)**

(ii) Examine **one** example of how human societies have always attempted to manage or control the natural processes which operate in river valleys. **(25 marks)**

**1997**

(i) Discuss, with reference to **three** characteristic landforms, the natural processes at work in river valleys. **(75 marks)**

(ii) Flooding in river valleys can be worsened by human activity. Examine one example of this. **(25 marks)**

**1994**

(i) With reference to erosional processes **and** to depositional processes, examine some of the ways in which rivers shape the landscape. **(75 marks)**

(ii) Human societies have always sought to control or manage river processes. Discuss this statement briefly, using **one** example which you have studied. **(25 marks)**

**1996**

(i) With reference to processes of erosion and processes of deposition, examine **THREE** ways in which rivers shape the Irish landscape. **(75 marks)**

(ii) Examine briefly **TWO** examples of human management of rivers. **(25 marks)**

**1992**

(i) Explain how the natural physical processes which are active in river valleys help to shape the landscape. **(75 marks)**

(ii) Examine the effect which a drop in base level can have on a mature river system. **(25 marks)**

**1990**

(i) Describe and explain, with reference to **THREE** typical landforms, how rivers shape the surface of the earth. **(75 marks)**

(ii) Examine using appropriate examples, how human activities interact with natural processes in river valleys. **(25 marks)**

**1988**

(i) Explain, with reference to **THREE** characteristic landforms, how rivers help to shape the earth's surface. **(75 marks)**

(ii) Examine briefly **TWO** ways in which a change in the relative levels of land and sea might affect the physical processes active in a river valley. **(25 marks)**

# Ice, Glacial Processes and Landforms

Fig. 12.1
Stein Glacier, Switzerland. Carefully examine the various aspects of this glacier as it creeps downhill.

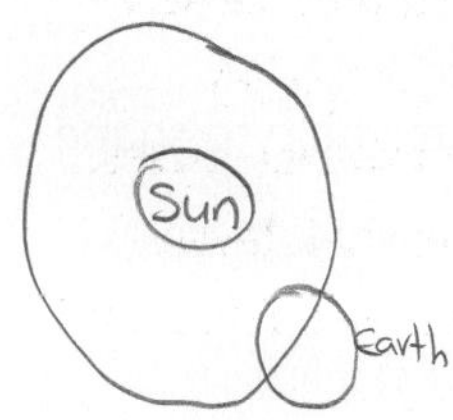

## Causes of Ice Ages

Ice ages may have occurred due to changes in the orbital motion of the earth. The coldest of the periods of ice may have occurred when three variations in the earth's orbital behaviour took place at the same time.

1. As the earth orbits the sun, its axis is **'tilted'**, and the northern hemisphere leans towards the sun in June and away from the sun in December. Small, slow changes occur in the tilt of the earth's axis. This can cause variations in the temperature range between summer and winter. Approximately every 41,000 years, the tilt reaches a more vertical position. Sunlight strikes the polar regions at a sharper angle, and the seasonal range of temperatures decreases.

2. Every 100,000 years, the orbit of the earth varies from a nearly circular orbit to a more elliptical one. When the planet is in this more elliptical orbit, it may be as much as 18,000,000 km (11,000,000 miles) further away from the sun than at other times. This '**eccentricity effect**' also decreases the seasonal range of temperatures.
3. Finally, it has been learned that the earth's axis '**wobbles**' like a spinning top, so it describes a circle once every 23,000 years. This process causes the northern hemisphere's summer to occur either when the earth is furthest from the sun in its elliptical orbit, or when it is closest.

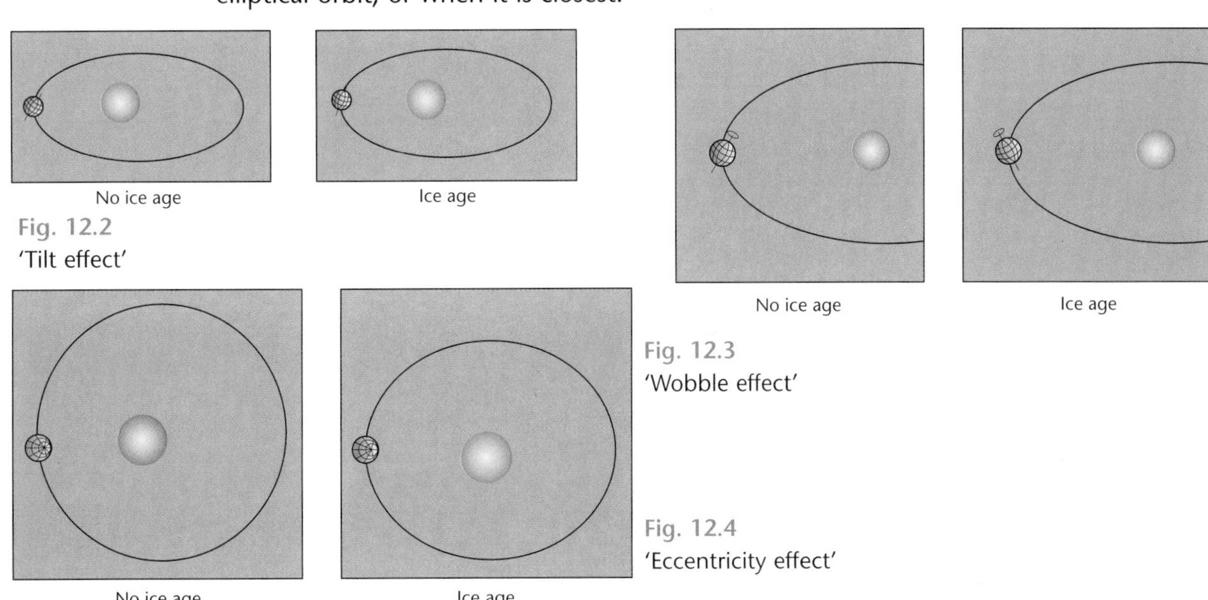


Fig. 12.2
'Tilt effect'

Fig. 12.3
'Wobble effect'

Fig. 12.4
'Eccentricity effect'

Fig. 12.5
Wall of glacier breaking up as it reaches the sea in Alaska, USA. This process is called '**calving**'.

**These three cycles** — occurring at intervals of 41,000, 100,000 and 23,000 years — apparently **join forces** to produce the longest and most severe **ice ages**. When the northern hemisphere's summer is cool, the accumulated snow and ice do not melt, leading to a glacial build-up. Any one of these is enough to cause some worldwide cooling and to provoke a 'mini-ice age', but when the three overlap, the great sheets of ice cover most of the globe.

It has been estimated that over a tenth of the earth's land surface is permanently covered with ice. The giant Antarctic and Greenland ice sheets exert major control over the earth's climate by extracting heat from the overlying air and surrounding ocean. For this reason, the ice sheets are closely monitored for climatic trends. Also studied closely for what they reveal of climatic trends are core samples extracted from the ice. They contain bubbles of *fossil air* — samples of past atmospheres trapped within the ice layers. The cores are, in fact, a reference library of the earth's ancient atmosphere; their layers record hundreds of thousands of years of climatic history. (See 'The Greenhouse Effect and Global Warming', p. 413.)

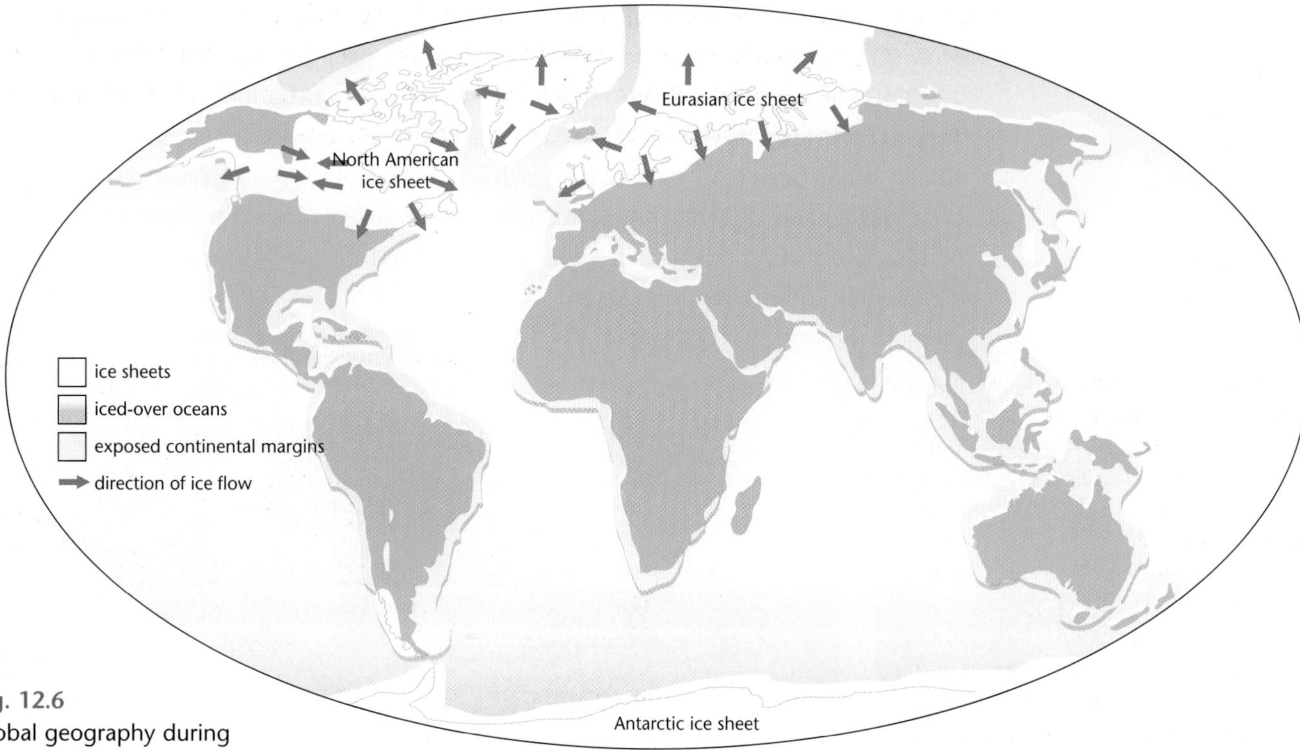


**Fig. 12.6**
Global geography during the last ice age. The growth of vast ice sheets lowered the worldwide sea level and exposed the continental shelves. Australia and Indonesia were one; Japan, Malaysia and many south-east Asian islands were joined to the mainland; and the Red, Black and Caspian seas were dry. Humans migrated across the exposed Bering Strait from Asia to settle in North America.

During the last ice age, which ended some 10,000 years ago, highlands and lowlands carried an extensive ice cover. Large areas at all levels were completely buried by a great thickness of ice, with the exception of exposed high mountain peaks. These exposed mountain peaks, called **nunataks**, were subjected to intense freeze and thaw action, which sharpened their edges to form pyramidal peaks, such as the Matterhorn in Switzerland.

To allow such masses of ice to form, the amount of snow that falls in winter must exceed the amount that can be melted away in the summer (**ablation**). The excess snow increases in thickness and gradually consolidates into ice.

Each additional snowfall adds to the weight of the accumulation and the lower ice crystals are compressed and compacted. Air, which was trapped between the ice crystals, is pushed out to form **firn** or **névé ice**. Further snowfalls intensify this process of **compaction** and **consolidation** until most of the air has been expelled. In this way, **blue glacier ice** is formed.

As ice increased in thickness in mountain areas during the ice age a dome of ice built up. As the centre of this dome rose, the margins moved outward under the influence of gravity and travelled down river valleys to form **glacier ice**.

As glaciers emerged from their valleys, some joined together to form **piedmont glaciers**.

Finally, all of these glaciers flowed onto the lowlands to form an **ice sheet** which covered the whole country.

As the ice advanced it picked up the weathered material which lay in its path and thus it came into contact with the underlying rock. Armed with this material, the ice advanced

**At its peak** about 18,000 years ago, the most recent advance of Pleistocene ice retreated to present locations by about 6000 BC. Meltwater raised sea levels about 100 metres.

**Processes of Glacier Movement**

- Plastic flow
- Sliding
- Melting
- Basal slip
- Rotational slip

across the landscape, scratching, scouring and polishing rock surfaces. Finally, with the onset of warmer conditions, the ice melted. Till or boulder clay was laid down across the lowlands by the ice sheets. Rivers, which flowed under or from the fronts of these melting ice sheets, deposited **fluvial** (river) **materials** of sand and gravel.

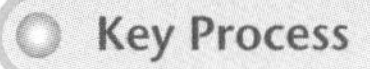


## Glacier Ice Movement

Fig. 12.7
**Main features of a valley glacier.** The glacier has been cut away along its centre line; only half is shown. Crevasses form where the glacier bed has a steeper slope. Arrows show direction of ice flow. A band of rock debris marks the boundary between the main glacier and a tributary glacier that joins it from a lateral valley.

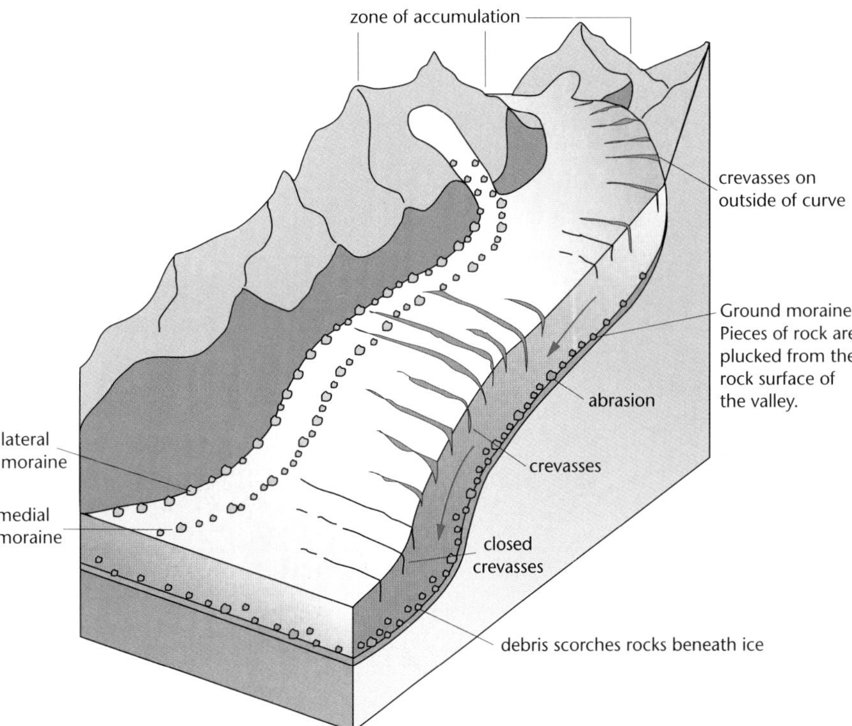


Crevasses do not extend below about thirty-five metres, because at that depth, the ice deforms under its own weight in the zone of **plastic flow**. In other words, rather than fracturing, the ice changes shape permanently and continuously in response to pressure generated by the weight of the overlying ice. In this zone, the ice flows, but very slowly. The squeezing is greatest at the head of the glacier where the ice is thickest. Restricted by the valley walls and floor, the ice is forced to flow downhill towards the glacier front, like toothpaste squeezed out of an open tube.

Glacial movement involves **sliding** and **melting**. Ice melts under pressure and refreezes as soon as the pressure is released. This is the principle behind making a snowball. As glacier ice moves downhill it alternately melts and refreezes in the downhill direction. Should the ice meet an obstacle in its path, it may melt along the uphill portion of the path where the ice is squeezed against the obstacle and refreeze downhill from the obstacle.

This melting supplies water on the valley floor and it mixes with the sand and gravel that is attached to the base of the glacier, creating a slippery slush. Like an oily skid, this slush allows the glacier to slide en masse downhill. This action is called **basal slip**. On occasion, some glaciers have been recorded moving downhill at up to twenty metres per day. Such a rapid movement is called a **surge**.

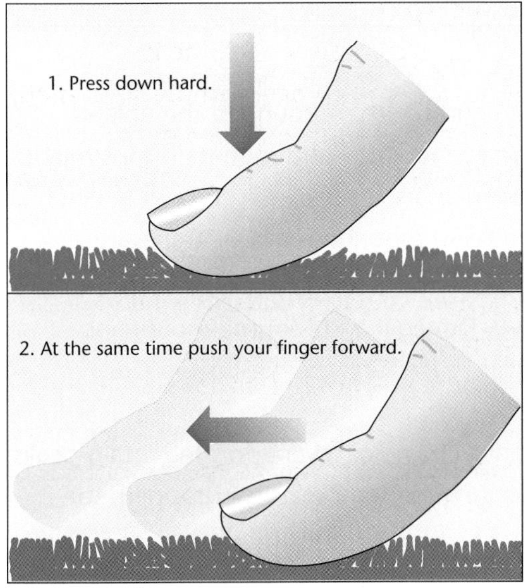

1. Press down hard.
2. At the same time push your finger forward.


Pressure and motion increase temperature and cause melting at base of glacier.

Fig. 12.8

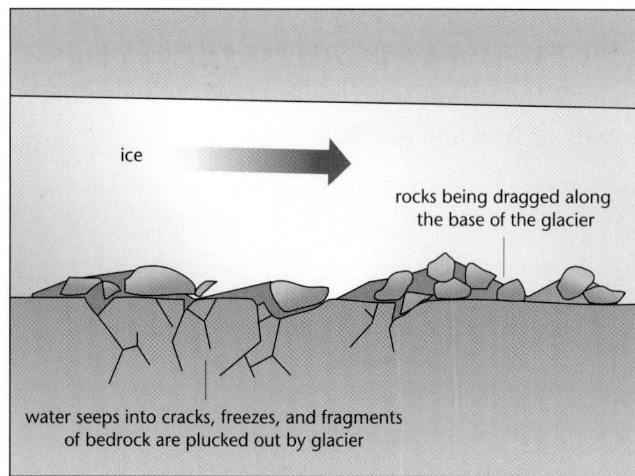


Fig. 12.9
Rock fragments and bedrock being plucked out and abraded by movement of a glacier

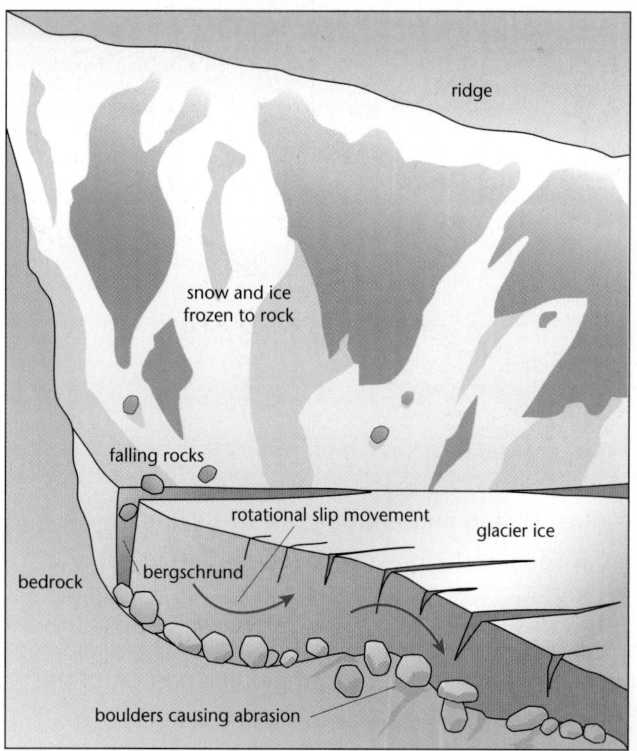


Fig. 12.10
Processes at work on and within a glacier

## Processes of Erosion

### Plucking

**Plucking** occurs because of the drag exerted by moving ice on the rock with which it comes in contact. The base and sides of a glacier may melt into the ground due either to the pressure of the ice or the heat which was caused by the friction of moving ice.

Meltwater flows into the joints and cracks of adjoining rocks. This water may then refreeze and cause the rock to adhere to the glacier. When the glacier moves on, it plucks chunks of these rocks from the bottom and sides of the valley. This process is especially effective in places where the rock is already weakened because of jointing or freeze-thaw action.

### Abrasion

The plucked rocks became embedded in the base and sides of the glacier. As the glacier moved, these rocks scoured, polished and scraped the surface over which they passed (much as rough sandpaper acts on timber), leaving deep grooves and scratches called **striations** on the rock landscape.

### Other Factors

The amount of plucking and abrasion often depends upon other factors.

1. **The weight of ice.** Erosion increases with the weight of overlying ice. Glaciers were often over 600 metres thick in Ireland during the ice age.
2. **Steep slopes.** Glaciers move faster on steep slopes, thus increasing their power of erosion.
3. **Resistance of rock.** The softer the rock, the greater the amount of erosion and the more rounded the upland peaks. Evidence of rounded hill and mountain tops may be seen in the sedimentary uplands of southern Ireland such as the Slieve Felim mountains in Co. Tipperary.

   Harder, more resistant rocks display steep sides and pointed peaks. Evidence of this may be seen in the igneous rocks of the Mayo, Donegal and Mourne mountains. Some mountains may rise to form peaks similar to the Great Sugar Loaf in Co. Wicklow, e.g. Croagh Patrick in Co. Mayo.
4. **Freeze-thaw.** Cracks in rock on high valley sides above ice level or on mountain peaks often filled with water during daytime. At night, when temperatures dropped, the water froze and expanded, breaking up the rock and causing rockfalls onto the glacier sides below. (See 'Mechanical Weathering', p. 113.)

# Zone of Accumulation
# Stages in the Development of a Glaciated Upland

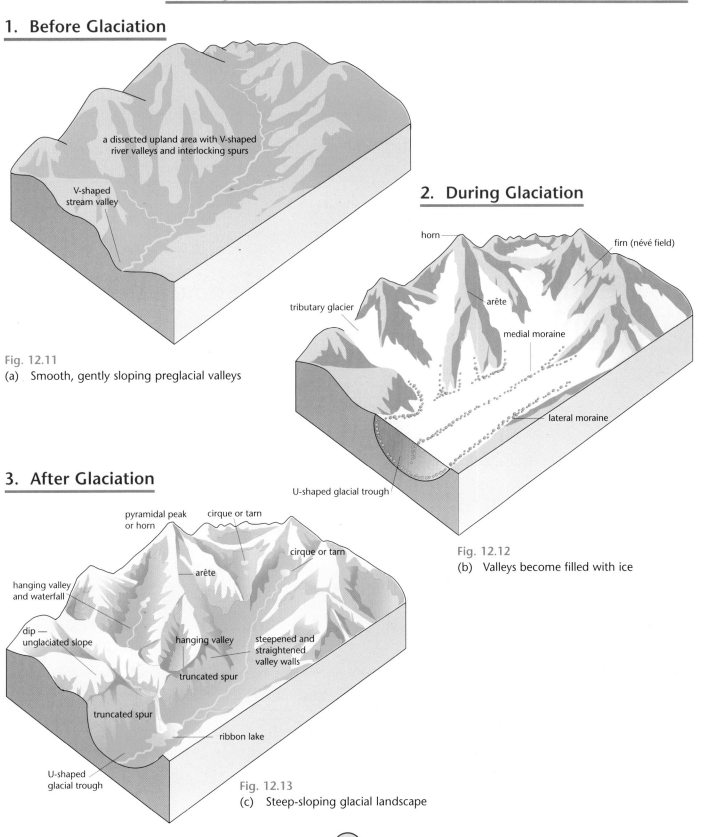

## 1. Before Glaciation

a dissected upland area with V-shaped river valleys and interlocking spurs

V-shaped stream valley

Fig. 12.11
(a) Smooth, gently sloping preglacial valleys

## 2. During Glaciation

horn

firn (névé field)

arête

tributary glacier

medial moraine

lateral moraine

U-shaped glacial trough

Fig. 12.12
(b) Valleys become filled with ice

## 3. After Glaciation

pyramidal peak or horn

cirque or tarn

cirque or tarn

arête

hanging valley and waterfall

dip — unglaciated slope

hanging valley

steepened and straightened valley walls

truncated spur

truncated spur

ribbon lake

U-shaped glacial trough

Fig. 12.13
(c) Steep-sloping glacial landscape

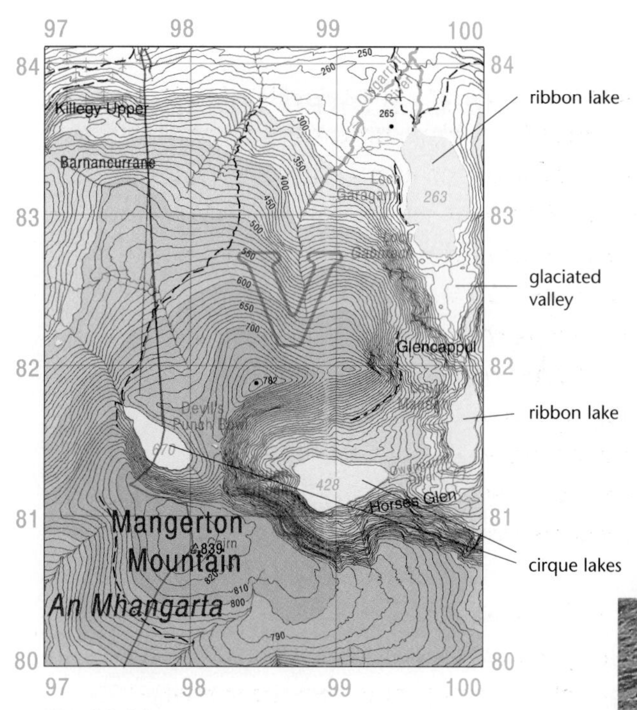


**Fig. 12.14**
Cirque contours. (a) In which direction are the cirques facing? (b) Explain why they regularly occur on this side.

## Landforms of Glacial Erosion

### Landform: Cirque

Examples: Coomshingaun in the east Comeraghs in Co. Waterford
The Devil's Punch Bowl in Mangerton in the Macgillycuddy's Reeks in Co. Kerry
Both of these cirques are in Ireland.

Cirques are sometimes called tarns, combes, cwms or corries. The largest is Walcott Cirque in the Antarctic continent, which has a backwall 3,000 metres high (10,000 feet).

**Fig. 12.16**
Cirque and cirque lake. Lough Nambrackderg in west Cork in Ireland.

**Fig. 12.15**
The process of rotational slip that creates a cirque

rockfalls from frost shattering fall into bergschrund
bergschrund
firn (névé field)
plucking steepens backwall
crevasses
heavily compacted snowfalls turn to glacier ice
rotational slip
bedrock
shattered rock causes abrasion and deepening of cirque hollow
rotational movement of ice causes over-deepening of cirque

Processes: Abrasion, Plucking, Freeze-thaw, Basal Sapping, Rotational Slip

### Formation

The upper end of a glaciated valley commonly consists of an amphitheatre-shaped, steep-sided rock basin. It is known variously as a cirque (French), a corrie, coire or coom (Gaelic), a cwm (Welsh), a combe or cum (in England) and a tarn (in Scotland).

Cirques have steep rocky walls on all sides except that facing down the valley. Cirque lakes regularly occupy over-deepened hollows at the base of these rock walls. Cirques are generally the source of ice for valley glaciers. (Figs. 12.14 and 12.16).

Cirques were formed when preglacial hollows were progressively enlarged generally on north- or north-east-facing slopes. The ice remained on these slopes longer than it did on other slopes. A patch of snow produced alternate thawing and freezing of the rocks around its edges, causing them to 'rot' or disintegrate. The weathered debris is transported by meltwater, thus forming a **nivation hollow**. This process is called **nivation** or **snow-patch erosion**. As snowfall accumulated, large masses of ice formed a **firn** or **cirque glacier**.

At this stage, the ice moves downslope and pulls away from the headwall of the cirque, to which some ice remains attached. This gaping crack or crevasse is called the **bergschrund**. The headwall of the cirque maintains its steepness from the meltwater, which seeps into cracks and, after alternate thawing and freezing, shatters the rockface (a process known as **basal sapping**). This action produces debris which falls down the bergschrund, freezing into the base of the ice field and acting as an abrasive. Ice movement pivots about a point situated centrally in the cirque, a process known as **rotational slip**. Through plucking and abrasion, this action increases the depth of the hollow, which often contains a lake when the ice finally disappears.

### Characteristics of a Cirque

1. Amphitheatre- or bowl-shaped hollow found mainly on northward-facing mountain slopes where ice remained for a longer time.
2. Steep rock cliffs form the headwall and sides of the cirque.
3. A **cirque lake** or **tarn** may occupy an over-deepened hollow or may be impounded by moraine debris.
4. Mountain lakes whose placenames begin with 'cum' or 'coom' or end in 'tarn' are cirque lakes.

### Landform: U-shaped Valley

Examples: Cummeenduff Glen in the Macgillycuddy's Reeks in Co. Kerry, in Ireland
Many valleys in the Lake District in England and Nant Gwynant valley and Nant Francon in Snowdonia in Wales
Yosemite Valley in Yosemite National Park, in the Sierra Nevada mountains in California in the United States

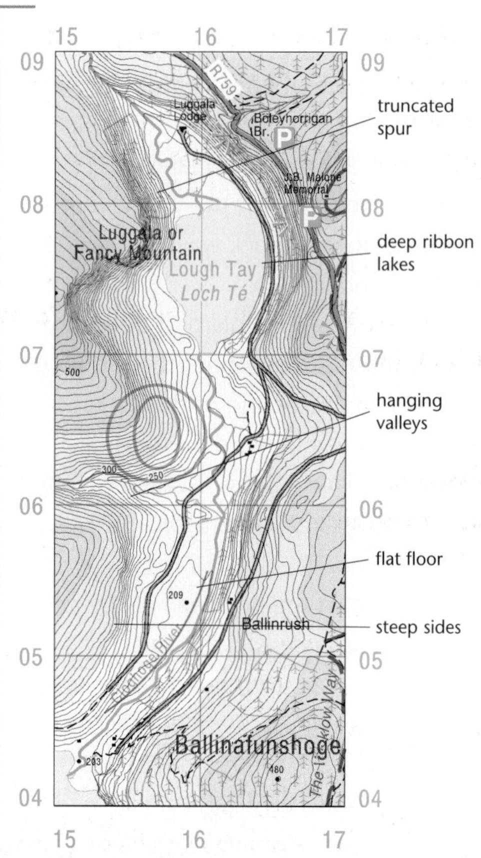


Fig. 12.17

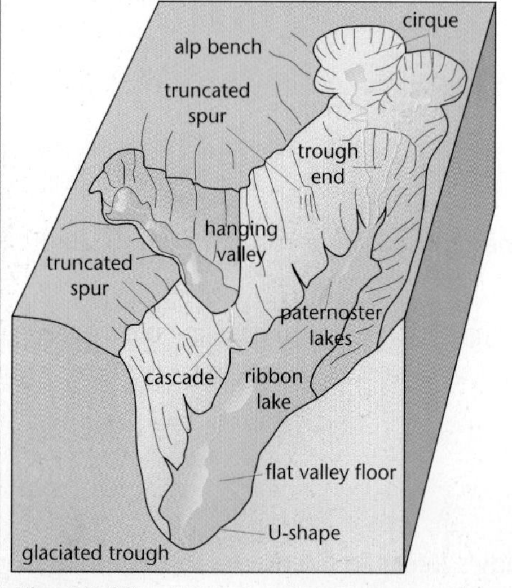


Fig. 12.18 **Glaciated trough.** Characteristic features in a deep glaciated valley.

### Formation

When glaciers moved downslope through preglacial river valleys, they changed their V-shaped profile into wide, steep-sided, U-shaped valleys. As the ice proceeded down-valley, it used material which it plucked away from the valley floor to increase its erosive power. Thus, gathered debris was used to increase vertical and lateral erosion in the valley. These processes of plucking and abrasion changed the preglacial V-shaped valley to a U-shaped glaciated valley.

Most of our mountain valleys were glaciated, e.g. Cummeenduff Glen in Co. Kerry and the Glenariff Valley in Co. Antrim, in Ireland.

A glacier is a solid mass of ice which moves down a valley. Due to its solid nature, it may have difficulty in negotiating a route through a winding valley which may also vary in width from place to place. However, a glacier overcomes this difficulty in a number of ways.

1. **Pressure** is exerted on a glacier as it passes through a narrow neck in a valley. Compression produces heat, causing some ice to 'melt' and allowing the glacier to 'squeeze' through, only to freeze again when the pressure is released. Elsewhere, obstacles in the glacier's path have a similar effect on the ice. Local melting on the upstream side allows the glacier to move over or around these obstacles as it moves downhill.
2. **Friction** between the base of the glacier and the valley floor causes melting, producing a thin film of meltwater which acts as a lubricant, so the glacier moves downslope.

   Well-developed glaciated valleys with flat floors are known as **glacial troughs**. Here, glacial erosion was intense due to the weight and pressure of the deep glacier. Some features of a glacial trough are: trough end, truncated spurs, hanging valleys, rock steps, ribbon lakes and paternoster lakes.

## Features within a Glaciated Valley

### Trough End

Deep glaciated valleys end abruptly at their upper end. This sharp change of slope is called a **trough end**. It occurs where **cirque glaciers coalesced** as they began their downhill journey.

### Truncated Spurs

As a glacier passed through a valley, its poor ability to navigate curves caused the ice to cut through all obstacles. Interlocking spurs were cut away to form truncated spurs.

### Hanging Valleys

Tributary valleys which were originally graded to the preglacial river valley are left 'hanging'. Their streams fall abruptly into the main glaciated valley in

cascades, producing waterfalls which are most noticeable during times of heavy rainfall.

The erosive power of the tributary glaciers in these tributary valleys was often unable to equal that of the main valley. Therefore, tributary valleys after the ice age were left hanging above the main valley floor. These display characteristics similar to the main glaciated valley.

### Fjords

When deep glacial troughs were submerged they formed parallel-sided inlets called fjords. Norway's western coast has many such inlets such as the Sogne Fjord.

## Landform: Ribbon and Paternoster Lakes

Examples: The Gap of Dunloe, in Co. Kerry in Ireland
Lake Windermere, in the Lake District in England
The Finger Lakes, in the USA



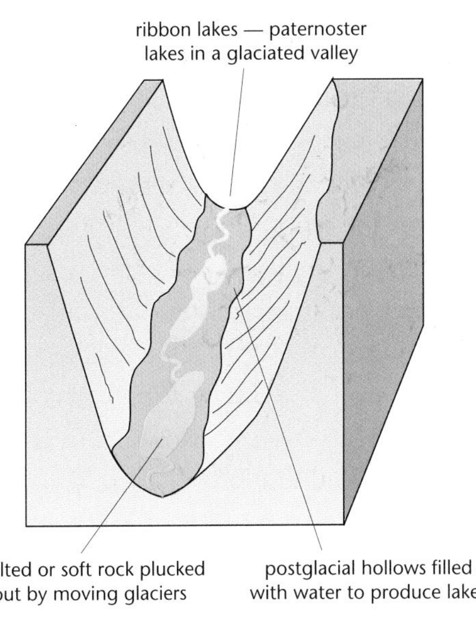


Fig. 12.19

ribbon lake
truncated spurs
ribbon lake
hanging valley

Fig. 12.20
The Gap of Dunloe, Co. Kerry

### Formation

### Rock Steps, Ribbon Lakes and Paternoster Lakes

The long profile of a glaciated valley may resemble that of a staircase. This is due to any or a combination of the following:

- the varied resistance of the bedrock
- velocity or thickness of the glacier
- patches of badly fractured bedrock

Fig. 12.21
A glaciated valley — the Gap of Dunloe in Co. Kerry. Identify:
(a) the paternoster lakes,
(b) a truncated spur and
(c) scree slopes.

'Rock steps' commonly occur where a tributary glacier joined the main valley and so the extra mass of ice was able to erode more vigorously. Over-deepened or unevenly scoured rock steps regularly contain small lakes which are joined by cascading waterfalls.

Long stretches of glacial valley floors may be scoured creating deep rock hollows. Again these may be patches of soft rock or badly fractured rock due to ancient earth movements. When filled with water, these basins may be found in isolation, when they are called **ribbon lakes**. As with the 'rock step' lakes, if they are found in a string, they are called paternoster lakes. Windermere, in the Lake District in England, is a ribbon lake. Luggala Lough and Lough Dan (see Ordnance Survey map extract, Fig. 12.17) are ribbon lakes in Co. Wicklow, in Ireland.

As glaciers moved through valleys, they passed over these fractured or soft rock patches. During cold spells or at night they may have stopped moving. At such times the liquid created by friction at the base of the glacier freezes and attaches the glacier to the bedrock (see Fig. 12.9, p. 200). Once the glacier moves again, chunks of shattered or soft rock are plucked from the valley floor creating hollows. These hollows increase in size as the glacier continues to erode the valley. When all the ice has melted from the landscape these hollows fill to form lakes.

### Fjords

Some glaciated U-shaped valleys were greatly deepened and straightened by ice. Once the ice age was over, sea levels rose again and some of these deep valleys which opened to the sea became flooded. The Norwegian coast is mainly a 'fjord coast'; Sogne Fjord and Hardanger Fjord are two examples. Killary Harbour in Co. Galway and Carlingford Lough in Co. Down are the two main Irish examples.

## Pyramidal Peak

A mountain peak in the shape of a pyramid is called a **pyramidal peak**.

### Formation

If three or more cirques cut back to back, a process known as **headwall recession**, the surviving central rock mass becomes a pyramidal peak. This peak is later sharpened by frost action.

Examples: Carrauntoohil and Brandon Mountain in Co. Kerry. One of the most famous examples of a pyramidal peak is the Matterhorn in Switzerland.

## Arête

An **arête** is a knife-edged ridge between two cirques.

When two cirques cut back to back or side by side (**headwall recession**), a sharp ridge forms between them. This ridge is an arête (Fig. 12.23). Example: Between Lough Nalackan and the Owennafeana Valley on Brandon Mountain in Co. Kerry.

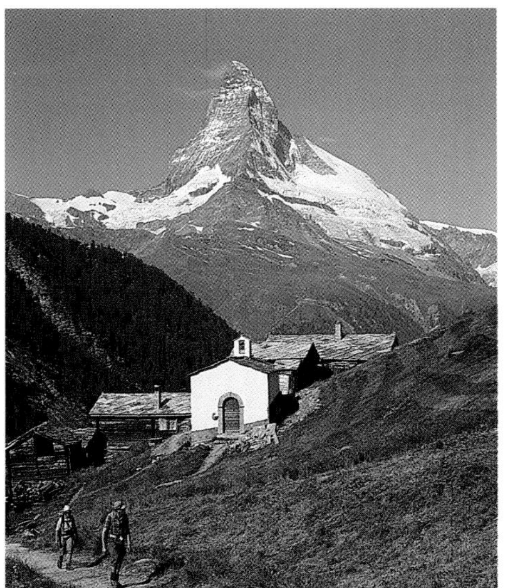

Fig. 12.23
The Matterhorn in the Alps is one of the best examples of a pyramidal peak. Arêtes are the steep-sided ridges which radiate from the peak.

## Roches Moutonées

Large rocky outcrops on valley floors provided obstacles to the movement of valley glaciers. Due to increased pressure at this point, local melting allowed the glacier to slip over and smooth out the upstream side of the outcrop. Once the pressure was released, the meltwater froze again and attached itself to the rocky outcrop. Plucking occurred when the ice advanced downslope, leaving the downslope side sharp and angular. Roches moutonées, as these features are known, were named by a French geographer in the 1880s because they resembled a type of wig which was fashionable at the time.

Fig. 12.22

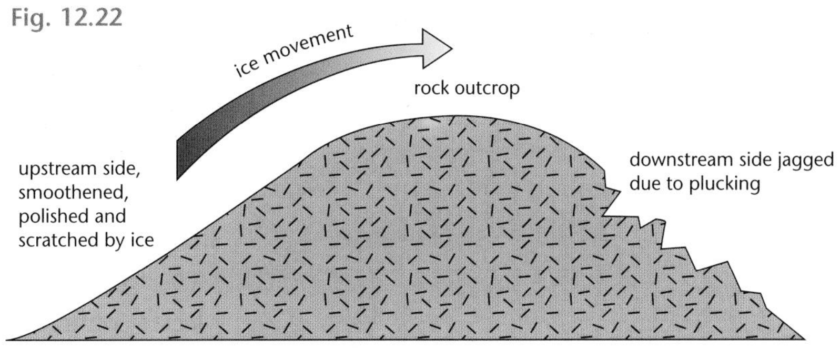

## Change of Slope

When a glacier did not entirely fill the preglacial valley, there is a noticeable change of slope, forming benches or 'alps' above the steep valley sides (Fig. 12.18).

Fig. 12.24
Converging Kaskawulsh glaciers in Yukon, Alaska, form medial moraines as lateral moraines combine (see 'Medial Moraine', p. 213)

## Knock-and-Lochan

Hollows formed in lowlands as ice sheets passed over areas of varying rock resistance. This resulted in numerous small rock basins containing lakes which are separated by low, rocky hills. This type of landscape is known as knock-and-lochan topography, as in Connemara, Co. Galway. The term is derived from Scottish lakes called 'lochans' and rocky hills known as 'knocks'.

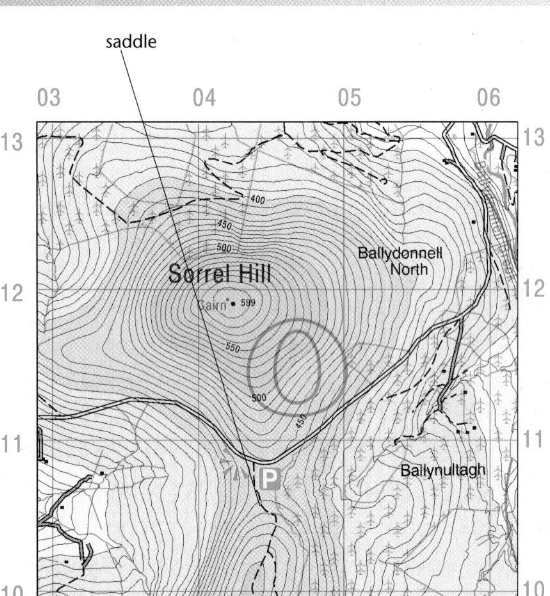


Fig. 12.25
A saddle between Sorrel Hill and Black Hill is used by a third-class road for access over the upland area

## Col/Saddle

In preglacial times, the source streams of rivers caused dips to form on hill and mountain ridges due to headward erosion. During the ice age, tongues of ice passed through these dips, or cols, and deepened and widened them to form saddles. Roadways take advantage of such gaps or passes through ridges and mountains to reduce the cost of road construction and shorten journeys (Fig. 12.25). In some instances, cols were eroded down to the level of the main valley floor to form a **glacial breach**. The Gap of Dunloe in Co. Kerry in Ireland is such a landform.

## Landform: Glacial Spillway

Examples: The Glen of the Downs in Co. Wicklow, in Ireland
Ironbridge Gorge in England

### Formation

Some mountain valleys were entirely blocked by moraines or ice masses lying across their former outlets. Meltwater from the ice and, in later times, rain-fed streams were dammed up within the valleys and formed lakes. These are called **proglacial** lakes. In time, the rising waters rose sufficiently high to flow over the lowest part of the moraine or the valley side. In such instances, spillways were cut by the escaping waters. Thus they are typically V-shaped. The processes of abrasion and hydraulic action were involved in their formation.

In Ireland, examples are found at the Scalp and the Glen of the Downs in Co. Wicklow (Fig. 12.26), at Keimaneigh near Gouganebarra in Co. Cork, at Dundonald in Co. Down and at Galbally in Co. Limerick.

Fig. 12.26
Glen of the Downs in Co. Wicklow

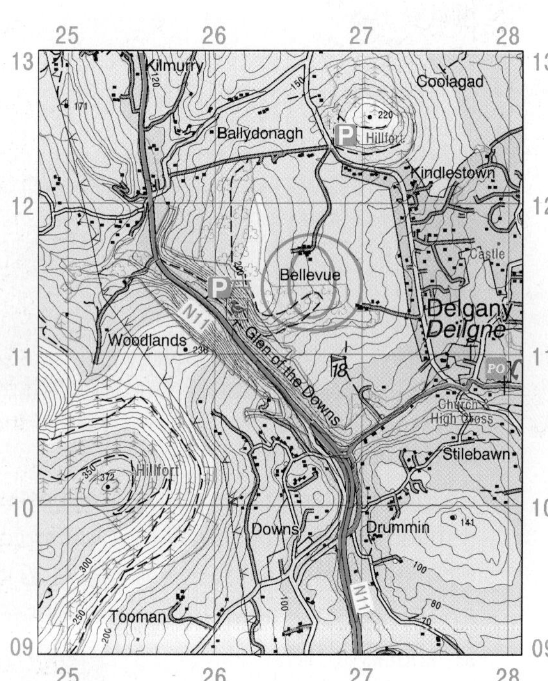


Fig. 12.27
The Glen of the Downs in Co. Wicklow, in Ireland, is a glacial spillway. Its steep sides and V-shaped profile indicate that it was formed by river erosion only.

**How to Describe a Glaciated Valley**

1. Name the valley. A valley gets its name from the river which flows through it. Example: the Blackwater river valley.
2. How high and how steep are the sides?
3. How wide is the floor?
4. Is the floor steeply sloping or is it level?
5. Is the valley floor flat?
6. Does the valley have a trough end? Is there a cirque above this feature?
7. Are there hanging valleys along the sides? Name or locate some of them.
8. Does the valley have truncated spurs? Locate them, describe how they were formed — very simply.
9. Are there ribbon lakes or paternoster lakes on the valley floor? Name them. Describe their formation — very simply.
10. Name the mountain peaks which border the valley.
11. In which direction is the valley sloping?

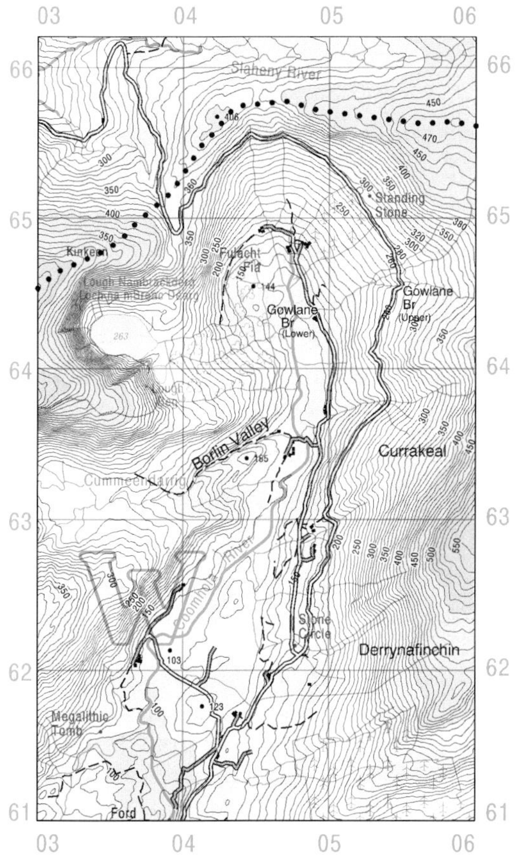


Fig. 12.28
An Ordnance Survey map extract of the Coomhola Valley in west Cork in Ireland

**Class Activity**

1. (a) Study the Comeragh region map extract (Fig. 12.29, p. 210). The grid references in the following list correspond with the glaciated landforms 1 to 10 mentioned below. Examine the grid references and then place each grid reference in the box beside the corresponding landform.

| | | | |
|---|---|---|---|
| S 325 105 | S 314 128 | S 296 104 | S 327 095 |
| S 288 088 | S 285 042 | S 326 125 | |
| S 304 116 | S 325 102 | S 279 069 | |

1. a cirque and cirque lake
2 a cirque without a lake
3 a rock cliff
4. an arête
5. rock steps with paternoster lakes
6. rock steps without paternoster lakes
7. a hanging valley
8. a glaciated valley
9. truncated spurs
10. a saddle

(b) Identify the pattern in the distribution of cirque lakes on the Ordnance Survey map extract and then account for this pattern.

(c) Identify the landform at grid reference S 325 110. With the aid of a sketch map, describe the processes involved in the formation of this landform.

(d) Describe three ways in which glaciers have changed this upland landscape.

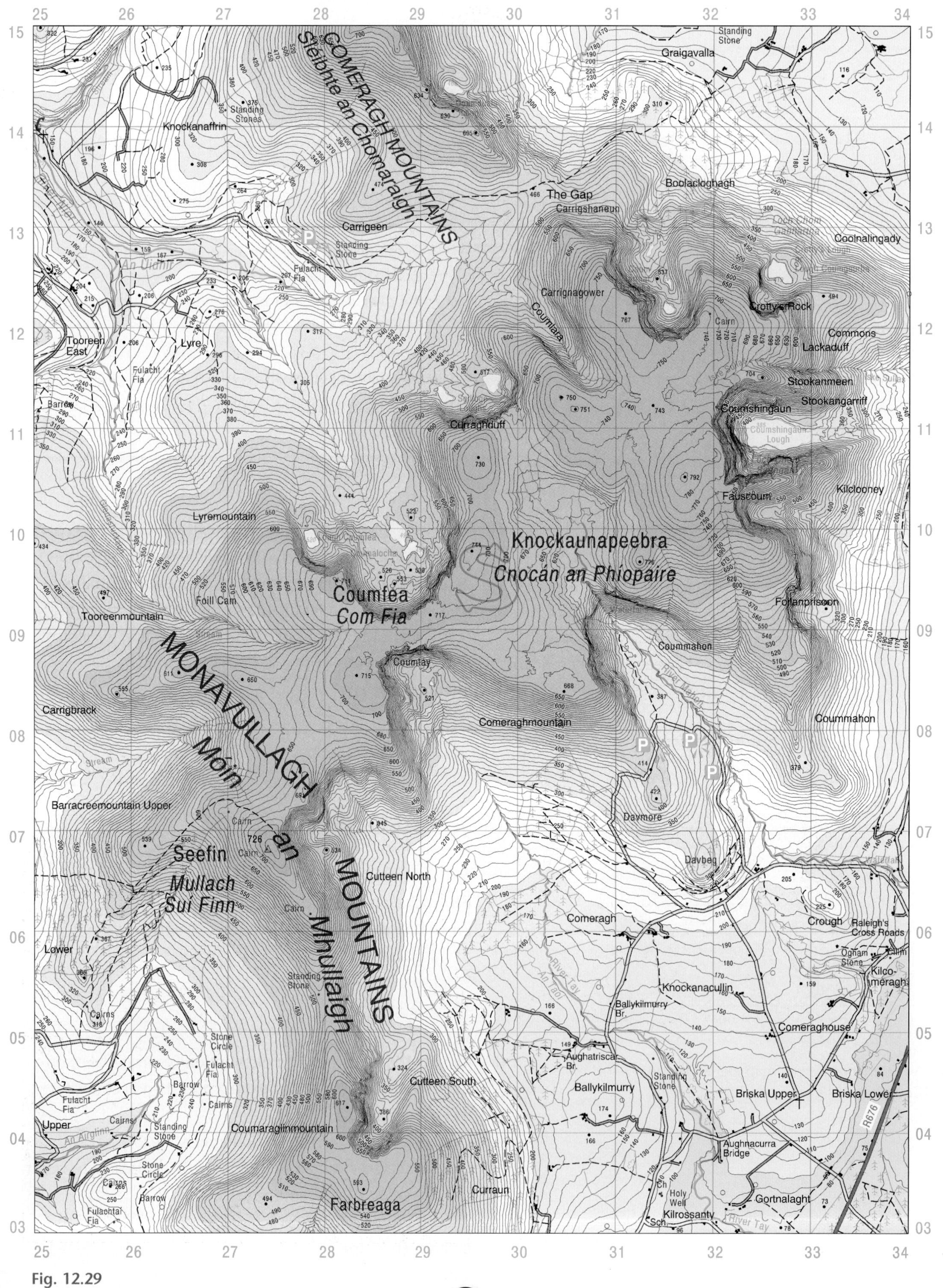


**Fig. 12.29**
Comeragh region

2. Draw a sketch map of the region shown on the Comeragh region map extract (p. 210).
   (a) Then on the map show and name its physical regions.
   (b) Choose any two contrasting regions from this sketch map. For each region that you choose, use the following headings to describe its characteristics:
      - relief (height and slope of the land)
      - drainage (patterns, rivers, direction of flow)
      - settlement (patterns, limit of settlement, location)
      - communications (types, influence of relief, density)
   (c) Study the Ordnance Survey map extract, Fig. 12.28 on p. 209.
   A third-class road winds its way over this upland area. With the aid of a sketch map describe how this routeway takes advantage of the relief of the area in its journey through the uplands.
3. Describe the shape, orientation and formation of Lough Nambrackderg (see Fig. 12.16 on p. 202).

## The Influence of Relief on Routeways

Study the map extracts below. Each extract is labelled. Each label/number highlights a particular influence of relief on routeways and how that routeway takes advantage of, or overcomes, the local relief.

For each numbered label, describe the way in which the development of communications is related to elements of the landscape.

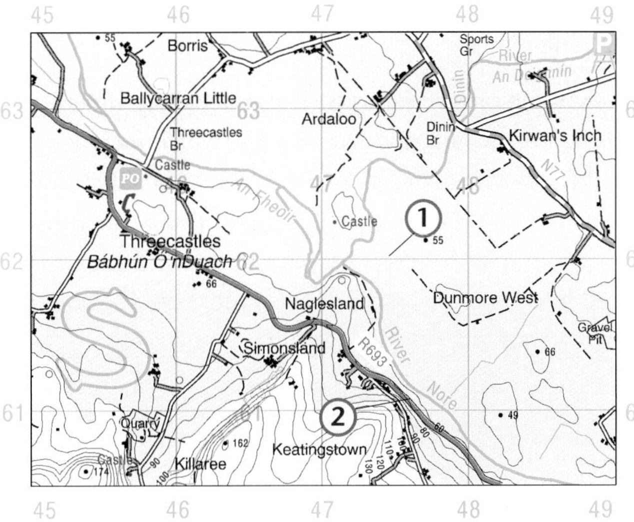


Fig. 12.30

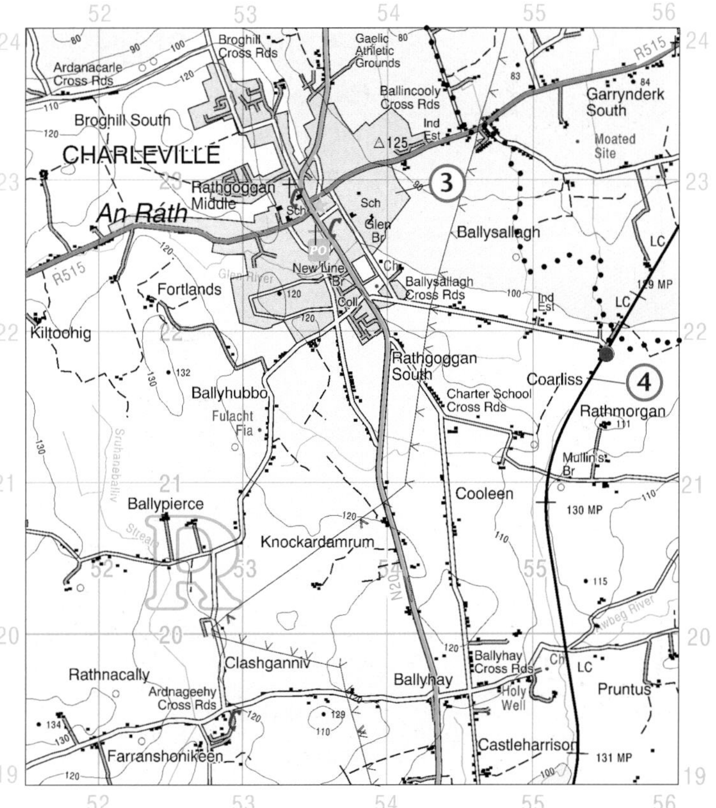


Fig. 12.31

Fig. 12.32

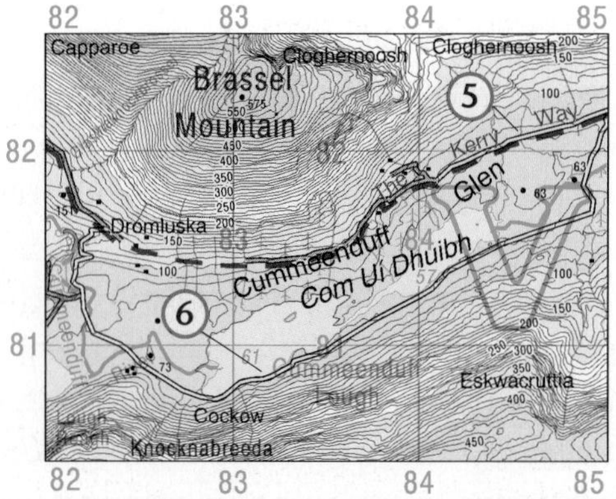

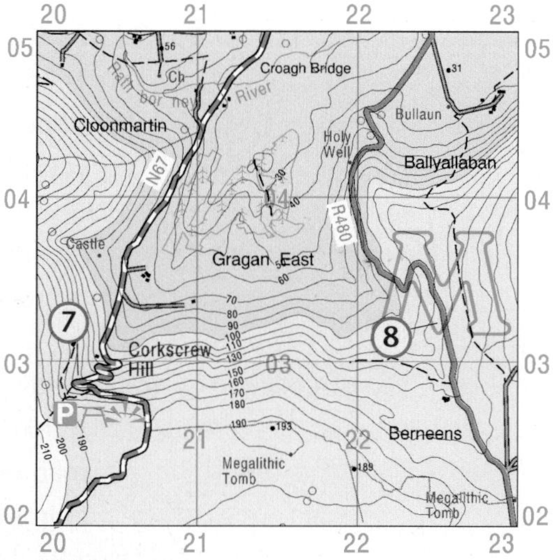

Fig. 12.33

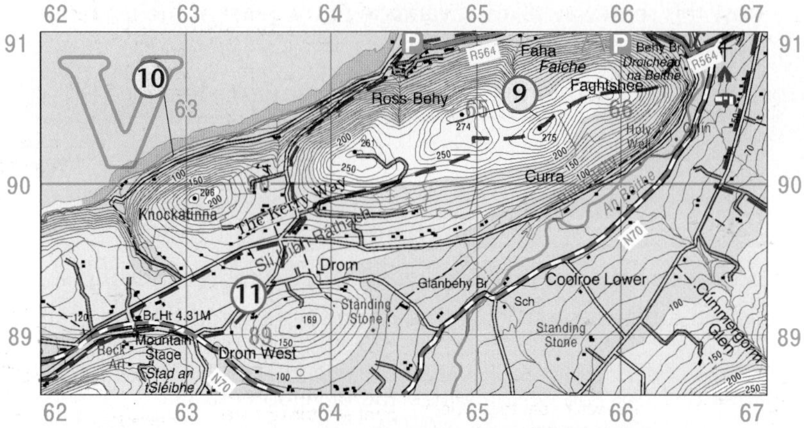

Fig. 12.34

# Zone of Ablation

## Landforms of Glacial Deposition

**Drift** is the term used to refer collectively to all glacial deposits. These deposits, which include boulders, gravels, sands and clays, may be **subdivided** into **till**, which includes all material deposited directly by ice, and fluvioglacial material, which is the debris deposited by meltwater streams. **Till** consists of **unsorted material**, whereas **fluvioglacial** deposits have been **sorted**. Deposition occurs both in upland valleys and across lowland areas.

In this section, however, till is divided into separate landforms: (a) moraine, (b) boulder clay, (c) drumlins, (d) crag and tail and (e) erratics.

## Moraine

### Landform: Moraine

Examples: East of Coomshinghaun Lake in Co. Waterford, in Ireland
Glengesh Valley in Co. Donegal, in Ireland
Lautenbrunnen Valley, near Interlaken in Switzerland
Mersey river valley in the central highlands of Tasmania

### Formation

Moraine contains rock fragments which range in size from large boulders to particles of dust. There are four main types of moraine: lateral, medial, recessional and terminal.

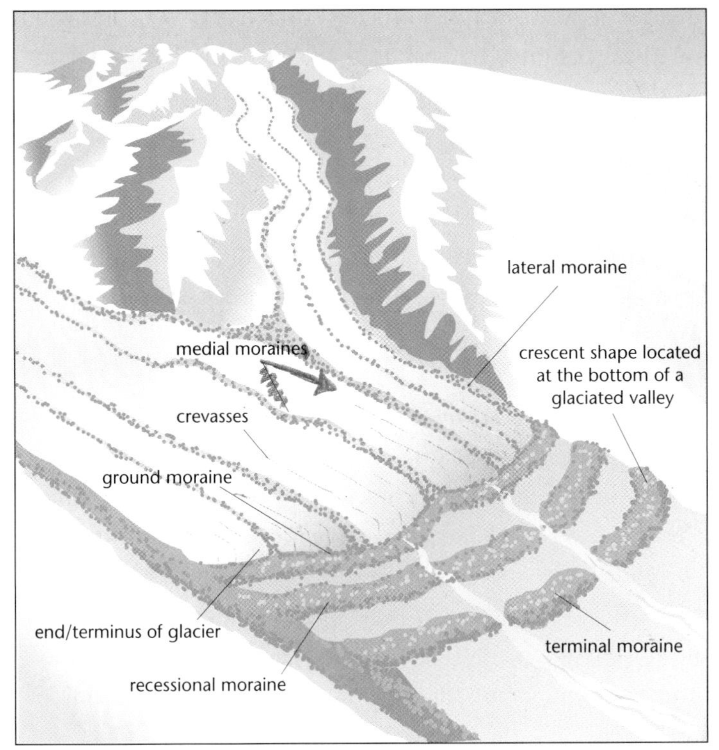


Fig. 12.35

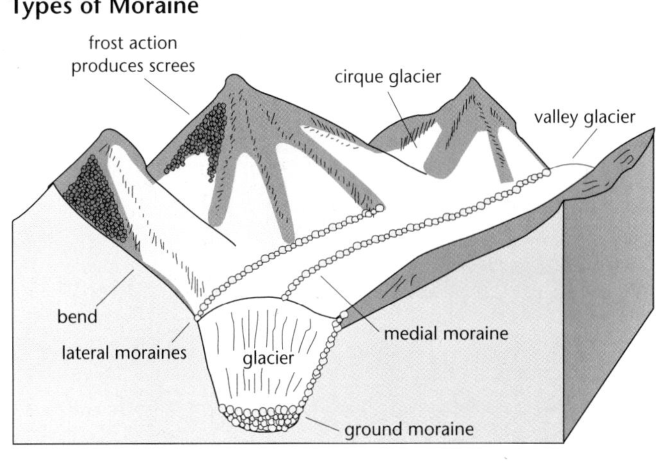


**Valley Glacier**



Fig. 12.36



### Lateral Moraine

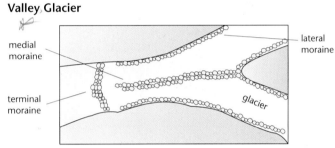

Long, sloping ridges of material left along valley sides after a glacier has melted are called **lateral moraine**. Freeze-thaw action is active on the ridges (**benches**) above glaciers and angular rocks of all sizes fall onto the glacier edges below. This material accumulates to form lateral moraine. Vegetation may cover this material in time. It may now be recognisable only by its lesser angle of slope than the valley walls or as a rocky, sloping surface along valley sides.

### Medial Moraine

A **medial moraine** is formed from the material of two lateral moraines after a tributary glacier meets the main valley. These lateral moraines join and their material is carried down-valley by the main glacier. It is laid down as an uneven ridge of material along the centre of the main valley. There may be many such medial moraines in a valley, the number of which varies according to the number of tributary glaciers.

### Terminal and Recessional Moraines

When glaciers stopped for a long period of time during an interglacial or warm spell, they deposited an **unstratified** (not layered) and crescent-shaped ridge of material across

valleys and plains. These deposits have an uneven surface and are composed of moraine. In some instances, they have caused lakes to form by impeding (preventing) drainage. In relation to upland areas, terminal moraines are found across the lower part or mouth of a valley, while recessional moraines are found at various places up-valley of the terminal moraine.

A terminal moraine runs north/south through the peninsula of Jutland. It divides the sandy outwash plain to the west from the boulder clay and tillage land to the east. It marks the front of the north European ice sheet where it stopped for a considerable length of time.

When terminal moraine deposits are exposed due to mining activity or recent slumping, they may be mistaken for esker deposits (see p. 218). A terminal moraine is exposed in Sinnot's quarry at Loughanavatta near Roscrea in Co. Tipperary in Ireland. There is no sorting of the material which is located near the entrance to the quarry, thus it is moraine. Even though these deposits are present here only 10,000 years, they are in the process of lithification (becoming rock). Even though they appear vulnerable to collapse, they do not crumble easily as they are bonded together by a lime cement. **Many** of the **boulders** in the moraine are **rounded** from **glacial action** much as they would be if they were water-borne.

Fig. 12.37
Unsorted moraine at Sinnot's quarry near Roscrea in Co. Tipperary. It has been partly lithified (become stone) and so does not crumble easily.

Fig. 12.38
Unsorted moraine at Sinnot's quarry near Roscrea in Co. Tipperary. It has been partly lithified (become stone) and so does not crumble easily.

## Landform: Drumlins

Examples: The Drumlin belt from Sligo to Strangford Lough, in Ireland
Dodge County in Wisconsin, in the USA
The Midland Valley, in Scotland
In the Solway Plain and the coastal plain near Lancaster in northern England

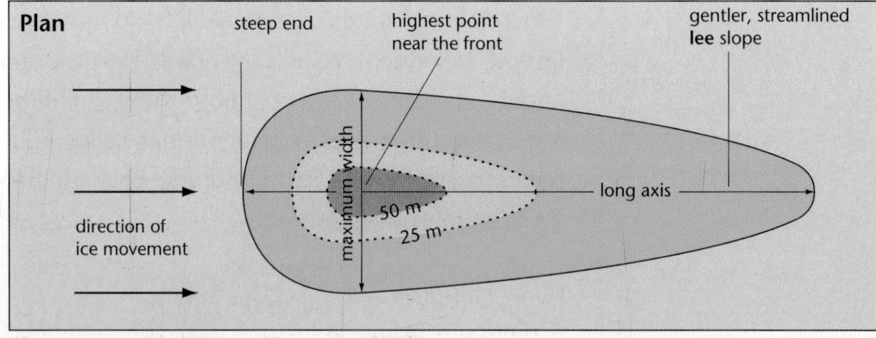


Fig. 12.39

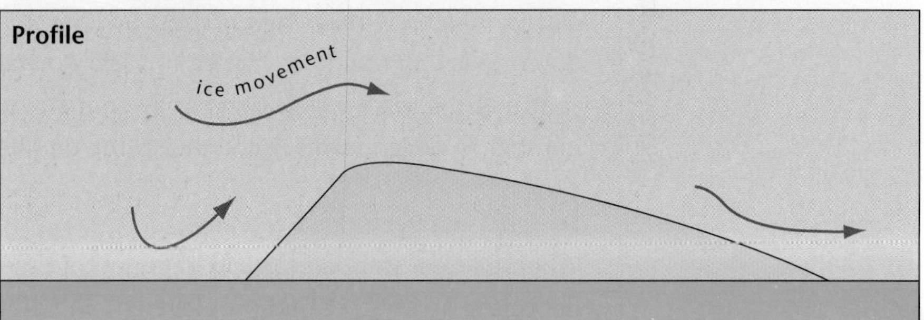


Fig. 12.40

Fig. 12.41
Drumlin landscape is sometimes referred to as 'basket-of-eggs scenery'

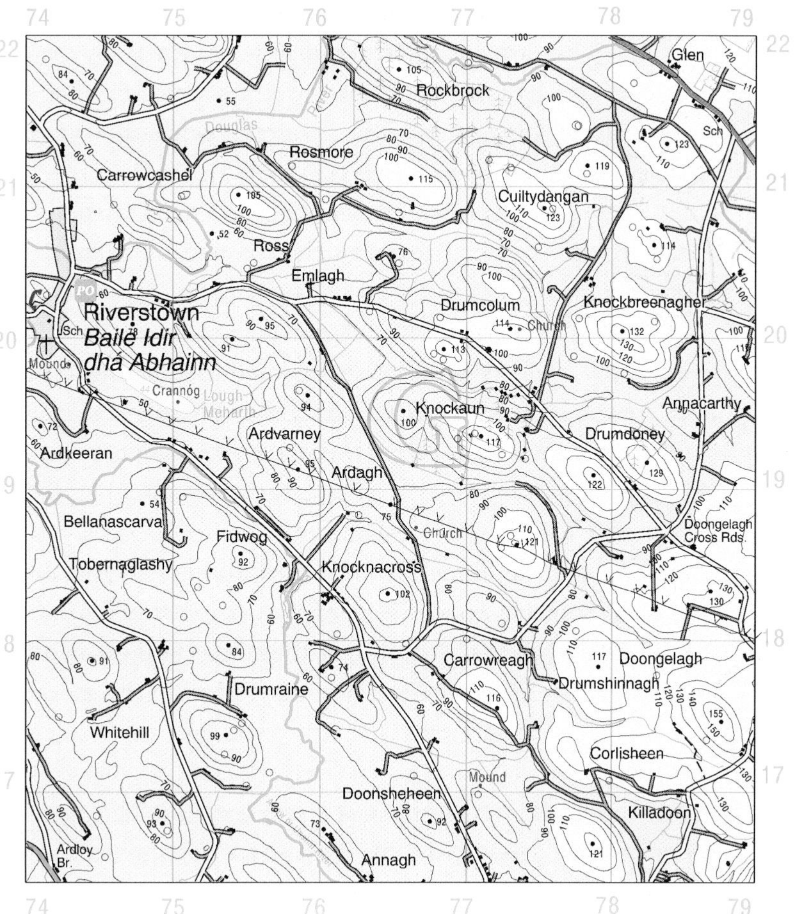


Processes: Plucking, Freeze-thaw, Deposition

## Formation

Drumlins are chiefly made from boulder clay. A drumlin consists of a layer of unsorted and unstratified material composed of rocks, pebbles, gravel, sand and clay. It represents the ground moraine of the stagnant ice sheet. When spread over extensive lowland areas, such deposits are termed **boulder clay** or **till plains**. In some areas, the boulder clay is over thirty metres deep and has been moulded into a variety of shapes. Till is largely derived by the process of plucking from the bedrock over which the ice passes. On the central plain in Ireland, it is formed from the carboniferous limestone which floors the plain and is often referred to as **limestone boulder clay**. Boulder clay deposits generally form undulating landscapes.

In some areas the boulder clay has been deposited as swarms of rounded hummocks, from small mounds just a few yards long and high to considerable hills a few kilometres or more in length and as much as one hundred metres high. Drumlins are found in eastern Denmark, Poland and in southern Germany. However, they are especially well developed in Ireland in a belt which stretches from Sligo in the west to Strangford Lough in the east. The term 'drumlin' is derived from Gaelic which means small hill.

**Drumlins** are rounded, oval-shaped, egg-shaped or whaleback-shaped hills. They usually occur in clusters or **swarms** forming **basket-of-eggs topography**. Their long axis

lies in the direction of ice movement. The steeper end represents the end from which the ice came. They are the result of moving ice depositing each clay mound because friction between the clay and underlying ground was greater than that between the clay and the overlying ice. Some drumlins were formed when boulder clay was deposited around obstacles such as boulders (see Fig. 12.41, p. 215).

In some areas, drumlins may impede drainage. Trapped water ranging from pools to oddly shaped lakes may form. Patches of marsh may form in other hollows due to waterlogging.

Drumlins in places such as Clew Bay in Co. Mayo and Strangford Lough in Co. Down were partly covered when sea levels rose (**eustatic movement** — the rise or fall of sea level) at the end of the last ice age some 10,000 years ago. A rise in temperature released vast volumes of water back to the sea when ice sheets melted.

### Crag and Tail

**Crag and tail** was formed when a hard, obstructive mass of rock (called the **crag**) lay in the path of oncoming ice. The hard crag protected the softer rocks in its lee from erosion, as the ice moved over and around the crag. On the downstream side, deposition was also encouraged to produce a tapering ridge of rock with glacial drift on the surface.

### Erratics

As ice moved from mountainous areas, large boulders were sometimes carried long distances and deposited as the ice melted. Sometimes, these boulders are perched in precarious positions and are referred to as **perched blocks**.

In this way in Ireland, Mourne granite was carried as far as Dublin, Wexford and Cork. Galway granite was carried as far away as Mallow, while Scottish ice dropped erratics in Monaghan and Dublin.

## Fluvioglacial Deposits

Fluvius is the Latin word for river.

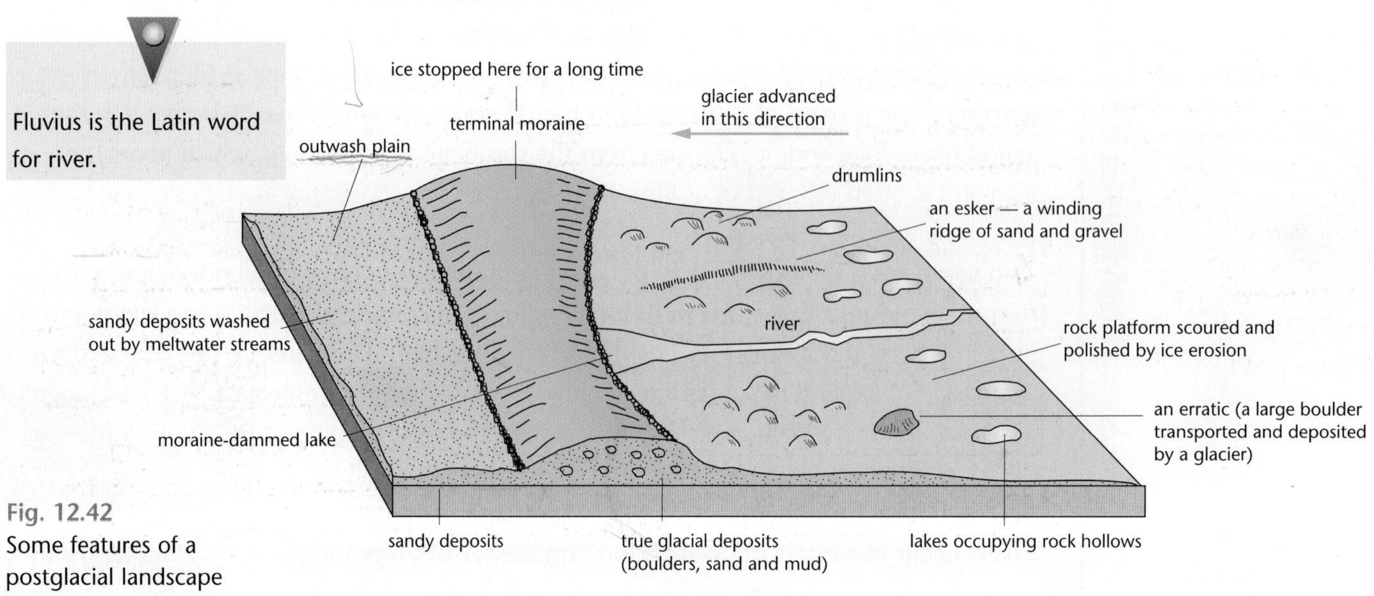


Fig. 12.42
Some features of a postglacial landscape

Towards the end of the ice age, vast amounts of meltwater were released from the melting ice as a result of rising temperatures. The many rivers which flowed from the melting glaciers carried large amounts of sands and gravels and deposited them to form the following **fluvioglacial features** (Fig. 12.42).

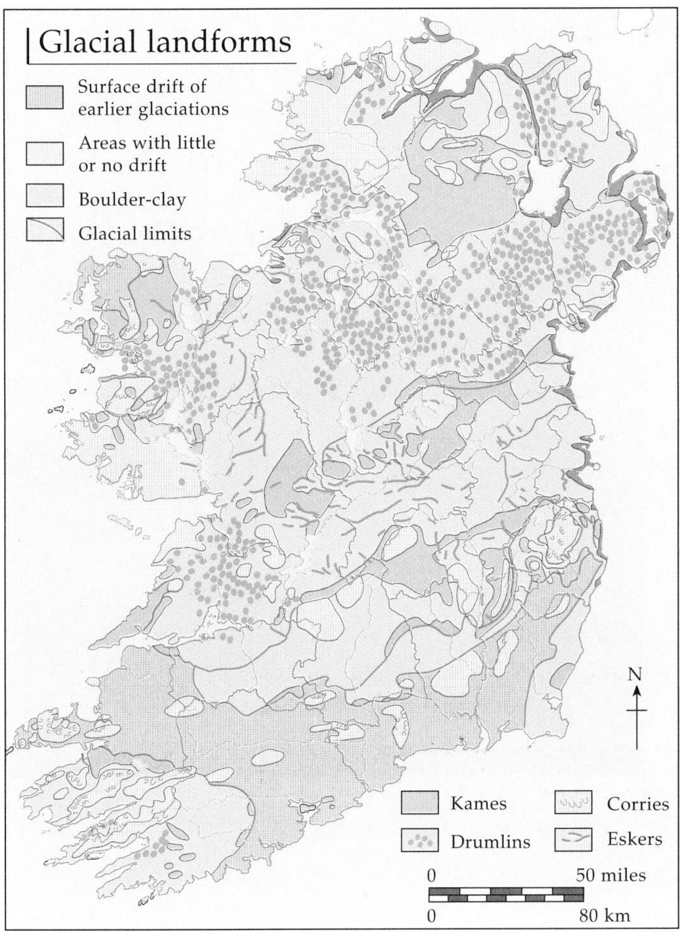


Fig. 12.43a
Glacial landforms. The erosional and depositional effects of the successive ice ages in Ireland were profound, more so than in Britain. Glaciers transported eroded material to the lowlands where it forms the widespread glacial drifts.

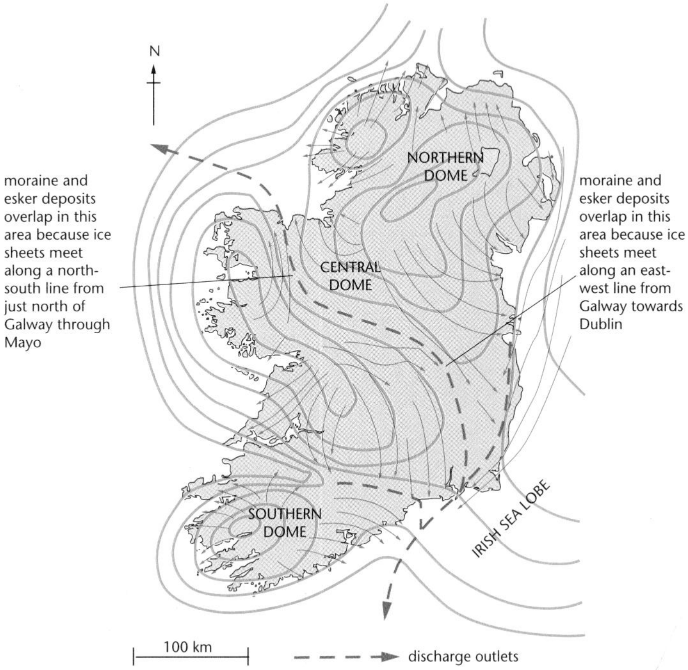

N
moraine and esker deposits overlap in this area because ice sheets meet along a north-south line from just north of Galway through Mayo
NORTHERN DOME
CENTRAL DOME
SOUTHERN DOME
IRISH SEA LOBE
moraine and esker deposits overlap in this area because ice sheets meet along an east-west line from Galway towards Dublin
100 km
discharge outlets

Fig. 12.43b
Irish ice generated a number of domes or small ice sheets, one in the west based on a line between west Galway and north Tipperary, one in the north based on a line between Leitrim and Lough Neagh and one in the south-west based on the Kerry/Cork mountains. Small ice caps also developed in the Wicklow Mountains, in Corca Dhuibhne and in the Macgillycuddy's Reeks.

## Outwash Plain

The melting ice sheet caused numerous rivers to flow from the front of the ice. This meltwater flushed sand, gravel and clay through the terminal moraine to form an **outwash plain** (Fig. 12.42). The Curragh in Co. Kildare in Ireland is such a feature. Heavy deposits of coarse material are generally deposited near the ice, while thinner and finer deposits are deposited further away. During spells of summer drought, such areas are inclined to scorch their vegetation cover due to the coarse or stony nature of the soil.

Large areas of the North European plain are covered by outwash sands and gravels. As the last ice advance stopped over Denmark and eastern Germany, rivers flowed westward across northern Europe. These rivers laid down vast amounts of sands and gravels. Today these deposits have been separated into individual heathlands by the north-flowing, postglacial and present-day rivers such as the Weser, the Ems and the Rhine to form the heathlands of Netherlands and northern Germany. Luneburg heath is one such area.

## Landform: Esker (Eiscir)

Examples: At Clonmacnoise in Co. Offaly
North and south of the River Brosna in counties Offaly and Westmeath
From Athlone in Co. Westmeath to Athenry in Co. Galway
From Athenry through north Galway and south Mayo to an area south of the Moy River

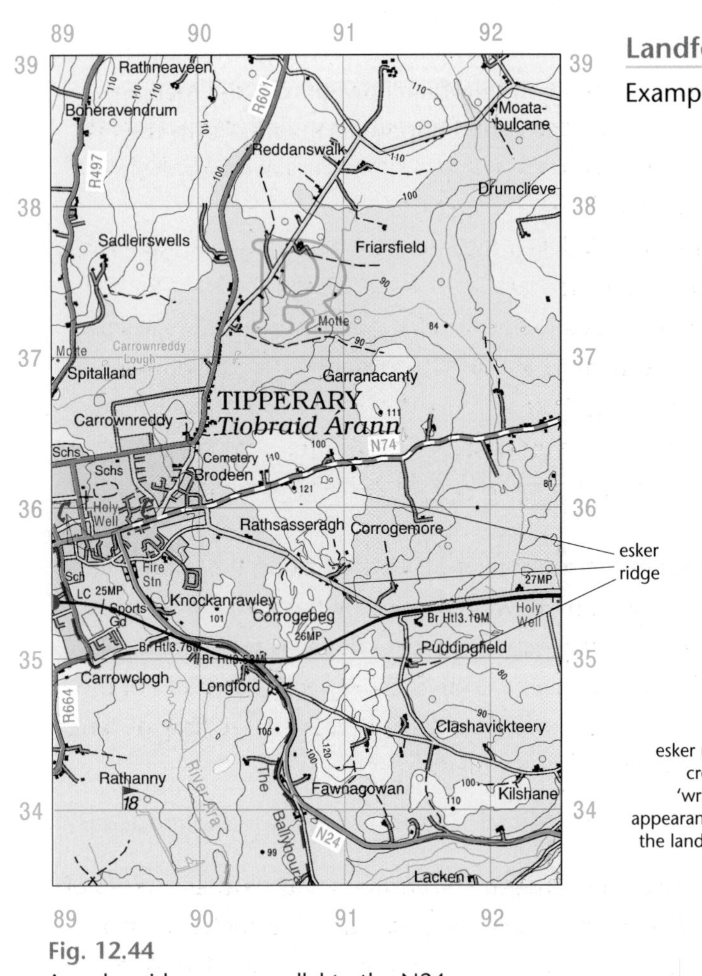


Fig. 12.44
An esker ridge runs parallel to the N24 roadway as it winds its way northward

Fig. 12.45
Esker ridges near Clonmacnoise in Co. Offaly

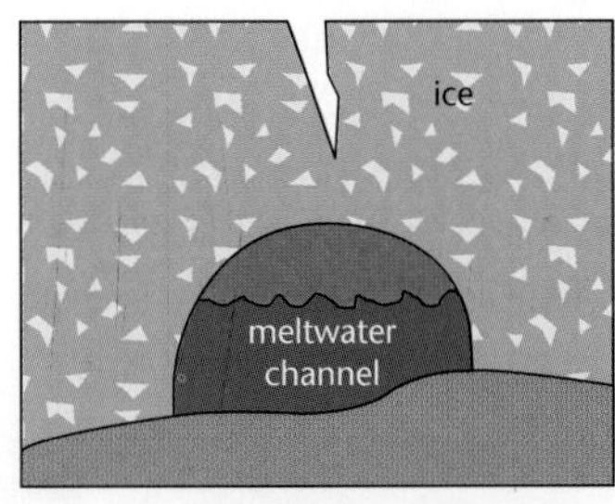


1. As ice melts, meltwater channels form under the ice.

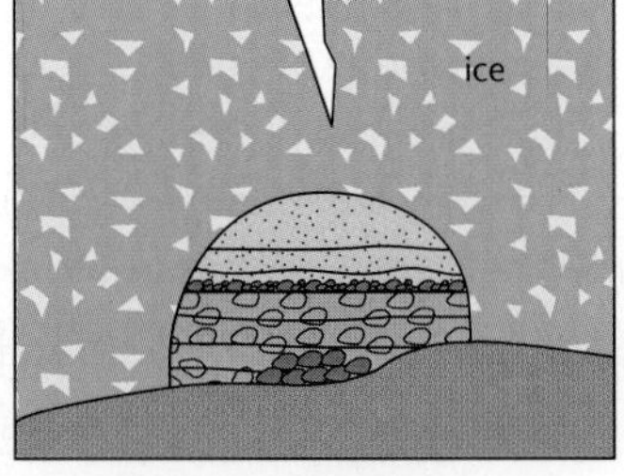


2. Sand, gravel and boulders are deposited, depending on the speed of meltwater flow.

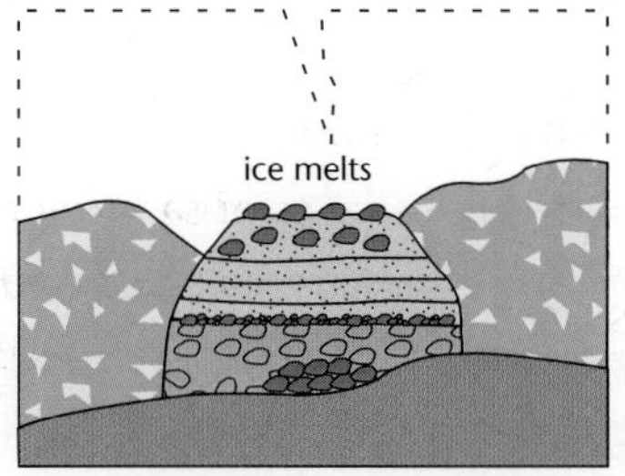


3. Meltwater channel fills with deposits as the ice melts.

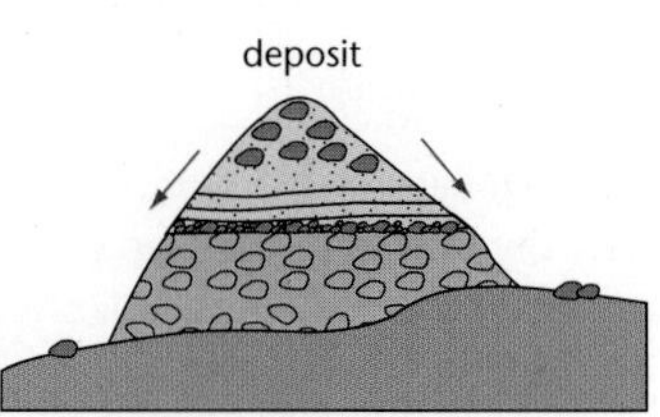


4. After the ice has melted, esker slopes stabilise, leaving a ridge of sand, gravel and boulders.

Fig. 12.46
The formation of eskers

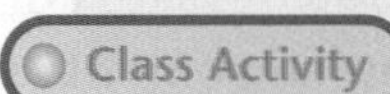

**Class Activity**

Due to differential weathering, an esker ridge appears as disjointed irregular hills on the Ordnance Survey extract, Fig. 12.44, as it winds its way northward through Tipperary town. Identify these hills and, with a marker, enclose the area occupied by these sandy ridges.

Processes: Suspension, Traction, Saltation, Hydrostatic Pressure, Deposition, Slumping

## Formation

Eskers are long, low and winding ridges of sand and gravel which are orientated in the general direction of ice movement. Sections through eskers have revealed alternate strata of coarse and fine deposits, representing times of rapid and slow ice melt respectively. They represent the beds of former streams flowing in and under ice sheets.

Changes in discharge routes sometimes led to a section of tunnel being abandoned by the main stream flow. It would then silt up with sand and gravel. When the ice ultimately disappeared, the tunnel fill would emerge as an esker, a ridge running across the country for several kilometres and bearing no relation to the local topography. The surrounding landscape may have a boulder clay covering. This gives rise to rich farmland that often stands in stark contrast to the sandy soils of an esker, which may display a poor-quality grass surface of coarse grasses and scrub. Eskers were formed as ice sheets retreated rapidly (Fig. 12.46).

Most river processes were involved in the movement of material along the subglacial channel. Material was carried in suspension; more was dragged by traction and bounced by saltation. Once the ice was gone, the newly exposed esker ridge was subjected to weathering processes such as slumping and gravity. Owing to the enclosed nature of the subglacial stream, hydrostatic (water) pressure was considerable, causing the flow of water to be rapid. Thus, these streams could carry a heavy load.

## An tSlighe Mhór — The Great Road

Eiscir Riada in Ireland marked a dividing line between the northern and southern halves of the country and enabled journeys to be made across the bogland of the centre of Ireland. This route was the most important factor for the construction of an tSlighe Mhór (old Celtic routeway), which stretched from Dublin to Galway.

Early roads were paved with stone, while wooden causeways (toghers) were constructed over bogland by the Celts.

## Kame Terraces

These are terraces of sands and gravels laid down by rivers which were trapped between a valley glacier and the valley side. Since glaciation, these terraces have undergone considerable change due to slumping and dissection by streams falling from the benches or higher valley sides.

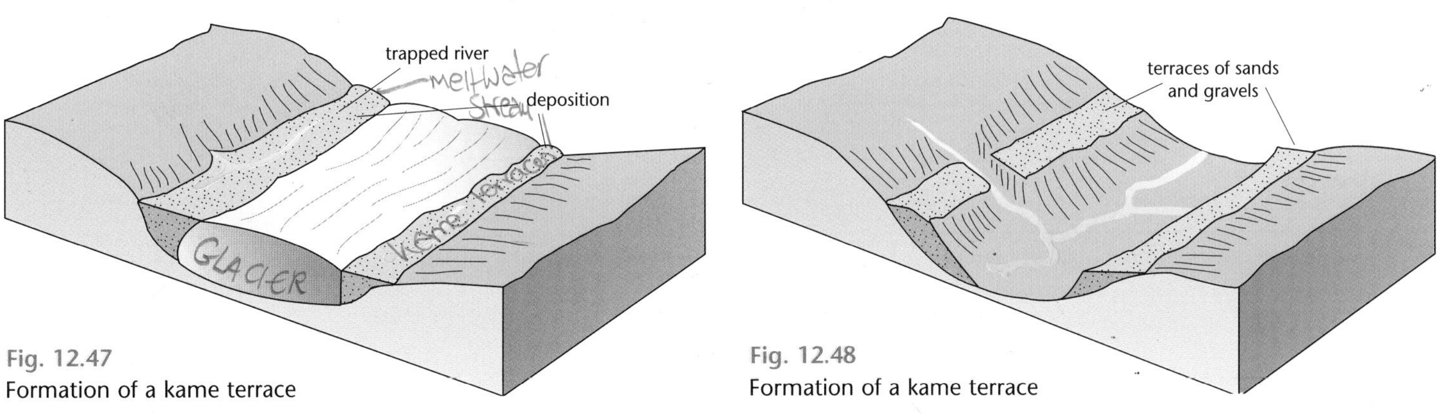


Fig. 12.47
Formation of a kame terrace

Fig. 12.48
Formation of a kame terrace

## Kames

These are undulating mounds of stratified sand and gravel laid down by rivers along the front of a long, stagnant and slowly melting ice sheet. They are basically alluvial cones and deltas laid down by meltwater streams falling from the ice sheets.

Kames are found in Ireland in the Lecale peninsula in Co. Down. This type of landscape is also found to the south of Big Delta, in Alaska.

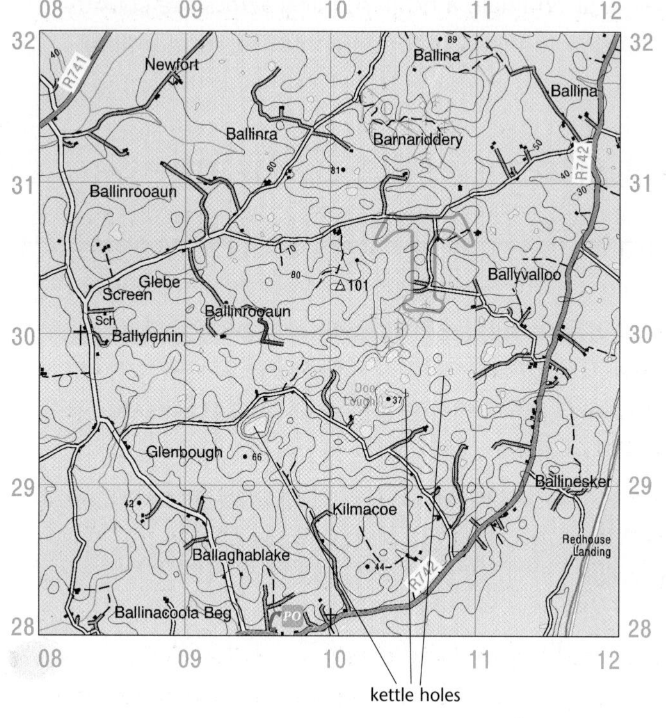


Fig. 12.49
Kettle holes dot this area near Curracloe and Blackwater in Co. Wexford

## Kettle Holes

As ice sheets retreated, large blocks of ice were often left isolated from the main ice mass. These were left buried or partially buried in outwashed sands and gravels. When these blocks of ice finally melted, hollows or depressions were left which often contain lakes. The word 'kettle' comes from the Kettle Range in Wisconsin near Lake Michigan, where these hollows are numerous. Kettle holes are often found in association with kames.

*Example:* South-west of Blackwater near Curracloe in Co. Wexford, in Ireland

## The Periglacial Fringe

The periglacial fringe was a zone or belt of considerable width which was not glaciated but was subjected to tundra conditions. As a result, this area was subjected to deep-freezing of the soils, subsoils and even bedrock. Such frozen subsoil is known as permafrost.

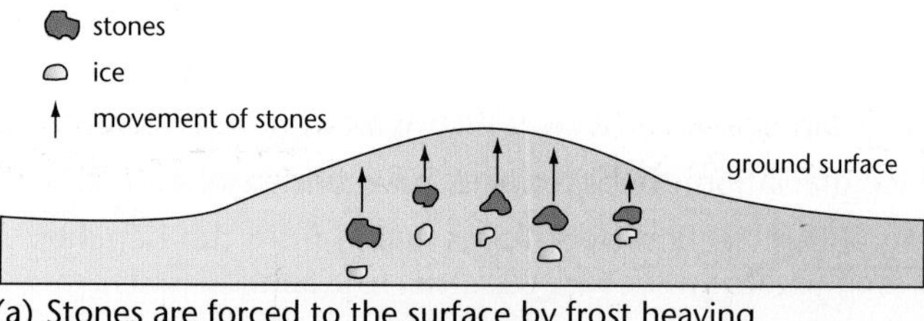


(a) Stones are forced to the surface by frost heaving

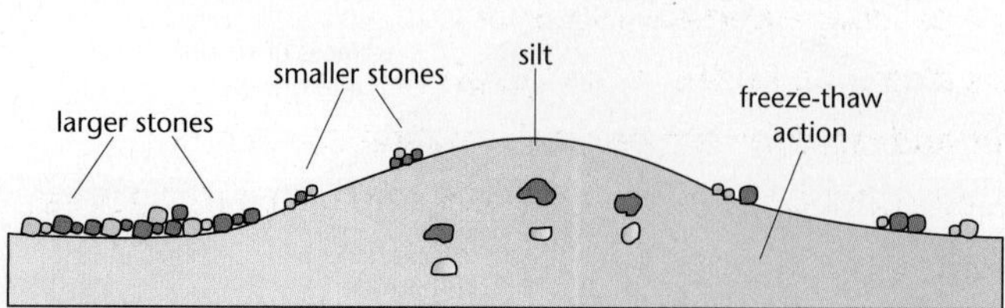


(b) Stones roll down into the hollows between mounds and material becomes sorted in size, with the finest deposits left in the centre of the polygon and on top of the mound

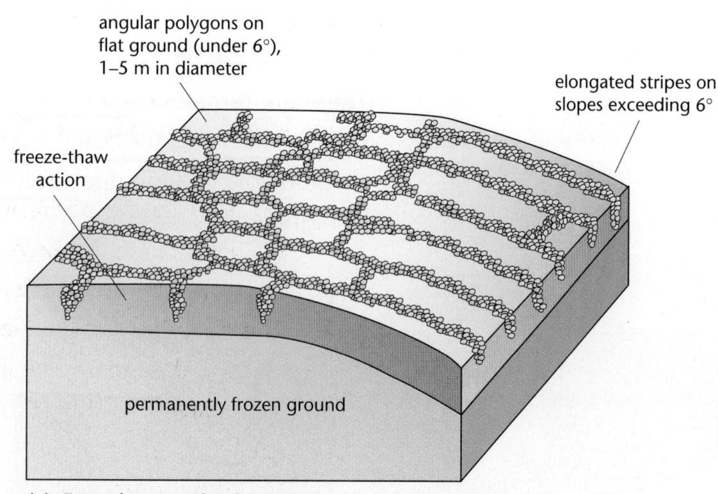


(c) Frost heave: the formation of polygons and stone stripes

Fig. 12.50

## Some Dry Valleys

During this cold spell, many rocks had their joints and fractures and bedding planes sealed by ice, and so were impermeable. On such rocks, run-off and erosion may have been much greater than under normal conditions. It is possible that many dry valleys of chalk and limestone uplands may have been formed in this way.

Freeze-thaw was an important form of weathering during periglacial conditions when the day-to-day temperature changes varied on either side of 0 °C. Jointed rocks would be especially prone to this form of breakdown. This would quickly produce quantities of scree for meltwater streams to remove.

Processes: Frost Heave, Freeze-thaw, Gravity

## Frost Heave — Stone Polygons

Frost heave includes many processes which cause fine-grained soils, such as silts, and clays to expand and cause stones to move to the surface. Stones lose heat more rapidly than soils. Consequently, the area under a stone becomes colder than the surrounding soil and ice crystals form.

Further expansion by the ice widens spaces in the soil, allowing more moisture to rise and freeze. The ice crystals, or even larger patches of ice which form at greater depth, force stones above them to rise until eventually they reach the surface. Where repeated freezing and thawing occurs, frost heave both lifts and sorts material to form stone-patterned ground on the surface. The heavier larger stones on flat ground move outward to form **stone polygons**. Where this process occurs on sloping ground, the stones will slowly move downhill under gravity to form elongated stone stripes.

Southern Munster in Ireland, and southern England were not covered by ice during the last ice advance. These areas were subjected to periglacial conditions at that time.

When deep stony soil was subjected to periglacial conditions, freeze-thaw action and the 'slushy' nature of the soil caused churning of the stones in the soil to form a folded pattern. Such patterns can be seen exposed at Baile na nGall on the Dingle peninsula, in Ireland.

---

Study the photograph of Bracklin Little in Co. Offaly, Fig. 12.51, p. 222. Then do the following:

1. (a) Classify the dominant glacial landform crossing the photograph.
   (b) Describe its surface characteristics.
   (c) With the aid of a sketch map, fully explain its formation.
2. There are a number of shades of brown on the photograph.
   Choose three different shades and describe the origin/formation of each.
3. There are a number of wrinkles/undulations in the area other than that in 1 above.
   (a) Suggest an origin for these.
   (b) What relationship, if any, exists between these undulations and at least one of the brown patches?
4. Describe how the physical landscape has influenced the development of communications in this area.

Fig. 12.51

Study the Ordnance Survey map of Clara/Tullamore, Fig. 12.52, p. 223. Then do the following:

1. (a) Locate the two examples of the main glacial landform of deposition on the map.
   (b) Explain how you identified and classified this landform type.
   (c) To what advantage have people used these glacial landforms?
   (d) Explain the processes that formed these landforms.
2. Why are buildings absent from the area in the vicinity of grid ref. N 255 305 and N 315 325?
3. Explain why a canal was constructed along the stretch of land at northing 25.
4. What were the physical advantages for choosing the site at Clara as an urban centre? Explain fully, using evidence from the map.
5. What pattern of drainage does the Tullamore River display?

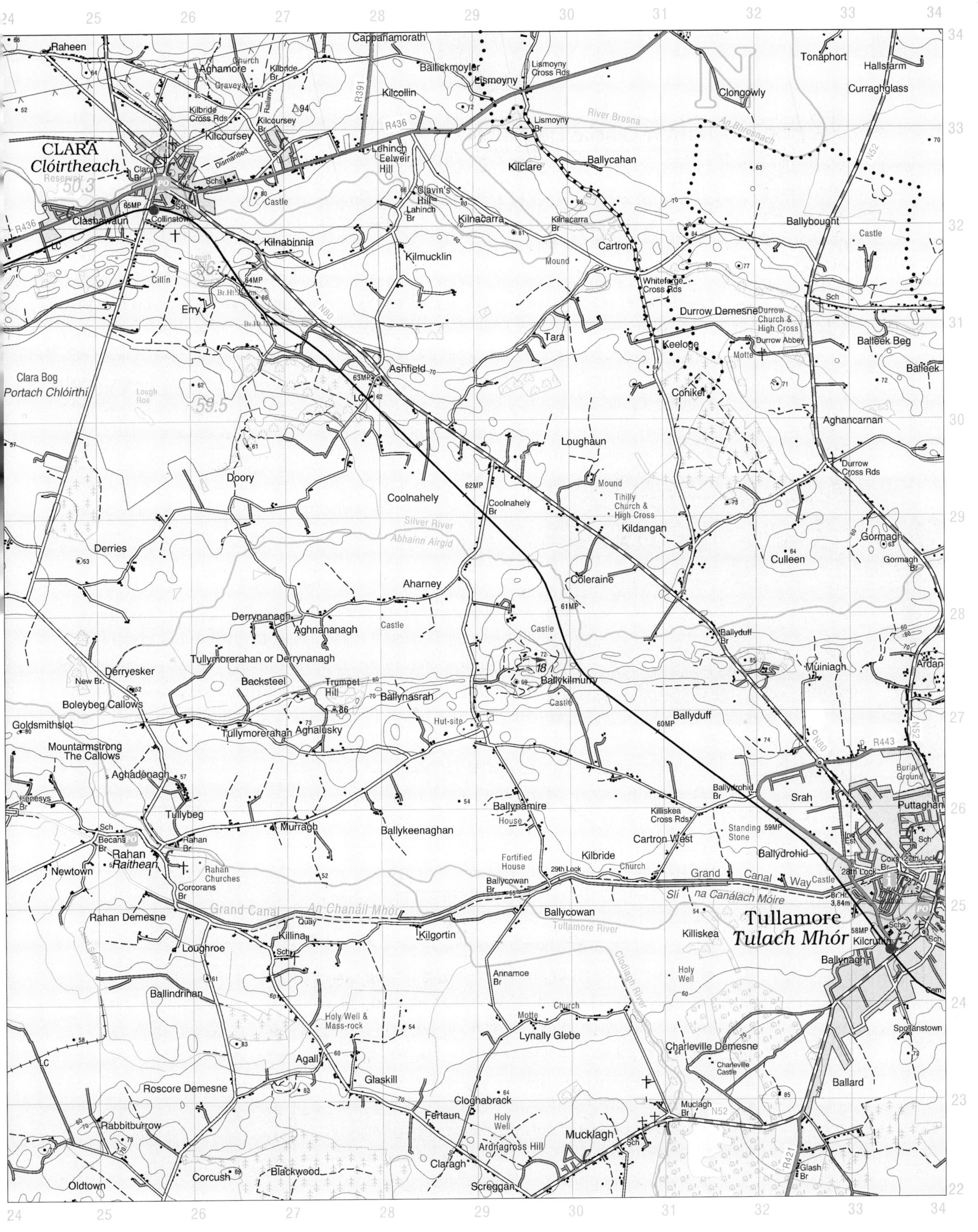


Fig. 12.52
Clara/Tullamore

# Dating Glacial Deposits

**varve**
Two layers of sediment deposited in one year's time in a glacial lake: one layer of light sand and silt deposited in summer and one layer of dark clay deposited in winter.

## 1. Varves

During the melting stage of a glacier, ice often impounded meltwater and prevented its escape. Lakes then developed at the edge of the ice margin. In the past, some of these lakes were enormous, as large as some of the bigger states in the USA, e.g. Kansas. When the ice had melted, the water drained, leaving behind stratified (layered) lake deposits, called **varves**. Each **varve** consists of **two layers of sediment, one coarse** and **one fine**.

Towards the centre of the lakes, coarse sediment graded into fine-textured, delicately layered varves. Each varve consists of a layer of light, silt-sized rock flour and a layer of dark clay. The rock flour represents a summer deposit, washed in by milky streams when melting was active. The clay settled slowly to the bottom during more peaceful winter conditions when the lake was frozen over and wind/wave turbulence was absent. Therefore, a single varve (two layers) records approximately one year of time.

Varve counts indicate that it took about twelve thousand years for a vast ice sheet to shrink from its maximum point of advance in what is now the United States back to Canada.

Fig. 12.53
Varves deposited in a glacial lake (Montana, USA). A light and dark layer together represents approximately one year. Geologists are able to date ice retreats by counting the varves in lake deposits left behind by the ice.

## 2. Glacial and Interglacial Ages

The beginning of the last ice age, the Pleistocene, has been dated to between 1.8 and 2 million years BP (before the present). During this time ice sheets advanced and retreated a number of times over northern Europe and North America. As a result, the Pleistocene is divided into glacial and interglacial periods.

## 3. Relative Dating of Glacial Deposits

The most important dating clues come from glacial till. Till deposits are stacked one on top of the other. The highest — and youngest — till deposit contains the least weathered rock fragments. Tills at progressively lower levels show progressively higher degrees of weathering.

Each till deposit represents an ice advance, the **oldest** on the **bottom**, the most **recent** on **top**. Between each till layer is an ancient weathered soil which indicates an interglacial period (warmer time). These ancient soils contain pollen grains, seeds, roots and buried logs, indicating that the soil nourished extensive forests. These soils may in turn be covered by glacial outwash sands and gravels and by non-glacial stream, lake or bog deposits — all of which are evidence of glacial melting and warmer climates.

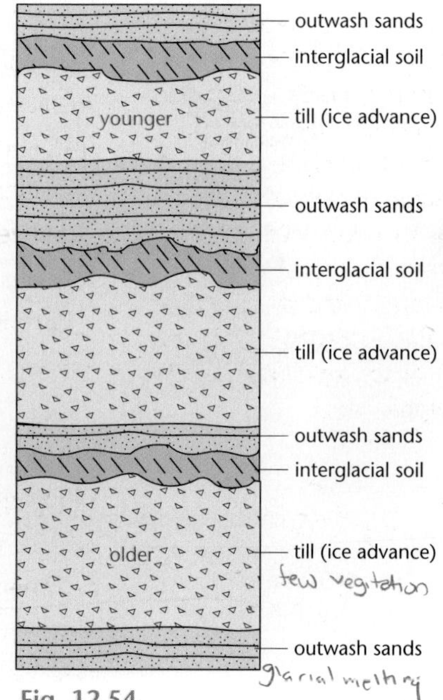


Fig. 12.54
Relative dating of glacial and interglacial deposits

# The Value of Glaciation to People

## Advantages

Fig. 12.55
Frozen archive of weather reports, the 50-metre face of the Quelccaya Ice Cap in Peru's Andes displays annual layers of snow separated by dry-season dust. A 165-metre-long ice core, taken at the ice cap's summit, holds 1,500 years of climatic data. Ice cores from Greenland and Antarctica trace conditions as far back as 150,000 years. But Quelccaya offers a rare record of equatorial weather patterns, such as El Niño, which can wreak global havoc.

### Scenery

Glaciated regions contain some of the most attractive landscapes in the world, such as the deep U-shaped glacial troughs. Their ruggedness, caused by the erosive action of the glaciers, has stamped a wild beauty on these areas, some of which are classified 'National Parks'. Steep cliff-like valley sides, cascading waterfalls, ribbon lakes and scoured rock surfaces are just some of the attractions.

Yosemite National Park in California and Glacier National Park in the United States, Glenveagh National Park in Co. Donegal and Glendalough in Co. Wicklow in Ireland as well as the Alpine Bernese Oberland in Switzerland are just some examples of these areas. Tourism, the fastest-growing industry, has its roots firmly based in glaciated landscapes, and accounts for a growing percentage of national economies worldwide.

### Boulder Clay

Farming thrives where boulder clay deposits are found on well-drained, sloping or undulating ground. Such clays will provide a deep and fertile soil, rich in minerals, especially if it has been eroded from a limestone landscape. In Ireland, well-known expanses of rich farmland, such as the Golden Vale, the Blackwater Valley in Cork and Waterford in Munster, the drumlin belt from Sligo to Strangford Lough, north Kildare and Meath in Leinster and the river valleys of Armagh in Ulster owe much of their fertility to the minerals which were deposited in their boulder clays.

The Sligo–Strangford drumlin belt was one of the most fertile areas in Ireland during the neolithic (young) period. After glaciation, the lowland soils of Roscommon and western Sligo had a very high lime content and were not as heavily wooded as the remainder of the country. Rich pasture provided excellent conditions for rearing cattle. The countless megalithic tombs found in this region are indicative of this farming activity. However, the lime content has since been leached from some soils by the heavy rains of the west of Ireland and many of these same soils are now acidic and less fertile.

Boulder clay represents the ground moraine of ice sheets. It covers large areas in eastern Denmark, northern Germany and Poland.

### Glacial Lakes and Lake-beds

Many glacial lakes have not survived because the ice sheets formed their margins, either in whole or in part, and when the ice melted, the lakes drained. Left behind were lake-bed sediments, as well as ancient beaches which mark their shorelines. Central Canada and Minnesota and North Dakota in the USA formed the lake-bed of a vast glacial lake called Lake Agassiz. As a consequence, this lake-bed now forms some of the richest farmland on the continent of North America.

Rich alluvium which once collected on the beds of these lakes now produces tillage crops or rich pasture for dairy and beef herds. In Ireland, the Glen of Aherlow in Co. Tipperary was once such a lake which drained southward through a glacial spillway towards Galbally, in Co. Limerick.

### Eskers, Road and Building Construction

These winding and stratified ridges of sands and gravels are found scattered over the central plain and east Galway and Mayo in Ireland, northern England and Scotland, and central Sweden. These deposits form the raw materials for the construction industry in the manufacture of concrete and pavement slabs. Quarrying is quite noticeable in Ireland in places such as the Roadstone quarry at Gooig in Co. Limerick and near Clonmacnoise in Co. Offaly as well as in many Alpine valleys. Sinnot's quarry near Roscrea in Ireland has both esker and moraine deposits.

In Ireland, in ancient times, eskers were used as routeways which were free from flooding, especially in the central plain. They stood above shallow lakes, which they impounded and in which peat deposits formed. Their sands and gravels are still in constant demand for road building purposes.

### Hydroelectric Power

Dams in glaciated valleys are also used to generate hydroelectric power. In the Alps in Italy, rivers that flow through glaciated valleys high up on the benches are dammed and their waters are funnelled through huge pipes to the hydroelectric stations at their base. These high 'heads' are ideal for electricity generation. Factories that need large quantities of cheap energy, such as electrometallurgical industries, generally locate near to these hydroelectric power stations. Such industries are located at the heads of Norwegian fjords and in Italian Alpine valleys.

## Disadvantages

### Disadvantaged Areas

Some regions of the world, such as the Scottish Highlands and western Ireland as well as large areas of Finland, have been denuded of their soil cover. In these areas, glaciers removed soil by the process of abrasion and carried it many kilometres, sometimes even up to 1,000 kilometres, and then dropped it as boulder clay or till across lowland areas. These highland areas are, therefore, able to support only a low population density and have, over the centuries, generally been areas of out-migration due to a lack of services and low income potential.

### Avalanches

High in the Swiss Alps today, many areas are still subjected to the processes of glaciation. Alpine glaciers wind their way slowly downslope. However, snow and ice accumulation on almost vertical slopes presents constant danger in potential avalanches, which regularly

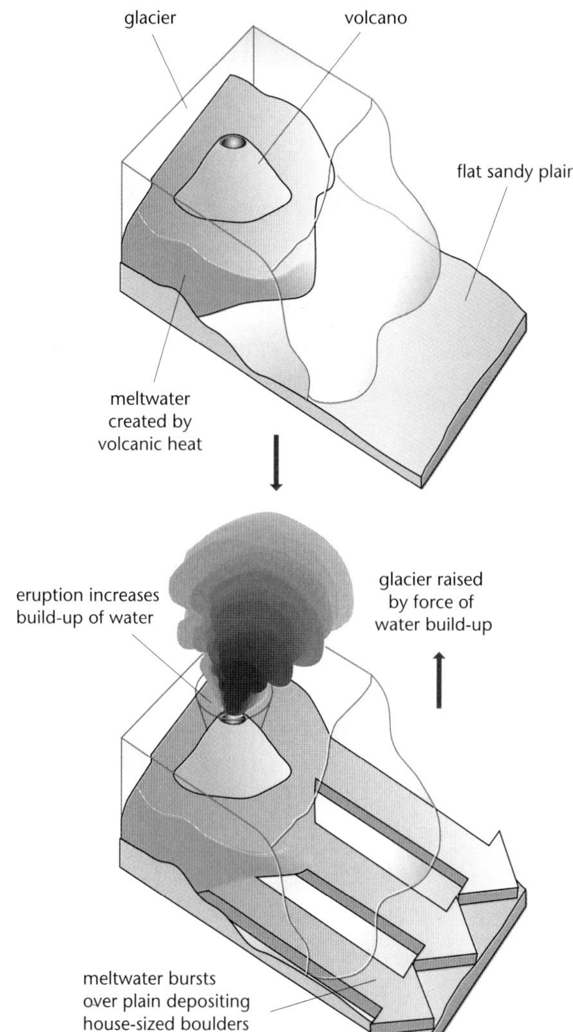


**Fig. 12.56**
In 1996 the Loki Volcano under Vatnajokull ice cap erupted. The hot magma created an underground lake which later burst through from under the ice and created a *Jokulhlaup*.

result in death. Avalanches in ski resort areas are the ones which regularly attract news coverage such as the avalanches in Austria in 1999, where thirty-four people died. Most occur as thaw sets in in early spring. Loud noises or vibration of any kind can create avalanches and send thousands of tonnes of snow and ice thundering down steep slopes, covering everything in their path.

## Glacier Bursts

Spectacular floods, called *Jokulhlaups* or glacier bursts, result from volcanic activity under glaciers. Huge meltwater lakes formed by a volcano's heat can suddenly burst from under the glacier in one of nature's most awesome displays of power. In 1918 Iceland's Myrdalsjokull glacier produced a *Jokulhlaup* with water flow estimated at three times that at the mouth of the Amazon River.

In October 1996 Iceland lay on full alert as a volcanic eruption under Europe's largest glacier, Vatnajokull glacier, threatened to burst through the ice cap and cause widespread flooding along the south coast. An eruption in 1938 at exactly the same place caused massive flooding.

Carefully study the Ordnance Survey map extract of the Derryveagh Mountains, Fig. 12.57, p. 228. Then do the following:

1. (a) What structural trend is evident in this area? Explain fully.
   (b) Account fully for this trend. (See 'Caledonian Fold Mountains, Fault-lines and Mineral Deposits', p. 74.)
2. Identify the glacial landforms at A, B, C and D. Explain fully how landforms A and B are formed.
3. (a) Identify, (b) locate and (c) describe the patterns of drainage evident on the map and (d) account for their formation.
4. Describe the course of the Bullaba River under the following headings: upper course, middle course and lower course.
5. With the aid of a sketch map, explain fully how the relief of this area influenced the development of the road network.
6. What characteristics of the natural physical landscape have led to this portion of this region being classified and protected as a National Park?

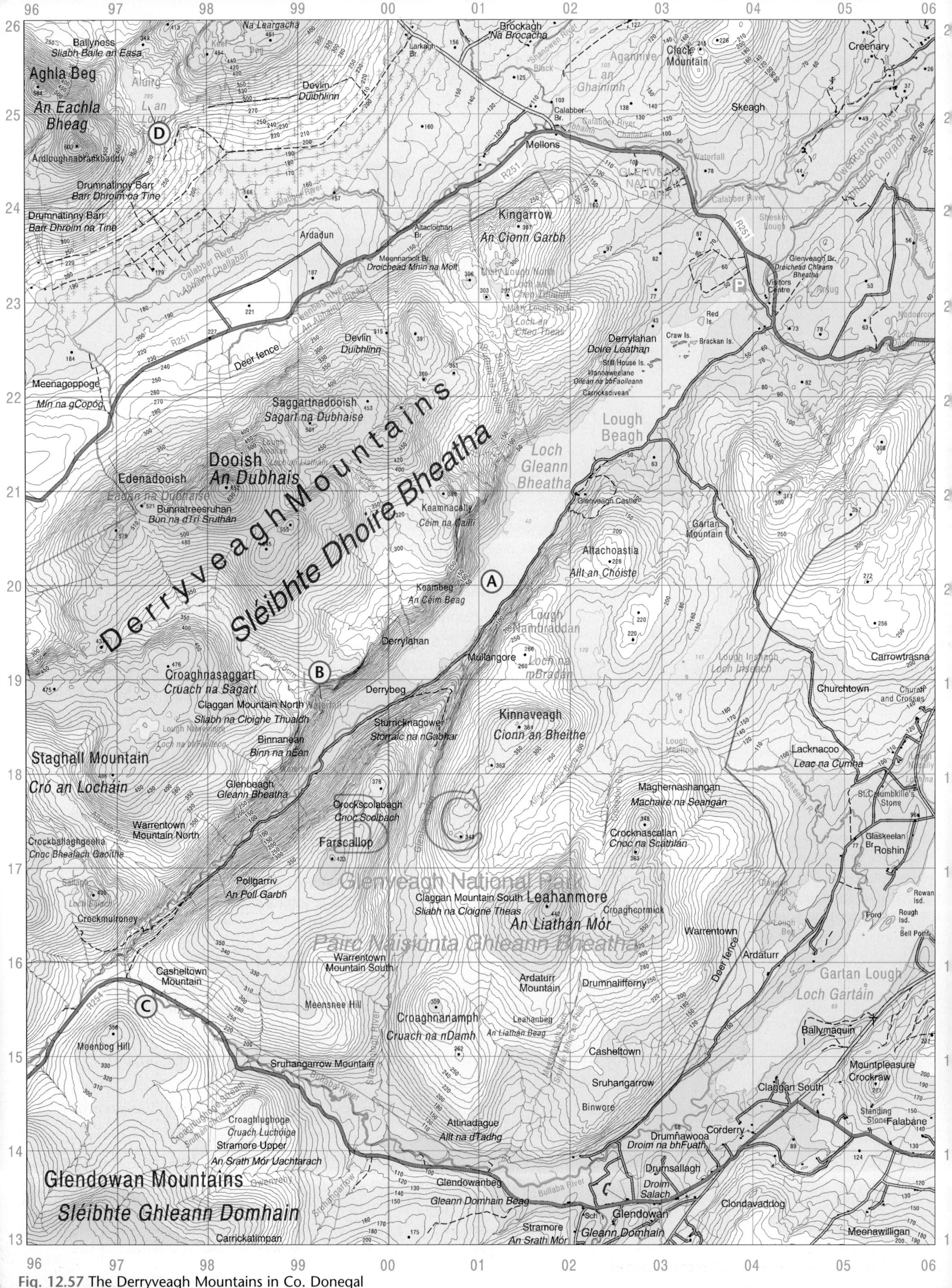


Fig. 12.57 The Derryveagh Mountains in Co. Donegal

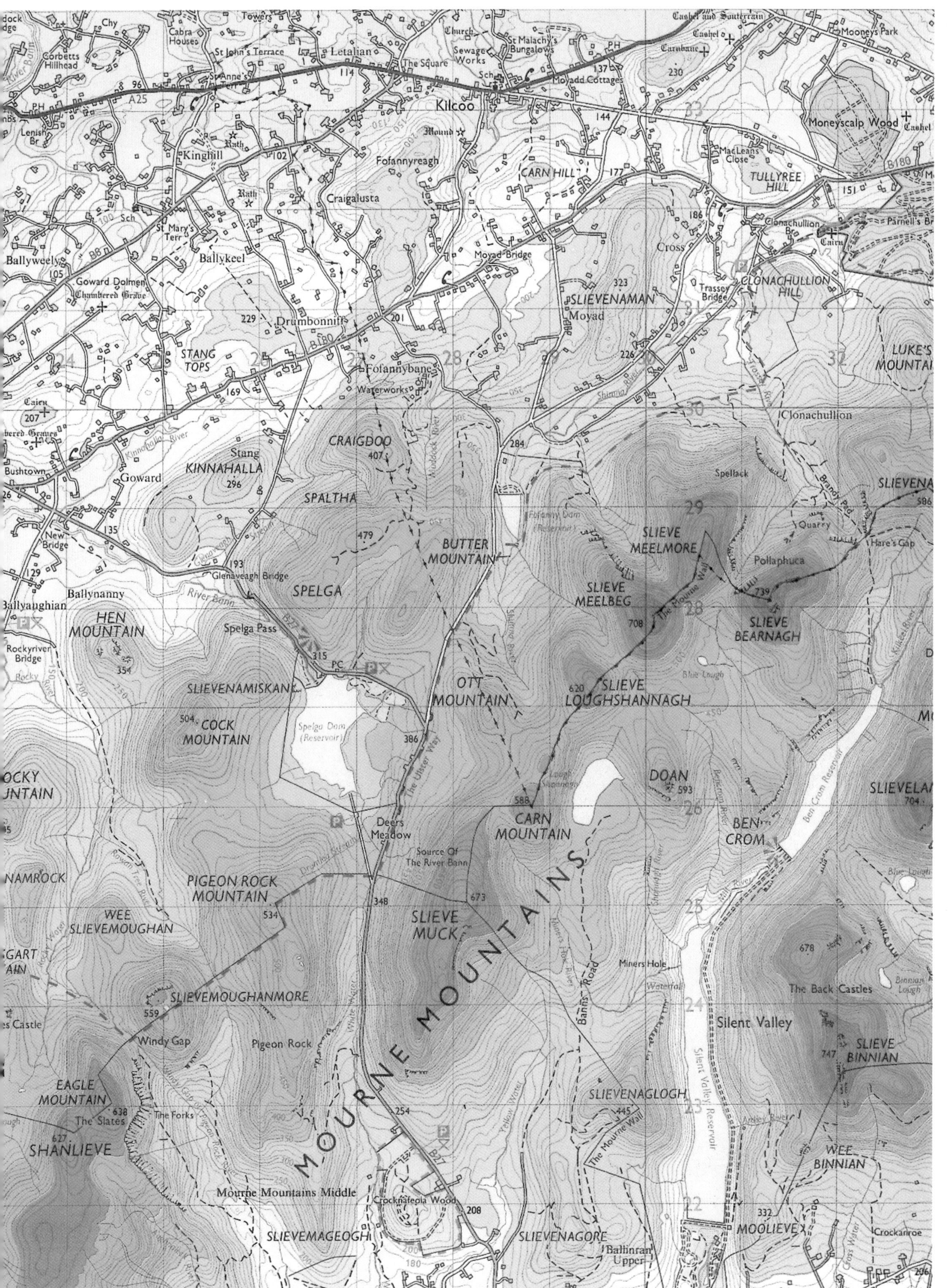


Fig. 12.58 The Mournes, Co. Down, Northern Ireland

Fig. 12.59
Ben Crom Reservoir, Mourne Mountains, Co. Down

## Map Questions — See Fig. 12.58, p. 229

1. Describe and explain, with reference to four typical landforms, how ice movement has shaped this physical landscape. In your answer refer to some landforms of erosion and deposition.
2. U-shaped valleys, hanging valleys, cirques and cirque lakes, saddles and drumlins are all landforms of glacial action. With reference to each landform, identify and locate one example and, with the aid of a sketch map, explain the processes which formed it.
3. Describe how the physical landscape has influenced the development of communications in the area.
4. Clean and sufficient sources of groundwater are becoming more difficult to find in recent years. With reference to the map, explain how local government has taken advantage of the natural landscape to overcome this difficulty.
5. The pattern of the highest points and ridges of the Mourne Mountains seems to display a definite trend. Identify this trend and account for its formation.
6. Classify the type of landscape in the northern portion of the map. Explain your choice fully, with the aid of a sketch map.
7. Identify one linear and one nucleated pattern of settlement north of northing 30. For each example, explain fully the reason for its presence at that location.
8. The northern half of the map displays a long history of settlement. Using examples from the map, explain this statement fully.
9. Names such as 'White Water' and 'Yellow Water' suggest fast-flowing rivers and landforms of young river valleys. Using evidence from the map, show how this area is affected by river erosion.

## Photograph Questions — See Fig. 12.59, p. 230

1. Is the feature at A a naturally formed or man-made landform or other? Explain fully, using evidence from the photograph to support your answer.
2. Glaciation has played a large part in the creation of this landscape. Choose two landforms created solely by the natural processes of glaciation. For each landform explain fully the processes involved in its formation. Identify certain characteristics of each of these landforms, which are clearly visible on this photograph.
3. The forces of denudation (weathering and erosion) are constantly at work reducing areas by erosion and increasing other areas by deposition. With reference to one example in each case, explain fully how the processes of weathering and erosion have influenced the formation of this region.
4. There is a whitish margin around landform A. What does this represent and what does it suggest about the weather condition of this region at the time this photograph was taken?
5. What evidence on the photograph suggests that much of this area is being eroded as a consequence of overgrazing?

## Photograph and Map Questions

1. Identify the landscape features at A, B, C, D, E and F.
2. Identify the spot on the map that each of the corners of the photograph represents. Join each of these spots. The enclosed area forms a square and represents the region in the photograph.
   (a) Name the river which flows into the lake at the most extreme south-west corner of the photograph.
   (b) Name the mountain in the north-east corner of the photograph and give its height in metres.
   (c) Name the lake on the south-eastern corner of the photograph.
3. What do the dark patches in the south to south-eastern part of the photograph represent?
4. What evidence on the map suggests that mineral extraction was carried on in this area in the past? Why, in your opinion, would this region have been chosen as having mineral potential? Explain fully. (See 'Mineral Formation at Subduction Zones', p. 41 and 'Caledonian Fold Mountains, Fault-lines and Mineral Deposits', p. 74.)
5. Using evidence from both the map and the photograph, show how this area displayed by the 'photograph' only is affected by river erosion.

## Fieldwork Activity

### Case Study: A Day Trip to the Galtees

Day trip to Lough Muskry in the Galtees. Access to the lake is by road and then pathway to the lake from Rossadrehid in the Glen of Aherlow. Trip takes a full day. Arrive at Rossadrehid at approximately 10 p.m. Finish at 4 p.m. (See Fig. 12.62.)

**Class Activity**

1. Find the hidden lake.
2. Ideal location for a fieldwork study on glacial landforms, for example cirque/cirque lake, moraines. Other: blanket bog and its vegetation, rockfalls.

A day trip to the Galtees — Trip C — glacial and river features

Fig. 12.60
**View A**
View from Knockastakeen in a south-south-east direction over Lake Muskry. This view shows the moraine material which impounds Lake Muskry.

**Fig. 12.61**

**View B**

View from the gully midway up the slope behind the small lake. This view shows the moraine material which impounds both lakes.

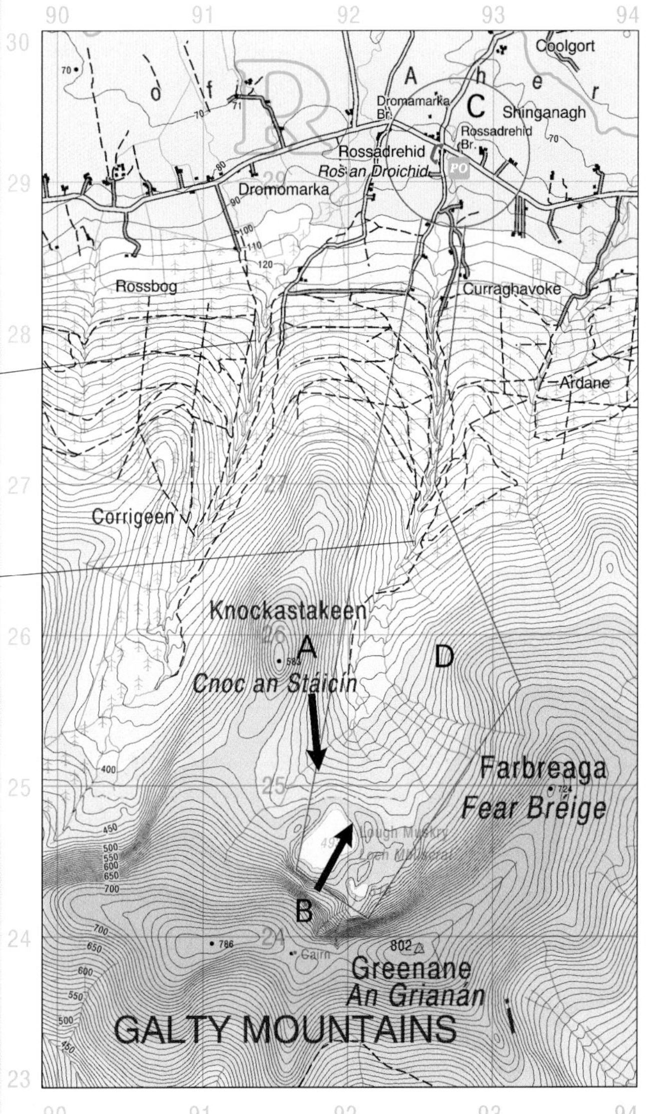

rockfalls

dry gullies

moraine impounding both lakes

Lake Muskry

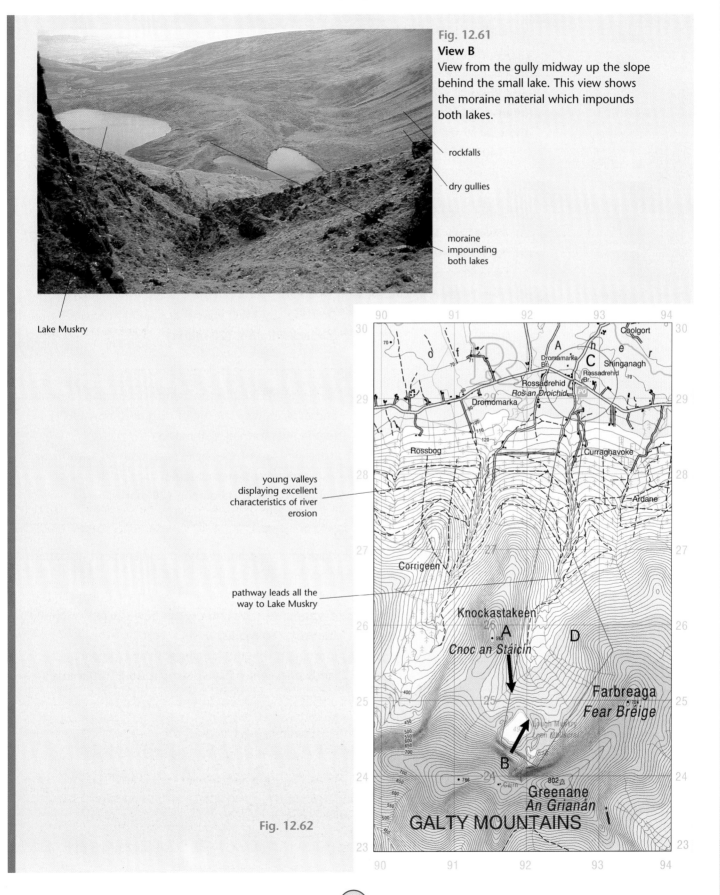

young valleys displaying excellent characteristics of river erosion

pathway leads all the way to Lake Muskry

**Fig. 12.62**

## Leaving Certificate Questions on Ice, Glacial Processes and Landforms

### Ordinary Level

**The Physical World**

**1998** — Explain the following:

(iv) Glacial action has resulted in many interesting landscape features in Ireland. **(40 marks)**

**1997**

Glaciated landscapes are shaped by erosion **and** deposition.

(i) Select any **THREE** features that result from the action of ice and in the case of each one you select:
- Describe and explain, with the aid of a diagram, how it was formed.
- Name a specific location where the feature may be found. **(60 marks)**

(ii) Describe any **TWO** ways that man has benefited from glacial landforms. **(20 marks)**

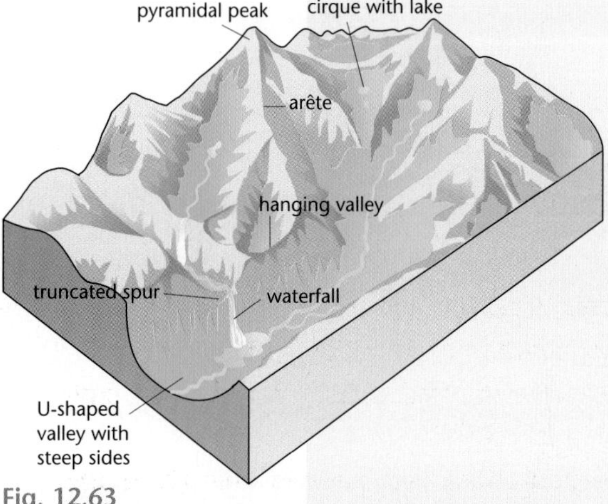


Fig. 12.63

**1995**

(i) Name any **THREE** features of glacial erosion. Describe and explain, with the aid of diagrams, how **EACH** feature was formed. **(60 marks)**

(ii) 'Glaciated landscapes have been both an advantage and a disadvantage to people.'
Briefly discuss this statement. **(20 marks)**

**1992**

This diagram shows an area which experienced glaciation in the recent geological past.

(i) With reference to landforms which are evident in the diagram, describe and explain how glaciation shaped the area. **(60 marks)**

(ii) Examine briefly the human economic potential of landscapes which have been glaciated. **(20 marks)**

Fig. 12.64

**1991**

(i) Describe and explain the formation of any **THREE** landforms which are produced by the processes of deposition by ice masses. **(60 marks)**

(ii) Examine briefly ways in which modern society makes use of landscapes affected by glacial deposition. **(20 marks)**

**1990**

There are many areas in Ireland today which show that erosion by moving ice played an important role in shaping our landscape in the past.

(i) Identify **THREE** landforms typical of such areas and explain the processes which produced them. **(60 marks)**

(ii) The melting of the ice masses at the end of the last glacial period released great quantities of meltwater. This meltwater had its own effects on the landscape. Explain the above statement. **(20 marks)**

Higher Level

**1998**

(i) Examine, with reference to processes of erosion **and** to processes of deposition, the formation of **three** landforms which result from glaciation. **(75 marks)**

(ii) There have been periods in the past when glaciation affected the earth's surface much more extensively than today.
Examine briefly some of the theories advanced to account for this. **(25 marks)**

**1997**

(i) Discuss **three** ways in which glaciation has shaped the Irish landscape. **(75 marks)**

(ii) Examine briefly how materials deposited by ice can be dated. **(25 marks)**

**1994**

(i) Identify **three** landforms produced by glaciation. In the case of **each** landform, describe and explain how it was formed. Use diagrams to support your answer. **(75 marks)**

(ii) Examine briefly how the results of glaciation of the landscape have been used to economic advantage by human societies. **(25 marks)**

**1992**

(i) Explain what basic conditions of climate and of topography are necessary in order to bring about a period of widespread glaciation of a landscape. **(30 marks)**

(ii) Examine how moving ice shapes the landscape. Refer in your answer to:
- Landforms formed beneath the ice and
- Landforms formed at or beyond the ice front. **(70 marks)**

**1991**

With reference to (i) moving ice and (ii) melting ice, examine how glaciation has helped in the past to shape the Irish landscape. **(100 marks)**

**1989**

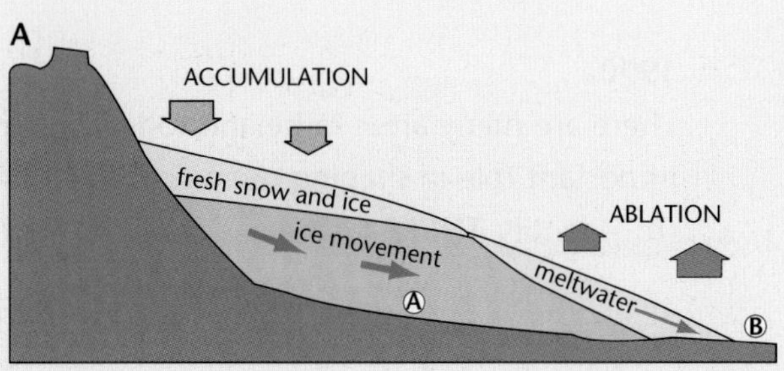


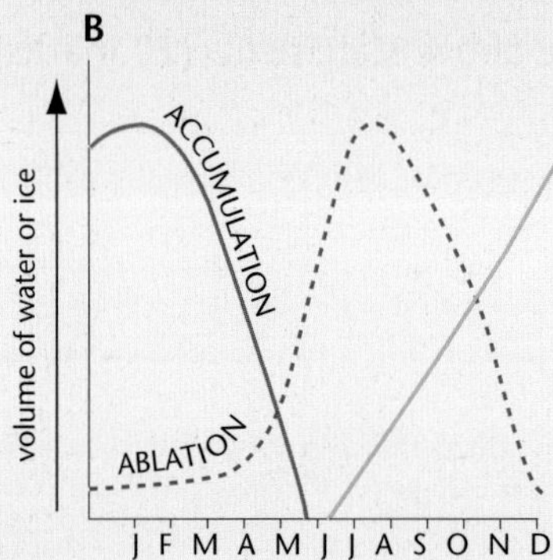


Fig. 12.65

(i) Examine diagram A above, which illustrates the glacier system. Describe the different physical processes which are active at point A and at point B. Explain briefly the formation of **TWO** characteristic landforms which would be produced at **EACH** point. **(80 marks)**

(ii) Examine again diagram A, and also graph B, which shows how rates of accumulation and ablation (melting) vary over a period of one year. Describe and explain what this information tells us about how the glacier system changes over time. **(20 marks)**

(b) The retreat of ice masses during a period of glaciation leads to the release of large quantities of meltwater.

  (i) Examine **THREE** ways in which this can affect landscapes. **(75 marks)**

  (ii) Explain **TWO** methods by which scientists attempt to date glacial deposits. **(25 marks)**

# The Sea, Coastal Processes and Landforms

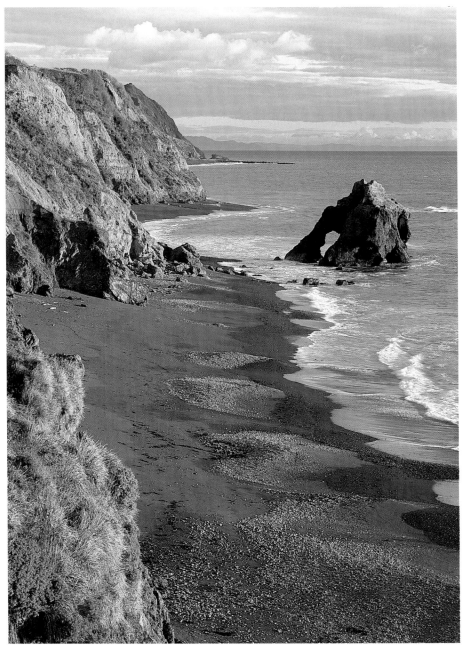

Fig. 13.1
The 'Needle' at sunset. Sinkyone Wilderness State Park, California, USA.

When referring to a particular stretch of coastline, correct terminology should be used.

**Coastline:** The margin of land bordering the sea, which is demarcated by either a cliff-line or the line reached by storm waves.

**Shore:** The area between the lowest tides and the highest point reached by storm waves.

**Beach:** That part of the shore with deposits of sand, shingle or mud.

The character of any coastline depends on a number of factors.

1. The **work of waves, tides and currents** which erode, transport and deposit material.
2. The **nature of the coastline**: whether the coastal rock is resistant or not; varied or homogeneous in character; type of coastline — highland or lowland and straight or indented.
3. The **changes** in the relative **levels** of land and sea.
4. **Human interference**: the dredging of estuaries, the creation of ports, the reclamation of coastal marshes, the construction of coastal defences against erosion such as groynes, dykes and breakwaters: the building of piers and promenades.

LANDSCAPES OF THE WORLD

# Marine Erosion

## Processes of Erosion

Fig. 13.2
The force of water from a power-hose removes dirt from a car

### 1·Hydraulic Action

The direct impact of strong waves on a coast has a shattering effect as they pound the rocks. Such waves breaking against the base of a cliff force rocks apart, making them more susceptible to erosion. Cliffs of boulder clay are particularly affected as loosened soil and rocks are washed away.

### 2·Compression

Air filters into joints, cracks and bedding planes in cliff faces. This air is trapped as incoming waves lash against the coast. The trapped air is compressed until its pressure is equal to that exerted by the incoming wave. When the wave **retreats**, the resultant expansion of the compressed air has an explosive effect, enlarging fissures and shattering the rock face (Fig. 13.3). ↳ cracks in the rocks.

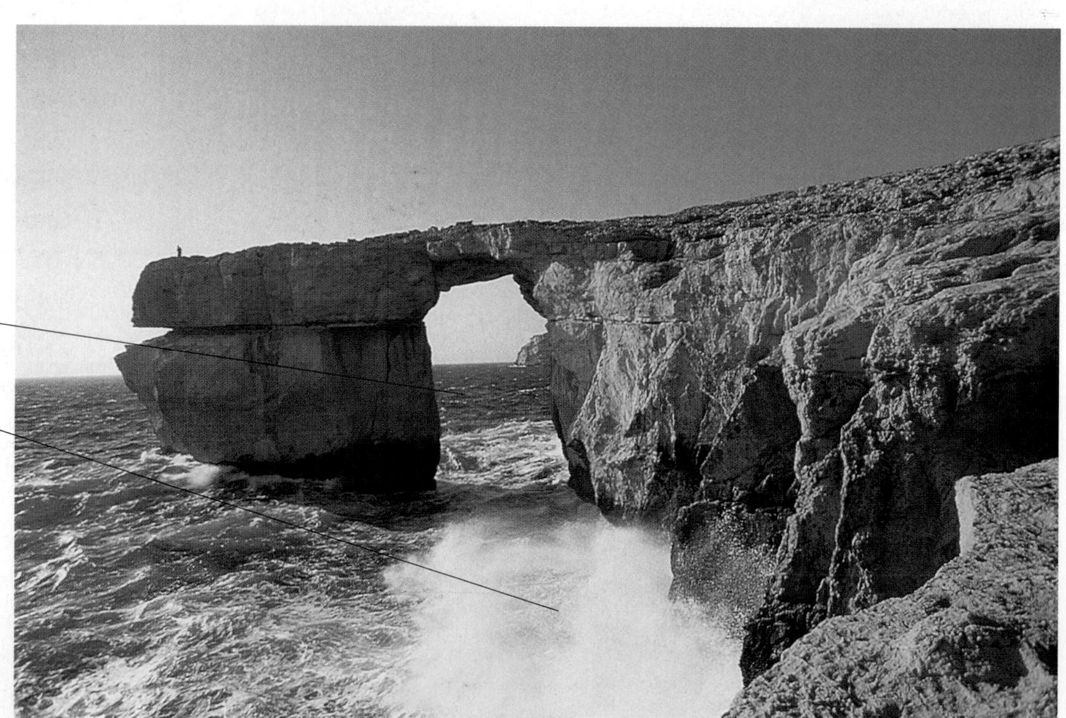
air is pressurised into cracks and crevasses in coastal rock
waves lash against the cliff face and trap air which is then compressed, causing hydraulic action

Fig. 13.3
Hydraulic action and compression in action on the Malta coastline

### 3·Corrasion and Abrasion

When boulders, pebbles and sand are pounded against the foot of a cliff by waves, fragments of rock are broken off and undercutting of the cliff takes place. The amount of corrasion is dependent upon the ability of the waves to pick up rock fragments from the shore. Corrasion is therefore most active during storms and at high tide when incoming waves throw water and suspended rock material high up the cliff face, and sometimes onto the cliff edge (see Fig. 13.4).

4. Attrition

Fragments which are pounded by the sea against the cliff and against each other are themselves worn down by attrition, creating sand and shingle.

5. Solution: $CO_2$ in the atmosphere

Fig. 13.4

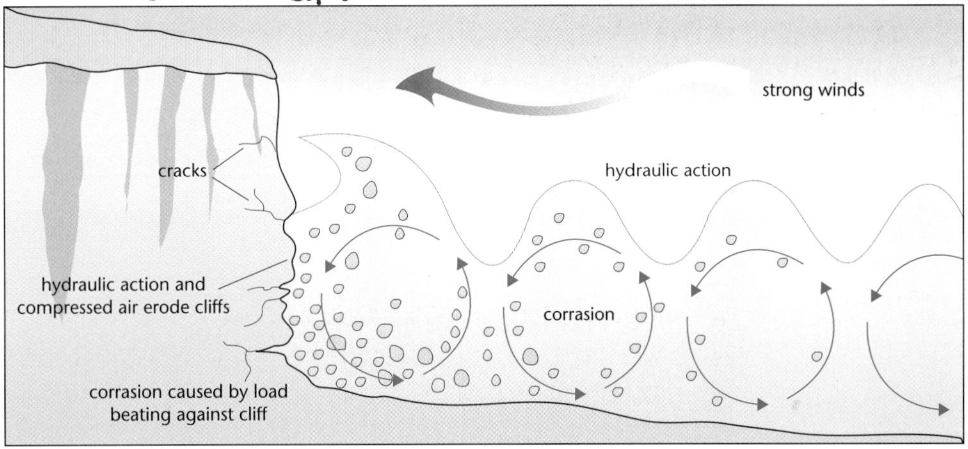

Fig. 13.5
Destructive wave action erodes carboniferous rock strata and exposes a fossilised tree trunk which was preserved in a thick layer of sediment created by a tidal wave on the Clare coastline in western Ireland

## Factors which Aid Erosion

### Destructive Waves

The power and size of a wave depends on wind speed and the **fetch**, i.e. the length of open water over which the wind blows. The stronger the wind and the longer the fetch, the stronger the waves and the greater the erosion.

Strong wind + long fetch =
strong waves = great destruction

Destructive breakers which pound the coastline have their greatest effect during storms. Because of their frequency (twelve per minute) and because of the almost vertical plunge of the **breakers** (breaking waves), the **backwash** is much more powerful than the **swash**. So these destructive waves dig up beach material or loose material near a cliff and drag it **seaward** (out to sea). The power of Atlantic waves on the western coast of Ireland is increased threefold during storms.

### Refraction

The speed of waves is reduced when waves approach a shore. The depth of water varies on shorelines which have alternate promontories and bays. In such cases, the water is shallower in front of the promontory than in the bay. As waves approach the shore, the shallower water off the promontory causes the wave to bend towards the headland, thus increasing erosion (see Fig. 13.7). This **wave refraction** or **bending** also occurs when waves pass the end of an obstacle such as a spit, creating a **hook**. Thus the process of refraction is involved in both erosion and deposition along a coastline.

Fig. 13.6
Wave refraction around a point of land off Palm Beach, Sydney, Australia. Here wave refraction creates triangular-shaped surf as the waves break along the shore.

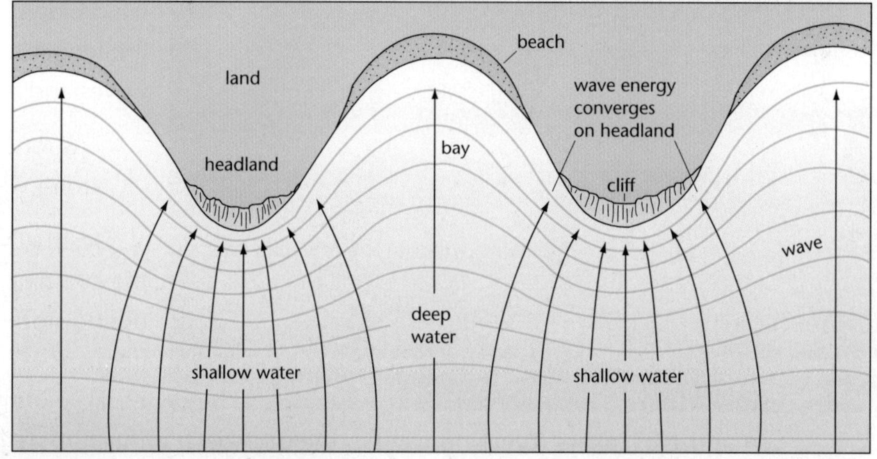


Fig. 13.7
Wave refraction on an irregular coast. Shallow water slows waves off headlands while the same waves move faster through the deep bays. Black arrows show transport direction of wave energy, concentrated on headlands, spread out in bays.

## Landforms of Erosion

On aerial photographs, **dark shadows** along a coastline often indicate the presence of cliffs and caves.

Cliffs are part of a coastline which are either sloping or veertical.

### Cliffs

Cliffs may be classed as either

1. active or inactive
2. sloping or vertical

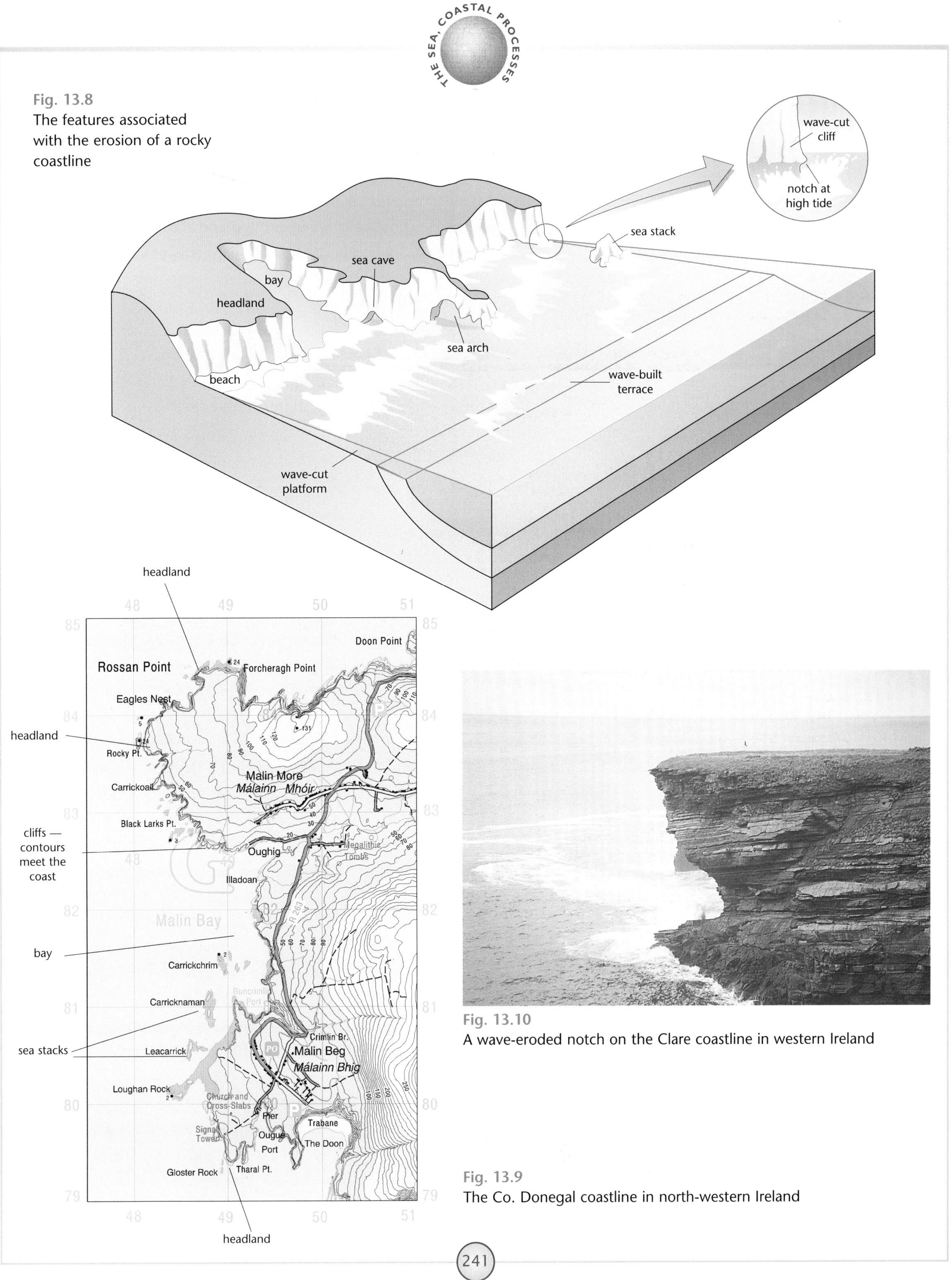


Fig. 13.8
The features associated with the erosion of a rocky coastline

Fig. 13.10
A wave-eroded notch on the Clare coastline in western Ireland

Fig. 13.9
The Co. Donegal coastline in north-western Ireland

Fig. 13.11
Sloping rock strata on a sea stack on the west coast of Ireland

1. **Active**

   Active — Cliffs are said to be active when they are vertical or near vertical and when free from vegetation, while waves rush at their bases and cause undercutting.

   Inactive — These cliffs are vegetated and stable. The sea floor, therefore, has a gentle gradient, which reduces wave action.

2. **Sloping Cliffs**

   In some places, rock strata dip seaward. Waves crash against the rock, removing soil cover. The angle of the rock strata and the slope of the cliff tend to coincide. Such cliffs may be seen at Slea Head in Co. Kerry, in Ireland. In other areas, the rock strata may dip landward, with the shape of the cliffs being determined by the slope of the rock strata.

### Vertical Cliffs

Vertical cliffs are generally found in areas where hard rock, such as old red sandstone, and limestone have horizontal strata of equal resistance.

Processes: Hydraulic Action, Abrasion, Compression, Weathering

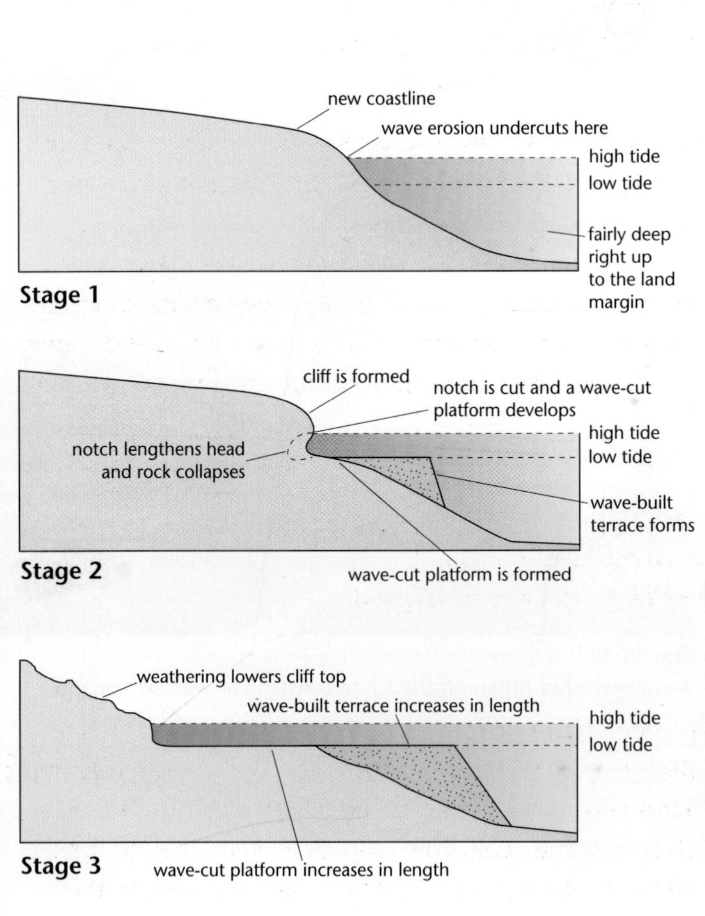


Fig. 13.12

### Landform: Cliff and Wave-cut Platform

Examples: Cliffs of Moher in Co. Clare, in Ireland
Big Sur coastline in California, in the USA
Cliffs of Dover, in southern England

As destructive waves lash against a steep coastline, a wedge of material is eroded, by the processes of **hydraulic action** and **abrasion**, to form a **'notch'**. This notch-line generally occurs along a weakness in the rock surface. It may not be horizontal, as it depends on the nature and dip (slope) of the softer rock layer.

As the notch is deepened, **undercutting** of the cliff face occurs. In time, the notch lengthens and deepens further and the **overhang** becomes too heavy, so it **collapses** and falls to the base of the cliff. These fallen rocks form protective material for some time, and when shattered create rubble for destructive waves to increase erosion and repeat the process.

- Strong waves breaking against the base of a cliff force rocks apart and make them more susceptible to erosion, with cliffs of boulder clay being particularly affected. This process is called hydraulic action.

Processes: Hydraulic Action, Abrasion, Compression, Weathering

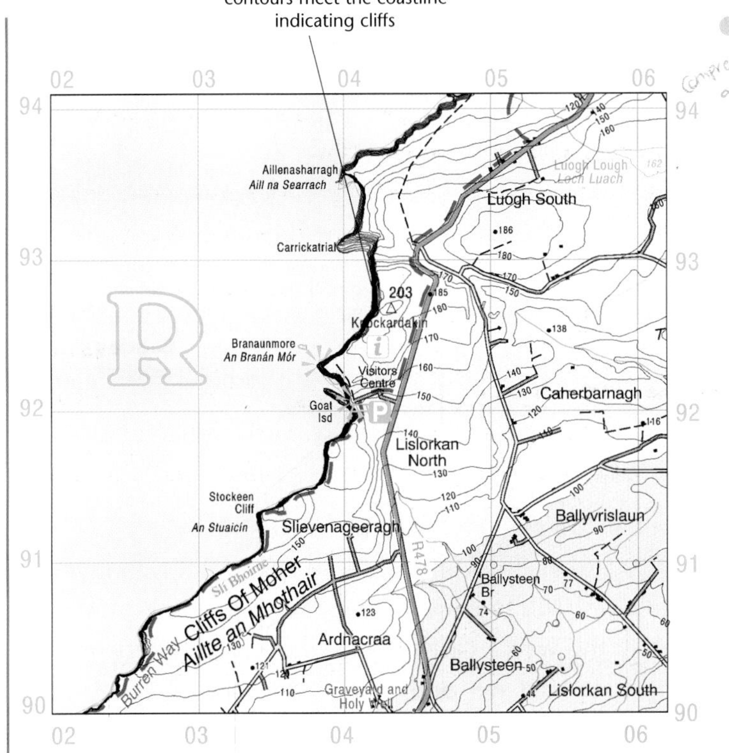


Fig. 13.13

- When waves crash against a cliff, air filters into **joints**, **cracks** and **bedding planes** in the cliff face. This air is trapped and the trapped air is **compressed** until its pressure is equal to that exerted by the incoming wave. When the wave retreats, the resultant expansion of the compressed air has an explosive effect, enlarging fissures and shattering the rock face.
- When boulders, pebbles and sand are pounded by waves against the foot of a cliff, fragments of rock are broken off, and undercutting of the cliff takes place (Fig. 13.10, p. 241). The amount of corrasion is dependent on the ability of the waves to pick up rock fragments from the shore. Corrasion is therefore most active during storms and at high tide, when incoming waves throw water and suspended rock material high up the cliff face and sometimes onto the cliff edge.
- Wind, rain and sometimes frost attack the upper part of a cliff face. These also cause erosion, and cause the cliff to retreat and finally reduce to a gentle slope.

### Wave-cut Platform

As a cliff face retreats, a wave-cut platform is formed at its base. This is a level stretch of rock often exposed at low tide, with occasional pools of water and patches of seaweed on its surface, and it displays a boulder beach in the backshore.

Generally the wider a wave-cut platform, the less the erosive power of waves, as shallow water reduces wave action, so the rate of coastal erosion slows down. Wave-cut platforms occur above present sea level in some parts of the country, such as at Dunquin in Co. Kerry and Annalong in Co. Down. These were formed when the sea was at a higher level than it is today.

Processes: Hydraulic Action, Abrasion, Compression

## Landform: Sea Stack

Examples: The Skellig Rocks off the west coast of Ireland
The Flagoni Rocks off the coast of Capri
The Needles, Isle of Wight in southern England

### Stages in the Formation of a Sea Stack

#### Stage 1 — Cave

Caves form in areas of active erosion, where there is some local weakness. A jointed or faulted zone with a regular outline may be visible at a cave entrance. The sea erodes more effectively at such places. Air filters into cracks and joints and bedding planes. This air is trapped as incoming waves lash against the coast. The trapped air is compressed until its pressure is equal to that exerted by the incoming wave. When the wave retreats, the resultant expansion of the compressed air has an explosive effect, enlarging fissures and shattering the rock face, thus deepening the cave.

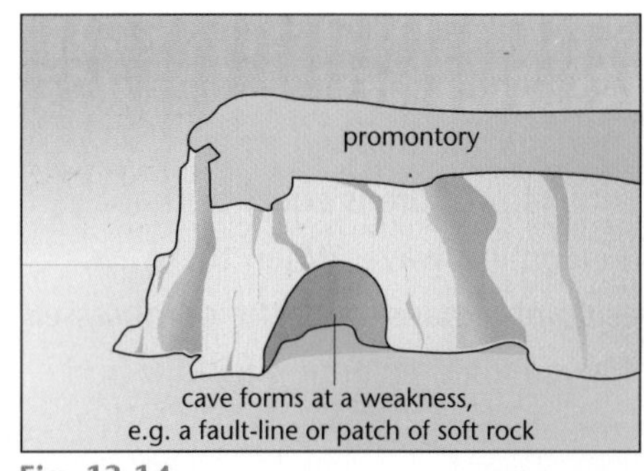


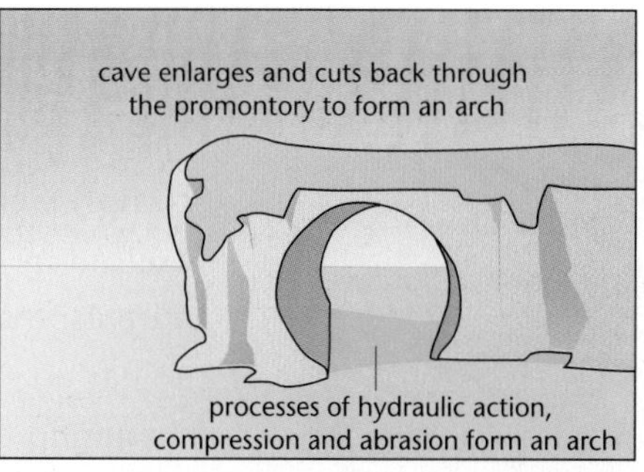


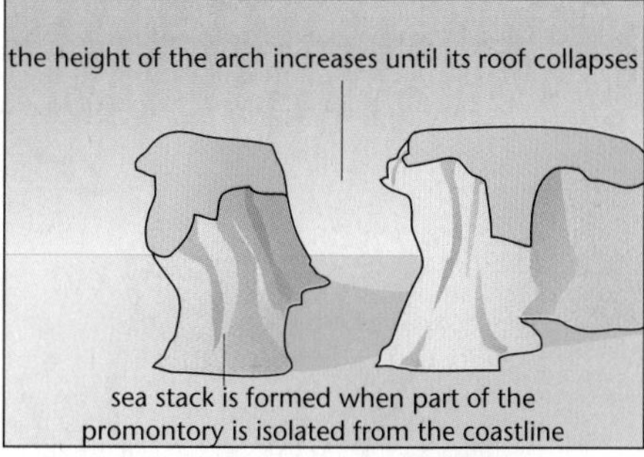


Fig. 13.14

Also aiding this process is the hydraulic force of the wave itself. This force erodes material from the cave just as a power hose does during a car wash.

### Stage 2 — Arch

In addition, boulders, rocks and pebbles are pounded by waves against the coastline, especially during storms. This material also enlarges and deepens a cave as it is thrown by the waves against the cave roof and sides. This process is called corrasion and is most active at high tide or during storms. In this way, a cave retreats into a promontory until it reaches the other side, thus forming a sea arch.

### Stage 3 — Stack

These processes continue until the arch roof becomes too wide or heavy and finally collapses into the sea. Part of the promontory is therefore cut off from the coastline and stands alone as a sea stack.

## Case Study: Blowhole at Bridges of Ross in Co. Clare, in Ireland

### Blowholes

Trapped and compressed air inside a cave may lead to the shattering of the cave roof. A chimney-like opening to the ground surface above may thus be formed some distance inland from the coastline. At high tide or during storms, powerful waves rush into the cave and force spray through the **blowhole** and into the air (Fig. 13.15).

Fig. 13.16
A fault-line is eroded in a wave-cut platform creating a fissure which ends in a blowhole at the Bridges of Ross in Co. Clare in Ireland

Fig. 13.15
Blowhole at end of fissure (Fig. 13.16). The cave roof has been shattered and the overlying boulder clay is eroded creating a blowhole.

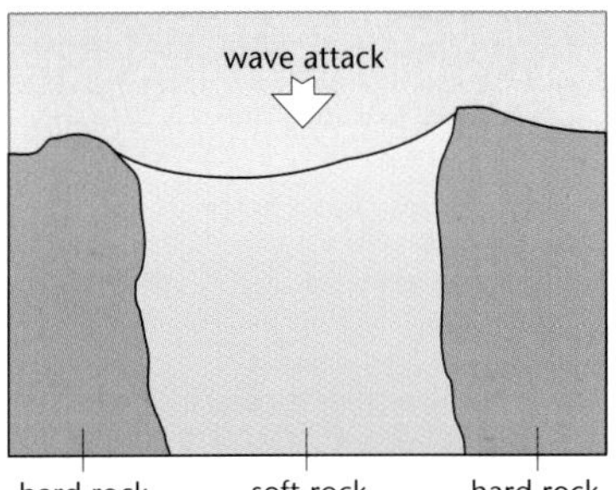


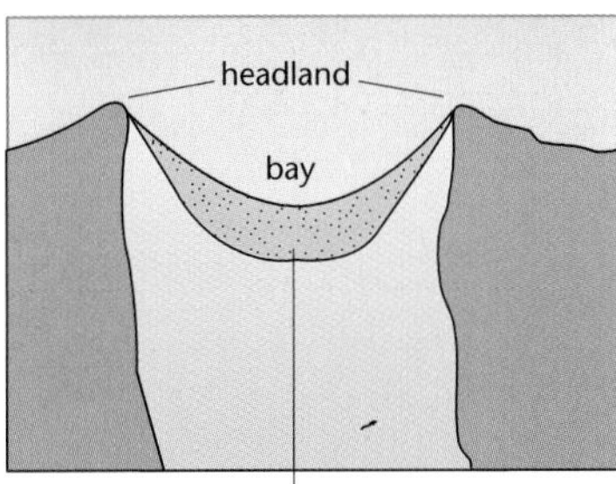


Fig. 13.17
Headlands and bay

## Bays

When a coastline is formed of alternate hard and soft rock, **differential erosion** occurs. The soft rock is worn away by marine processes at a faster rate than the hard rock. In such places, the soft rock is replaced by a **bay**, while the hard rock stands out as a **headland** (Fig. 13.17).

### Tip

Use these questions as a guide only. Explain your description fully.

### How to Describe a Coastal Area

1. Divide the coastline into lowland coastline and elevated coastline.
2. *Low-lying coast*
   (a) What is the general direction of the coastline?
   (b) Are there submarine contours indicating the depth of water offshore?
   (c) Is the coastline straight or is it broken into bays?
   (d) Is it a coastline of erosion or deposition?
   (e) Is it backed by sand dunes with a beach along the shore?
   (f) Are there lagoons, sand spits or marshes present?
   (g) Is there a coastal plain running parallel to the coast? If there is, describe its characteristics.
   (h) Is it a coastline of submergence or emergence? Explain.
3. *Elevated coast*
   (a) What is the general direction of the coastline?
   (b) Are there submarine contours indicating the depth of water offshore?
   (c) Are there any erosion features indicated by features or names?
   (d) If cliffs are present, give their height(s).
   (e) Is there a rock symbol indicating erosion?
   (f) Is the coastline straight or broken?
   (g) Is it a ria or fjord coastline?

Study the photograph of Malinbeg on p. 246 and then use the spaces provided to identify the following landforms of coastal erosion A to E.

A ______________________________

B ______________________________

C ______________________________

D ______________________________

E ______________________________

Fig. 13.18

E
C
D
B
A

this neck of land is a sand bar created by longshore drift
beach created by longshore drift
silting often occurs on leeward side of a bar
golf courses are regularly located on sand dune areas as the surface is dry throughout the year

Fig. 13.19
Features of coastal deposition: sandbar, Barra Da Tijuca beach, Rio de Janeiro

## Marine Deposition

On reaching a shore, waves are said to **break**. The way in which this occurs is of fundamental importance to coastal processes. Shallow water causes incoming waves to steepen, the crest spills over and the wave collapses. The turbulent water created by breaking waves is called **surf**. In the landward margin of the **surf zone**, the water rushing up the beach is called the **swash**; water returning down the beach is the **backwash**. The swash moves material up the beach and the backwash **may** carry it down again.

## Processes of Deposition

### Constructive Waves

**Constructive waves** or **spilling breakers** break slowly with a powerful swash up the beach. The swash is spread over a large area and much of it percolates through the ground, with little water returning down the beach as the backwash. Hence, there is a **net gain** of material up the beach, and so the title constructive waves. If, on the other hand, the backwash is more powerful than the swash, any loose material is dragged seaward and the coastline is eroded further.

Fig. 13.20

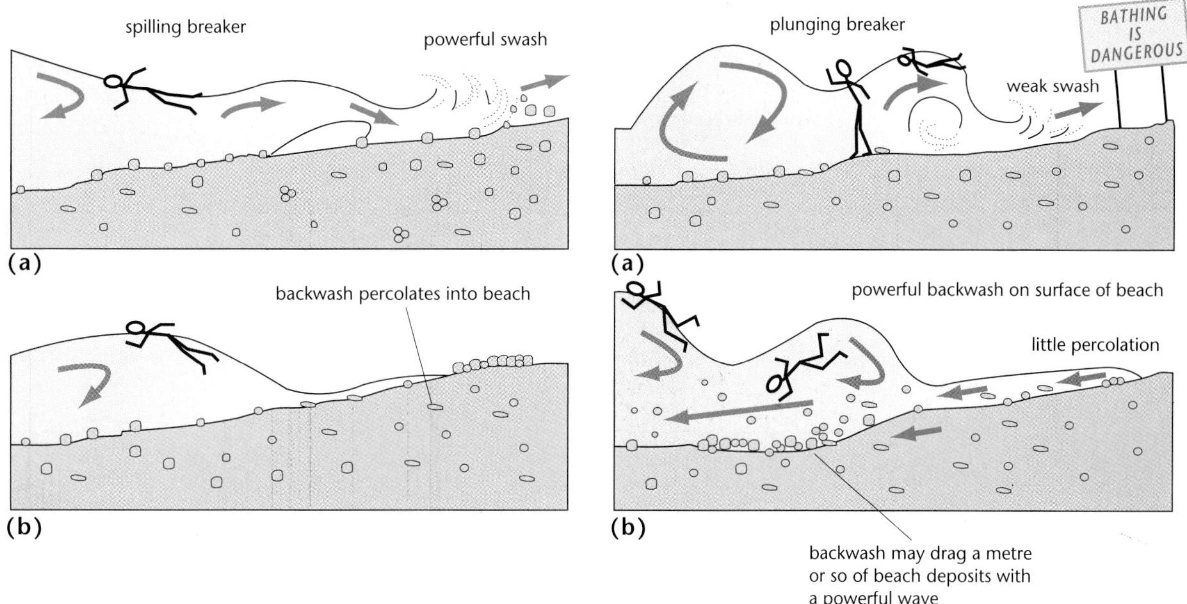


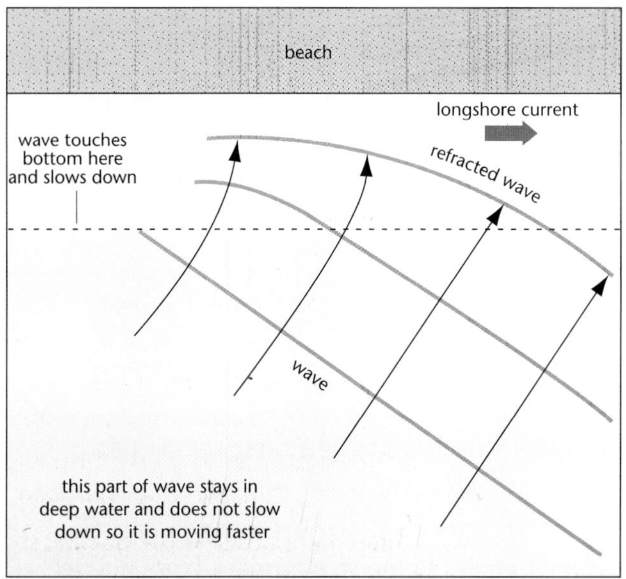


Fig. 13.21
Wave refraction changes the wave direction, bending the wave so it becomes more parallel to shore. The angled approach of waves to shore sets up a longshore current parallel to the shoreline.

### Wave Refraction

Most waves approach a coast at an oblique angle. One end of a wave breaks first, and then the rest breaks progressively along the shore. This angled approach of a wave towards shore may change the direction of the wave. One end of the wave reaches shallow water first, touches the bottom and slows down. The remainder of the wave continues at its deep-water speed. As more and more of the wave touches the bottom, more of the wave slows down. Therefore, as the deep end of the wave moves faster than the shallow end, the wave crest bends and becomes more parallel to the shore (see Fig. 13.21).

### Longshore Drift (Littoral Drift)

Longshore drift refers to the movement of sediment, such as sand and shingle, along a shore. When waves break obliquely onto a beach, pebbles and sand are moved up the beach, at

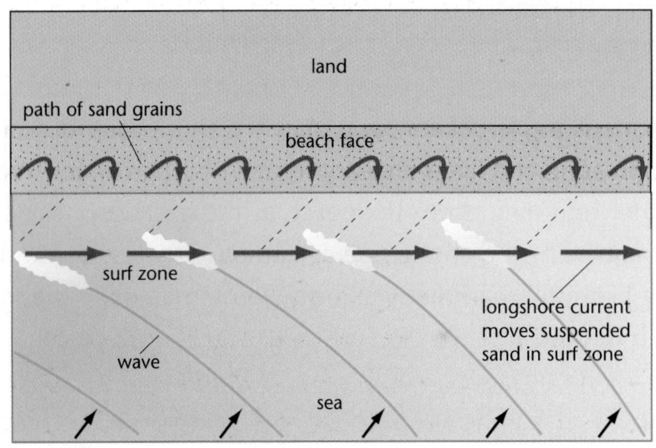


Fig. 13.22 The process of longshore drift

the same angle as the waves, by the swash. The backwash drags the material down the beach at right angles to the coast, only to meet another incoming wave, when the process is repeated. In this way, material is moved along the shore in a zigzag pattern.

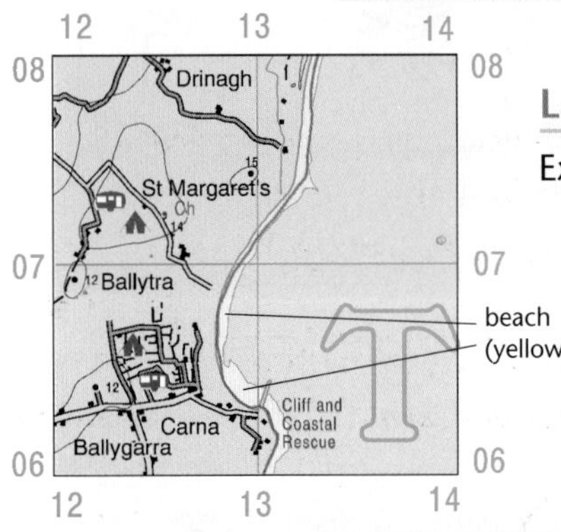

Fig. 13.23

## Landform: Beach

Examples: Malibu beach in Los Angeles, in the USA
The Algarve coastline in Portugal, such as Alvor and Lagos
The Spanish coastline on the Mediterranean and the many pocket beaches on the Amalfi coast in Italy
Curracloe in Co. Wexford, in Ireland
Dollymount Strand in Dublin, in Ireland

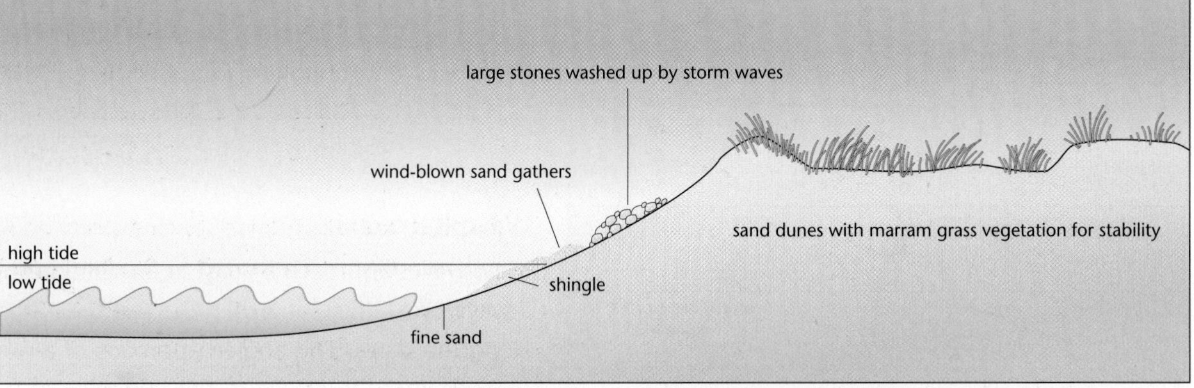


Fig. 13.24

### Processes: Longshore Drift, Deposition, Constructive Wave Action

Beaches are formed by the process of longshore drift. This process refers to the movement of material (sand and shingle) along a shore. On reaching a shore, waves are said to break. Shallow water causes incoming waves to steepen, the crest spills over and the wave collapses. When constructive waves break, they break slowly, and their ability to move material reduces.

The **spilling breaker** has a **powerful swash** which rushes, obliquely, up the beach. It spreads over a large area, and drops most of its load as the water percolates through the sand. Little water returns down the beach as the backwash. The backwash drags the remainder of its load down the beach at right angles to the shore. Hence there is a net gain of material up the beach and so the title '**constructive waves**'.

The swash and backwash create a **zigzag** movement of material **along the shore**. This movement is called **longshore drift**.

### Other Beach Landforms

**beach cusps**
A series of ridges of material near the high water mark on a beach. They are separated by rounded, evenly spaced depressions which run at right angles to the shore. They are the result of a powerful swash and backwash.

**runnels and ridges**
Broad, gently sloping ridges of sand and shallow, gently sloping depressions which run parallel to the shore on the seaward edge. At low tide, the runnels may contain long pools of water which are trapped by the ridges as the tide recedes. They are formed by constructive waves.

On an upland coast, a beach may be just a loose mass of boulders and shingle under a cliff, while a bay between headlands generally has a crescentic beach at its head and is called a bayhead beach or a pocket beach. Many such beaches are found on the western coasts of Ireland, Scotland and Wales. Many others may be found on the Portuguese coastline and Greek and Newfoundland coasts.

A beach is defined as the accumulation of material between low tide level and the highest point reached by storm waves.

The most typical beach is one with a gently concave profile. The landward side is backed up by dunes, which are succeeded by a stretch of shingle and sand and sometimes rock which is covered by seaweed. This rock represents the underlying wave-cut platform.

An ideal beach profile has two main parts:

1. **The backshore:** This is composed of rounded rocks and stones, as well as broken shells, pieces of driftwood and litter thrown up by storm waves. This part of the beach has a steep gradient and is reached by the sea during the highest tides or during storms.
2. **The foreshore:** This is composed of sand and small shell particles. It has a gentle gradient and is covered regularly by the sea each day.

## Landform: Sand Spit

Examples: Inch Strand near Dingle in Co. Kerry, in Ireland
Strandhill in Co. Sligo, in Ireland
Cape Cod in Massachusetts, in the USA
Borth Point at the Dovey estuary, in Wales

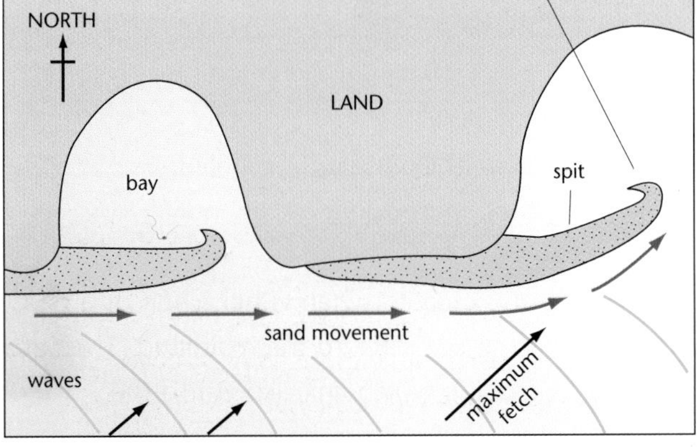


Fig. 13.25
Longshore drift of sand can form spits

A **spit** is formed when material is piled up in linear form, but with one end attached to the land and the other projecting into the open sea, usually across the mouth of a river estuary. Spits generally develop at places where longshore drift is interrupted and where the coastline undergoes a sharp change of direction such as at river estuaries and bays, or between an island and the shore.

In such places, longshore drift is unable to continue its zigzag movement of sediment along the shore and it deposits this material at the seaward point of the coastal beach in the lee of the headland. Where material continues to be deposited at such locations, sand spits will form.

In the diagram, Fig. 13.25, the prevailing winds and maximum fetch (the length of open ocean over which the wind blows) are from the south-west, and so material is carried eastward by longshore drift.

As the spit continues to grow, storm waves throw some larger material, such as pebbles, shingle and larger stones, above the high water mark. This allows

Processes: Longshore Drift, Wave Refraction, Deposition

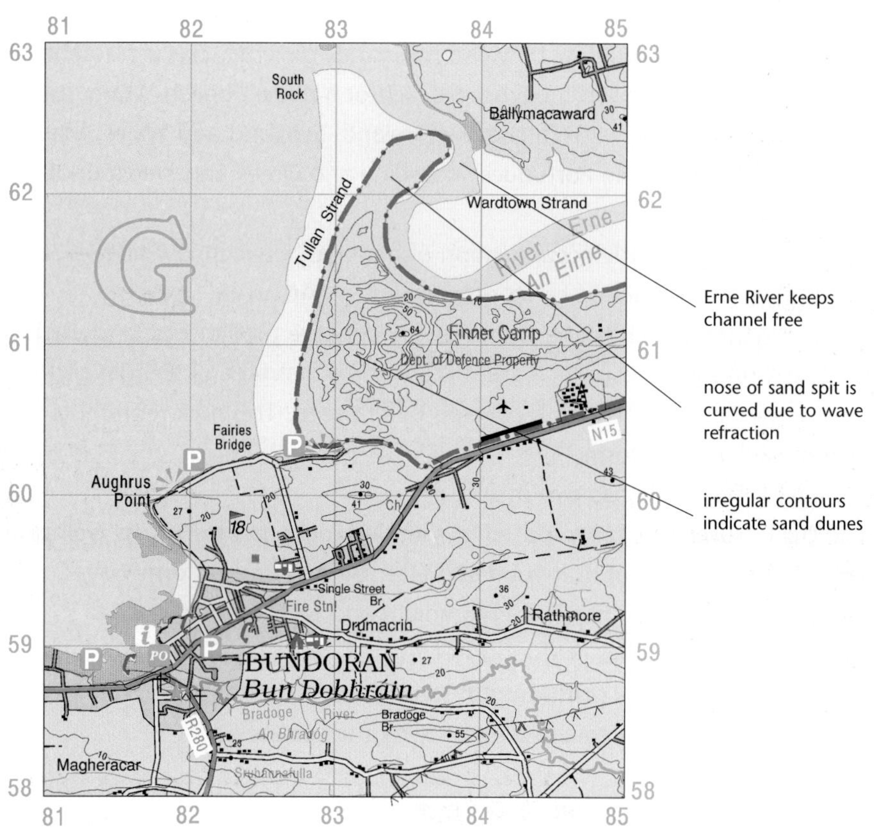


Fig. 13.26
Ordnance Survey map extract displaying sand spit, sand dunes and beach deposits

the growing spit to become more permanent.

Sand spits display a regular form. On the seaward edge, spits display a long wide beach, backed by sand dunes which are generally stabilised by marram grass. This grass is a rough, coarse variety with a well-developed root system. The root mass binds the sand particles, thus reducing or preventing erosion. These sand dunes formed from dry beach sand which was blown up the beach by strong sea breezes. Banks of these sand particles accumulate to cover a sizeable area.

The sand dunes on a sand spit give way, on the landward side, to coastal marsh with reeds, grasses and silt.

The seaward end of a sand spit is often curved or hooked. This curving is due to waves advancing obliquely up the shore, causing them to swing round the end of the spit. This process is called wave refraction (see p. 247). Sand dunes develop on the landward side of these spits.

The curved hook may change shape due to a change in wind direction coming from the second most dominant source.

## Landform: Tombolo

Examples: Omey Island and Beach in Connemara, in Co. Galway in Ireland
Inishkeel, near Portnoo in Co. Donegal, in Ireland

Processes: Longshore Drift, Deposition, Wind Action

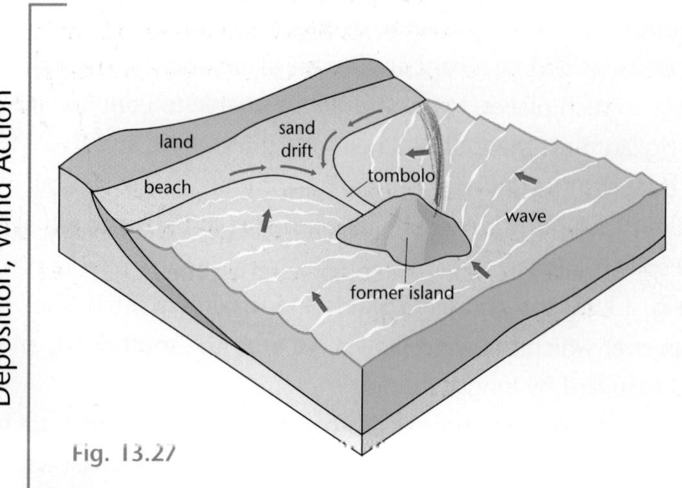


Fig. 13.27

This landform is created when either a spit or a bar links an island or sea stack to the mainland. The Sutton tombolo in Dublin, which links Howth to the mainland, and the Fenit tombolo in Co. Kerry in Ireland are two examples.

Should two tombolos join an island, from opposite directions, they may enclose sea water between them, thereby creating a lagoon. Tombolos are regularly formed along rugged coastlines, such as the west coast of Ireland.

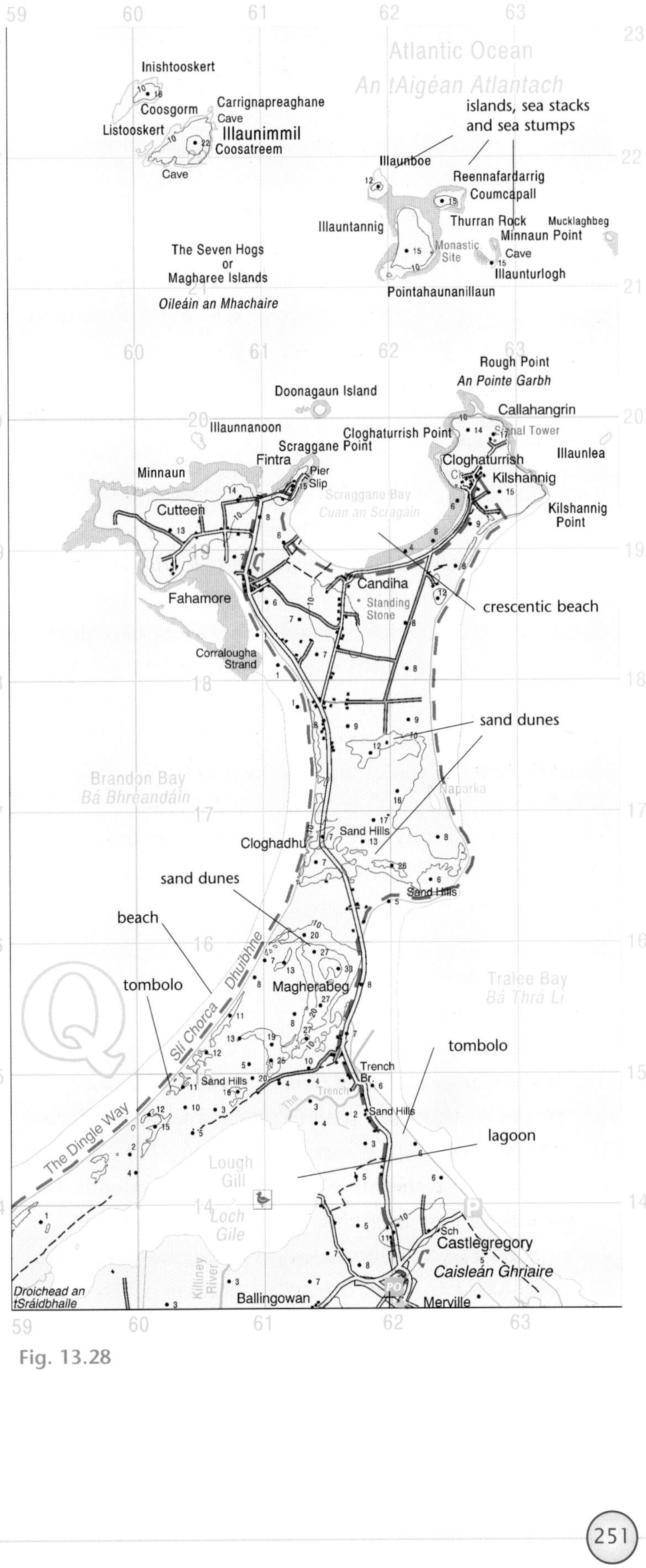

Fig. 13.28

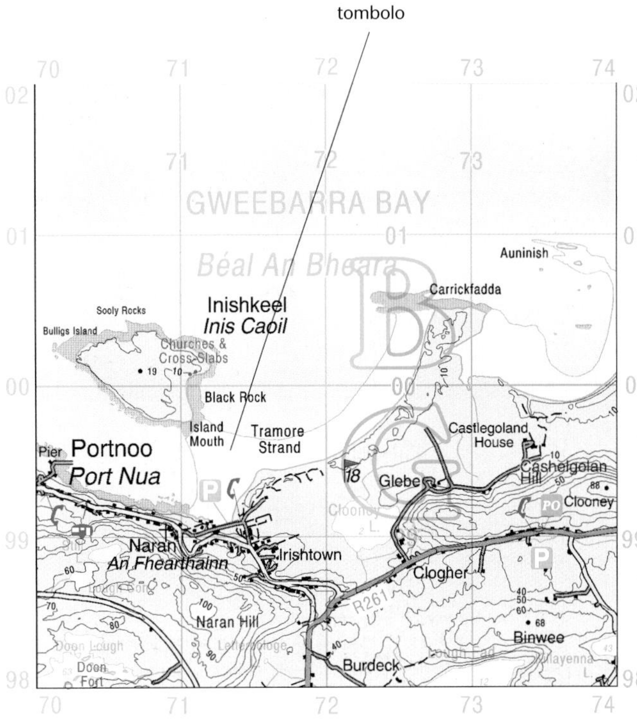

Fig. 13.29
Tramore Strand joins Inishkeel Island to the mainland in Co. Donegal

## Landform: Sandbar/Lagoon

Examples: Baymouth Bar enclosing Tacumshin Lake and Lady's Island Lake in Co. Wexford, in Ireland
Loe Bar in Cornwall in England, which is composed of shingle. Coney Island and Jones Beach are the seaside areas for New York City. They are **offshore bars**.

**Barrier island** bars form a continuous coast along the Florida coastline. Hotels and other coastal developments are constructed on these bars (see Fig. 13.80, p. 276).

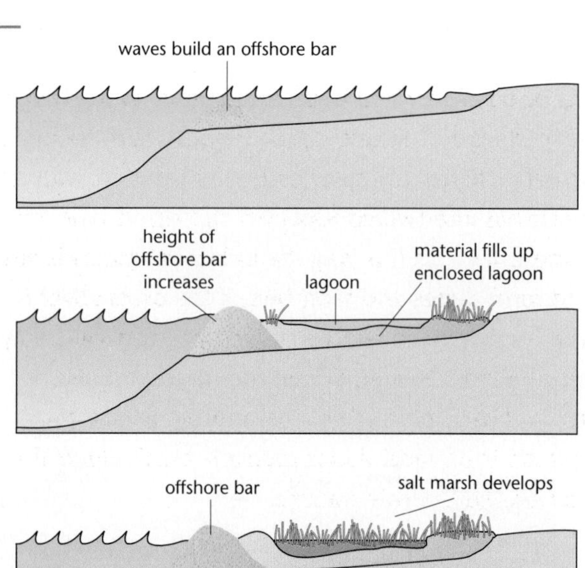


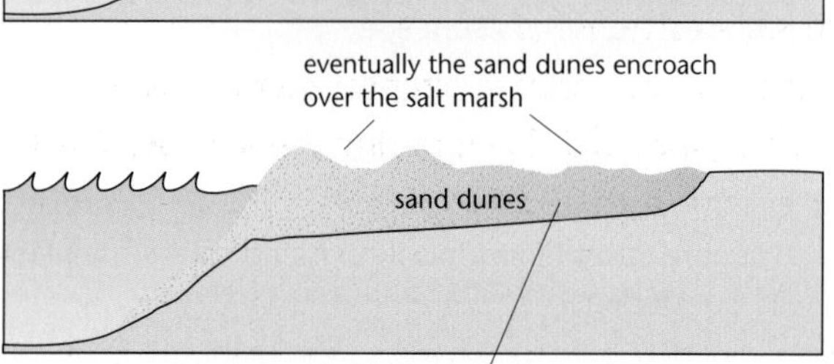


Fig. 13.30
Life cycle of a lagoon

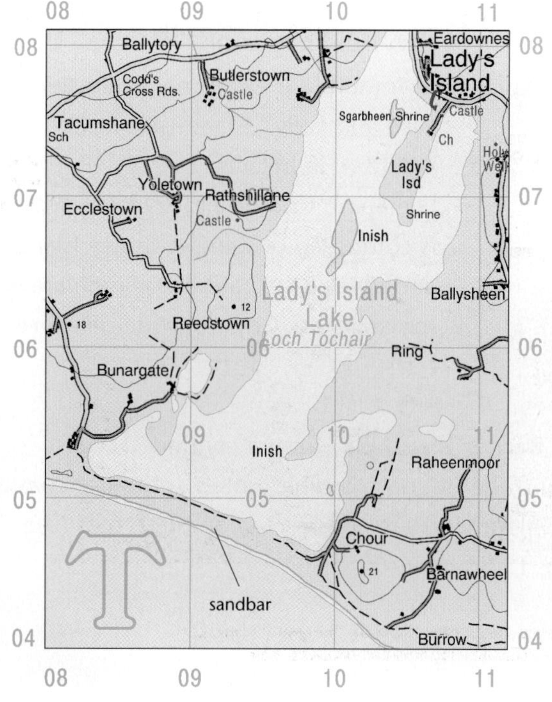


Fig. 13.31
A baymouth bar cuts off Lady's Island Lake from the sea in Co. Wexford, in Ireland

## Formation

When a bay is cut off from the sea, as is the case with Lady's Island Lake (Fig. 13.31), and Tacumshin Lake in Co. Wexford, in Ireland, a landform called a **lagoon** is formed. It is created when waves build a bar above water and across a bay or parallel to the coastline. Waves wash sand into the lagoon, and rivers and winds carry sediment into it. The lagoon eventually becomes a **marsh**. Finally, the combined work of waves, rivers and winds turns the marsh into an area of **sand dunes** (see Fig. 13.30).

## Sand Bars

There are two main types of sand bar.

### 1. Offshore Bar

These are ridges of sand lying parallel to a shore and some distance out to sea (Fig. 13.30). On gently sloping coasts, large ocean **waves** (breakers) **break**, dig up the seabed and throw the loose material forward to form a ridge of sand. Once the ridge is formed, the bar increases in height by **constructive wave action**. **Longshore drift** may lengthen the bar at both ends. If submerged, such ridges are dangerous to shipping and can cause ships to run aground.

Offshore, these ridges are pushed along in front of the waves until, finally, they may lie across a bay to form a **baymouth sand bar**. Lady's Island Lake and Tacumshin Lake in Co. Wexford, in Ireland, were originally bays of the sea before they were cut off by such a sandbar (Fig. 13.31).

### 2. Baymouth Bar

A sandbar may also form as a result of the growth of a sand spit across a bay. This type of landform is called a **baymouth bar**. Sometimes, where **tidal or river scouring** takes place, a bar may be **prevented from sealing off** the bay completely (see Fig. 13.33, p. 253).

Offshore barrier island bars are relatively uncommon in Britain and Ireland. However, they are globally widespread, accounting for 13 per cent of the world's coastlines. They are readily recognisable on maps of the Gulf of Mexico, the northern Netherlands, and southern and western Australia. Bartragh Island in Killala Bay in Co. Mayo in Ireland is an offshore bar.

**Alternative Theory of Barrier Island Bar Formation**

This theory suggests that rises in postglacial sea levels may have partly submerged older beaches or ridges of glacial drift that were formed during the ice age (when sea levels were lower). The upper parts of these beaches now appear above sea level as barrier island bars.

## Sand Dunes

**Sand dunes** form on the landward side of beaches when large expanses of sandy beach are exposed at low tide. The surface sand dries out. Strong coastal winds blow the dry sand onshore, forming small mounds which are quickly colonised by vegetation, normally **marram grass** (Fig. 13.32). Marram grass has intertwined roots which help to bind the loose sand, while the grass itself traps more sand. In this way the individual mounds are stabilised, growing larger and joining to form dunes and sand hills. If the grass cover is destroyed by fire, pathways or excessive human interference, the unprotected sand may be blown away and may damage fields and crops inland. Sometimes marram grass or coniferous trees are planted deliberately to break the force of the onshore winds, thus stabilising the sand dunes. Conifer plantations on sand dunes regularly occur along the west coast of France such as near Royan on the Gironde estuary.

Examples: Along much of the lowland parts of the west coast of Ireland such as at Lahinch in Co. Clare and Strandhill in Co. Sligo

The Landes area on the south-west coast of France

The sandy countryside of the Landes occupies almost 15,000 square kilometres, covered by marine sands laid down during the ice age. The largest dune, the Dune de Sabloney (the largest moving dune in Europe), exceeds 90 metres.

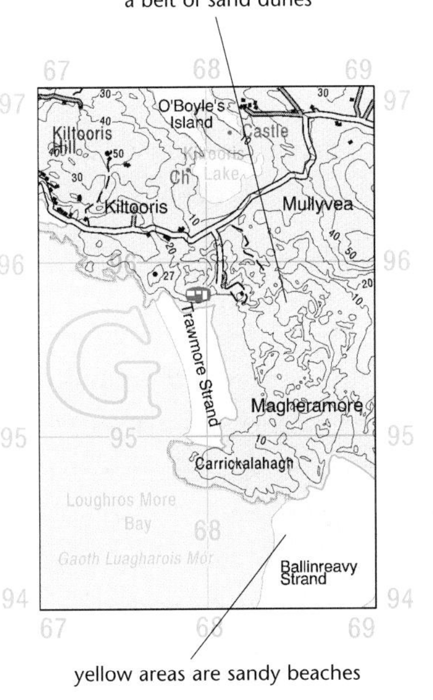


Fig. 13.32
Sand dunes on the Donegal coastline in Ireland

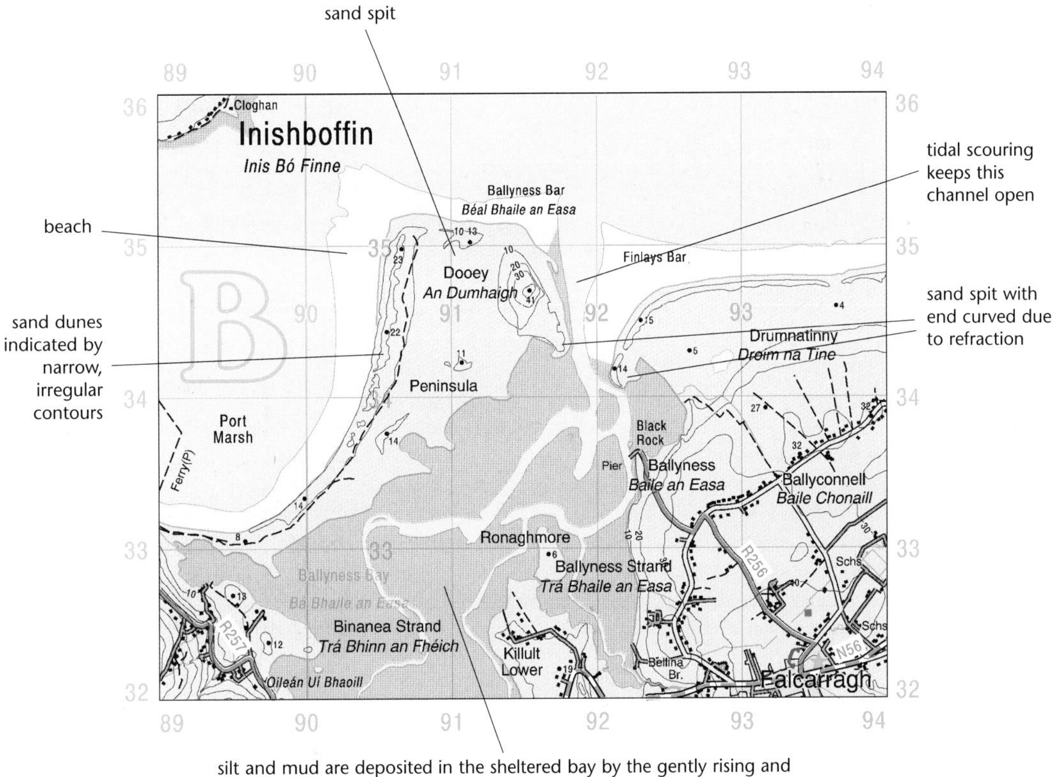


Fig. 13.33
Coastal features in Co. Donegal, in Ireland

## Changes of Sea Level

Throughout the world, wave-cut platforms and beaches are found well above sea level, while other places are submerged by the sea. This suggests that changes in sea level have occurred along many coastal areas at various times. These changes may be the results of:

1. eustatic movement
2. isostatic change
3. tectonic activity

### 1. Eustatic Movement

Changes in the level of the sea relative to that of the land may have a great effect on the form of a coast. A vertical rise or fall of a few metres on a low-lying coast can produce huge changes in the shape of a coastline.

Sometimes, a worldwide and uniform rise or fall of sea level occurs. This is known as **eustatic movement**. The most important eustatic movement is associated with postglacial changes (see Fig. 13.35).

During the ice age, large volumes of water were stored in the form of ice sheets which covered large areas of Europe, Asia and North America. It is estimated that sea level was then as much as ninety metres lower than present sea levels. Once the ice age ended, the melting ice sheets returned the water to the oceans. Thus sea levels rose again and many low-lying areas were submerged.

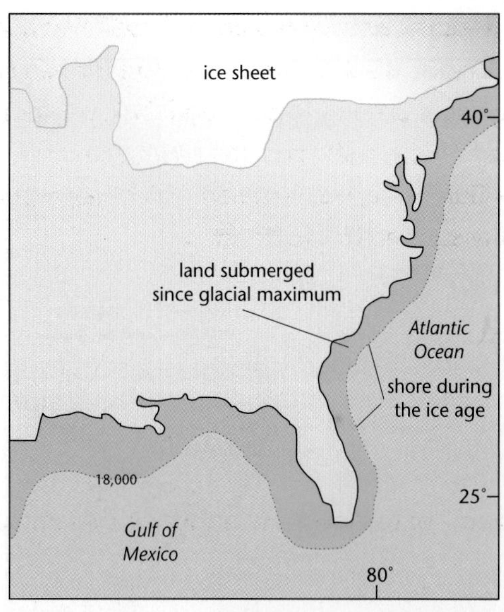


Fig. 13.34

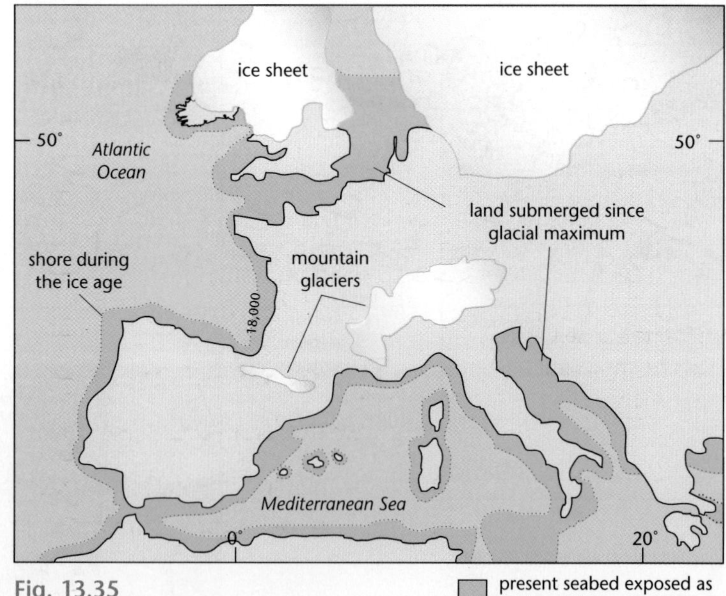


Fig. 13.35

### 2. Isostatic Change

If the concept of isostasy (see p. 13) is correct, we should expect that if weight is added to a continent, it will be depressed; conversely, if weight is removed then it will rise. For example, during the last ice age, continental ice sheets caused downwarping of the earth's crust. Since the ice melted some 10,000 years ago, uplifting of as much as 330 metres has occurred in the Hudson Bay region of Canada, where the thickest ice had accumulated. Uplift is still taking place around parts of the North Sea and the Baltic Sea coasts.

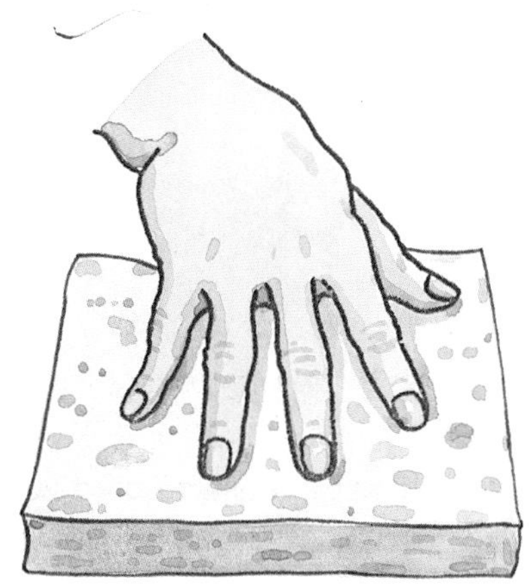

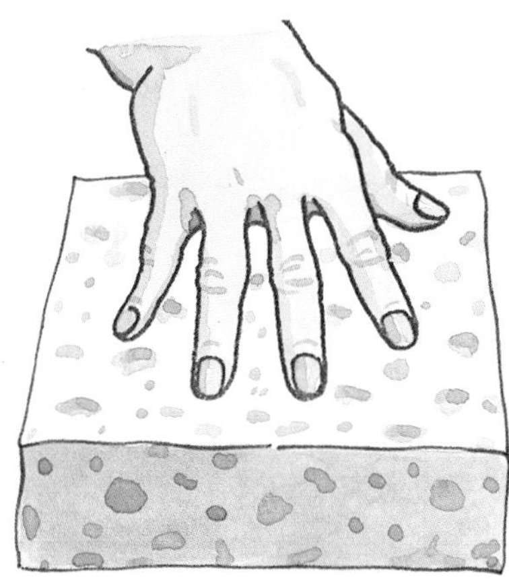

Fig. 13.36
When pressure is applied by a hand to a sponge, the sponge is squeezed and reduces in volume. If this pressure is released the sponge rises and expands. Similarly, when great thickness of ice blanketed a landscape the surface was pressed down and squeezed. When the ice melted, the ground rose again.

Fig. 13.37
Caves formed at sea level are now dry and well above sea level

### 3. Tectonic Activity

: can cause sea levels to rise/fall and change the form of a coast.

Some coastal areas may be submerged due to faulting. The downward movement of a land area may allow the sea to invade and submerge the area. Local **warping**, or tilting of the earth's crust, may also affect the shape of a coastline.

Tectonic activity has lifted these caves above water (Fig. 13.37). The caves were formed when they were partially submerged. This photograph was taken by the author on the south coast of Capri in the Gulf of Naples, a region which has experienced many such earth movements.

## Types of Coastlines

### 1. Concordant Coastlines

When a coastline is roughly parallel to upland or to mountain ranges located immediately inland, it is said to be **concordant**.

#### (a) Submerged Upland Coast

When a coastline with upland ridges and valleys parallel to the sea is submerged by the sea, a **Dalmatian coastline** is formed (called after the Dalmatian coast of Croatia on the Adriatic Sea, Fig. 13.38). The higher parts of the upland ridges tend to stand out as islands parallel to the coast, while the lower parallel valleys form long **sounds**. Cork Harbour in Co. Cork, in Ireland, is a small-scale example of such a coastline. Here, the sheltered

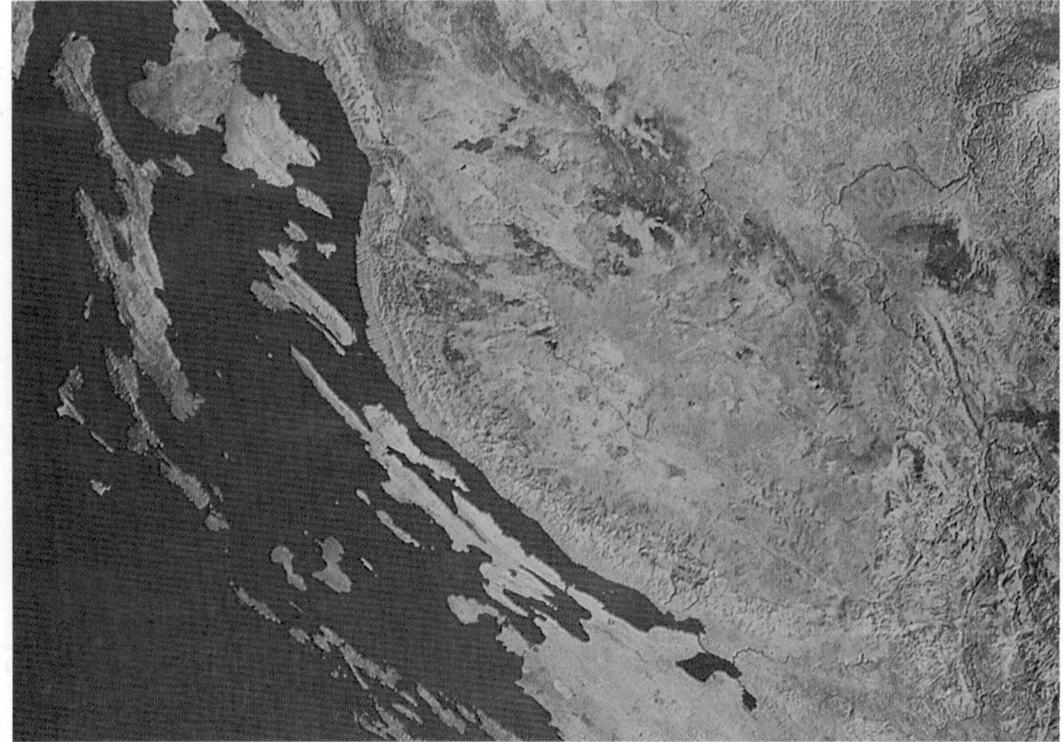

Fig. 13.38
The folded ridges and valleys of the Dalmatian coast in Croatia, on the Adriatic Sea, were flooded by the sea. The upland ridges appear as a series of linear islands running parallel to the coast.

sounds provide long coastal stretches for many activities such as water sport facilities and industrial wharfage, while the bays provide a facility for mariculture, fishing and recreation. San Francisco Harbour and the southern Chile coastline are also concordant coastlines.

### (b) Submerged Lowland Coast

Sea levels rose immediately after the ice age. Melting ice sheets returned vast volumes of water to the seas and oceans. In places like **Clew Bay** in Co. Mayo, in Ireland, which was low-lying land, sea water flooded large areas, forming a concordant lowland coast.

When submergence of a lowland area occurs, extensive areas of land are covered by the sea. This happens because slopes are gentle, and a very slight depression allows the sea to flood a large area. Thus a slight positive change of sea level can convert a drumlin countryside such as Clew Bay into a number of rounded, gently humped islands (see map, p. 215 for drumlin landscape). In such areas, deposition is actively creating offshore bars, spits, coastal lagoons, marshes and tombolos.

Where such submergence takes place in a river valley, broad shallow estuaries are formed. Such estuaries display tracts of marsh and mudflats which are exposed at low tide with a maze of creeks and winding shallow inlets. An example of this is the Shannon estuary in western Ireland.

The lower portions of many river valleys were submerged by the rise in sea level that followed the end of the ice age. Chesapeake Bay and Delaware Bay, on the east coast of the United States, are estuaries and excellent examples of such drowned river valleys.

## 2. Discordant Coastlines

When upland and lowland areas meet the coast more or less at right angles, resulting in long headlands and bays, the coastline is said to be **discordant**.

### (a) Submerged Upland Coast

(i) **Rias:** Rias represent a discordant coastline of submergence. They are formed when submergence affects an upland area where the ridges and valleys meet the coastline more or less at right angles. The rias are funnel-shaped inlets, decreasing in depth and width as they run inland. A river which was originally responsible for the formation of the valley flows into the head of the inlet. Rias are common in southern Ireland from **Dingle Bay to Roaring Water Bay**. The narrow low-lying edges of the rias in the south-west have a covering of glacial till, represented by the many drumlins which

### Wave Action Alters a Submerged Highland Coast

**1. In the beginning**

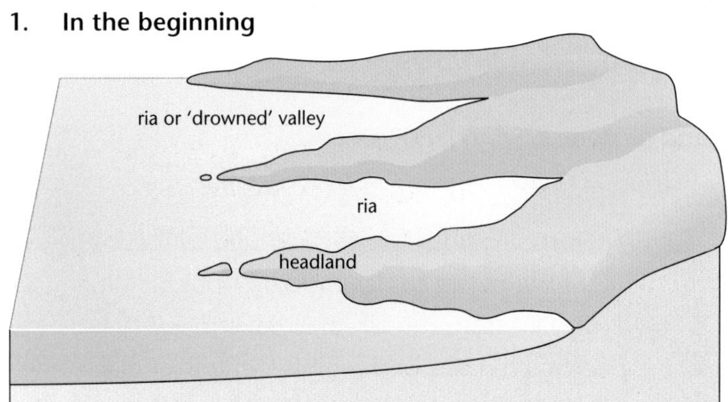


Fig. 13.39
Irregular coast of rias and headlands. Headlands are cut back by wave erosion, and cliffs, caves and stacks form.

**2. Stage of youth**

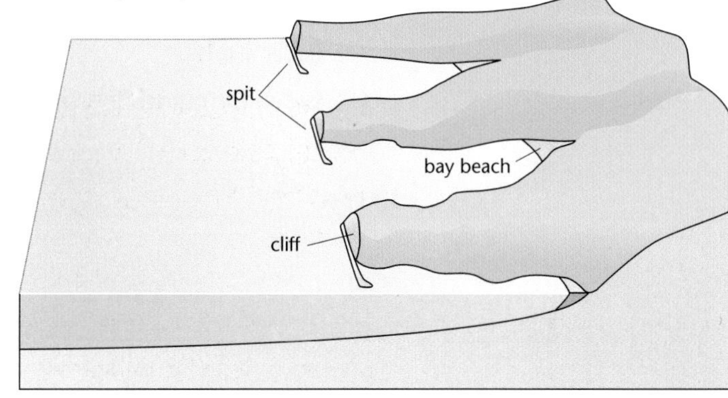


Fig. 13.40
Wave deposition is more important than erosion. Spits and bay beaches are formed. The coast becomes straighter, since erosion of the headlands still goes on.

**3. Stage of early maturity**

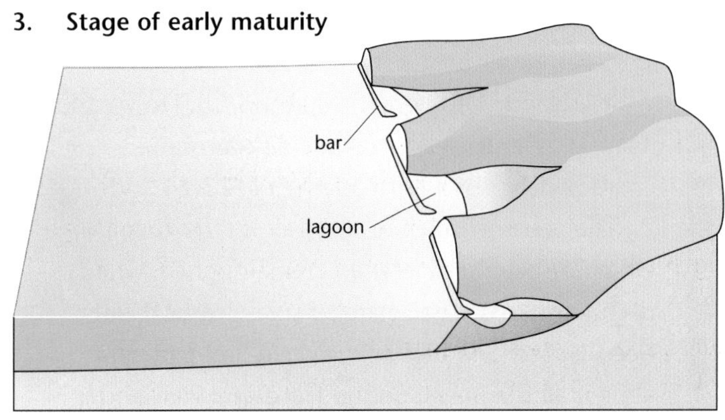


Fig. 13.41
Erosion is still cutting back the headlands. Bars extend across the bays, which are now turned into lagoons. These fill in with sediments and marshes form.

**4. Stage of late maturity**

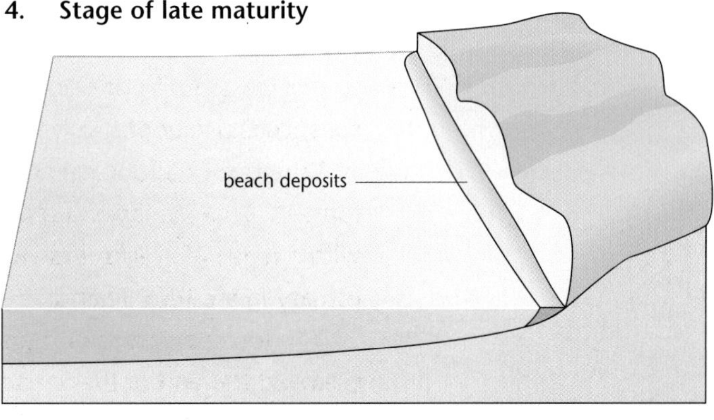


Fig. 13.42
The coast is cut back beyond the heads of the bays and is now almost straight

form islands, such as those in **Bantry Bay**. Some of these inlets such as Bantry Bay are exceptionally deep and form natural deep-water and sheltered harbours which can accommodate even the largest ocean-going bulk tankers (see photograph of Durrus in Co. Cork, below).

### Formation

As Greenland drifted away from Europe, the west side of Ireland was deprived of support. The east–west valleys tilted seaward and the sagging allowed the sea to flood their ends to give us the indented coastline of the west and the ria coast of the south-west, on the Kerry and west Cork coastlines.

Study the photograph below of Durrus in west Cork in Ireland (Fig. 13.43) looking seaward over Dunmanus Bay. Then answer the following:

1. The photograph shows a submerged coastal area. Is this a coastal lowland or a coastal upland? (In other words is the land that surrounds the sea water low-lying land or upland?)
2. What characteristics indicate that this area was submerged?
3. From your list of characteristics in question 2 above state whether this is a ria or fjord inlet?
4. Is the farmland rich grazing land or poor grazing land? Discuss.

Fig. 13.43

(ii) **Fjords:** Fjords were formed due to the submergence of deep glacial troughs. When viewed from the air, they have many characteristics of the higher edges of glaciated valleys, including vertical or very steep and parallel mountainsides, hanging valleys, and truncated spurs which slope steeply to the water. Islands may appear at the entrance to such inlets, representing the exposed higher parts of otherwise submerged terminal moraines.

In Ireland, Carlingford Lough in Co. Down and Killary Harbour in Co. Galway display some characteristics of a fjord inlet (Figs. 13.44 and 13.45).

The western coast of Norway is a fjord coastline which displays all the characteristics of drowned glacial troughs. Their parallel sides, vertical cliffs, deep waters and highland terrain are evident in almost all the inlets. Sogne Fjord and Hardanger Fjord are just two examples.

Fig. 13.44

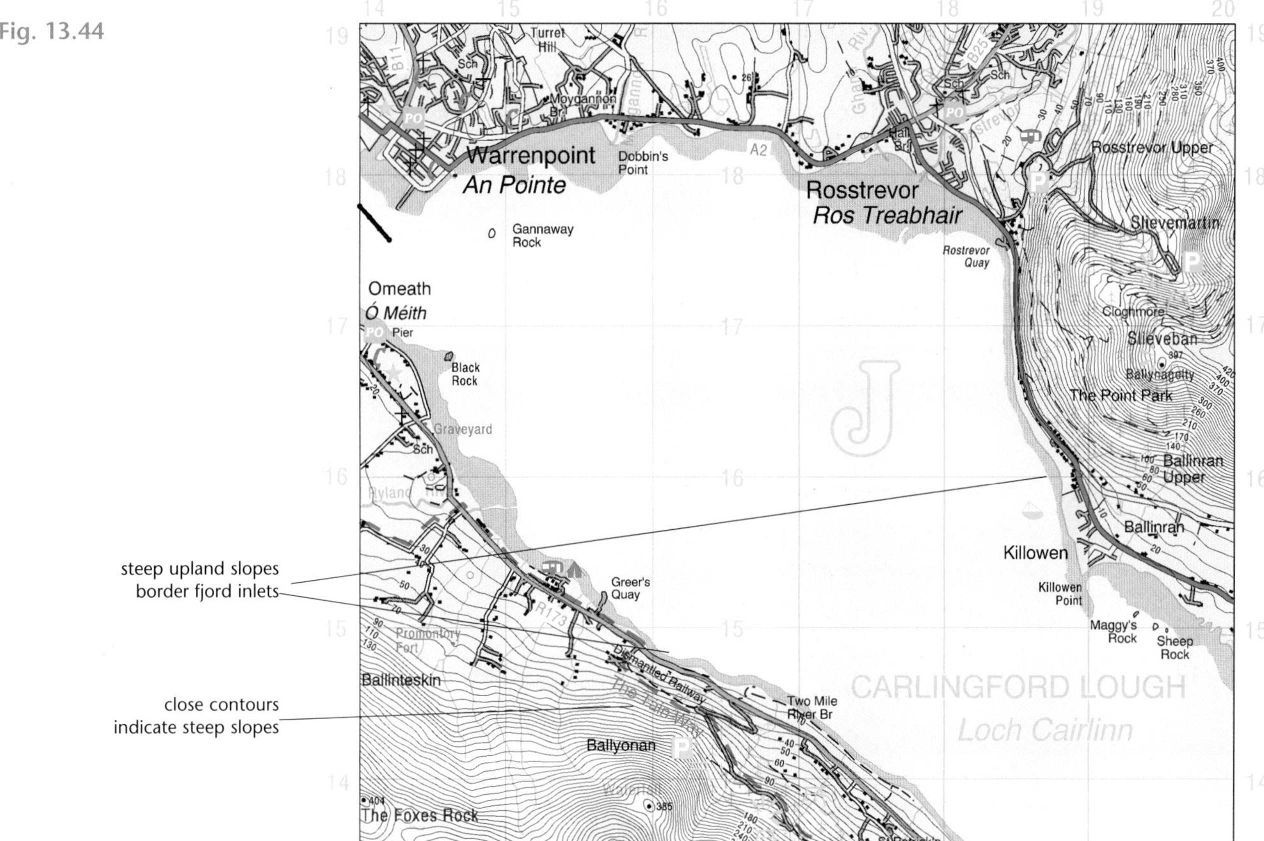


## (b) Emerged Upland Coast

A coastline which has experienced a fall of sea level (negative) or a rise of the land is known as a **coastline of emergence**. The chief feature of such a coast is a raised beach or cliff-line now found well above the present shore. The old coastline may be noticed as a

Fig. 13.45

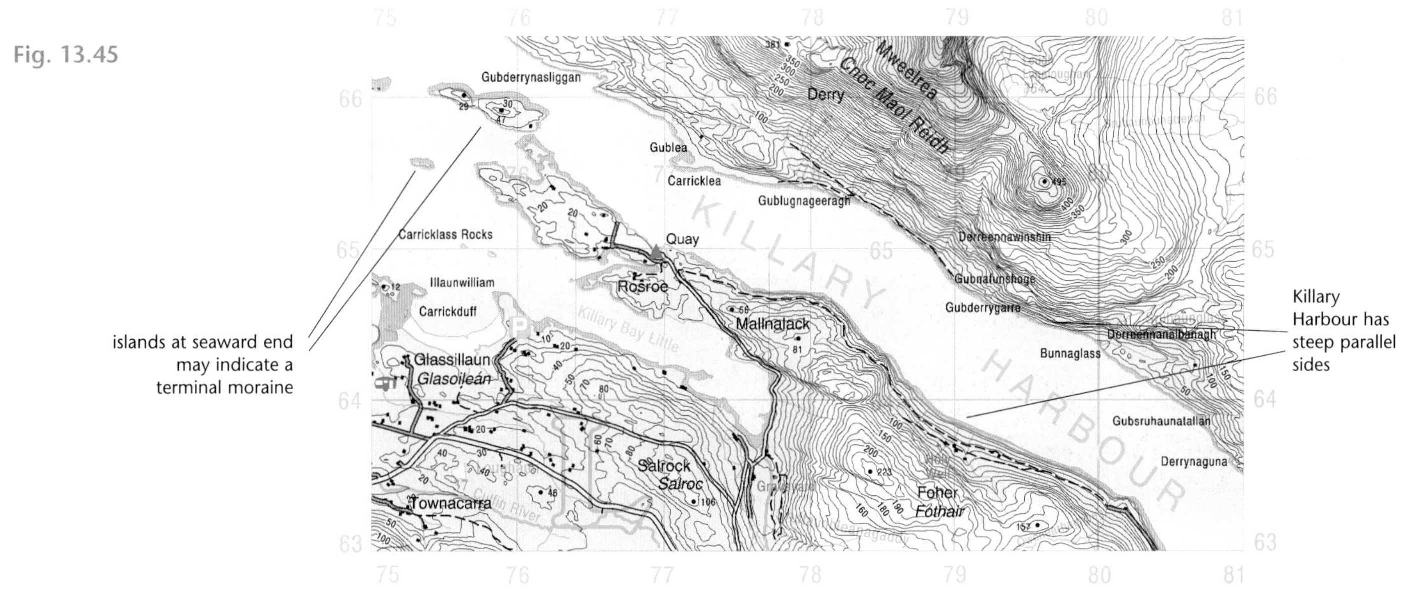

distinct notch in the land which slopes seaward. Caves which are backed by a cliff and fronted by a level wave-cut platform, and in some cases sea stacks now located high and dry above the sea, are evidence of a sea level of another time. This wave-cut platform may have a covering of beach material such as sand or shingle which is now camouflaged by a carpet of grass. Such level platforms near to the sea are often used as locations for coastal routeways, as at Black Head in the Burren area of Co. Clare or near Annalong in Co. Down, in Ireland.

Fig. 13.46
The Norwegian coastline has numerous fjord inlets which provide sheltered waters for fishing and steep south-facing sides for settlements and the growth of apples and other fruits. Notice the snowline in the background.

Fig. 13.47
Emerged highland coastal features

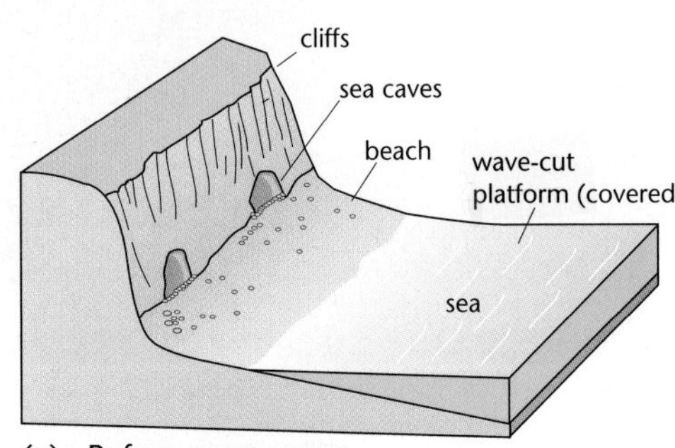


(a) Before emergence

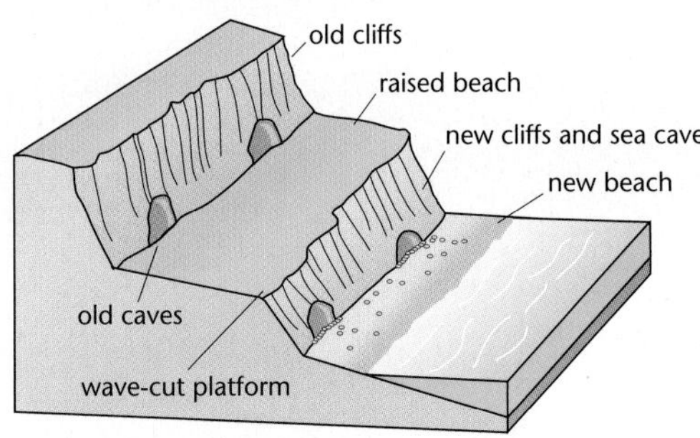


(b) After emergence

Fig. 13.48
An abandoned cliff-line. The smooth cliff face indicates wave action.

Fig. 13.49
Bloody Bridge, Northern Ireland

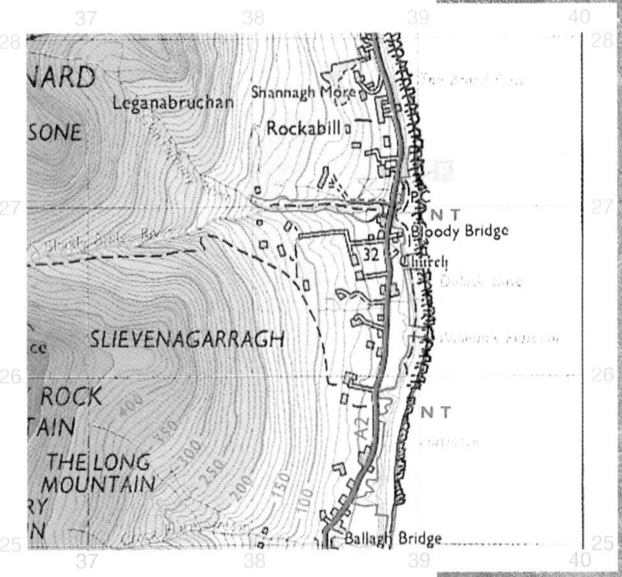

Fig. 13.50

Study the photograph and the map extract of Bloody Bridge/Mourne Mountains in Co. Down, in Ireland. Then do the following:

1. Classify this coastal area.
2. What evidence on the photograph suggests that sea levels in this region have not always been stable? Develop your answer fully.
3. Land quality has often influenced farm structures/boundaries. Discuss this statement using evidence on the photograph to support your answer (see enclosures, p.333).
4. Soil development in any particular area depends on a number of geographical factors. Discuss this statement using evidence from the photograph to support your answer.
5. Describe the course of the Bloody Bridge River which flows from east to west.
6. Rural land use varies greatly as a consequence of specific geographical factors. Discuss.

**periglacial**
An area near to an ice sheet which suffered intense freeze-thaw conditions causing a churning of the soil.

## (c) Emerged Lowland Coast

This forms when a part of a continental shelf emerges from the sea and forms a coastal plain. The coast has no bays or headlands and deposition takes place in the shallow water offshore, producing offshore bars, lagoons, spits and beaches. Examples of this type occur along the south-east coast of the USA and the north coast of the Gulf of Mexico. In such areas the development of ports is difficult.

In the south-east of the United States, the slope of the emerged plain is so slight that a 1-metre rise in sea level causes the coastline to retreat by 1,000 metres or more. Sea level rise is responsible for the numerous barrier islands that rim the east and Gulf coasts.

The islands and coastal bars were formed, probably, as beaches on the continental shelf during the ice age, when sea levels were lower. After the ice age, the rising sea levels flooded the land behind the beaches, isolating them from the coast.

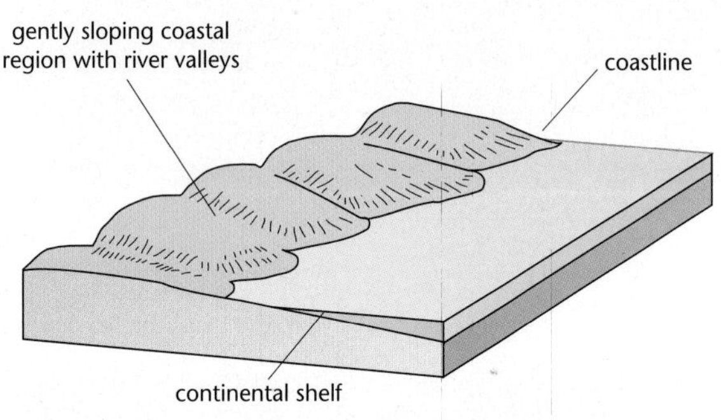


Fig. 13.51
Before emergence

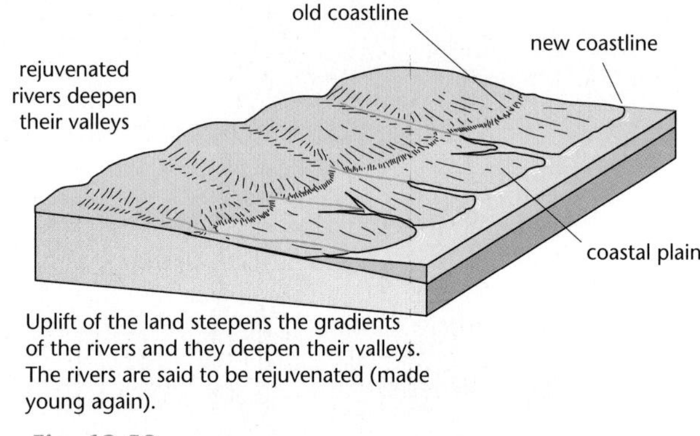


Fig. 13.52
After emergence

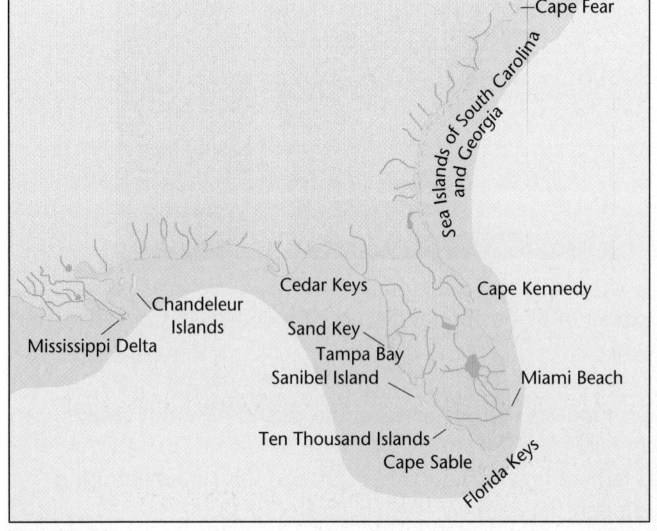


Fig. 13.53
The east and Gulf coasts of the United States. Rising sea level has created a complex coast of embayments, barrier islands, spits and beaches.

Fig. 13.54
A raised beach at Baile na nGall on the Dingle Peninsula, in Ireland

Fig. 13.55 Ardglass in Co. Down

Study the vertical photograph of Ardglass, in Co. Down above and the map extract and photograph, Figs. 17.58 and 17.59, p. 368. Then do the following:

1. Name the bay A, the headland B, the coastal landforms, C, D, E, F.
2. Explain fully the processes involved in the formation of landforms C and G. Use labelled diagrams.
3. Identify the landform at G. Explain, with the aid of labelled diagrams, the varied processes involved in its formation.
4. Draw a sketch map. On it mark and label
    - the coastline
    - the road network
    - two areas of coastal erosion
    - two areas of coastal deposition
    - a disused railway line
    - five areas of different rural land use
5. Draw an enlarged sketch map of Ardglass town. On it mark and label
    - the coastline
    - three piers, three harbours and three areas of secondary industry associated with the main function of the town
    - two churches, two castles
    - five other areas of different urban land use

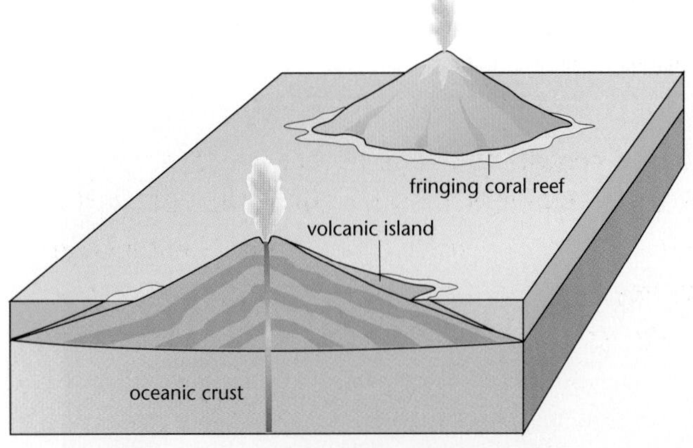


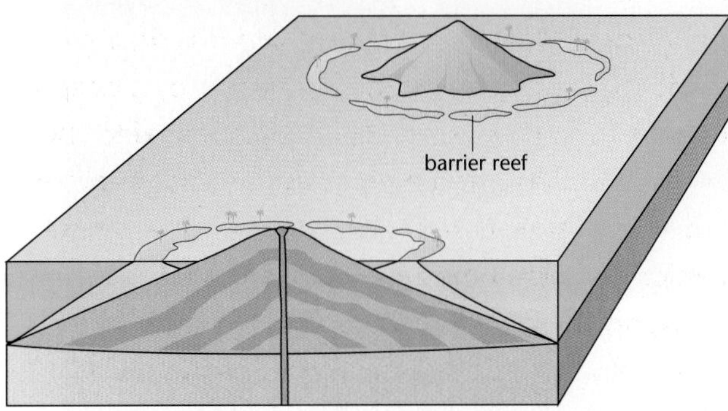


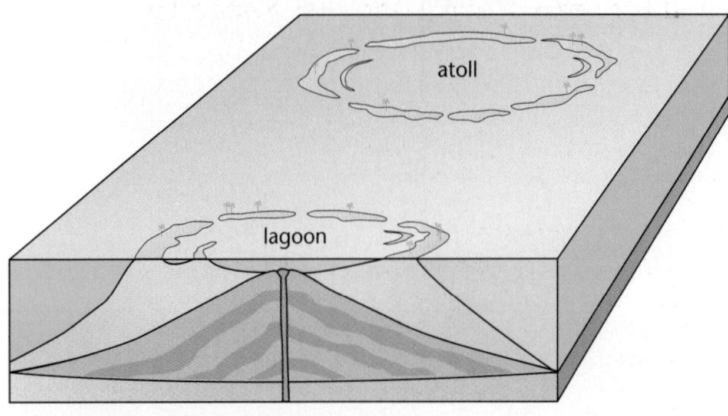


**Fig. 13.56**
Formation of a coral atoll due to the gradual sinking of oceanic crust (floor)

# Coral Coasts

## Coral Reefs and Islands

### Nature of coral

It is a limestone rock made up of the skeletons of tiny marine organisms called **coral polyps**. The tube-like skeletons in which the organisms live extend upward and outward as the old polyps die and new ones are born. Coral polyps cannot grow out of water and they are therefore formed below the level of low tide. They thrive under these conditions:

1. sea temperature of about 21°C (70 °F)
2. sunlit, clear salt water down to a depth of about 55 metres (180 feet)

Extensive coral formations develop between 30°N and 30°S, especially on the eastern sides of land masses where warm currents flow near to the coasts, such as the east Australian coast. They do not develop on the western coasts in these latitudes because of the cool currents which flow along these coasts, e.g. the Benguela Current in south-west Africa and West Australian Current.

### Types of Coral Formation

Coral masses are called **reefs** and there are three types:

1. **Fringing Reef:** a narrow coral ridge separated from the coast by a lagoon, which may disappear at low water.
2. **Barrier Reef:** a wide coral ridge separated from the coast by a wide, deep lagoon.
3. **Atoll:** a circular coral reef, which encloses a lagoon.

### Structure of a Coral Reef

Most reefs are fairly narrow and the coral ridge lies near to low water level. The seaward edge is steep and pieces of coral are broken off by wave action and are thrown up onto the ridge, where they form a low mound. On the landward side of this the breaking waves deposit sand in which the seeds of plants, such as coconut, readily germinate. Coral atolls in the Pacific are of this type.

## Types of Reefs

### Fringing Reef

This type of reef consists of a ridge of coral which is connected to, and is built out from, a coast. The surface of the ridge is usually flat or slightly concave and its outer edge drops away steeply to the surrounding sea floor. A shallow lagoon usually occurs between the coast and the outer edge of the reef.

Fig. 13.57
Fringing reef in Tahiti

### Barrier Reef

This reef is similar to a fringing reef except that it is situated several miles off the coast and is separated from it by a deep water lagoon. The coral of a barrier reef is often joined to the coast although the lagoon may be too deep for coral to grow on its bed. Some barrier reefs lie off the coasts of continents, e.g. Great Barrier Reef along the east coast of Australia. Others occur around islands forming a continuous reef except for openings on the leeward side. In some cases, fringing reefs develop on the inner side of lagoons which lie between a barrier reef and the coast of the island that it encircles.

Fig. 13.59 (a)
Barrier reef at low tide

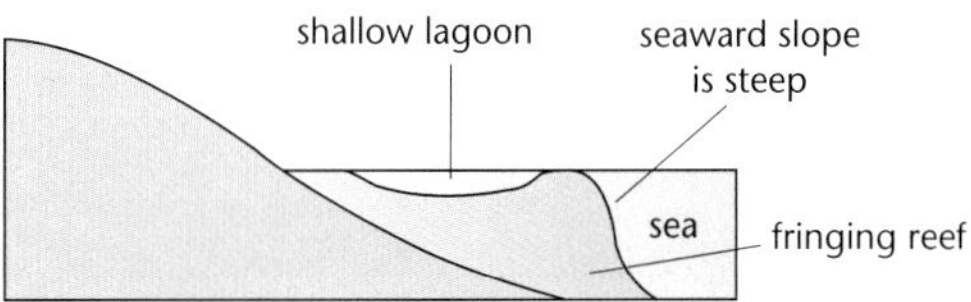

Fig. 13.58 (a)

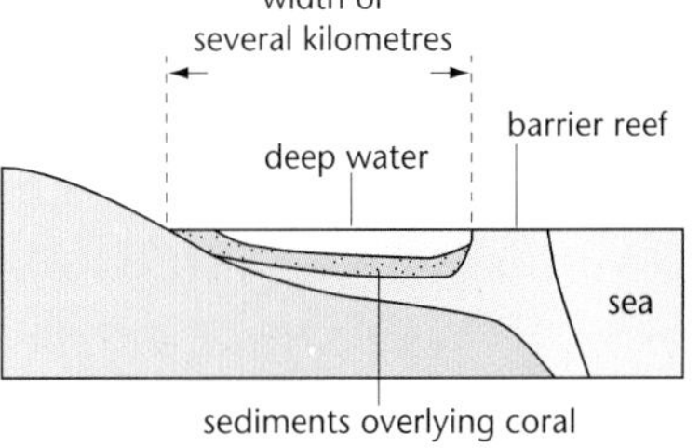

Fig. 13.58 (b)

Fig. 13.59 (b)
The underwater coral world with its many colours on the Great Barrier Reef in Australia

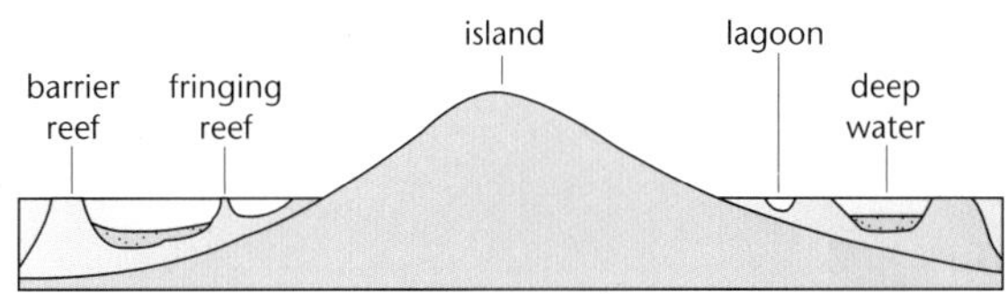

Fig. 13.58 (c)
Both types of reef can occur together

Fig. 13.60
Pacific atoll and surrounding reefs. Rangiroa in Polynesia, one of the world's largest atolls.

## Atoll

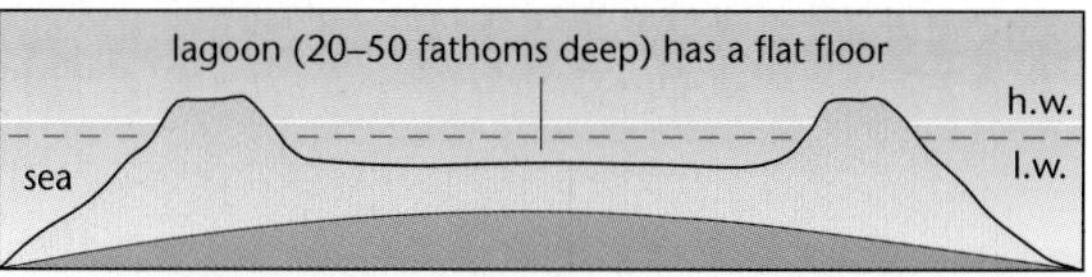


h.w. high water l.w. low water

Fig. 13.61

Figs. 13.61 and 13.62 show drawings of a typical atoll. Atolls are particularly common in the Pacific and Indian Oceans. Some are very large, e.g. that of Suvadiva in the Maldives. The lagoon of this atoll is 64 kilometres (40 miles) across and its reef extends for 190 kilometres (120 miles). The Gilbert and Ellice Islands of the Pacific are all atolls.

In 1831 the naturalist Charles Darwin set out aboard the British ship HMS *Beagle* on its famous five-year expedition that circumnavigated the globe. One outcome of Darwin's studies during the voyage was his hypothesis on the formation of coral islands, called atolls.

However, atolls present a difficulty. How can corals, which require warm, shallow, sunlit water no deeper than 45 metres, create landforms that reach thousands of metres to the ocean floor.

There are three main theories on the formation of coral reefs and atolls.

1. Sir John Murray's theory states that coral reefs and atolls formed on submarine hills or plateaus within 55 m (180 feet) of the surface.
2. Reginald Daly's theory suggests that a rise in sea level was responsible for their formation, and that coral growth kept pace with the rising waters.
3. Sir Charles Darwin's theory stated that because coral requires shallow, sunlit waters no deeper than 45 m (150 feet) to live, coral reefs form on the flanks of sinking volcanic islands. As an island slowly sinks, the corals continue to build the reef complex upward. Thus atolls are thought to owe their existence to the gradual sinking of the oceanic crust.

   American scientists made extensive studies of two atolls (Eniwetok and Bikini) that were going to be sites for testing atomic bombs. Drilling operations at these atolls revealed that volcanic rock did indeed underlie the thick coral reef structure. This finding was a striking confirmation of Darwin's explanation.

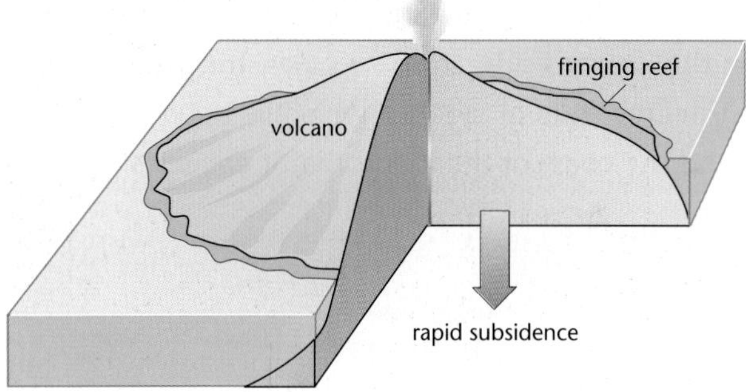


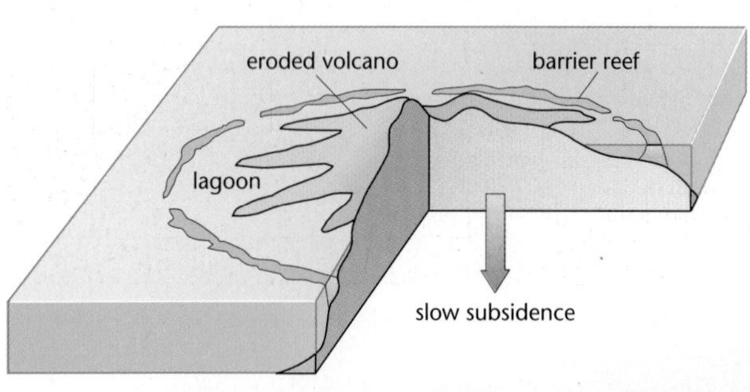


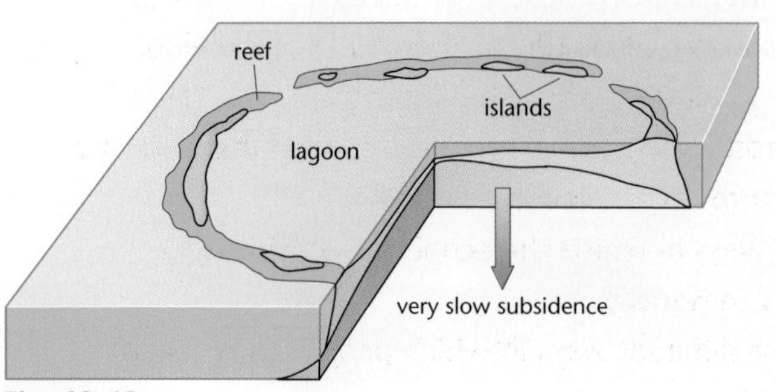


Fig. 13.62
Evolution of an atoll from a subsiding volcanic island

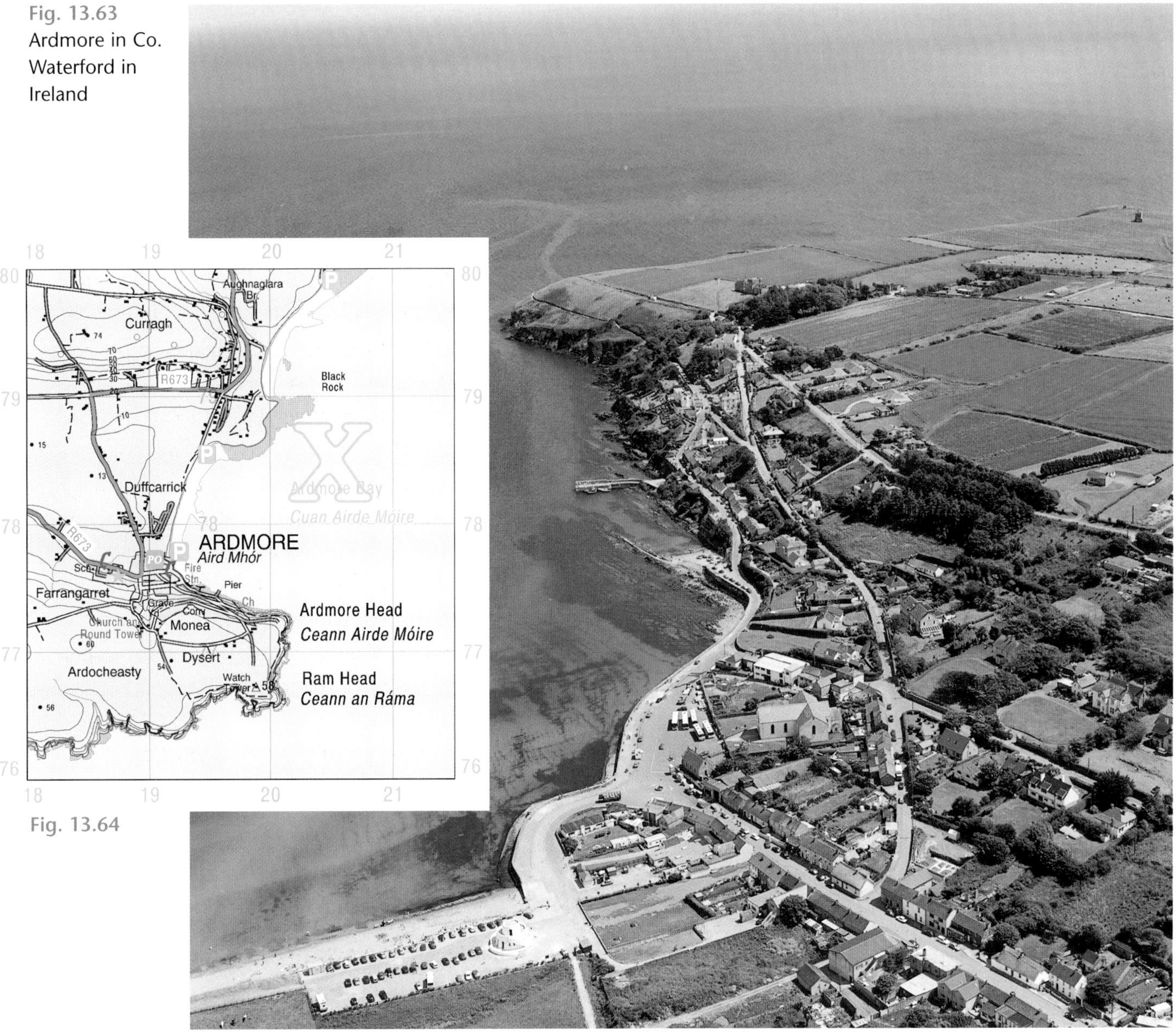

Fig. 13.63
Ardmore in Co. Waterford in Ireland

Fig. 13.64

Study the photograph of Ardmore in Co. Waterford, in Ireland, Fig. 13.63, and the corresponding Ordnance Survey map extract, Fig. 13.64. Then do the following:

1. In which direction was the camera pointing when the photograph was taken?
2. Describe how the processes of erosion and deposition have shaped this coastline. Refer in your answer to one landform in each case.
3. Describe and briefly explain two ways in which the economic activities of settlements such as this may vary seasonally.
4. 'Conflict is inevitable between the different ways in which people make use of coastal areas like this.' Examine this statement.
5. Describe in detail the variation in building size and function in the settlement.

Fig. 13.65
Killybegs in Co. Donegal

Study the photograph of Killybegs in Co. Donegal, in Ireland, Fig. 13.65, and answer the following questions:

1. Classify the settlement according to its main function. Explain your answer with evidence from the photograph.
2. Draw a sketch map of the area shown and on it mark and name the following features:
   (a) the coastline
   (b) three major piers
   (c) the road pattern
   (d) four different land uses
3. 'The settlement's main function has a dominating effect on all activities in the area shown, so much so that it becomes a way of life in which all community groups are involved.' Discuss this statement with reference to the photograph.
4. What evidence in the photograph suggests that this is a prosperous and expanding settlement?

Fig. 13.66
Ballycotton in Co. Cork in Ireland

Fig. 13.67
Ordnance Survey map extract of Ballycotton area in Co. Cork in Ireland

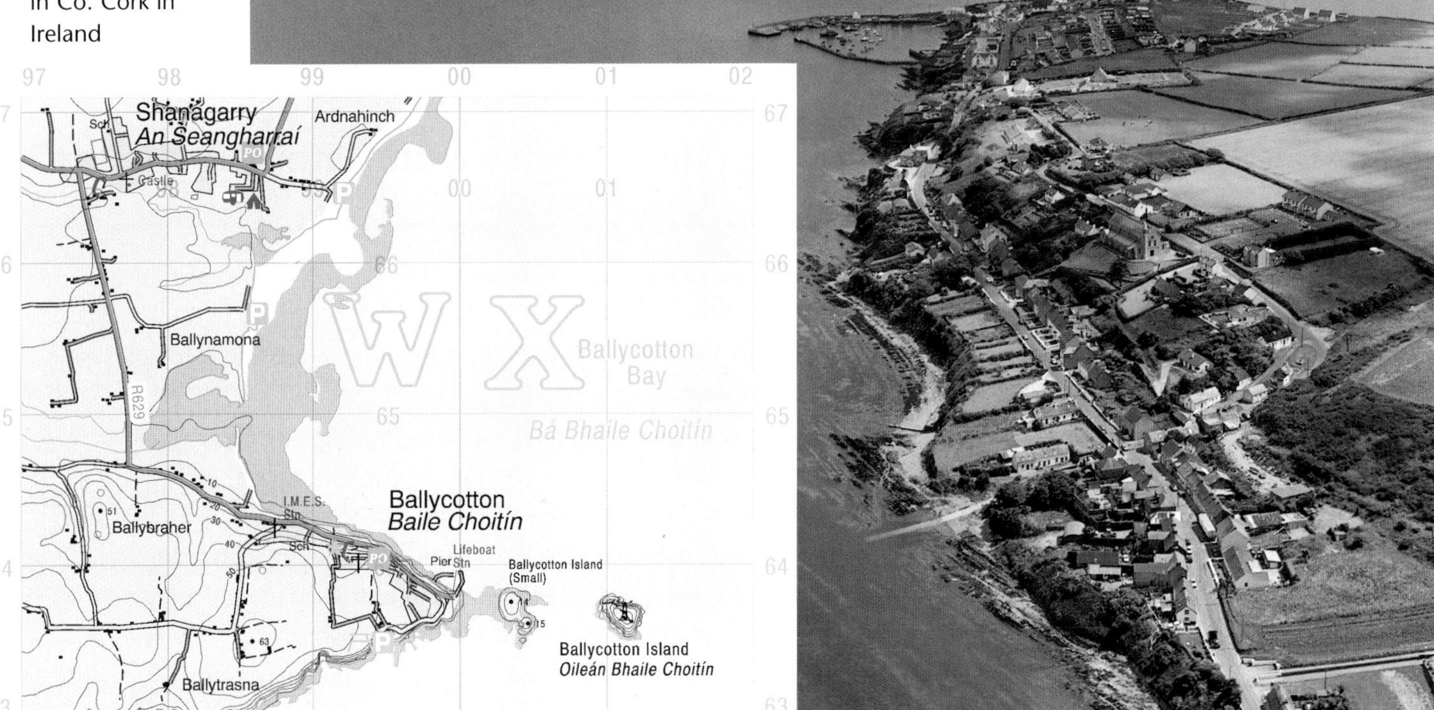


Study the photograph of Ballycotton in Co. Cork, in Ireland, Fig. 13.66, and the corresponding Ordnance Survey map extract, Fig. 13.67, and then do the following:

1. In which direction was the camera pointing when the photograph was taken?
2. Draw a sketch map of the area shown in the photograph and on it mark and label
   - (a) the coastline
   - (b) the street pattern of the village
   - (c) seven areas of different land use
3. This coastal area indicates that coastal erosion is severe at Ballycotton.
   - (a) What serious implications could this have for the village in the future?
   - (b) What efforts could be made to reduce the destructive effects of the sea at the village (see pp. 273–8)?
   - (c) What knock-on effect, if any, has the erosive power of the sea at Ballycotton had on the area to the north of the village?
   - (d) What knock-on effect could the efforts at (b) have on the area mentioned in (c)?
4. From evidence on the photograph can you suggest whether or not this village has a sewage treatment plant? If not, suggest what effect this could have on the local environment.

Study the photograph of Dunmore East in Co. Waterford, in Ireland, Fig. 13.70, p. 271, and the corresponding Ordnance Survey map extract, Fig. 13.69. Then do the following:

1. Classify this settlement according to its main function. Explain your choice using evidence from the photograph and map extract.
2. This settlement is located on the south-east coast of Ireland. Outline the main advantages of this region for a settlement with a function such as mentioned in question 1 above.
3. Growth of urban settlement in coastal areas throughout the world regularly changes the quality of the local environment over time. With reference to the map, to the photograph and to other areas that you have studied, explain how this may occur.
4. Outline the changes made in this area to create a safe environment for this settlement's function.

Fig. 13.68
How salt is removed

## Human Interference with Natural Processes in the Seas and Oceans of the World

### Positive Interference

#### A Source of Fresh Water

Many parts of the world lack plentiful supplies of fresh water for domestic or industrial activities.

The process of extracting fresh water from salty sea water is called desalination. Desalination is a costly and complex process, which requires large amounts of electricity. The water produced can cost up to ten times the normal price, providing 'natural' purified domestic water supplies. However, a new plant will be built on the Norfolk coast to supply one of the driest areas of Britain. If it proves cost-effective, many more will be built in coastal areas. This may provoke furious reaction from environmentalists. Many of these plants may be sited in beauty spots such as river estuaries and undeveloped shorelines and thus may be a source of conflict between the public, environmentalists and governments.

#### A Source of Energy

#### Tidal Power

Tides have been used as a source of power for centuries. As early as the twelfth century, waterwheels driven by the tides were used to power flour and saw mills. During the seventeenth and eighteenth centuries, most of Boston's flour was ground at a tidal mill.

Tidal power is harnessed by building a dam across the mouth of a bay or river estuary which has a large tidal range (the difference between the level of high tide and low tide).

Although estimated total tidal power potential throughout the world represents only a relatively small proportion of world energy requirements, its development would save the

Fig. 13.69

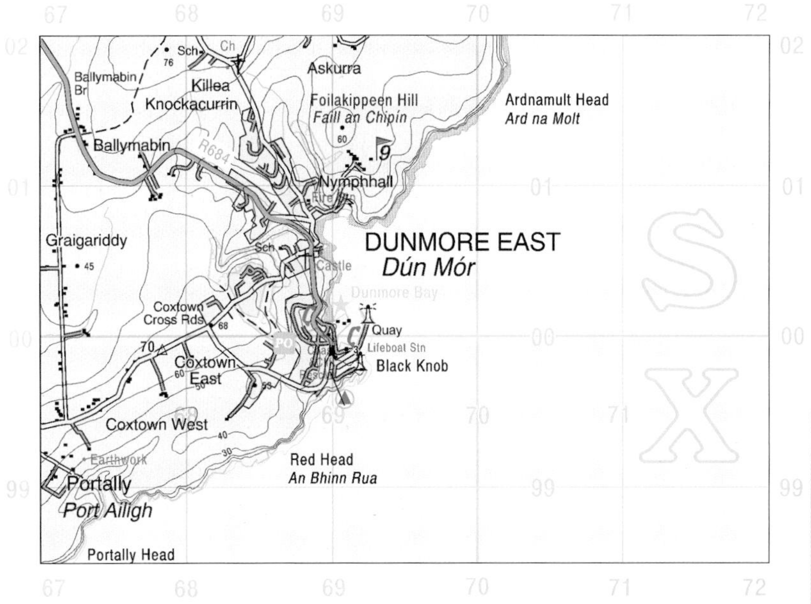

Fig. 13.70

use of a sizeable amount of fossil fuels. Tidal projects worldwide have been estimated to have a potential energy output equivalent to more than a billion barrels of oil a year.

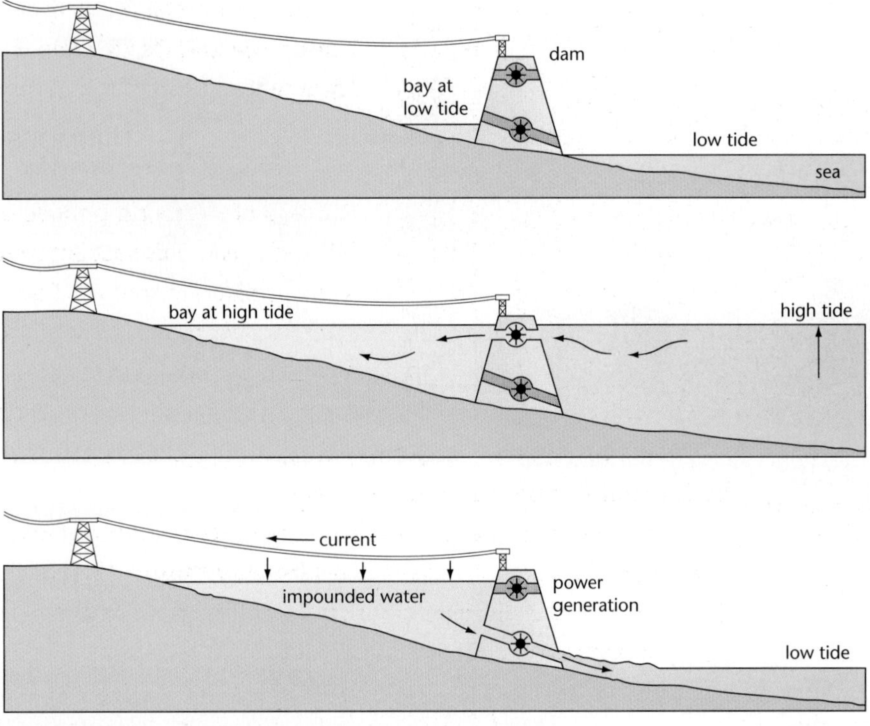


Fig. 13.71
The tidal power station at La Rance in Brittany, France, opened in 1966, consists of a dam blocking the 750-metre-wide (2,460 ft) estuary of the River Rance

## A Source of Minerals

### Bedrock

The bedrock of the sea floor in shallow water areas, such as the continental shelves, provides us with vast supplies of oil and natural gas. Britain, Norway, the United States, Venezuela, Kuwait and many more countries are huge producers of these minerals.

### Sea Floor

Manganese nodules are stone oval-shaped mineral deposits which lie on the sea floor near to mid-ocean ridges. They consist of concentric layers rich in copper, cobalt and nickel as well as manganese wrapped around a hard rock core (nucleus). Geologists think that the nodules formed as a result of reactions between cold sea water and hot fluids rising from fractures along mid-ocean ridges. Though strewn throughout the ocean floor, they are concentrated mostly in the vicinity of mid-ocean ridges.

**Class Activity**

Explain how the processes of rising and falling tides create electricity. Use Fig. 13.71 to develop your answer.

## A Source of Land

### Land Reclamation and the Struggle against the Sea in the Netherlands

Reclamation of land from the sea in the Netherlands has continued over hundreds of years. Approximately 50 per cent of the country's total land area consists of polderlands on which over 60 per cent of its fifteen million population live. In the past, windmills were used for powering water pumps. Over time these were replaced by steam pumps and in the twentieth century diesel and electric pumps allowed reclamation on a much larger

Fig. 13.72
Almere, an 'overspill town' in the Netherlands built on reclaimed land from the sea

scale than before. Rather than just defending existing land from flooding, the Dutch reclaimed land from the seabed and created large freshwater reservoirs for water supplies.

The most ambitious actions to date have been the **Zuider Zee** and **Delta projects**. Outline the main advantages of each of these projects (see Class Activity).

## A Source of Food

The seas of the world provide over ninety million tonnes of fish each year. This vast amount has increased from about twenty million tonnes in 1950. Much of this increased catch is a consequence of huge 'factory ships' and modern technology, such as sonar and monofilament nets. Most of this catch is caught on the continental shelves.

The waters of the continental shelves rarely exceed 185 metres (600 feet) in depth, and so sunlight can penetrate to the seabed.

Because of this and because many rivers drain into the seas, these areas are rich in **plankton**. Plankton are created by **photosynthesis** (the process by which a plant makes food). This activity is generally confined to the upper 60 metres of the sea, where sunlight, oxygen and carbon dioxide are available.

Thus the continental shelves are major fishing areas and form a vital part of some countries' economies, such as Ireland, Britain, Canada, China and Japan.

The warm currents of these continental shelves, such as the North Atlantic Drift and the Gulf Stream in the North Atlantic and the Kuro Siwo in the Western Pacific, allow year-round ice-free waters for fishing in these areas.

**Class Activity**

Explain fully how the Dutch have created polderland. Focus on the Zuider Zee project. Use embassy publications as a base for your research.

## Coastal Protection

### Creation of Some Coastal Defences

Due to recent increased environmental awareness, it is recognised that the coastline is a valuable natural resource, which needs careful and sensitive management. This is especially so in the case of small island countries, such as Ireland and Britain, where the coastal zone has a direct and major influence on the economic and social welfare of the countries.

### Groynes

A groyne is a thin, parallel-sided barrier built at right angles to a beach or coast. One end is attached to the shore on the beach, usually not below the low water line. Groynes are built for the purpose of trapping sand that is moving parallel to the shore. Normally, beach sediment (sand) is carried by the process of longshore drift (sometimes called littoral drift) along the shore. This sediment is trapped at the updrift side of the groynes. At the downdrift side due to the reduction in sediment passing the groyne, some local erosion occurs.

The location and spacing of the groynes must be analysed to ensure that the correct level of sediment retention is obtained to provide a balance between the need for

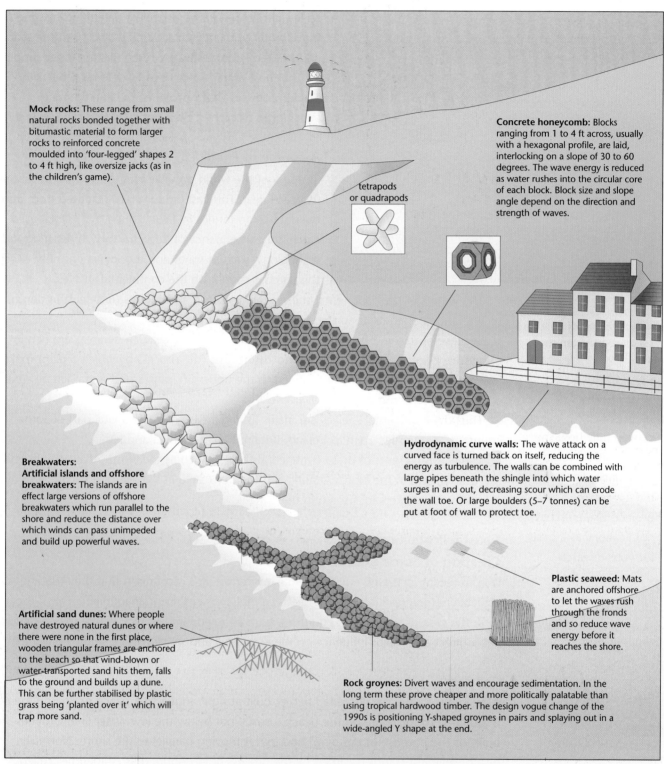


**Mock rocks:** These range from small natural rocks bonded together with bitumastic material to form larger rocks to reinforced concrete moulded into 'four-legged' shapes 2 to 4 ft high, like oversize jacks (as in the children's game).

tetrapods or quadrapods

**Concrete honeycomb:** Blocks ranging from 1 to 4 ft across, usually with a hexagonal profile, are laid, interlocking on a slope of 30 to 60 degrees. The wave energy is reduced as water rushes into the circular core of each block. Block size and slope angle depend on the direction and strength of waves.

**Breakwaters:**
**Artificial islands and offshore breakwaters:** The islands are in effect large versions of offshore breakwaters which run parallel to the shore and reduce the distance over which winds can pass unimpeded and build up powerful waves.

**Hydrodynamic curve walls:** The wave attack on a curved face is turned back on itself, reducing the energy as turbulence. The walls can be combined with large pipes beneath the shingle into which water surges in and out, decreasing scour which can erode the wall toe. Or large boulders (5–7 tonnes) can be put at foot of wall to protect toe.

**Plastic seaweed:** Mats are anchored offshore to let the waves rush through the fronds and so reduce wave energy before it reaches the shore.

**Artificial sand dunes:** Where people have destroyed natural dunes or where there were none in the first place, wooden triangular frames are anchored to the beach so that wind-blown or water-transported sand hits them, falls to the ground and builds up a dune. This can be further stabilised by plastic grass being 'planted over it' which will trap more sand.

**Rock groynes:** Divert waves and encourage sedimentation. In the long term these prove cheaper and more politically palatable than using tropical hardwood timber. The design vogue change of the 1990s is positioning Y-shaped groynes in pairs and splaying out in a wide-angled Y shape at the end.

**Fig. 13.73**
The creation of coastal defences

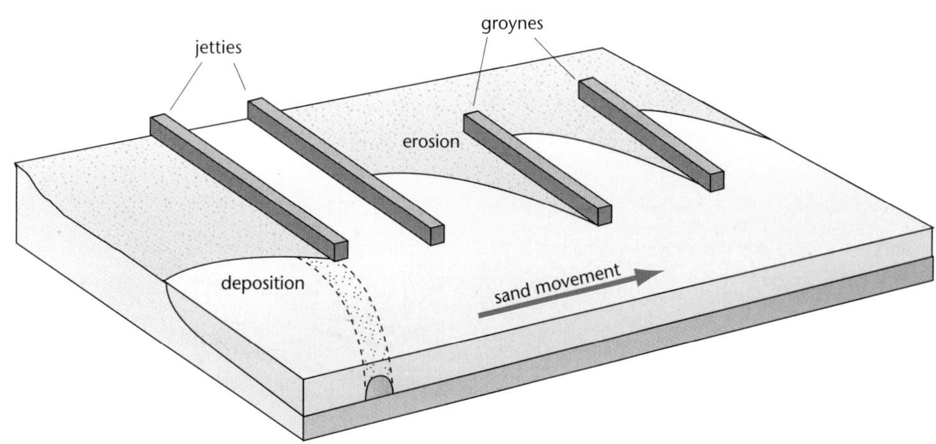


Fig. 13.74
Jetties, groynes and breakwaters interrupt the movement of sand by beach drift and longshore currents. Beach erosion often results downcurrent from the site of the structure.

Fig. 13.75
Tetrapods are used as a method of reducing wave action. They protect the pier from erosion. These are used at Kilmore Quay in Wexford to protect the pier.

protection of the area and the requirements of the zones which are downdrift of the breakwater system. After careful monitoring, these groynes may be adjusted if erosion is too great on the downdrift side.

Groynes may have many shapes, such as T, L and Y (sometimes referred to as fishtail). Each shape may be chosen for its effect on any individual area. Rock groynes are used at Rosslare Strand in Co. Wexford, in Ireland, to reduce erosion of the spit and beach (see Fig. 13.81, p. 277 and Fig. 13.84, p. 278 and Fig. 13.85, p. 279).

## 2. Jetties

The construction of jetties may result in the interruption of longshore drift sediment. Typically, sediment will back up against the upshore wall and forms a curved underwater bank of sand, called a deflection shoal, off the inlet mouth. **Sand bypassing** is the removal of this sand accumulation by dredging or pumping via a pipeline and transferring it elsewhere.

Fig. 13.76
Rock breakwaters are used to protect the fishing vessels and Pontoon restaurant from wave action at Sorrento in Italy

## 3 Breakwaters

Offshore breakwaters are long parallel mounds of rubble, or rock, which are built parallel to the shore for the purpose of reducing wave energy. In some cases, sunken shipwrecks or other materials are used. Again, as with groynes, breakwaters are designed to suit each individual site of coastal erosion. The effect of wave change around and behind the breakwater is to set up new currents which trap sediment in the lee of the structure (see Fig. 13.77).

The sediment deposited in the lee of the breakwater may result in either an increase in the beach width at the shoreline or a sandbar reaching to the breakwater in the form of a tombolo. The formation of the tombolo will depend on local factors, such as tidal current, storm frequency or shape of breakwater.

Fig. 13.77

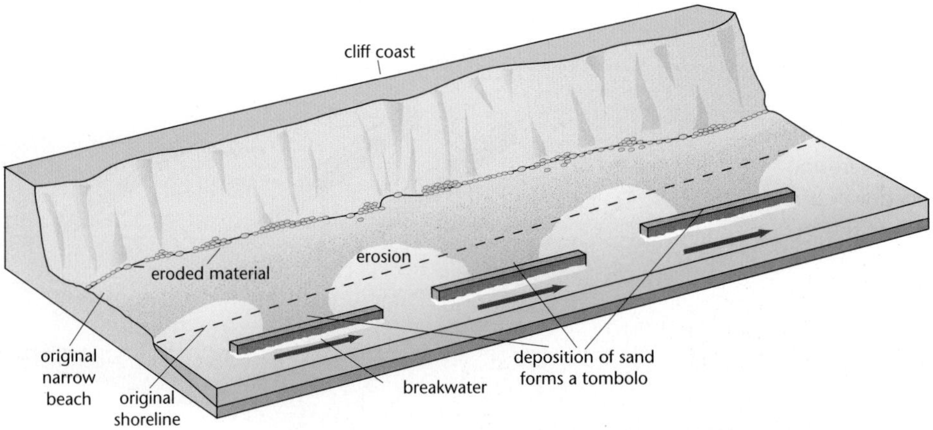

cliff coast
eroded material
erosion
original narrow beach
original shoreline
breakwater
deposition of sand forms a tombolo

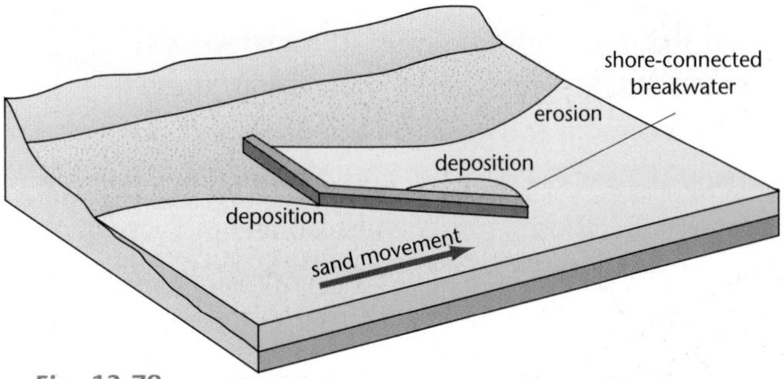

shore-connected breakwater
erosion
deposition
deposition
sand movement

Fig. 13.78

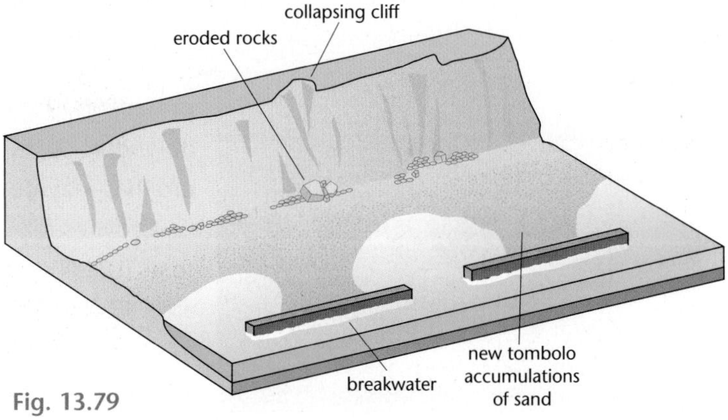

collapsing cliff
eroded rocks
breakwater
new tombolo accumulations of sand

Fig. 13.79

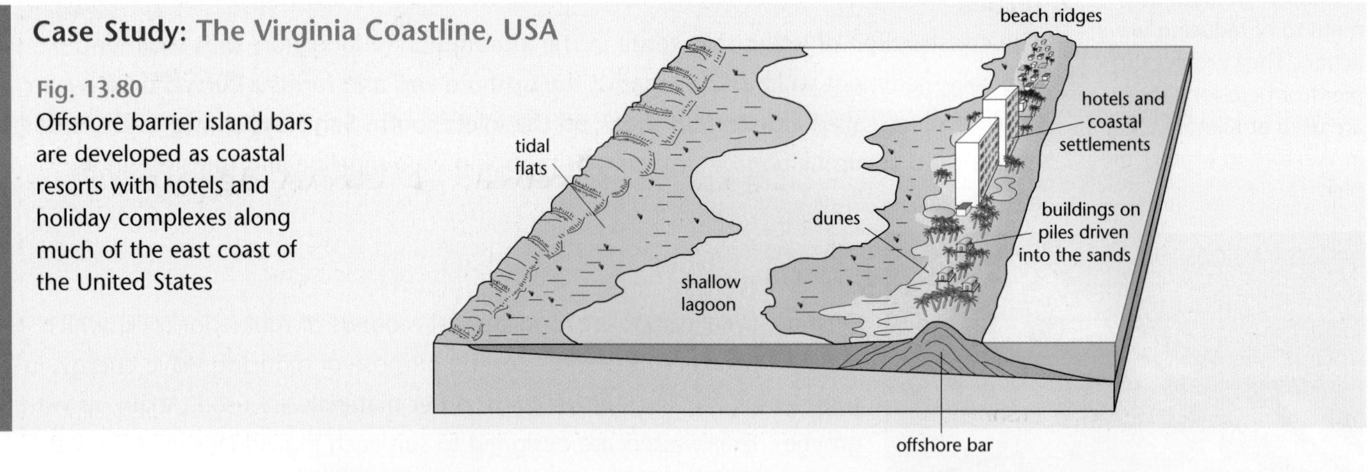


## Case Study: The Virginia Coastline, USA

Fig. 13.80
Offshore barrier island bars are developed as coastal resorts with hotels and holiday complexes along much of the east coast of the United States

**Home Activity**

Study the diagram, Fig. 13.80 and then do the following questions:

1. Identify two disadvantages of coastal development for areas such as that displayed in the diagram. For each disadvantage you give, explain fully the implications for such development on a coastal environment.
2. Identify two advantages of coastal development for areas such as that displayed in the diagram.

Fig. 13.81
Groynes have been built perpendicular to the shoreline of Eastbourne in Sussex in England, to prevent excessive loss of sand by longshore drift at this popular resort area

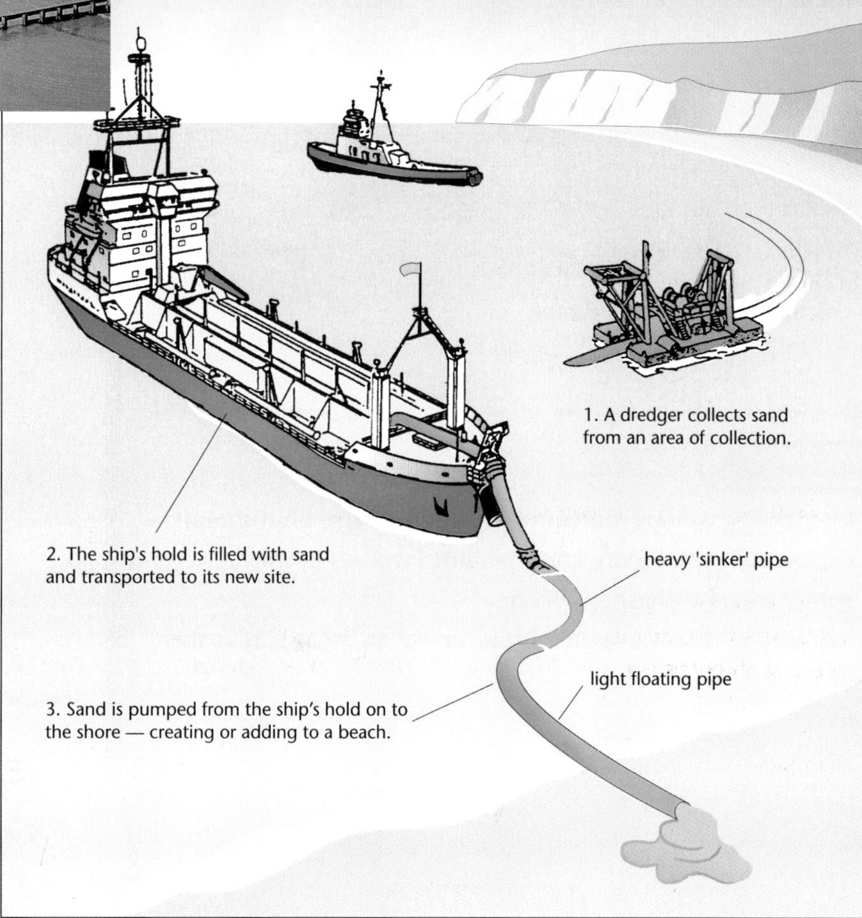

1. A dredger collects sand from an area of collection.
2. The ship's hold is filled with sand and transported to its new site.
heavy 'sinker' pipe
light floating pipe
3. Sand is pumped from the ship's hold on to the shore — creating or adding to a beach.

Fig. 13.82
Beach nourishment

## Beach Nourishment

Beach nourishment is simply the pumping of sand onto an eroded beach to change wave movement so as to increase deposition. Source material for beach nourishment is normally obtained by offshore dredging although in certain cases land sources are used for small schemes.

It is important that the source material is selected with great care to ensure that the grain size and its distribution are appropriate for the site. In certain circumstances, such as increased wave action, the use of a coarser sediment than the original beach material may be appropriate.

Fig. 13.83
A coastal fishing village in Tunisia

## Class Activity

Identify the various sources of pollution on this photograph.
Explain fully the theory and methods involved in the construction of these dwellings.
Explain the disadvantages of tourism for an area such as this in Tunisia.

## Case Study: Rosslare Strand in Co. Wexford, in Ireland

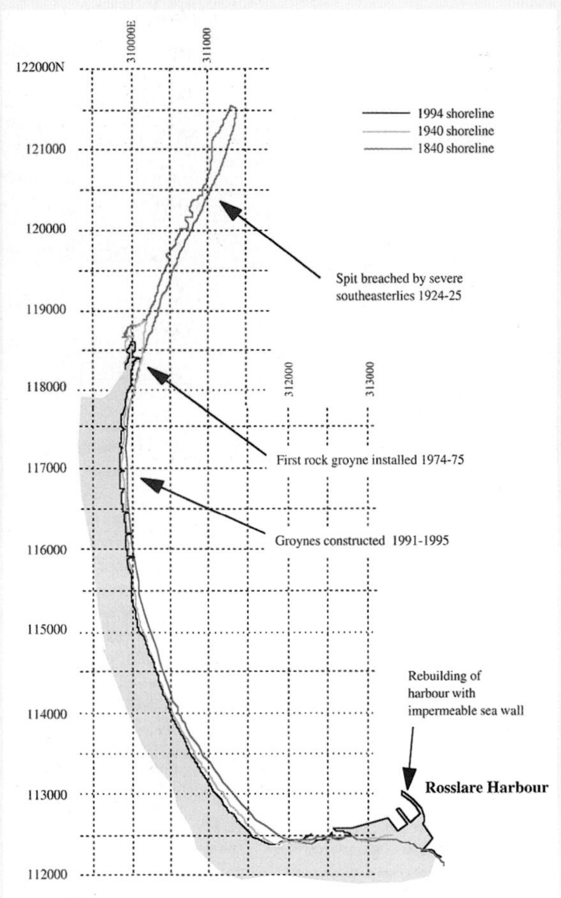


Fig. 13.84
Shoreline as shown on OS maps 1994, 1940 and 1840

## Class Activity

Carefully examine the Ordnance Survey map extract of Rosslare in Co. Wexford, Fig. 13.85 on p. 279, and the diagram, Fig. 13.84 above. Then do the following:

1. Draw a sketch map of the area shown on the map extract. On it mark and label:
   - the coastline
   - Rosslare Harbour and pier
   - Rosslare Strand and beach
   - four third-class roads
   - one national primary road, two regional roads
   - linear settlement along a third-class road
   - a sand spit
   - a cliff
   - a headland
2. List the tourist facilities available in this area.

Fig. 13.85

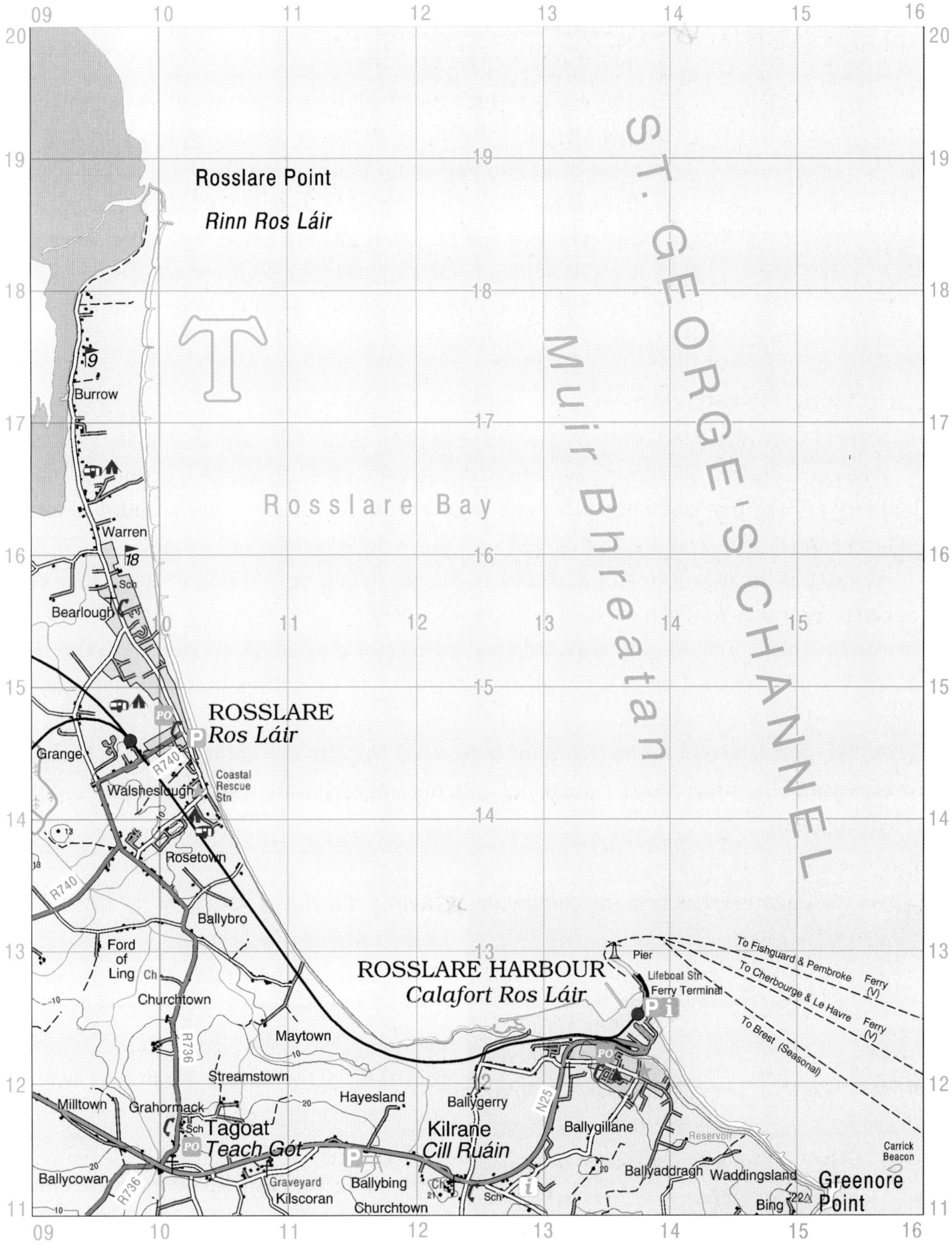

3. Natural processes as well as human interference are responsible for the creation/development of this area as a tourist destination. Discuss — in your answer refer to two natural processes and two human interferences.
4. Coastal erosion as well as man's interference over the years has influenced the present shoreline at Rosslare Strand. Discuss fully.

## Negative Interference

### Overfishing — The Oceans

Factory trawlers are a major cause of depletion of some fish species and a serious strain on 70 per cent of the world's fish stocks. These trawlers with their high-tech equipment are capable of trailing nets of up to one kilometre long and of hauling in 400 tonnes of fish in a single trawl. They have crews of as many as 100. Sophisticated fish-finding sonar, spotting aircraft and precision satellite-based navigation leave little to luck. Once found, schools of fish are hauled aboard for conveyor-belt sorting, gutting, filleting and freezing. Cruises can last for days or months, in all kinds of weather and in every corner of the oceans.

### Case Study: By-catch Waste

To combat overfishing in EU waters, quotas have been set by the EU member states.

As a consequence of licensing systems, trawlers with a licence to catch pollock or mackerel must dump any haddock or hake that come aboard. These **rejected fish**, called by-catch, **amount** to an estimated **twenty-seven million tonnes a year**, more than 25 per cent of the total caught worldwide.

By-catch is the main cause of depletion of the North Sea herring population. Large industrial trawlers licensed to catch sprat only for fishmeal haul in huge amounts of young herring which must then be dumped overboard as by-catch. While young herring are a delicacy in Holland, they are of no interest to the sprat trawlers, which are mainly Danish. Yet the Danish by-catch amounts to more than three times the new quota allowed to the Dutch, who would eat the young fish.

The EU must take part of the blame for overfishing. As with the farming system, it supports the large producer at the expense of the small trawler/traditional fishermen. Withdrawal of smaller vessels is promoted while larger vessels are encouraged.

### Shrimp Farming

#### Loss of Mangrove Forests

One-quarter of the world's shrimp is raised in ponds in fifty countries throughout the world. Shrimp farming is particularly damaging to the mangrove forests in the tropics. Mangroves are narrow strips of coastal forests which survive in tidal swamps by filtering sea water and which protect coastal areas from erosion. Much of the world's shrimp supply is raised in ponds gouged out of these thick mangroves. Here, sheltered from the open sea, the shrimp spend six months living in high-density ponds — as many as 500,000 per hectare — before being harvested and processed for export.

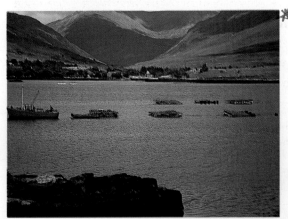

**Fig. 13.86**
Mariculture cages in Killary Harbour in Co. Galway. Does this activity affect this rugged rural landscape in any way? Discuss.

#### Creation of Disease

High shrimp density eventually fosters disease. Within three to six years, ponds exhaust their usefulness. Disease-carrying bacteria settle into the shrimp waste, poisoning the pond against further use. The growers then move along the coast, destroying more mangrove forest and rice fields to make way for more shrimp ponds.

## Leaving Certificate Questions on the Sea, Coastal Processes and Landforms

Ordinary Level

**Coastal Landscapes**

**1998**

(i) Name **THREE** landforms that result from the action of the sea.
Explain, with the aid of a diagram, how **EACH** feature was formed. **(60 marks)**

(ii) Describe and explain **TWO** ways of preventing coastal erosion. **(20 marks)**

**The Physical World**

**1997** — Explain the following:

(i) Erosion by the sea has resulted in distinctive landscape features. **(40 marks)**

(ii) Marine erosion has created spectacular coastal landscapes. **(40 marks)**

**Coastal Landscapes**

**1996**

Coastal landscapes are shaped by erosion and deposition.

(i) Select any **THREE** landforms that result from action by the sea and, in the case of each that you select:
- Describe and explain, with the aid of a diagram, how it was formed.
- Name a specific location where the feature may be found. **(60 marks)**

(ii) Describe any **TWO** ways by which human activity attempts to protect coastlines from erosion. **(20 marks)**

**1995**

(c) Sea cliff, sea arch, beach, lagoon, blowhole, sand spit.

(i) In the case of **EACH** of the above coastal features, state whether it is the result of erosion or deposition. **(18 marks)**

(ii) Select any **THREE** of the features listed above and in the case of **EACH**:
- Name a specific location where the feature may be found.
- Describe and explain, with the aid of a diagram, how it was formed. **(48 marks)**

(iii) In recent years, coastal erosion has caused enormous damage to coastal areas. Describe **TWO** methods used to limit this damage. **(14 marks)**

**1993**

(b)

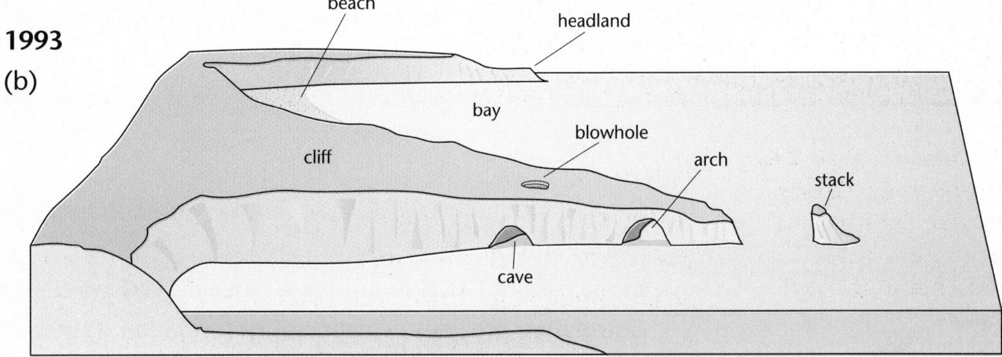


Fig. 13.87

The diagram (Fig. 13.87) shows an area where marine erosion has been active.

(i) With reference to any **THREE** major landforms which are evident in the diagram, describe and explain how marine erosion has shaped them. **(60 marks)**

(ii) Materials produced by erosion at one part of a coastline are transported to other parts and deposited. Human action is sometimes taken in order to interfere with these natural processes. Explain why. **(20 marks)**

**1990**

(b) (i) With reference to **THREE** typical landforms, describe and explain how the sea shapes the coastline. **(50 marks)**

(ii) There is often a conflict between the different ways in which modern society uses coastlines. Explain this statement, referring to **TWO** human activities.

**Higher Level**

**Physical Section**

**1996**

(i) Examine the processes which influence the formation of any **THREE** landforms found along the Irish coast. **(75 marks)**

(ii) 'Coastlines are subjected to much human use and abuse.'
Discuss this statement, with reference to Ireland. **(25 marks)**

**1991**

(b) (i) 'The coastal zone is undergoing constant change, resulting from both natural physical processes and the pressures caused by human activity.'
Examine this statement, referring to examples you have studied. **(75 marks)**

**1990**

(c) The shaping of coastal landforms involves interaction between three forces: Waves, Currents and Tides.

(i) Examine the statement above, with reference to **THREE** landforms which are typical of coastal regions. **(75 marks)**

(ii) Most coastlines contain landforms which show evidence of change in the relative levels of land and sea over time.
Describe and explain the formation of any **TWO** such landforms. **(25 marks)**

**1987**

(b) On a rough outline sketch-map of the world, indicate the distribution of (i) volcanic islands and (ii) coral islands.
In the case of **EACH** island-type, explain **TWO** ways in which the pattern of distribution provides evidence of its formation. **(100 marks)**

## Photograph Question

### 1992

Examine the aerial photograph which accompanies this paper [Fig. 13.70, p. 271] and answer the following:

**NB: DO NOT use tracing paper when answering this question.**

(i) On a sketch map, based on the photograph, mark and label:
- The coastline
- The network of roads
- **FOUR** different land-uses which are related to this coastal location. **(30 marks)**

(ii) With reference to landforms which are evident in the photograph, describe and explain how the processes of erosion and of deposition affect coastlines. **(35 marks)**

(iii) 'Conflict is inevitable between the different ways in which people make use of coastal areas like this.'
Examine this statement. **(35 marks)**

### 1989

(a) Examine the aerial photograph which accompanies this paper [Fig. 13.88], and answer the following:

Fig 13.88

(i) Draw a sketch-map based on the photograph and on it mark and label:
- **TWO** areas where coastal erosion is evident.
- **TWO** areas where coastal deposition is evident.
- **THREE** areas of different land-uses which are related to the coastal location of this settlement. **(35 marks)**

(ii) With reference to **ONE** landform in **EACH** case, describe and explain how the processes of erosion **AND** of deposition operate in coastal areas such as this.

(iii) 'The interaction between human activities and natural processes in areas such as this can have negative effects on the coastal environment.'
Examine the above statement.
**(25 marks)**

# Deserts, Wind Processes and Landforms

## Deserts

Fig. 14.1
A distinctive pattern in rocks in Navajo Sandstone, Colorado Plateau, Arizona, USA. These rocks were once sand dunes. Each layer (stratum) is easily identified by its own bedding plane.

Desert conditions are temporary phases over various parts of the earth's surface. Many of today's deserts were once centres of thriving civilisations. For example, the Northern Sahara Desert once produced grain for the Roman Empire. In other places, animal drawings of bison and gazelle in desert caves indicate a surrounding land of tropical savannah grassland. At the end of the ice age, the present 'rain belt' of temperate latitudes extended further south, covering present deserts. As the ice sheets withdrew, so did the rain belt and the desert edges advanced poleward.

### Important Aspects of Desert Landscapes

- There is a partial or almost complete absence of vegetation cover. This creates an unprotected surface for denudation (weathering and erosion).
- **Water is the dominant agent of erosion.**
- Wind is also an agent of erosion.
- Mechanical and chemical weathering are active.

### Plate Tectonics and Deserts

While studying the processes at work in deserts, it is important not to forget that the processes associated with the theory of plate tectonics have also played their part in desert formation.

### Some Influences of Plate Tectonics in the Formation of Deserts

- The movements of the earth's tectonic plates have brought the continents to their present locations, where diverging winds (high pressure areas) have made rainfall scarce.
- Mountain ranges, formed as a consequence of colliding plates and subduction, form barriers to rain-bearing winds that lead to rain shadows.
- The locations of the continents create loops in ocean currents that force cold currents

to flow parallel to the west coasts of South America and Africa. This causes the formation of the Atacama Desert in Chile and Peru and the Namibian Desert in Namibia in Africa.

- Faulting, compression and uplift of the earth's crust have changed the elevations of some areas, which gradually leads to climatic change. For example, prior to the uplift of the Sierras in Nevada and eastern California in the USA, the climate of that region was much like the humid and swampy conditions of present-day central Florida and Louisiana. Sediments deposited in Nevada at that time reflect this environment. Later sediments deposited during the uplift of the mountains record the changes to the present-day desert conditions.

Only about 29 per cent of the earth's surface is dry land and only a very small proportion of this is fertile and capable of being used agriculturally. About one-third of the earth's dry land surface experiences desert or semi-desert conditions. This does not include the polar and sub-polar lands, which are sometimes called the 'cold deserts'.

Lack of abundant, consistent rainfall is the dominant factor of all deserts. Its quantity, frequency or absence is responsible for the variety, distribution or absence of vegetation and for the factors that make each desert region unique.

Although it seldom rains in a desert, **running water is the dominant agent of erosion**. Rain, when it does come, falls in heavy downpours. Such rainstorms create flash floods which may be very destructive and deadly.

The location of most deserts is related to descending air. When air sinks, it is compressed and warmed, much like a bicycle pump flex which heats as a tube is pumped. Such conditions are just the opposite of what is needed to produce clouds and rain. Consequently, these regions are known for their clear skies, sunshine and ongoing drought.

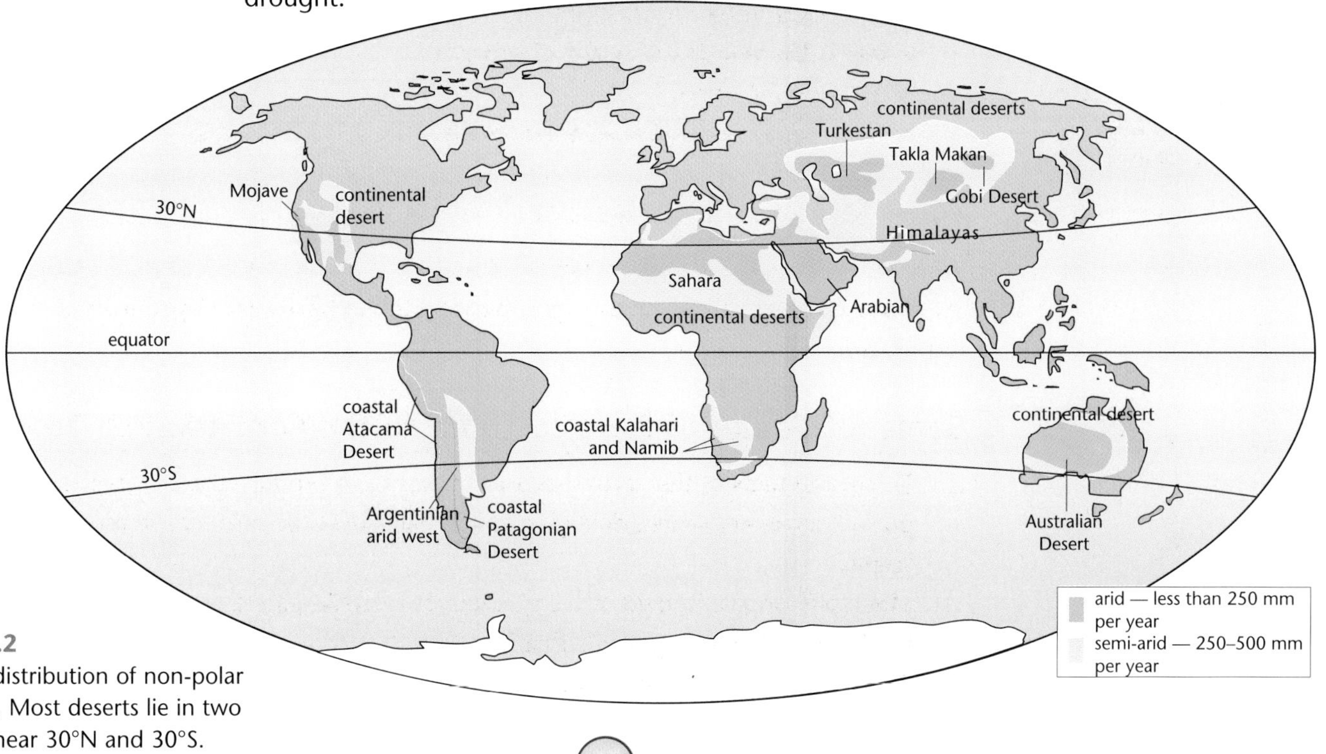



**Fig. 14.2**
World distribution of non-polar deserts. Most deserts lie in two bands near 30°N and 30°S.

# Main Causes of Deserts

## 1. Descending Air

Subtropical high-pressure regions between 15° and 30° north and south of the equator are dry areas. The descending air is compressed and warmed and becomes increasingly capable of holding more and more moisture. Thus, the air near the surface is very dry. This creates clear skies.

## 2. Rain Shadow

Within the tropics, the prevailing winds are the North-East Trades and the South-East Trades. Where they meet mountain barriers, such as the Andes in South America, the Drakensberg in South Africa, and the Great Divide in Australia, they are forced to rise and lose their moisture on the eastern sides. When they descend, they are compressed, heated and dry.

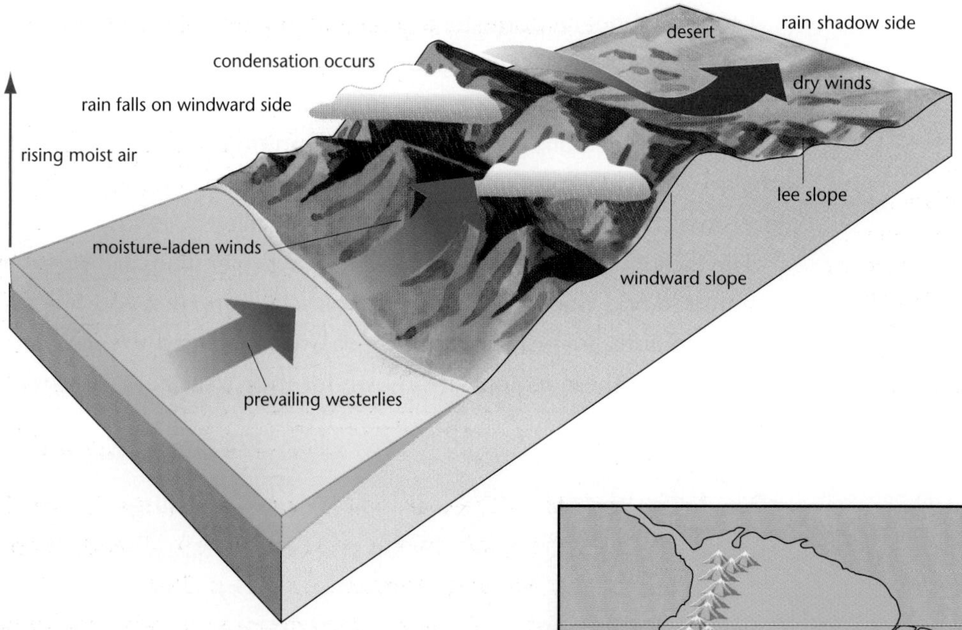

**Fig. 14.3**
Rain shadow effect

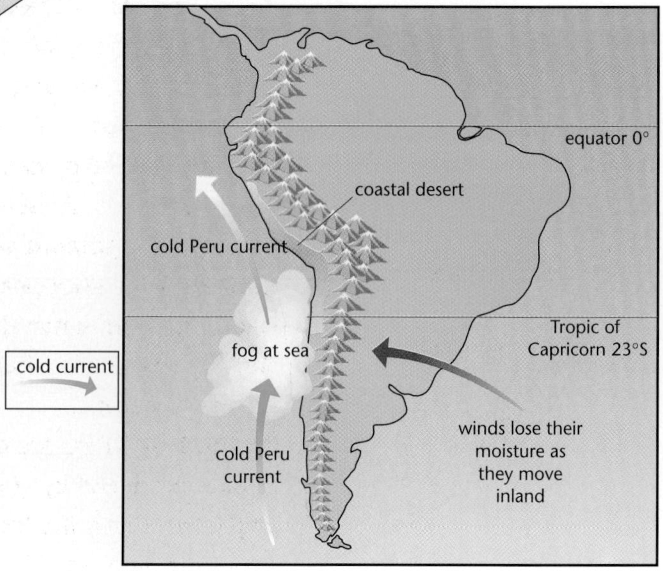

rising moist air
rain falls on windward side
condensation occurs
desert
rain shadow side
dry winds
lee slope
moisture-laden winds
windward slope
prevailing westerlies

cold current
equator 0°
coastal desert
cold Peru current
fog at sea
Tropic of Capricorn 23°S
winds lose their moisture as they move inland
cold Peru current

**Fig. 14.4**
South American coastal Atacama Desert

## 3. Cold Water Coastlines

Cold ocean currents and upwelling of cold ocean waters on western continental coastlines give rise to deserts. Winds which blow over these waters are cooled. On reaching land they are heated and so retain their moisture.

## Desert Processes and Landforms

1. **Sandy Deserts** — These have various names, such as erg in the Sahara and koum in Turkestan. They are vast, almost horizontal sandsheets or undulating plains of sand and are produced by wind deposition. Outstanding examples are found in the Sahara and the great sandy desert of north-western Australia.
2. **Stony Deserts** — Horizontal sheets of smoothly angular boulders and pebbles cover the surface. Extremes of heat, cold and light dew cause expansion, contraction and shattering of the bedrock. Finer sands are blown away. These deserts are called serir in Libya and Egypt, reg in Algeria and gibber plains in Australia. 'Desert pavement' also refers to this type of desert surface.

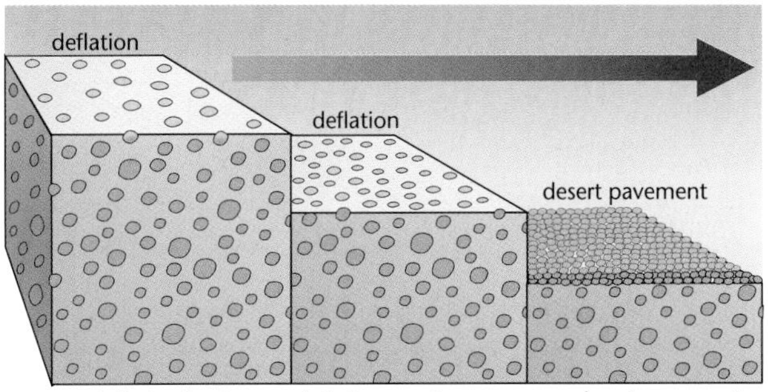

Fig. 14.5
Formation of desert pavement. Deflation lowers the surface by removing sands and silt. Coarse particles of stones and rocks are gradually exposed and interlock to form a layer and protect the surface from further erosion.

3. **Rocky Deserts** — These are called hammadas in the Sahara. They consist of smooth bare rock surfaces, windswept and almost clear of sand. These bare surfaces may be interrupted by yardangs and zeugens (see p. 290).
4. **Desert Plateaus** — These are deep depressions regularly formed by earth movements. They are crossed by canyon-like valleys of rivers whose water is derived from beyond the desert lands. These depressions have steep plateau edges, isolated **mesas** and **buttes** and **piedmont fans** at the foot of steep slopes (see p. 299).
5. **Rock Peaks** — Rock peaks, with their steep, craggy faces cut into and divided by wadis, rise from a mantle of angular rock wastes or screes. The chaotic peaks of Sinai and the Tibesti and Ahaggar ranges of the central Sahara display this spectacular eroded landscape.
6. **Badlands** — These are developed in semi-desert regions mainly as a result of water erosion produced by violent rainstorms. They occur in upland areas. As there is little vegetative cover, the torrential rains create extensive gullies and ravines which are separated by steep-sided ridges. Excellent examples of these deserts occur in North America, extending from Alberta in Canada to Arizona in the United States.

Fig. 14.6
Dust storm in eastern Namibia

## Wind (Aeolian) Processes

Winds are able to support, erode and deposit materials much like rivers and seas.

## Wind Transport

Wind moves material in three ways:

- **Suspension:** In extraordinary storms, wind is capable of carrying rock particles several centimetres in diameter. These can be lifted to heights of a metre or more. In most regions, however, the largest particles of sediment that can be suspended in the air are grains of sand.
- **Surface Creep:** Wind blowing across beds of sand moves the grains in a rolling motion. This process is called surface creep.
- **Saltation:** In strong desert winds, small sand grains are lifted off the surface into the air. They are carried forward along an arcuate path landing a short distance downward. This process is similar to the saltation process in rivers where pebbles are bounced along the river-bed. As these sand grains are bounced along, they strike other grains.

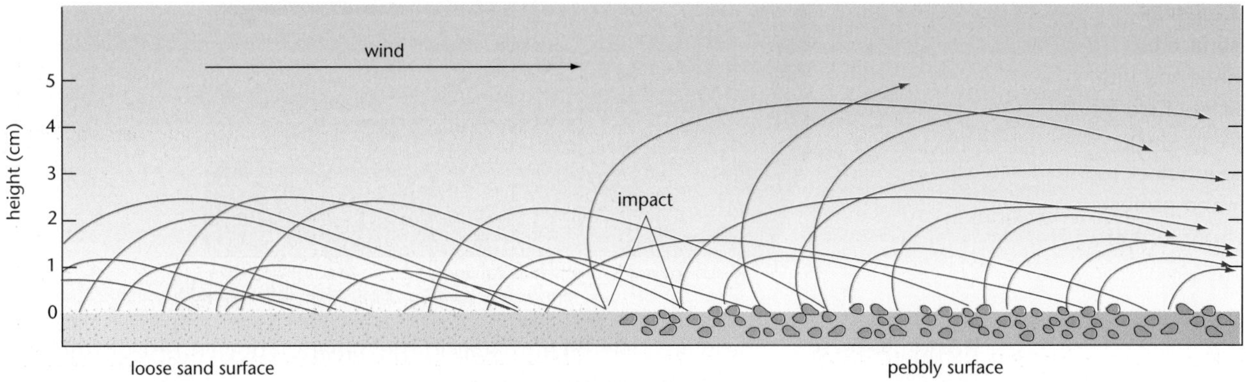


Fig. 14.7
Strong wind causes movement of sand grains by saltation. Impacted grains bounce into the air and are carried along by the wind as gravity pulls them back to the land surface where they impact other particles, repeating the process.

## Processes of Erosion

There are three main processes of wind erosion: 1. deflation, 2. corrasion (abrasion) and 3. attrition.

### 1. Deflation

Deflation is the picking up and blowing away of loose dust and sand. This process provides most of the wind's load. The finest grains are swept high, while the larger grains are carried closer to the ground. Larger particles may be bounced along the ground by the process of saltation.

### 2. Corrasion (Abrasion)

This is the 'sandblasting' effect of wind carrying a load of hard sand (quartz) grains. This process can smooth and polish rock when it is of uniform hardness. Large individual rocks,

Fig. 14.8 (a)
Rocks polished by wind-borne minerals in Sinai Desert in Egypt

Fig. 14.8 (b)
'The Devil's Marbles', exfoliation domes, Northern Territory, Australia

broken off by mechanical weathering but too heavy to be moved by the wind, are worn on the windward side. These rocks are called **ventifacts**. A ventifact with three wind-eroded surfaces is called a **dreikanter**.

### 3. Attrition

The particles of sand which are moved by the wind collide with each other or with larger particles on the ground. This process reduces these particles into tiny rounded grains.

## Landforms of Wind Erosion

### Landform: Rock Pedestal

Examples: Lut Desert in Iran
Death Valley in California

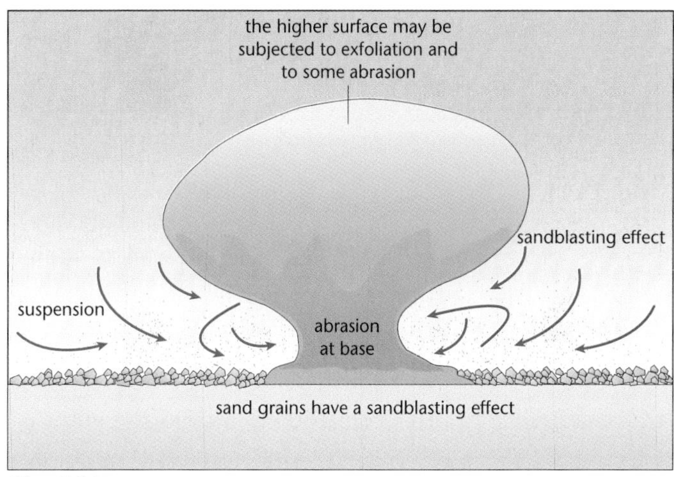


Fig. 14.9
Gara (plural gour)

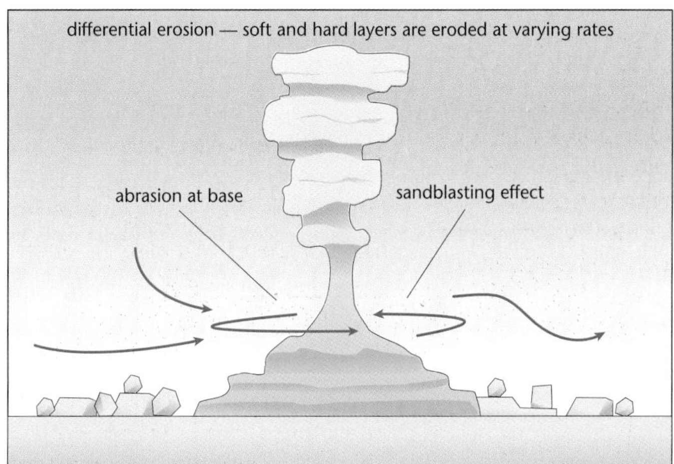


Fig. 14.10
Rock pedestal

Processes: Suspension, Abrasion, Saltation, Exfoliation

#### Formation

Rock pedestals are formed in deserts as a consequence of abrasion. When winds blow over desert surfaces, they pick up grains of sand and hold them in suspension. These sand grains are then used by the winds to erode large rocks which lie in their path. As sand particles in suspension and heavier particles by saltation are moved just above desert surfaces they erode the base of rocks which lie in their path. This undercutting process is most marked about half a metre above the ground.

The upper surface of the rock is weathered by the mechanical process of exfoliation. The expanding of the rock by day and contracting of the rock by night together with moisture, such as dew, shatters the rock surface. This creates a

Processes: Suspension, Abrasion, Saltation, Exfoliation

rounded surface as curved 'shells' of the rock fall to the ground. A gara is formed in this way.

Large blocks of rock with bands of hard and soft layers (strata) are eroded at different rates. The soft rock erodes quickly while the hard rock is more resistant and erodes more slowly. The rock pedestals formed as a consequence of this process take many shapes, each depending on the composition and shape of the original rock.

The bases of these blocks may be eroded to such a degree that the rock seems about to topple over. These perched rocks are called rock pedestals.

## Landform: Zeugens

Examples: Egyptian Desert, Arabian Desert

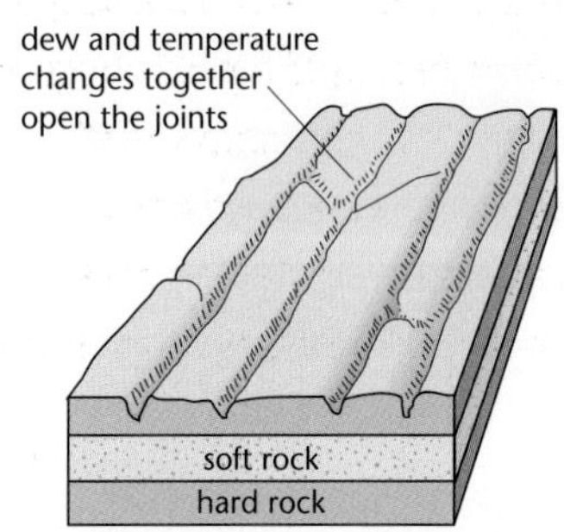


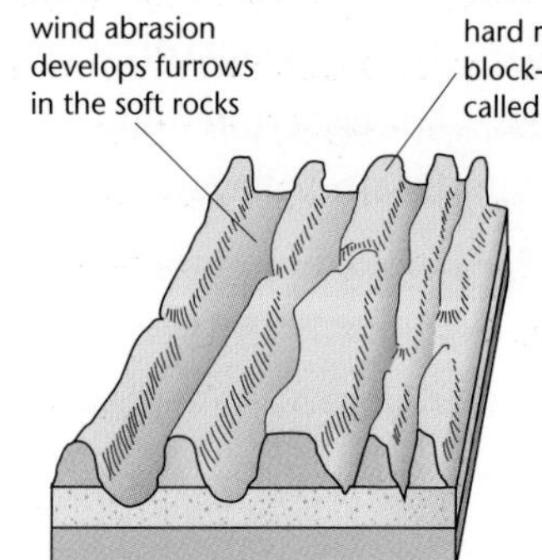


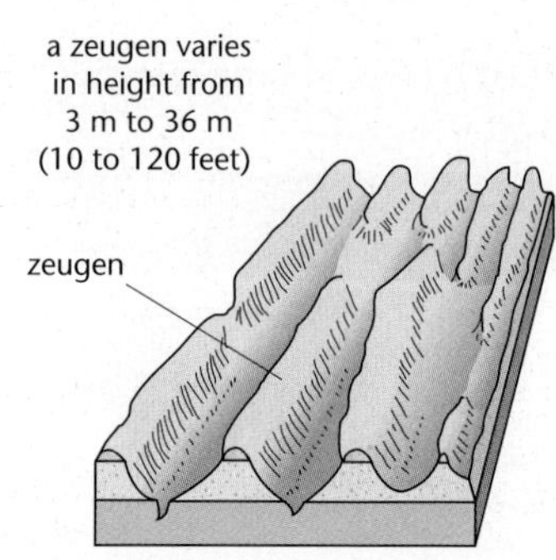


Fig. 14.11
(a) Weathering opens up the joints
(b) Wind abrasion continues the work of weathering
(c) Wind abrasion slowly lowers the zeugens and widens the furrows

Processes: Suspension, Abrasion

Zeugens are formed where a hard layer of surface rock is underlain by a soft layer. Weathering attacks the joints on the hard surface layer. When the joints are widened sufficiently, the wind, saltation particles and suspended sand erode by the process of abrasion until separate tabular masses, called zeugens, are left standing upon the softer underlying rock. This erosion process continues until they are completely undercut and gradually worn away. Such zeugens often stand up prominently thirty metres or more above surface level.

## Landform: Yardangs

Examples: Atacama Desert in Chile
Takla Makan and Gobi Deserts in central Asia
Western Egypt

Processes: Abrasion, Saltation, Deflation

Yardangs are extensive ridges of rock, separated by grooves (troughs), which lie parallel to the direction of the prevailing winds. ('Yardang' comes from the Turkish word *yar*, meaning steep bank.) As with zeugens, the processes of abrasion and saltation erode the soft rock bands faster than the hard bands. The prevailing winds, which are the most regular winds in an area, use the wind-blown sand and dust to further deepen and

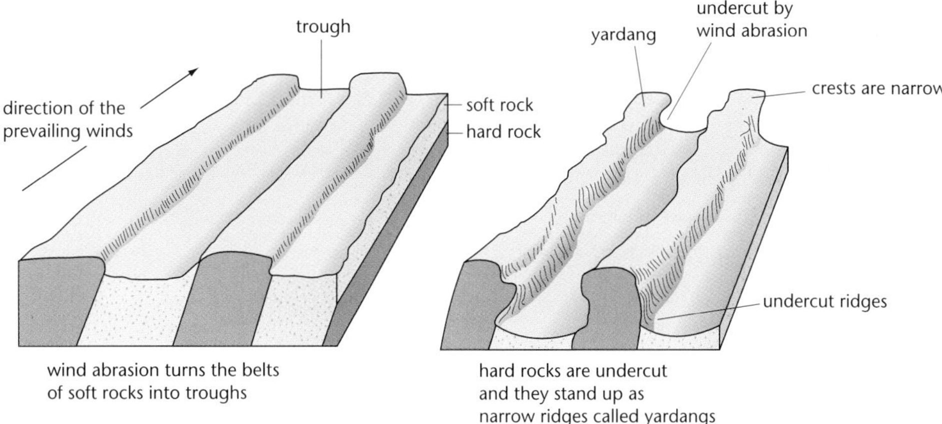

Fig. 14.12

broaden the depressions between the hard bands. These processes of abrasion and deflation finally create a 'ridge and furrow' landscape of well-worn ridges, with blunt rounded fronts facing the prevailing winds and with narrowing 'keel-like' crests (like an inverted ship's hull).

Typically, yardangs are sharp-crested and carved from hard, compacted sediments or from highly weathered crystalline rocks. Individual yardangs range up to a few tens of kilometres long and up to 100 metres high.

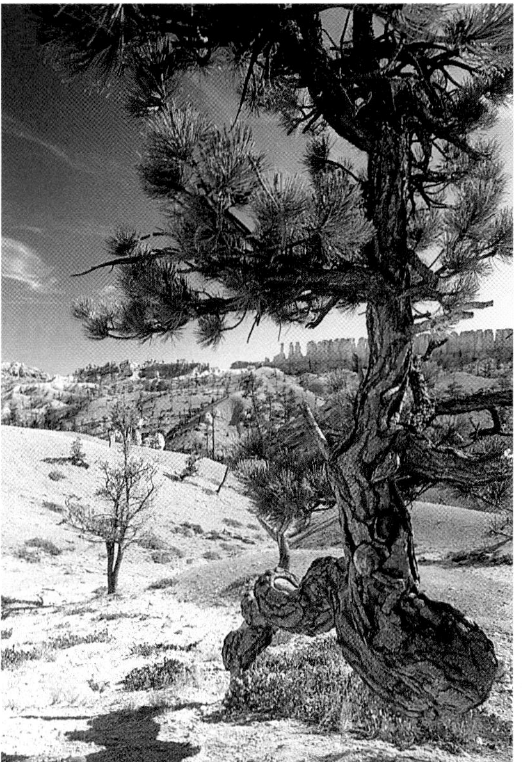
Fig. 14.13
A long root system which reaches the water-table allows some plants to survive in a desert

## Deflation and Deflation Hollows

Deflation is the progressive removal of fine material by the wind, leaving pebble-strewn '**desert pavements**' or '**reg**' or '**gibber plains**'. Throughout much of the Sahara and especially in Sinai in Egypt, vast areas of these pebble and stone landscapes were formed by deflation. The pebbles and stones were carried as part of the loads to these areas of rivers and flash floods from surrounding upland. They were deposited with clay, silt and sand on the plains. Since then, the wind has removed these fine materials, leaving the coarse particles too large to be moved. The pebbles and stones interlock like cobblestones which protect the material underneath from further erosion. (See 'Stony Deserts', p. 287.)

## Landform: Deflation Hollows

Example: Qattara Depression in Egypt

Processes: Earth Movements, Faulting, Deflation, Abrasion, Saltation

Small saucer-shaped depressions and larger basins created by wind erosion or tectonic movements (faulting) or a combination of both are numerous throughout desert lands. Thousands of these basins occur in the Great Plains region of North America from Canada to Texas. Some are only a metre or so deep and no more than two kilometres long. However, others, such as the Qattara Desert in Egypt, are very large. This depression is 122 metres below sea level and 250 kilometres long. It contains salt marshes and the sand eroded from it forms a zone of dunes on the lee side.

Fig. 14.14
Some depressions produced by wind deflation reach down to water-bearing rocks. A swamp or an oasis then develops.

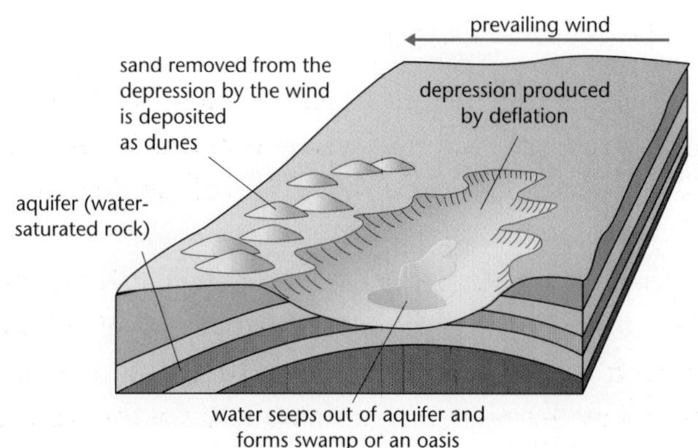
prevailing wind
sand removed from the depression by the wind is deposited as dunes
depression produced by deflation
aquifer (water-saturated rock)
water seeps out of aquifer and forms swamp or an oasis

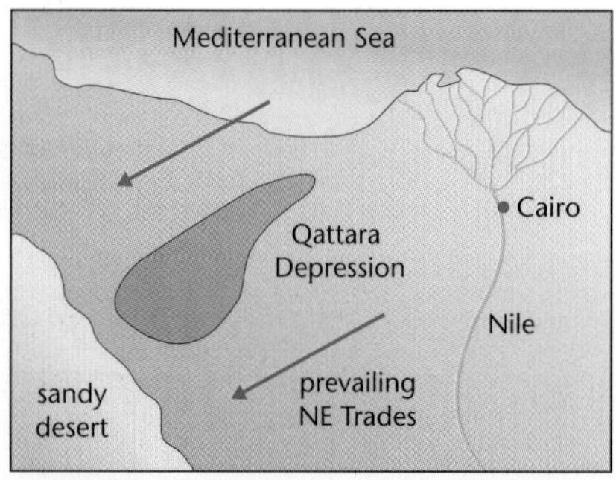
Mediterranean Sea
Cairo
Qattara Depression
Nile
sandy desert
prevailing NE Trades

The Qattara Depression is 130 metres below sea level. It has salt marshes and the sand excavated from it forms a zone of dunes on the lee side.

Fig. 14.15
Plate tectonics and deflation hollows.
The formation of some depressions may first be caused by faulting. The soft rocks thus exposed are eroded by the wind.

(1) Deflation hollows in hot deserts

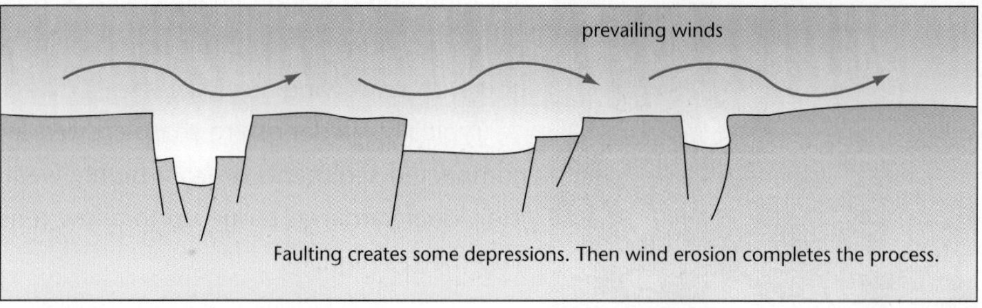
prevailing winds
Faulting creates some depressions. Then wind erosion completes the process.

(2) Deflation hollows in hot deserts

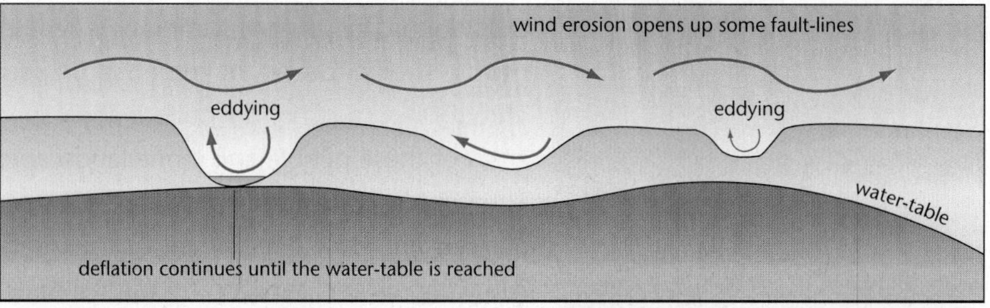
wind erosion opens up some fault-lines
eddying
eddying
water-table
deflation continues until the water-table is reached

(3) Wind erosion smoothens the surface

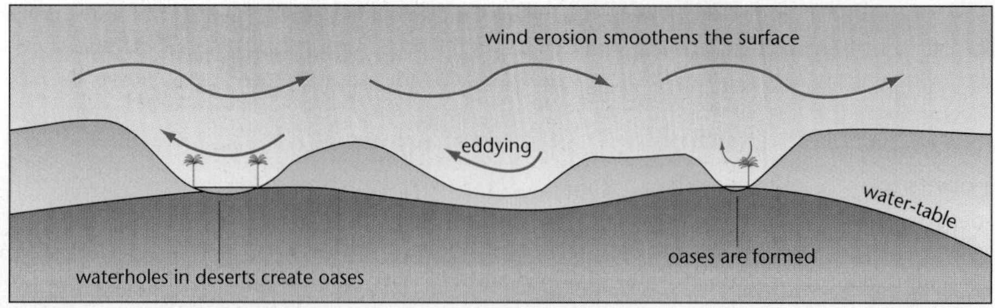
wind erosion smoothens the surface
eddying
water-table
oases are formed
waterholes in deserts create oases

The formation of a depression may first be caused by faulting. Once this has occurred the softer rocks exposed by this movement are attacked by abrasion, where the sand in suspension and saltation is used by the wind as an eroding agent. The hollow is enlarged and deepened by eddy action of the wind. Once the process is initiated, it becomes self-perpetuating. Deflation continues until the water-table is reached, thereby creating an oasis.

## Wind (Aeolian) Deposition

Sand is constantly on the move in a desert region. The lightest particles can even be carried thousands of miles from their desert source. Other particles are moved along the surface and create a hissing movement, especially as they tumble down the face of a sand dune.

Sand movement in areas outside of desert regions is best displayed on a beach at the seaside where there is a constant breeze. Towels spread out on the dry sand are quickly invaded by sand particles moving either along or up the beach. The type and amount of material carried by the wind depends on the size of the particles and on the strength of the wind.

Factors which determine landforms created by wind deposition include:

1. the nature of the surface over which the material is moved, for example
   — does it consist of deep sand?
   — is it bare rock?
   — is it a stony area?
2. the presence of an obstacle
   In some instances the presence of an obstacle, such as a bush or a rock, may create a mound where the wind is forced to slow down. Some of its load is dropped and the mound continues to grow.
3. the direction and strength of the dominant wind
   If there is a prevailing wind in the region it may create a particular type of landform, such as a seif dune. Seif dunes align themselves parallel to the wind (see p. 296).

Fig. 14.16
(a) Sand dunes in Namibia are created by wind.
(b) Soil erosion in KwaZulu Natal is created by water.

4. The presence of vegetation may influence the effect of the wind. Trees reduce wind speed. Drought-resisting plants, such as eucalyptus, creosote bushes and palm trees, help bind desert sand particles together and protect oases.

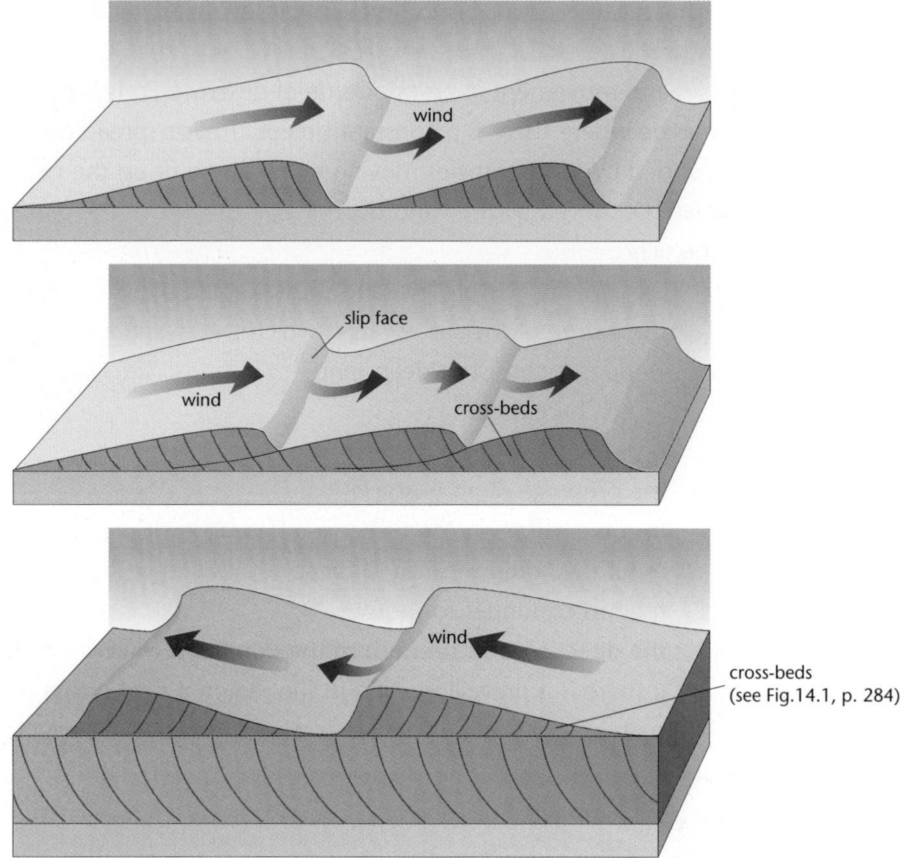


**Fig. 14.17**
Changing wind patterns are reflected in the bedding planes which form some sandstone rock

## Landforms of Wind Deposition

### Landform: Barchan Sand Dunes

Examples: Libya, Egypt, Namibia, north-west Australia, Colorado in the USA

Processes: Deposition, Saltation, Suspension

#### Formation

Dunes are sand heaps where winds that carry abundant sand lose speed because of an obstacle, depression or friction with the land surface. In time, the sand itself acts as a barrier to the wind, trapping newly arrived sand. Even though sand dunes are found in a variety of shapes and sizes, most of them are variations of a single plan — they have a gentle windward slope up which the sand moves, and a steep downhill or lee slope down which the sand falls from the crest.

There are many types of sand dunes, e.g. transverse, seif, barchan and star.

Barchan dunes, otherwise known as crescentic dunes, derived their name from the deserts of Turkestan. They have a crescentic shape and lie across the direction of the wind, with their 'horns' trailed out in the direction towards which it is blowing.

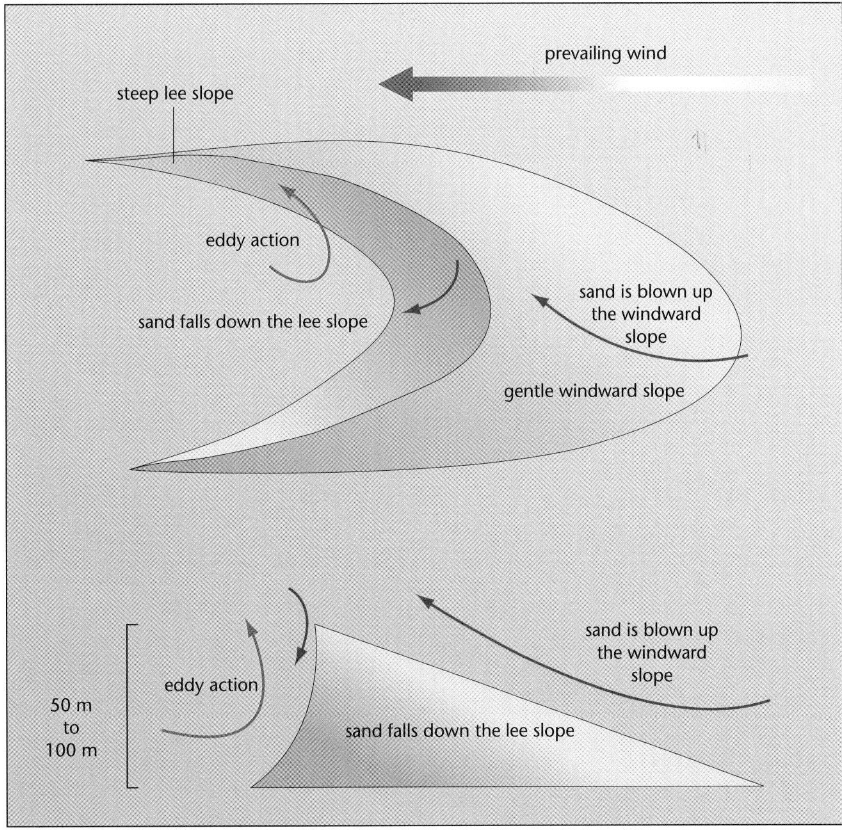


Fig. 14.18
Barchan dune

Patches of pebbles or sudden wind fluctuation may cause the accumulation of a low heap of sand, around which the barchan dune forms. They occur on fairly level open surfaces. The direction of movement of barchan dunes is dependent upon the wind. In Turkestan, for example, the wind changes seasonally, blowing alternately southward and northward, and the dunes change likewise, their horns swinging right around. Barchans are sometimes found as isolated hills. Generally, however, they occur in swarms, sometimes as a regular series, more often as a chaotic, ever-changing pattern of partially joined ridges at right angles to the prevailing wind.

As sand accumulates to form a barchan, movement is slowest at the centre of the dune, since more energy is needed to move sand up the gentle windward slope to the crest. At the edges, where the dune is lower, the movement is faster. Thus 'horns' develop, and a crescent shape forms. Barchans vary in height from 1 to 30 metres (1 to 100 feet).

Sand dunes are found in coastal areas in western Europe. They are especially large and extensive in the Landes area in south-west France, the best-known example being the dunes of Pilaf. In most coastal regions, marram grass binds and stabilises the dunes.

## Landform: Seif Sand Dunes (Longitudinal Dunes)

Examples: West Australian Desert
Thar Desert in India
Sahara Desert south of the Qattara Depression
Iranian Desert

Processes: Saltation, Suspension, Eddying, Deposition

### Formation

Seif or longitudinal dunes are long, symmetrical and narrow. Seif (pronounced 'safe') is derived from the Arabic word for a 'sword', which such dunes resemble. They may be up to 200 metres high and 300 kilometres long. The long axis of a seif dune lies parallel to the prevailing wind.

Seif dunes may be formed by the joining of a line of barchan dunes where cross-winds exist in addition to prevailing winds. The prevailing winds blow along the depressions (valley-like) between the seif dunes, keeping them free from sand. Whirlwinds or eddies build up the sides of the dunes once they become established. It is these eddies which drop the sand grains and thus build up the dunes.

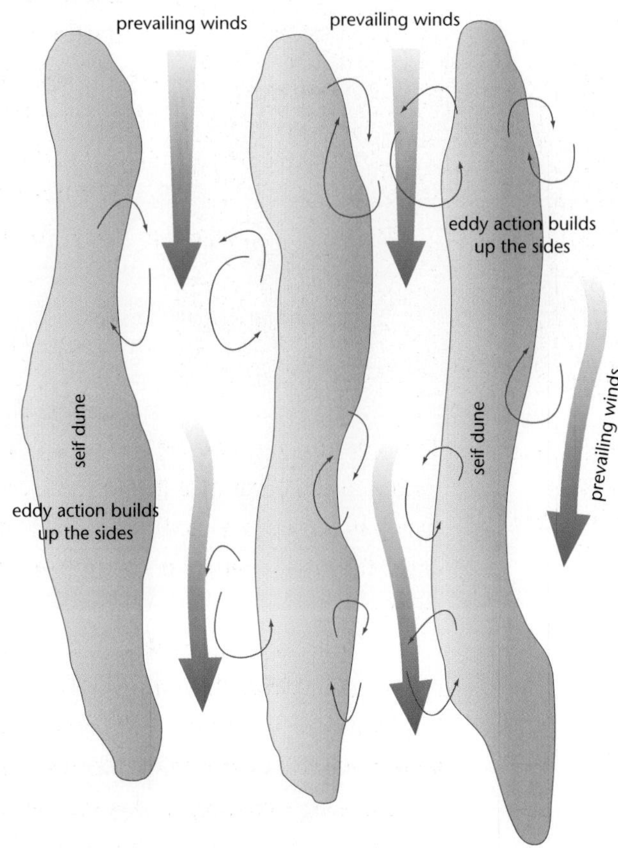


Fig. 14.19

## Landform: Loess Plains

Examples: North-west Paris Basin (called Limon in France)
Borde Region in northern Germany
central Belgium — North-west European Plain
Central Plains in the USA (called Adobe)
Loess Plateau of central and north-west China
the Pampas of Argentina

Processes: Deposition, Suspension, Transportation

Loess (from the German for 'loose') is a thick deposit of fine-grained dust consisting largely of **silt** but regularly accompanied by **fine sand** and **clay** particles. Loess is almost the perfect soil, in so far as it has a balance of fine sand, silt and clay. It is friable (crumbly) when tilled, with a porosity of 60 per cent. It is deep with no stones, ideal for root crops and cereals.

In China this material covers almost half a million square kilometres to a depth of 100 to 300 metres. It occurs at all elevations from near sea level to about 2,500 metres. Wind from the Gobi Desert carried the silt and clay that formed these deposits. Because loess is very cohesive (sufficiently sticky) it is easy to dig into and excavate and has the ability to stand as a vertical cliff without slumping. For centuries, the Chinese have dug cave-like homes in loess cliffs. Volcanic ash has a somewhat similar texture and was also excavated for homes. The last one in Tenerife in the Canary Islands was vacated in 1972.

Fig. 14.20
Some rivers which are fed by glaciers in far-off mountains create canyons by vertical erosion

Loess represents the accumulation, over a long time, of fine material carried by outblowing winds from dry, high-pressure areas. These areas may be deserts of varying kinds, such as the vegetation-free land along the fronts of melting ice sheets during interglacial or postglacial periods.

The Hwang-Ho (yellow river) in China erodes arid loess areas. Its heavy load of loess particles has given the river its name.

## Landforms of Water Erosion and Deposition in Deserts

Some desert rivers derive their water from their sources beyond the desert boundaries. The source of the River Nile lies in the Ethiopian Highlands. Meltwater from the ice-capped summits provides a continuous flow of water. The Nile Valley is defined by vertical edges, indicating that there is no lateral erosion, and that the valley is created by vertical erosion alone.

The Colorado, the well-known river of the south-western USA, has cut an immense canyon, the Grand Canyon, across the Colorado Plateau. While the river mainly concerns itself with vertical erosion, the immense depth is also due to the uplift of the surrounding land while the river retained its own level.

However, generally in deserts, most rivers, some permanent, some intermittent, do not flow out of the desert to the ocean. This is especially true of the Basin and Range region in the western United States. This region of 800,000 square kilometres has more than 200 mountain ranges which rise 900 to 1,500 metres above the basins that adjoin them. Over time, these regions undergo great physical change.

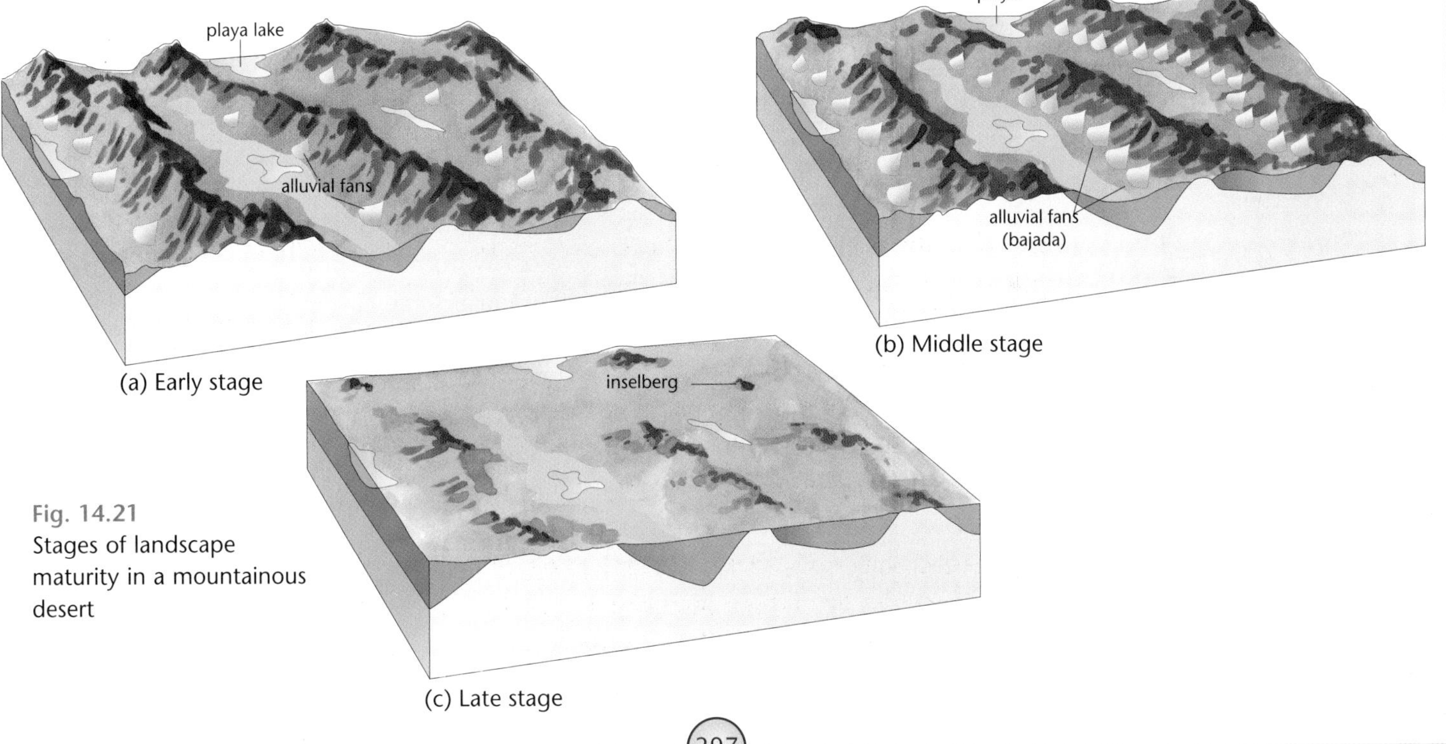


Fig. 14.21
Stages of landscape maturity in a mountainous desert

## Early Stage

The mountain ranges are attacked by the forces of denudation. Flash floods erode the mountains and create alluvial fans at ravine exits. Playas and playa lakes (salt lakes) form in the basins.

## Middle Stage

Alluvial fans join to form **bajadas** (see p. 299). Pediments (see p. 300) form and the mountain ranges are lowered. Playa sediments accumulate on the basin floors. The difference in altitudes between the mountain heights and playa elevations reduces.

## Late Stage

Basins fill with sediment. Pediments reduce the mountain ranges to isolated hills and inselbergs. Bajadas create gentle slopes between the pediments and the basin sediments.

### Landform: Mesas and Buttes

Example: Monument Valley National Park in Arizona

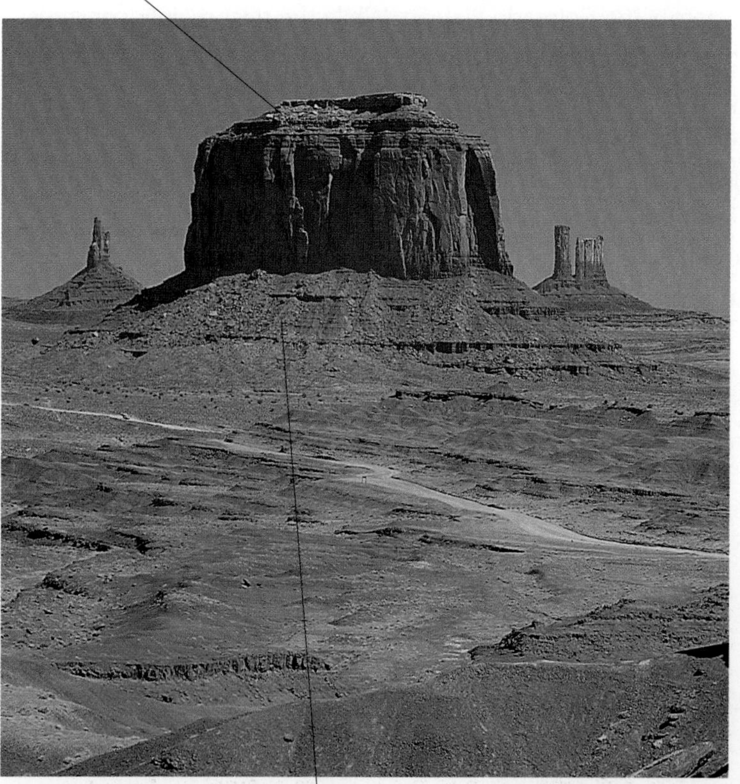


**Fig. 14.22**
Mesas and buttes in the Monument Valley National Park, Arizona, USA

Exfoliation, Freeze-thaw

In some areas, plateaus have been dissected by exotic rivers as they derive their waters from distant sources outside desert areas. The vertical erosion of some of these rivers has kept pace with slow uplift of land over a long period of time, forming canyons. Their sides, the plateau edges, are steep and regularly vertical and cliff-like. If rock layers are horizontal and vary in hardness the plateau edges may have a stepped appearance. Some sections of the plateaus, or remnants, form mesas (large sections) and buttes (small sections).

Mesas and buttes are flat-topped and rise vertically from the surrounding pediment with an angular scree cover from desert weathering. They are generally capped with a layer of resistant rock. Some mesas in Arizona have summits large enough to have been used as defensive village sites by the Hopi Indians. Buttes are simply smaller versions of similar landforms. The most spectacular mesas and buttes are in Monument Valley National Park in Arizona.

Fig. 14.23
Block diagram of an arid landform pattern

dissected plateau
canyons
interior drainage
mesa
playa lake (salt lake)
butte
alluvial fan

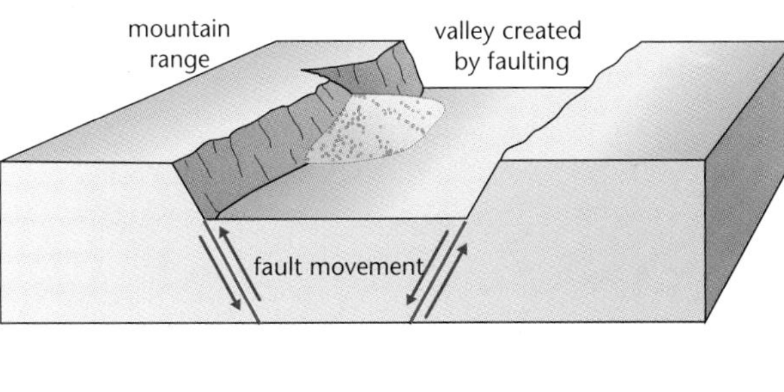


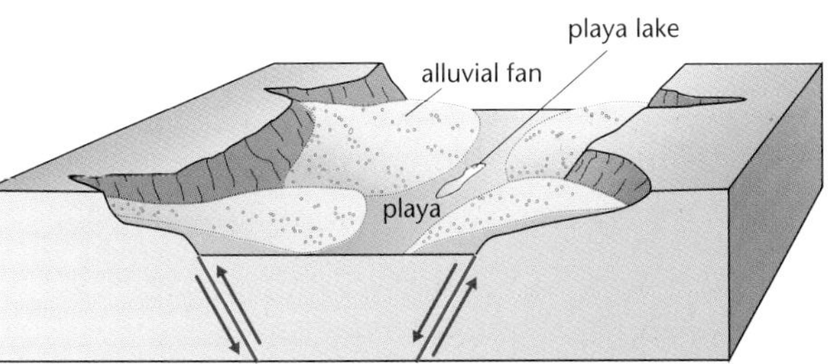


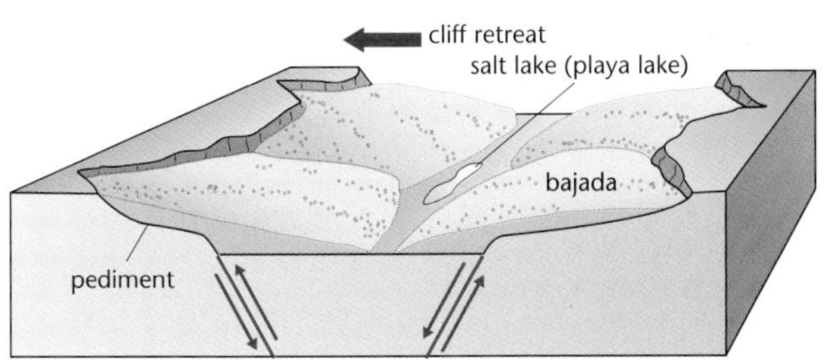


Fig. 14.24

## Landform: Alluvial Fans and Bajadas

Example: Death Valley in California

Processes: Transportation, Deposition, Run-off, Suspension

Alluvial fans are fan- or cone-shaped bodies of alluvium, rock debris, pebbles, sand and silt deposited by flash floods where stream channels widen as they flow out of steep mountain valleys, canyons or wadis. Over the years, **fans enlarge**, eventually **joining** with fans from adjacent canyons to form a **bajada**. These landforms are especially common in arid and semi-arid lands. A bajada is much more extensive than an individual alluvial fan and may have a gently rolling surface caused by the merging of adjoining fans.

Arid regions generally lack permanent rivers and they are characterised as having interior drainage. Rather than flowing to the sea, these rivers typically flow to low regions in the deserts where their waters seep and disappear into the deposits of earlier floods or evaporate from shallow lakes called playa lakes. Playa lakes are temporary, lasting only for a few days after a rainstorm. After the lake water evaporates, a thin layer of fine mud remains on the dry lake floor. This mud dries and cracks in the sun. If the run-off contained a large amount of dissolved salts, the playa may have a coating of dry white salt instead of mud. Lake Eyre in central Australia and the Great Salt Lake, Utah in the USA are examples of salt lakes.

In the United States, southern Oregon, western Utah, all of Nevada, southern California, Arizona and New Mexico all display patterns of interior drainage. This pattern is radial. However, unlike radial drainage in wet upland regions where the rivers flow outward, these rivers flow into the centre, where their waters evaporate.

Fig. 14.25
Salt lake in the Atacama region of Chile

## Landform: Pediment

Example: Mount Graham Pediment, south-eastern Arizona

Processes: Weathering, Erosion

A pediment is a gently sloping surface cut into the solid rock of a mountain. It may be a bare rock surface or it may be covered with a thin veneer of gravel. A pediment develops uphill from a bajada as the mountain front retreats. It is believed that as the steep mountain front erodes, it retreats uphill, maintaining a relatively constant angle of slope. A pediment is an erosional surface. An abrupt change in slope marks its upper limit.

It is difficult, sometimes, to distinguish a pediment from the surface of the bajada since both have similar slopes and gravel cover. But their origin is different. The pediment is a rock surface created by erosion of the steep mountain front. The bajada, on the other hand, is a surface created by wadi deposition and may be underlain by hundreds of metres of sediment.

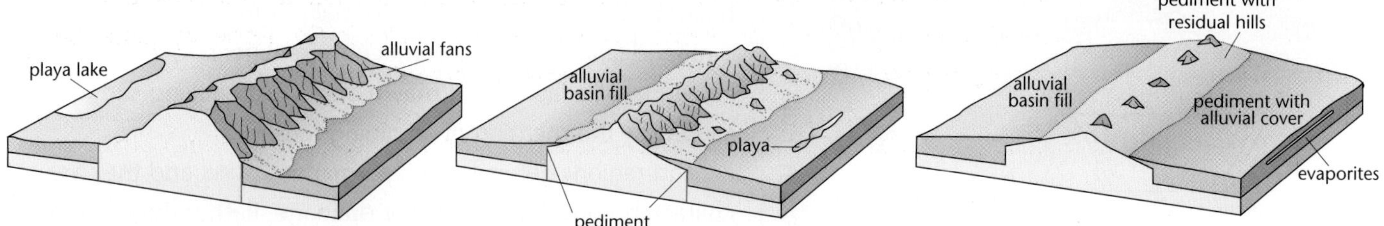


Fig. 14.26
Pediment development in a mountainous desert

## Landform: Wadis or Arroyos

Examples: Arabian Desert
Australian Desert
Death Valley and Monument Valley in the United States

Processes: Traction, Suspension, Deposition, Braiding, Erosion

As explained in chapter 11, the main processes of water erosion are hydraulic action and abrasion. In a desert it may not rain for many months or even years, but when it does rain it falls in torrential downpours. These short-lived rainstorms create equally short-lived flash floods.

With no vegetation to slow run-off, these torrents carry immense loads of silt, sand, pebbles and stones. As they flow across desert areas, they continue to add to their load. In some instances, they carry so much solid matter that they are more like mudflows than rivers.

Fig. 14.27
A dry wadi near Yiti in Oman

As they slow down, they drop their load quickly. Some **individual layers** may be as much as **three to four metres deep or more**.

**Wadis or arroyos are deep, steep-sided ravines which are cut by flash flood waters in desert areas.** The frequency of flash floods is generally once each year in semi-desert areas and once in ten years in extreme desert areas.

When it does rain in desert and semi-desert areas, the hard desert rock surface is initially unable to absorb this instant water supply and so it flows evenly across the land in a 'sheet flood'. Soon after, it finds its way to small channels which join to form narrow, vertically sided gorges, called wadis. They quickly fill with rapidly flowing water, containing stones, rocks and mud in its load. In times of drought the wadis appear as interesting, pebble-strewn dry river-beds. However, in reality, these harmless dry valleys can be treacherous. In an instant, even on a dry day, they can change to a raging torrent, whose water may derive from a distant thunderstorm.

The enormous energy of a flash flood enables large boulders to be moved by traction and enormous amounts of material to be carried in suspension. Once the waters abate, the river's load reduces and material chokes the river's channel to cause braiding. River channels such as these provide the raw materials of conglomerates. In Ireland, some of these conglomerates may be seen on the Old Red Sandstone mountains of southern Ireland. They are easily recognisable on the Galtee Mountains in Co. Tipperary, either on the high ridges or as erratics and moraine material near Lake Muskry.

## Leaving Certificate Questions on Desert, Wind Processes and Landforms

### Ordinary Level

**The Physical World**

**1997** — Explain the following:

(ii) Misuse of the soil can lead to soil erosion. **(40 marks)** [See chapter 19.]

**1992** — Explain the following:

(iv) By variations in the processes of transportation and deposition, wind produces several different types of sand dunes. **(40 marks)**

### Higher Level

**1993**

(a) (i) Describe how wind shapes and modifies landscape features in hot desert regions. **(40 marks)**

(ii) 'Desertification is a major problem of our time.'

With reference to African deserts explain the main causes and consequences of desertification. **(60 marks)** [See chapter 19.]

# Settlements and Settlement Patterns

Fig. 15.1
Enniskillen, Co. Fermanagh

**Tip**
If one is asked to (a) **study the buildings** in a town and (b) **identify** its urban functions or (c) write an **account** of the **historic and economic development** of the town, then the **characteristics** of the **chosen buildings** should first be **listed** in order to classify them, e.g. circular, tall, tapering stone building with a conical roof represents a round tower.

## Buildings, Structures and Functions of Settlements

The following is a list of urban functions which can be identified by recognising certain town buildings and structures. The number of buildings which may be identified in any particular photograph will depend on the height of the camera above the ground and the angle at which the photograph was taken.

## Functions, Buildings and Structures

| | |
|---|---|
| **Religious** | Round tower, church, church in ruins, abbey, abbey in ruins, convent, cathedral, graveyard |
| **Defence** | Motte, castle, tower, town gate, town walls, military barracks |
| **Market** | Town square, Y-shaped junction in streets, market house, town located in fertile plain, focus of routes, fair green |
| **Port** | Quay, dock, warehouses and cranes along the quay, ships or boats, boat slip, drawbridge, mooring posts, canal, canal lock |
| **Manufacturing** | Mill, mill-race, weir, factory, plant works, industrial estate |
| **Commercial** | Bank, hotel, post office, coal yards, timber yard, Georgian buildings in Central Business District (CBD), railway station |
| **Financial** | Bank, post office, building society |
| **Legal** | Courthouse (often with Grecian columns), Garda station, jail |
| **Administrative** | Multi-storey buildings for offices, Georgian buildings in Central Business District (CBD) |
| **Medical** | Hospital — generally a large flat-topped building in its own grounds on the outskirts of a town initially, but now maybe surrounded by housing estates |
| **Educational** | Schools — recently constructed schools are large buildings in their own grounds. Can be distinguished from other buildings by associated structures such as tennis courts and sports fields. |
| **Holiday resort** | Beach, caravan park, hotel, golf course, youth hostel, marina with boats moored, pier |
| **Recreational** | Hall, cinema, ballroom, park, ball alley, sports ground or stadium |
| **Retail** | Large ground-floor windows, different shades of colour between ground and upper floors, parking outside buildings (especially in villages) |

**Tip**

In settlement chapters 15–18 incl., just develop a **general understanding** of settlement development and change over time. It is **not essential** to remember every single factor about each individual settlement.

**Tip**
Facts to learn before continuing: see pp. 303 and 304.

## Settlement Characteristics

| Period | Street Names, Pattern/Word | Defence/Enclosing Settlement | Diagnostic Major Buildings |
|---|---|---|---|
| Pre-Christian farmers 4000 BC to 600 BC | Dolmen, wedge megalithic tomb, fulacht fia, barrow, cairn, wedge, stone circle, passage grave. | Circle of stones. Circular mounds. | Copper mines. Stone circle. Cairn of stones. Standing stones (gallauns). Circular hollow. |
| Celtic 6th century BC to 6th century AD | Promontory fort. Rath, Cashel, Lis, Dún, Doon. Ogham stones, togher. | Circular earthen bank. Circular stone wall. | Ring-fort. Stone fort. Cluster of trees in a field. |
| Early Christian (E) 6th–11th centuries | Street called after local saint. Circular. Radial. Kill, Well, Ch. | Circular earthen/stone bank. | High cross. Round tower. Early church. Native Irish monastery. |
| Viking (V) 9th and 10th centuries | High Street. Olaf Street. Coppergate Alley. Irregular. Gable end of houses onto street. Winetavern, fishmongers. | Earthen/stone bank/wall. None visible above the ground today. | None visible above the ground today. |
| Norman (N) 12th and 13th centuries | Watergate Street. Castle Street. Linear, chequered. Long, narrow burgage plots onto streets, sometimes still the same size today. Market place or market cross. | Strong, fortified stone walls. Bridging point of river. | Castle or tower house. Fortified town house. Church. Religious house. Hospital and/or leper house close to town. Town gate. |
| Plantation (P) 16th, 17th, 18th centuries | The Diamond. Sometimes similar to Norman. 2 streets joined by a square or diamond. Burgage plots — as Norman. Irishtown. Englishtown. Towns with landlord's name — Mitchelstown, Charleville. | Strong, fortified stone walls. Gun emplacements. Ornamental gardens, woods near castle. Fair green. | Fortified town house. Church. Market house. Religious house. Town square. Demesne. |
| Georgian (G) 18th and 19th centuries | Georges Street. Wellington Place. Geometric plan. Wide streets. Buildings often laid out around a crescent or green. | None. | Typical Georgian houses. Custom house. Industrial and commercial buildings, warehouses, monuments. Square or rectangular town square. |
| Later (L) | Canal Street. Factory Terrace. Varied plan. | None. | Varied. Depending on town's function, e.g. railway station, canal hotel, Victorian buildings, factories, etc. |

# Pre-Christian Settlement Patterns

## Stone Age Settlements

Stone Age people in Ireland may be divided into two groups:

- the first **settlers**
- the first **farmers**

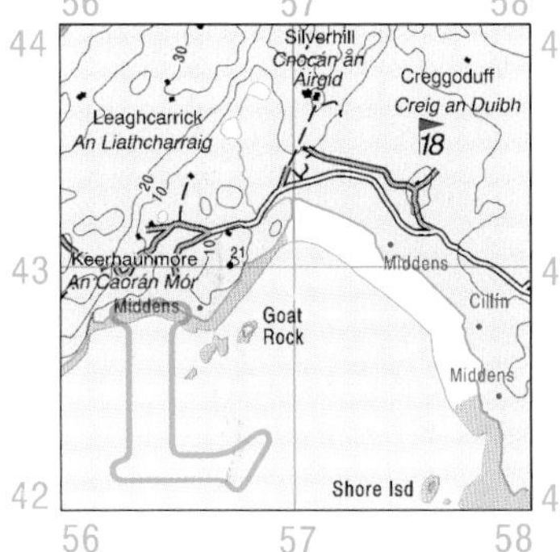


Fig. 15.2

Middens are heaps of shells in sand on the seashore which represent gathering and eating of seafood by early settlers

### The First Settlers

Ireland's first settlers were **hunters, fisherfolk and food gatherers**.

They were mesolithic people, which means they belong to the Middle Stone Age. ('Meso' means middle and 'lithos' means stone.) No evidence of these people appears above the ground.

In Pre-Christian Ireland, elevated sites, particularly hilltops, were chosen for defence. The earliest known defensive settlement in Ireland is at Lough Gur in Co. Limerick. It may have been inhabited as early as 4000 BC.

### Early Settlements and the Need for Water

A constant, **clean water supply** was needed for drinking, washing and cooking, so waterside locations such as coastal estuaries, riverside sites and lakeshores were chosen for Stone Age settlements.

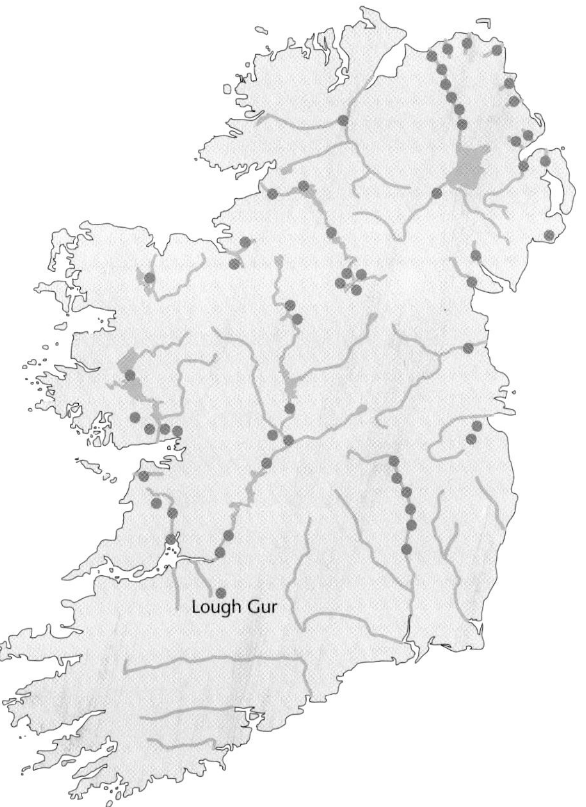


Fig. 15.3
The location of settlement sites of Ireland's early settlers. Riverside and lakeshore settlement sites were popular.

Fig. 15.4
Study the Ordnance Survey map extract. Then find what needs did Lough Gur and the surrounding area fulfil for its early settlers.

## Early Settlements and the Need for Defence

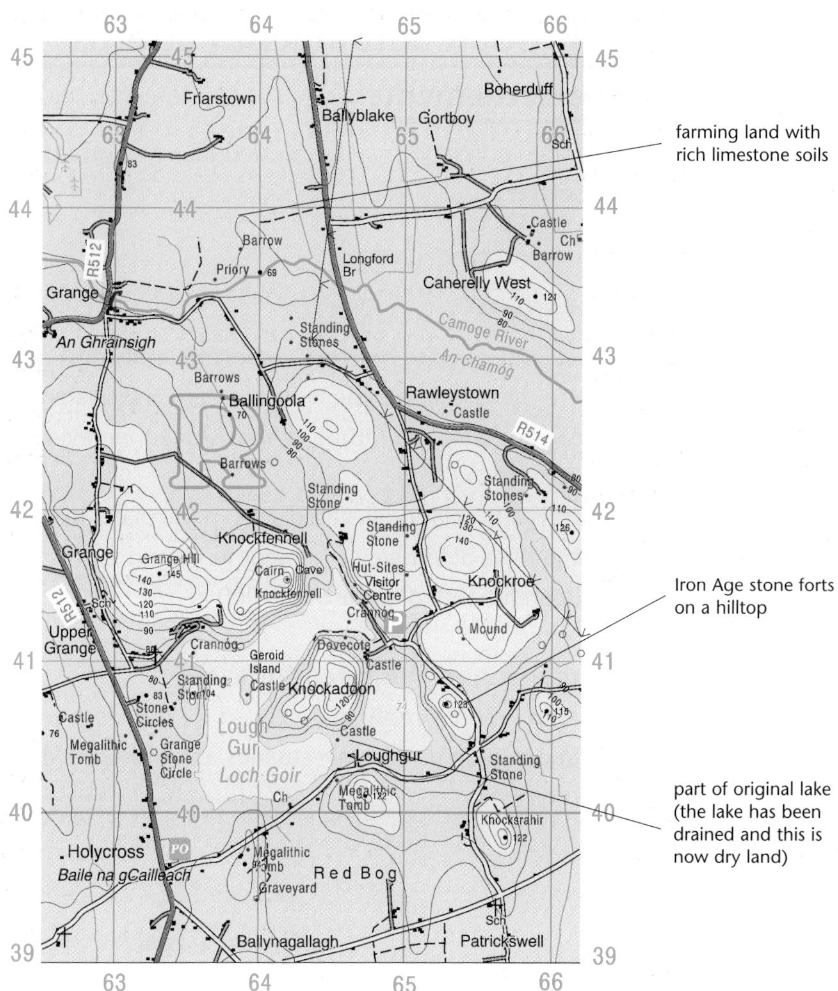


Fig. 15.5
The combined distribution of court tombs, wedge tombs, portal tombs and passage tombs in Ireland

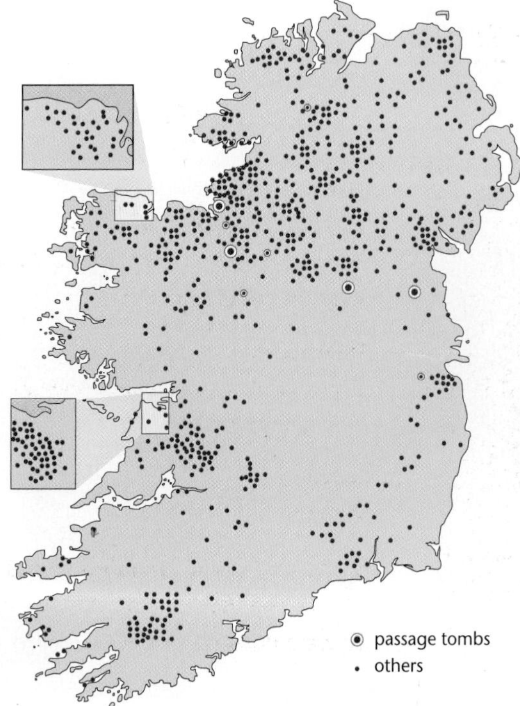


### The First Farmers

The first farmers or neolithic (New Stone Age) people came to Ireland about 7,000 years ago (5000 BC). By 3000 BC, farming was practised throughout most of the country. These farmers came to Ireland either directly from France or Spain, or more likely from Britain.

These neolithic people laid the foundations of the agricultural economy that persists in Ireland to the present day.

At the time of their arrival, most of Ireland was covered with dense forest. River estuaries were sought out and these channels provided the earliest and often the only means of exploring the country.

The first farmers left very little evidence of their dwellings on the landscape. What they did leave, however, were large graves and places of religious worship made of large stones. These features are called **megaliths** ('mega' = large, 'lithos' = stone). While many of these were destroyed in the past, reliable estimates place the

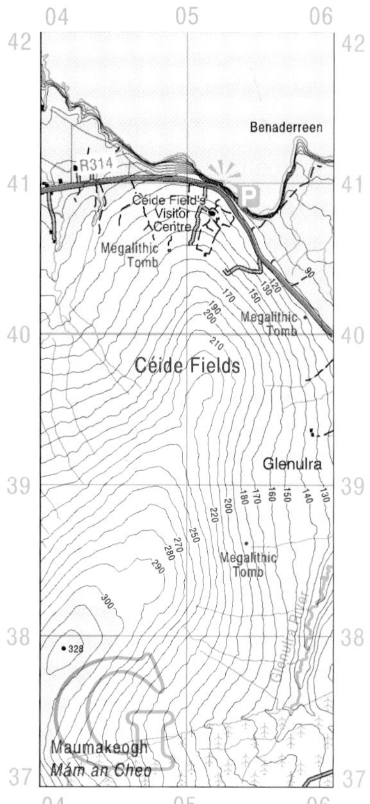


Fig. 15.6
Evidence of pre-Christian farming in an upland area in north-west Mayo

surviving number of megaliths at about 1,200. There are many variant forms, but it is possible to classify them into four main types — **court**, **portal**, **passage** and **wedge tombs**.

After glaciation, the lowland soils of Roscommon and western Sligo had a very high lime content and were not heavily wooded. Rich pasture provided excellent conditions for rearing cattle. The lime content has since been leached from some soils by the heavy rains of the west of Ireland and many of these same soils are now acidic. Countless megaliths are found in this region.

Numerous burial tombs are found in **upland areas** or **'hilly lowland'**. This suggests that these sites, with their **lighter and easily worked soils**, were preferred to the heavy clays of valley flood plains. Early farmers grazed cattle on the Burren, especially in winter, since this area of Co. Clare had a better topsoil cover then than it has today. Deforestation by early farmers left the light soil cover exposed to rain and strong sea winds and much has since been eroded.

Some of the first farmers sought out areas with plenty of elm and oak, which indicate rich and fertile soils. These areas were cleared of trees and agriculture was practised. The earliest evidence of farming in Ireland comes from Cashelkealty on the Beara Peninsula in Co. Kerry and is dated to 5834 BC.

It is not certain where these farmers landed or in which part they first practised farming. It is suggested that they landed on the west coast and then moved across to the east, though others believe the opposite. What is known, however, is that their farming practices were quite advanced. In north-western Mayo, for instance, the **Céide Fields** (Fig. 15.6) excavations show that farmland was divided by these early neolithic people into fields with dry stone wall boundaries. When the climate became wetter, peat growth rates rose and the fields were abandoned. These ancient fields and walls are now covered by blanket bog.

## Some Copper and Bronze Age Settlements

Fig. 15.7
Some Copper and Bronze Age settlements in Co. Waterford

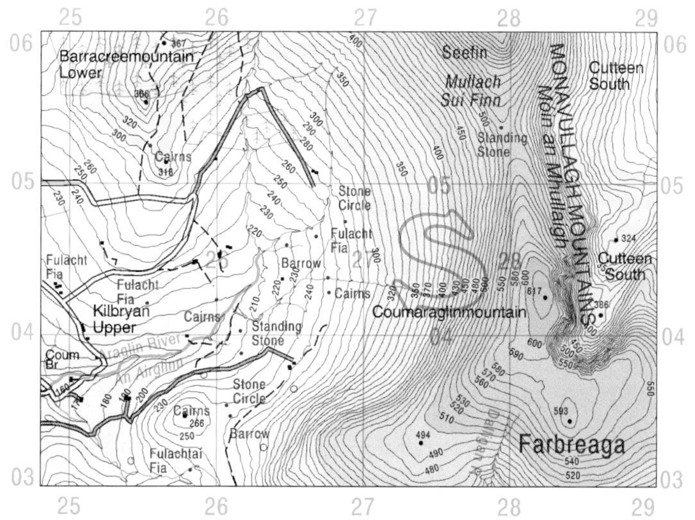


**cairns**
Some bronze age tombs were covered by mounds or stones. These mounds are called cairns.

**fulachta fia**
Cooking sites located near sources of clean water.

**barrows**
Burial sites.

**stone circles and standing stones**
These were used to date the seasons or were ritual sites or both.

Fig. 15.8
A stone circle, Drombeg, Co. Cork. Most stone circles date from the Bronze Age; however, some date from the Iron Age.

### Evidence of Copper and Bronze Age People in Ireland

Many other groups entered Ireland after the first farmers. Some of these new cultures include those that brought the knowledge of smelting, moulding and working of various metals. Copper was the first metal used for domestic and defence equipment. While domestic copper utensils were efficient, copper weapons were not, as they were too soft and could retain their edge for only a short time. Bronze was then adopted for knives, shields and daggers but it too lost to iron when the Celts invaded.

### People and Minerals in Prehistoric Ireland

Copper mining activity was carried on in some parts of Ireland from about 1400 BC onward. Copper deposits were mined in such places as Avoca in Co. Wicklow, Bunmahon in Co. Waterford and Ross Island in Lough Leane in Killarney. However, by far the most copper was mined during prehistoric times at Mount Gabriel in west Cork. Estimated copper output at this time for west Cork and Kerry is 370 tons. This large output suggests that copper was exported to Britain and the Continent during the Copper and Bronze Ages (see Fig. 15.9).

Copper was extracted from the mine by building a fire against the ore rock. Once the rock was heated, water was spilt on it, causing it to cool rapidly. It shattered and was then removed for smelting.

Later, during the Bronze Age, copper was alloyed with tin (90 per cent copper, 10 per cent tin) to make bronze.

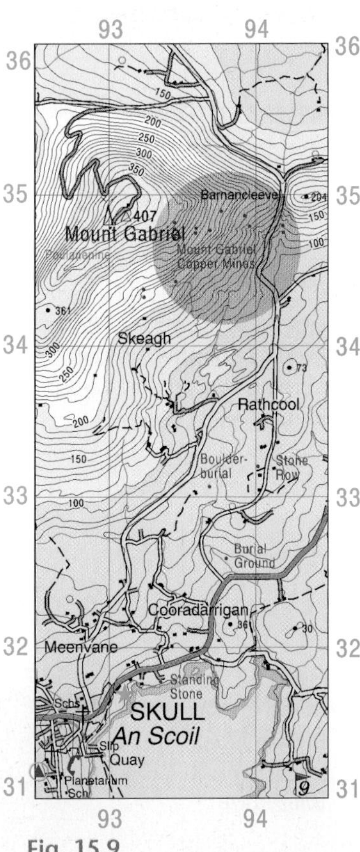


Fig. 15.9

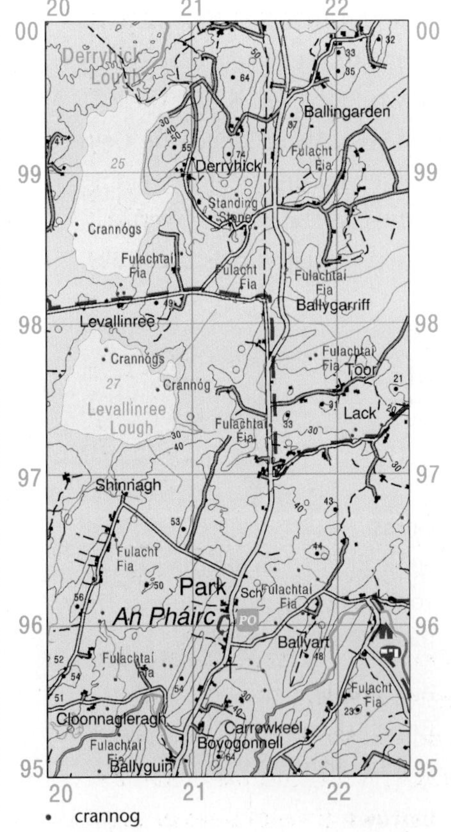


- crannog
- ring-fort

Fig. 15.10
Evidence of Celtic settlements on Ordnance Survey maps

## Some Iron Age Settlements

### The Celts

From about 600 BC to AD 250, Ireland was invaded by numerous farming tribes called **Celts**. They hunted and fished as well as keeping some cattle, which they enclosed within their ring-forts at night for safety. Cattle were the most common animals but pigs, sheep and horses were also kept. Celtic people grew crops such as rye, barley and oats, while corn was grown and ground into flour with the use of quern-stones.

Just learn a general account of these settlement eras and be able to recognise their associated names on an Ordnance Survey map.

These Celtic tribes often went to war against each other and so they chose easily defended sites on which to build their settlements. They encircled the settlements with a defensive wall or walls. The materials used to build these walls varied from place to place. Where stone was absent, high circular banks of clay were built.

Celtic defensive settlements were called: **Rath** (Rathluirc), **Lios** (Lisdoonvarna), **Dún** (Dundalk), **Caher** (Cahir), **Cashel** (Cashel). Folklore often referred to these places as fairy-forts, and it was this name that often ensured their survival.

**Rath** (Rathluirc),
**Lis** (Lismore),
**Cashel** (Cashelkealty),
**Dun** (Dundalk),
**Caher** (Caherciveen).

## Evidence on the Landscape

### Hill-fort

This is a circular or oval bank or fence which encircles a hilltop. It is somewhat similar in appearance to a ring-fort except that it is usually larger. Its defensive walls were built of stone or earth or both. In the case of Navan Fort in Co. Armagh, the earliest defensive wall was made of timber posts which were set vertically in the ground, with lighter timbers interwoven horizontally between them. This type of fencing is known as post and wattle.

### Identifying Celtic Settlement on Ordnance Survey Maps

The names of some defensive Celtic settlements can be recognised today in their place names.

Fig. 15.11
Evidence of Celtic settlements on photographs.
A royal ceremonial hill-fort at Cashel, Co. Cork. This site was used in the late Bronze Age, Iron Age and into the Middle Ages, presumably for periodic meetings and ritual purposes.

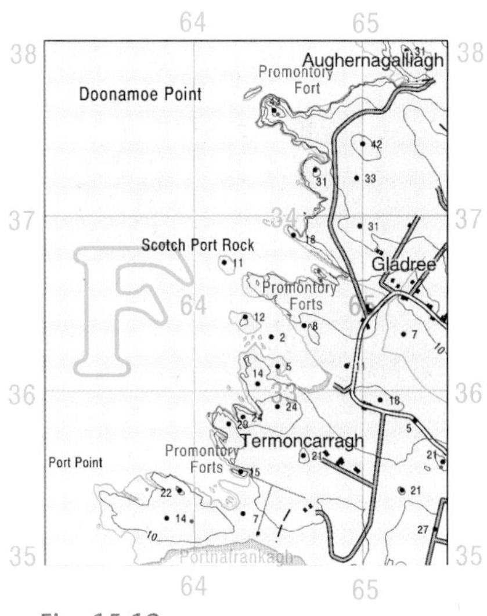


Fig. 15.12
- Promontory forts. These are forts on a cliff edge for defensive purposes.

circular walls of stone or earth provided protection from attackers

a cliff provided defence on this side

Fig. 15.13
Cahercommaun stone fort in the Burren of Co. Clare — a promontory fort on an inland site

### Ring-fort

Ring-forts represent enclosed farmsteads dating to early Christian times, AD 400 to 1300. They are referred to as 'fort' on Ordnance Survey maps and indicated by a red circle. A ring-fort may often be visible from the air only as a circle of trees or even a cluster of bushes in a field.

Caher, Dun or Cashel indicates a **'stone ring fort'**, e.g. Caherconnell in the Burren in Co. Clare, Dun Aengusa on the Aran Islands, Cashel in Co. Tipperary.

### Promontory Forts

These forts were built by Iron Age people on a coastal promontory (narrow neck of land) which had steep cliffs on three sides. They could be attacked on only one side, and so were defensive coastal settlements.

Some promontory forts were built on inland cliff edges such as Cahercommaun fort in the Burren in Co. Clare. (See Fig. 15.13 on p. 309.)

**ath**
*Áth* = ford, e.g. Athenry: *Áth na Ríogh* = Ford of the Kings.

**bun**
End; mouth of, bottom of, e.g. Bunratty: Mouth of the Rathy River.

**cashel/caher**
Circular stone fort.

**cappa/agh**
A plot of land for tillage.

**carrick**
A rock.

### Crannogs

Crannogs were forts built in a lake. First used in Ireland in the Bronze Age, they continued to be used well into Celtic times. They are mostly found in the lakes of the drumlin landscape north of the central plain. Foundations of stone, brushwood and clay formed the base of these island forts. The defensive walls were built of strong upright poles with light sticks interwoven horizontally between them and the dwellings were made of wattle and daub. The forts were joined to the mainland by pathways that could easily be dismantled if the settlement was attacked. Others had causeways, submerged pathways which were known only to the people in the fort.

### Togher (Toher)

A togher was a wooden routeway built across a marshy area during Celtic times. It regularly features as a place name.

Carefully examine this Ordnance Survey map extract, Fig. 15.14. Then do the following:

1. Coastal upland regions such as this have attracted settlement over thousands of years. With reference to the map explain clearly why this area was so attractive to various settlement groups and identify evidence that supports this trend.
2. Settlement in areas such as this may be temporary or permanent depending on the seasons of the year. What evidence on the map supports this view? Give three fully explained statements.
3. The deposition of sand and shells can change the shape of a coastline over time. With reference to two landforms explain this statement fully. Support your answer with evidence from the map.
4. The processes of coastal erosion and deposition in places such as this sometimes operate together to alter the shape of a coastal area. With reference to two landforms explain this fully.

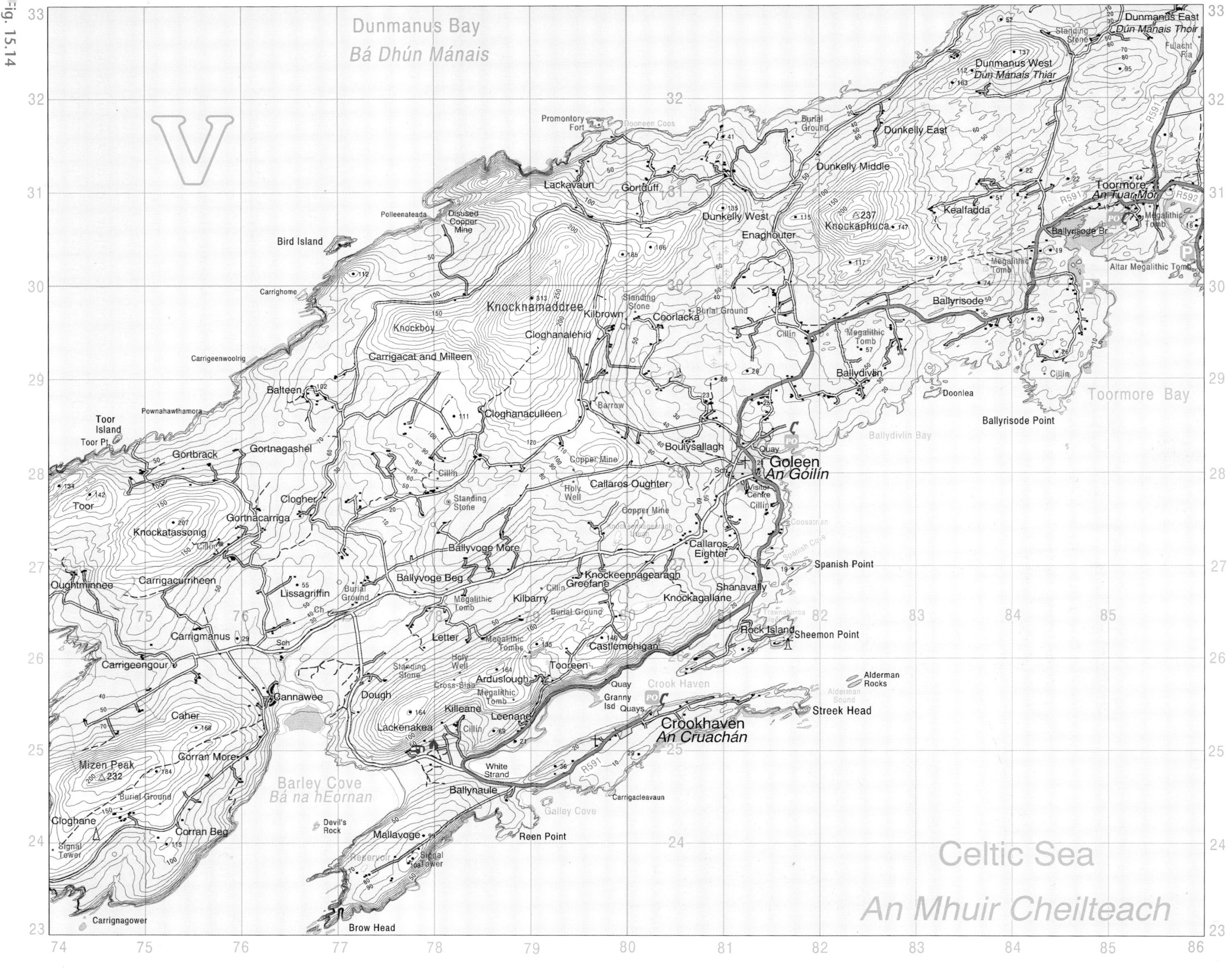


Fig. 15.14
Mizen Head region

# Early Christian and Norman Settlements

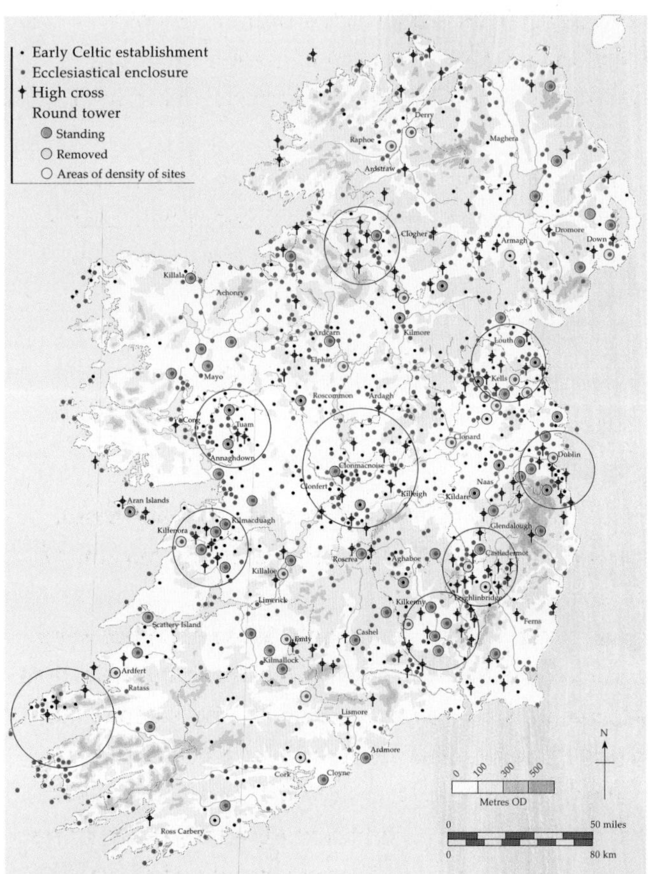


Fig. 16.1
The distribution and densities of early Celtic monasteries in Ireland

Every Irish town or village has its own set of characteristics, some of them unique to individual settlements or regions. In this book, we are generally interested only in those features or structures which are noticeable from the air or marked on maps. By examining such features as location, street plan or buildings, we should be able to reconstruct the **historical and geographical development** of a settlement.

## Early Christian — The Native Irish Monasteries

### Background

Sites chosen by the early Celtic missionaries were located in isolated, scenic areas where an appreciation of the tranquillity and natural beauty of the land formed an integral part of Irish monastic life. These early clerics were known as **hermits**. Later, as monastic congregations increased, people were unwilling to live in such isolated places. The need for fertile land eventually developed so that the increasing numbers could be supported. Settlements therefore tended to be located in places which were more accessible and fertile.

These early settlements attracted many young people of both sexes to study the Scriptures. Many were employed to illuminate manuscripts, producing such works as the Book of Kells. Others worked as the stonemasons who built churches and carved Celtic high crosses, as metalworkers making sacred vessels such as the Ardagh Chalice, as poets or labourers.

However, none of the great religious foundations from this period subsequently became a location for a major urban centre. Some did develop into smaller urban centres such as Kells in Co. Meath, Downpatrick in Co. Down, Armagh in Co. Armagh and Kilkenny in Co. Kilkenny.

The larger monasteries were built within circular enclosures. The buildings consisted of churches, cells, a round tower, a scriptorium and library. Other buildings, such as craft workshops, were built outside the walls. **Where monastic settlements grew into towns, they sometimes retained a circular and radial street plan pattern, representing the enclosing wall and the original routeways** leading into the settlement, which can still be seen at such towns as Kells, Armagh, and Lusk in Co. Dublin.

**cillín/kil**
Small early Christian Celtic church.

**clon/cluain/cloon**
A meadow.

**corn**
Round hill.

**dare**
Áth Dara, Adare — The Ford of the Oak Tree.

**disirt/dysert**
Hermitage.

**domhnach**
A church.

**down/don/dun/doon**
Fort.

**drom**
Ridge-like hill.

Place names can help to identify the origins of some settlements. When 'Kil' or 'Cill' is combined with a place name, it usually refers to 'ceall' or 'cill', a church; for example, Kilbride — Cill Bhríde — the church of Bríd. Similarly the word 'monaster' in 'Monasterevin' — Mainistir Eimhín — the monastery of Eimhín.

In the period between the collapse of the Roman Empire and the Renaissance, education standards fell throughout much of Europe. It was at this time that monastic communities (Fore Abbey had 3,000 people) in Ireland offered the opportunity of a quality education to those from Europe who could afford it. As a consequence of this demand, early Christian monasteries in Ireland became centres of learning and formed Ireland's first urban settlements.

Irish monastic settlements grew from simple isolated cillín to pre-urban clusters with considerable wealth and power. Their influence spread throughout Europe. Towns such as St Gall in Switzerland and Bobbio in northern Italy have their origins in Irish missionary settlements.

## The Early Monasteries

### Wooden Churches — Fifth to Eighth Centuries

The earliest churches and enclosing fences were made of wood (**wattle**). In many cases, evidence of their existence has disappeared from the landscape. However, the remains of some enclosing stone walls are still visible from the air.

The first record of a stone church dates from AD 789, at Armagh. Stone churches were similar in design to the wooden type. They are easily recognised from the air by their small size. Other notable features are rectangular, one-roomed buildings with very steep gables, indicating the former existence of a high, pitched roof similar to the roofs of earlier wooden churches. The earliest stone churches were roofed with thatch or wood, while later structures were roofed with stone. Ornamentation was absent from these early stone churches. The doorways of some had Romanesque design. These rounded arches are found only in the churches of the native monasteries. Some patterns of decoration in these arches indicate a European influence.

No associated settlement

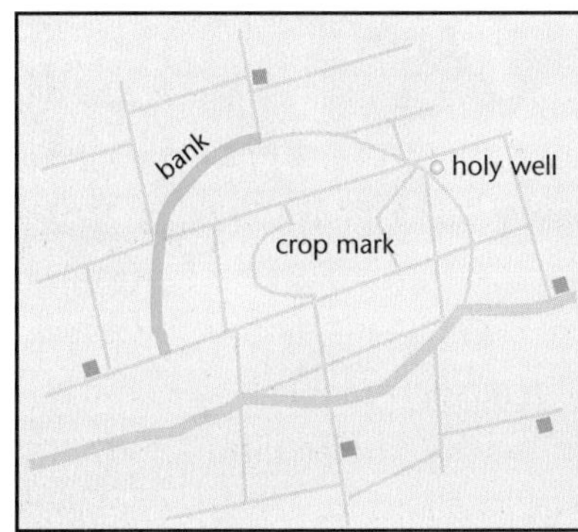


Enclosure around village or within town

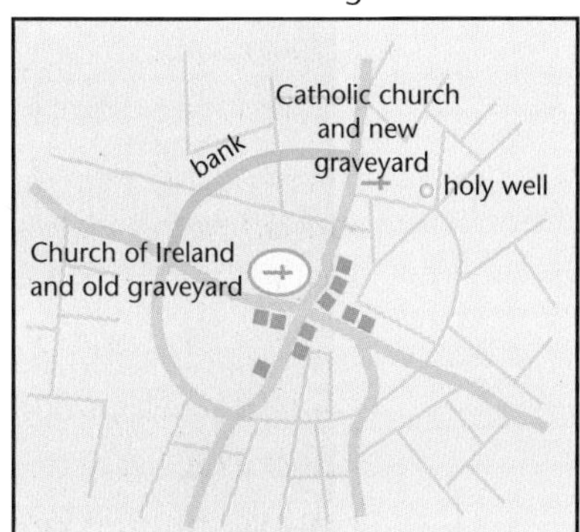


Fig. 16.2
Some evidence of monastic enclosures on the modern landscape. Can you suggest why the Protestant church is located in the town centre? The town began as an early Christian monastery.

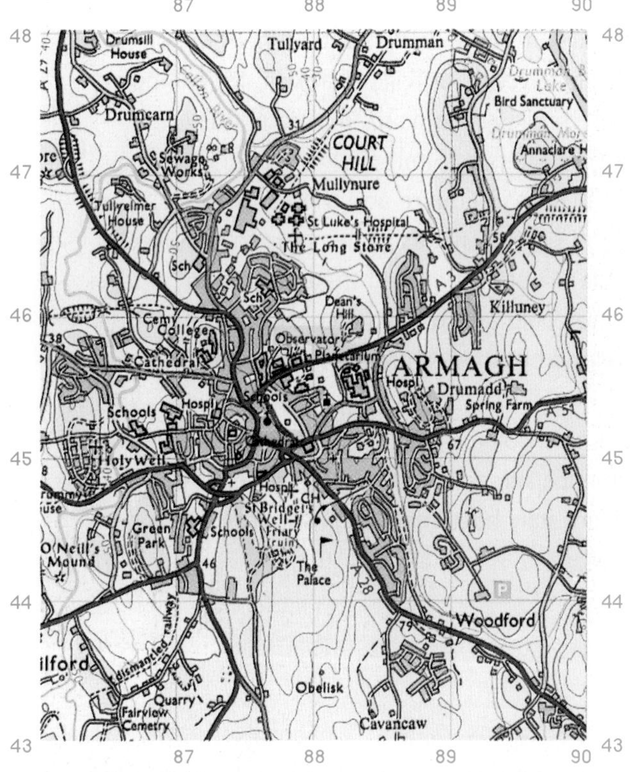


Fig. 16.3
Armagh

Carefully examine the Ordnance Survey map extract of Armagh town in Co. Armagh, Fig. 16.3.

1. What patterns do the Class A roads form with reference to Armagh town centre?
2. What pattern do the streets form in Armagh town centre? (See street pattern, pp. 312 and 383.)
3. As a consequence of these observations, do these patterns suggest anything about the possible origin of the town?
4. Could any other factor/s have influenced the town's street/road patterns besides that at 3 above? (Tip — the physical landscape — glacial deposition.)
5. Describe the location of Armagh town (i.e. site and situation).

Fig. 16.4
An early Christian stone church

Fig. 16.5

Carefully examine this photograph of Armagh town and learn all about early Christian settlement covered so far in this book. Then do the following:

1. Give an approximate date for the origin of this settlement. Explain your answer fully, using evidence from the photograph.
2. (a) What pattern, if any, did the initial routeways of this settlement form?
   (b) How have they influenced the present street pattern?
3. What dominant function of this settlement today has its origins in earlier times? Explain fully.
4. Draw a sketch map of the area. On it mark and label
   - the street/road network
   - a flyover
   - three churches
   - two hospitals
   - four schools
   - an observatory and a planetarium
   - two factories
   - two housing estates

Fig. 16.6

Carefully examine the photograph of Gouganebarra in west Cork, in Ireland, Fig. 16.6. Then do the following:

1. What evidence on the photograph suggests that this area had an early monastic settlement?
2. What landscape characteristics existed here that attracted early Christian monks to this location? Use the following headings to explain your answer: site, location, situation of site, soils.
3. Account for the distribution of vegetation shown in the photograph.

**Curragh**
Marsh, moorland.

**Cush**
Foot (at the foot of . . . ).

**Derry**
Oak wood or grove.

**Derryard**
High oak grove.

**Donagh**
Church.

**Donaghcloney**
Church of the meadow.

**Dooish**
Black hill.

**Dooks**
Dunes

**Doo Lough**
Black lake.

**Dromada**
Long ridge.

**Dromahair**
Ridge of the two demons.

**Dysert**
Isolated place, hermitage.

**Dysert O'Dea**
Hermitage of O'Dea.

## The Vikings

The coming of the Vikings in the late eighth and early ninth centuries introduced a period of raiding which involved monastic sites all around the Irish coast and along inland rivers. This was soon followed in the early ninth century by the building of fortified Viking strongholds at such places as Dundalk, Dublin, Wicklow, Wexford, Waterford and Limerick. Trading bases were also set up in Cork, Kinsale and Youghal. The Vikings introduced organised trade to Ireland. Even after their defeat by the Normans, their trading expertise made them valuable members of coastal urban society.

### Round Towers

**Round towers** were erected on well-established monastic sites. Some of these monasteries may have been built as early as the sixth, seventh or eighth centuries. Irish monasteries at this time held considerable influence and power over their individual 'hinterlands'. They held lay and monastic wealth and became patrons of artists and masons. Thus they were vulnerable to attack from enemies. The towers were built to protect the monks and their followers from raids by the Vikings and by neighbouring Irish tribes. **These towers or bell houses are tall, slender, tapering structures, sometimes more than thirty metres high and generally with a conical roof.** They had a number of storeys, with each floor made of wood. Light was provided by one window on each floor, except on the top storey, which may have had four or more openings to act as a belfry (Fig. 16.7). While the towers were intended to protect their occupants, the opposite was often the case, as the building design acted as an excellent chimney if the structures within were set alight.

The Norman settlers often set camp at already established monastic centres. Thus new rulers retained old centres of power.

Fig. 16.7
Clonmacnoise monastery on the River Shannon

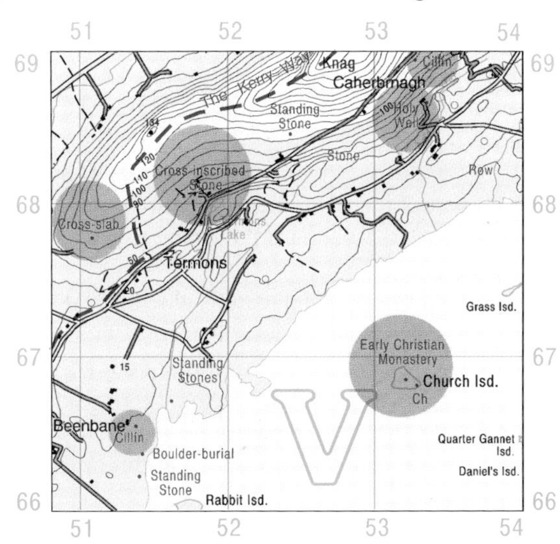


Fig. 16.8
Monastery, Cillín, Holy Well, Ch. and Cross- inscribed Stone indicate early Christian monastic settlements

**Fig. 16.9** Clonmacnoise in Co. Offaly

Carefully study the photograph, Fig. 16.9, and the Ordnance Survey map extract, Fig. 16.10, and then answer the following:

1. Identify evidence of two ancient settlements and state in which part of the photograph they are located.
2. Give a detailed account of the characteristics of each settlement.
3. (a) Give fully explained reasons why these settlements were developed at these sites. Use evidence from Figs. 16.9 and 16.10 to support your answer.
   Tip: (i) river/glacial deposition, (ii) riverside, (iii) flood plain, (iv) fuel.
   (b) Many settlements that began at this time in Ireland have survived and are thriving urban centres today. Give at least **ONE** fully developed point why the settlement sites on the photograph failed to survive.
4. There are brown patches in the background of the photograph.
   (a) What are they?
   (b) Why did these landforms develop in these places? Use evidence from the photograph to support your answer. (Tip — see chapter 2.)
5. At which stage of maturity is the river in the centre of the photograph? Give evidence to support your answer.
6. At which time of year was this photograph taken? Refer to three different aspects on the photograph to support your answer.
7. What type of landscape does the photograph represent? (Tip — lowland, highland, etc.) Explain.
8. In which direction was the camera pointing when the photograph was taken?

**Fig. 16.10** Clonmacnoise in Co. Offaly ▶

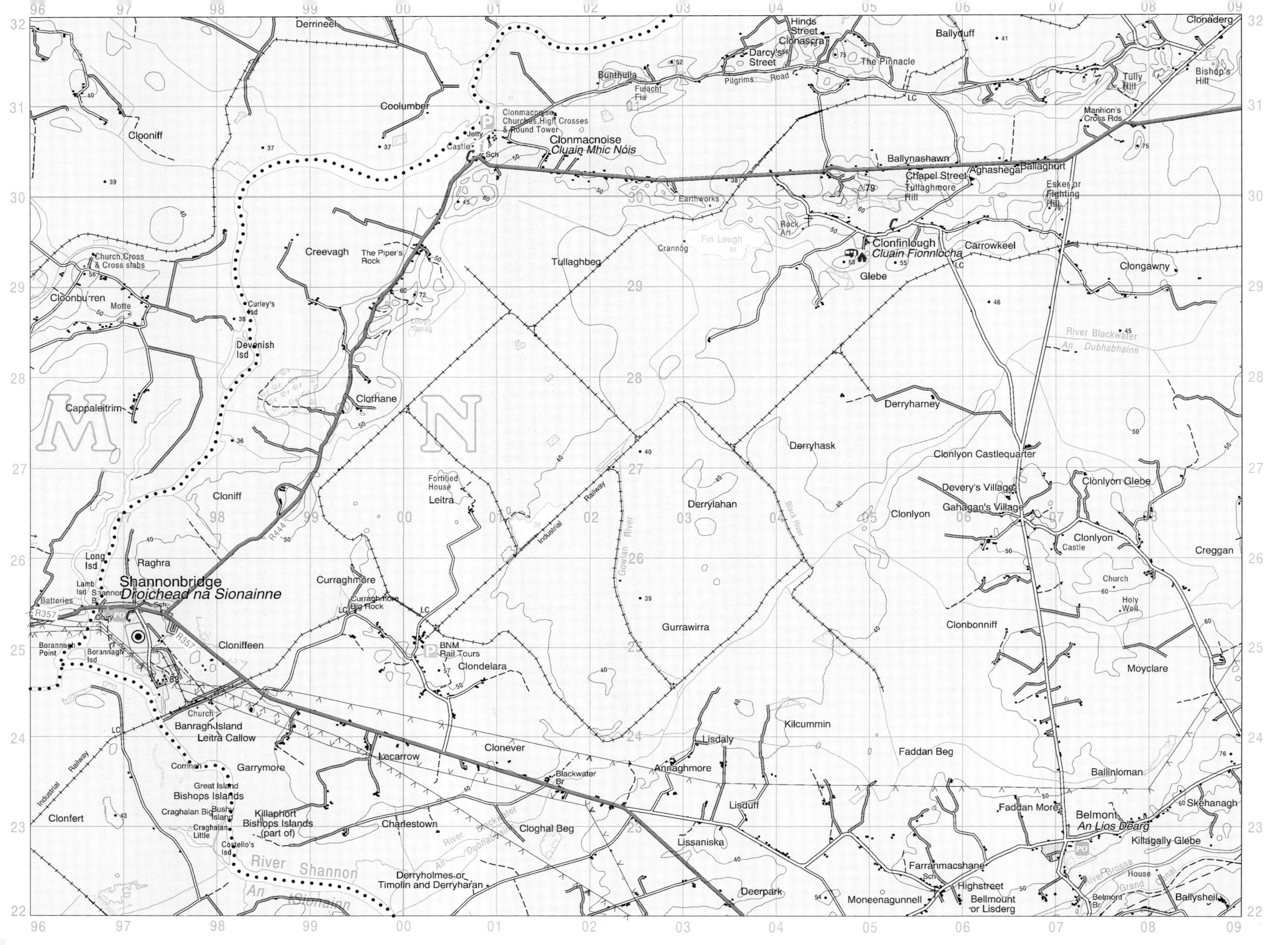

Derrineel
Hinds Street
Clonascra
Darcy's Street
The Pinnacle
Ballyduff
Clonaderg
Tully Hill
Bishop's Hill
Bunthulla
Fulacht Fia
Pilgrims Road
Coolumber
Clooniff
Clonmacnoise Churches, High Crosses & Round Tower
Jetty
Castle
Sch
Clonmacnoise
Cluain Mhic Nóis
Mannion's Cross Rds
Ballynashawn
Chapel Street
Aghashega
Ballaghurt
Tullaghmore Hill
Esker or Fighting Hill
Earthworks
Rock Art
Fin Lough
Crannog
Tullaghbeg
Church, Cross & Cross slabs
Creevagh
The Piper's Rock
Clonfinlough
Cluain Fionnlocha
Carrowkeel
Glebe
Clongawny
Cloonbu'ren
Motte
Curley's Isd
Devenish Isd
Lough Ranag
River Blackwater
An Dubhabhainn
Cappaleitrim
Clorhane
Derryharney
Derryhask
Clonlyon Castlequarter
Fortified House
Leitra
Cloniff
Industrial Railway
Derrylahan
Devery's Village
Gahagan's Village
Clonlyon
Clonlyon Glebe
Castle
Creggan
Black River
Gowlan River
R444
Long Isd
Raghra
Shannonbridge
Droichead na Sionainne
Lamb Isd
Batteries
Quay
PO
R357
Curraghmore
Curraghmore Big Rock
LC
Church
Holy Well
Clonbonniff
BNM Rail Tours
Gurrawirra
Boranagh Point
Borannagh Isd
Cloniffeen
Clondelara
Moyclare
Church
Banragh Island
Leitra Callow
Kilcummin
Lisdaly
Clonever
Lecarrow
Faddan Beg
Corrinch
Garrymore
Annaghmore
Ballinloman
Great Island
Bishops Islands
Blackwater Br
Lisduff
Faddan More
Belmont
An Lios Dearg
Skehanagh
Clonfert
Craghalan Big
Bushy Island
Killaphort
Bishops Islands (part of)
Craghalan Little
Charlestown
Cloghal Beg
Costello's Isd
Lissaniska
Killagally Glebe
River Shannon
An tSionainn
Derryholmes or Timolin and Derryharan
Farranmacshane
Highstreet
Bellmount or Lisderg
Moneenagunnell
Deerpark
River Brosna
House
Grand Canal
Belmont Br
Ballysheil

Study the Ordnance Survey map extract of North Co. Dublin, Fig. 16.11. Then do the following:

1. Draw a sketch map of the area shown on the map and on it mark and label
   - the coastline
   - the main towns
   - the main transport routes
   - one specific area which you think should be zoned for residential settlement only
   - one specific area which you think should be zoned for industry only
   - two areas which you think should be zoned as green belts (see 'Green Belts', p. 392)
   - three areas where coastal erosion is evident
   - three areas where coastal deposition is evident
2. Carefully study the text on p. 312, Fig. 16.2 on p. 313 and the photograph on p. 314 and the Ordnance Survey map extract of North Co. Dublin, Fig. 16.11. Then identify the pattern which the streets/roads form at Lusk.
   Does this pattern have an association with the town's origin? (See Fig. 17.88, p.383.)
   Explain your answer and support it with evidence from the map under the following headings: transport routes, structures.
3. Suppose that you are the industrial development **planning engineer** for this north Dublin region. Two industries wish to construct their factories near grid reference O 238 526.
   One of these factories is a heavy industry, the other a light manufacturing industry. Each company promises to employ 1,000 people. The products each company proposes to manufacture have a guaranteed long-lifetime market.
   (a) Choose a suitable site for each industry. Give a six-grid reference for each location. At this time suppose there is widespread unemployment in this region and a general election is pending.
   (b) In the case of each industry, outline **three advantages** and **disadvantages** of this location. Then decide whether each planning application will be successful or not. In each case, give **two fully developed points** explaining your decision.
4. Global warming is expected to cause ice-cap melting at the poles. It is also expected to change global climates.
   (a) With reference to this part of north Co. Dublin, suggest ways in which these factors could affect the region.
   (b) Suggest one/two ways in which this region might prepare for these changes. See chapter 13, pp. 273–6.

Fig. 16.11 ▶

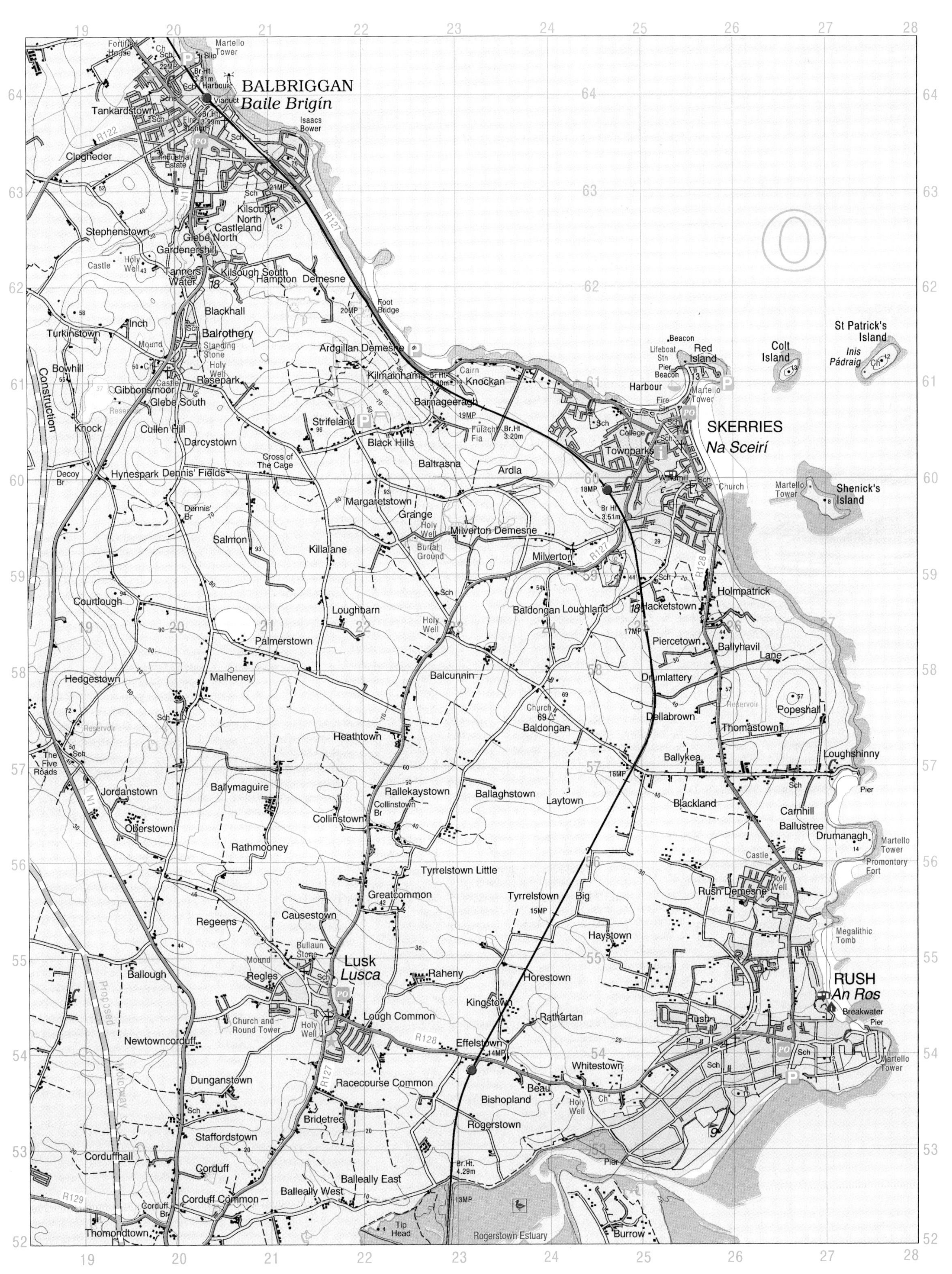

BALBRIGGAN
Baile Brigín
Fortified House
Martello Tower
Slip
Br.Ht. 3.81m
Harbour
Viaduct
Br.Ht. 3.99m
Fire Station
Tankardstown
Isaacs Bower
R122
Clogheder
Industrial Estate
R127
Kilsough North
Castleland
Stephenstown
Glebe North
Gardenershill
Castle
Holy Well
Tanners Water
Kilsough South
Hampton Demesne
Foot Bridge
Blackhall
Inch
Turkinstown
Balrothery
Mound
Standing Stone
Bowhill
Construction
Ardgillan Demesne
St Patrick's Island
Inis Pádraig
Colt Island
Beacon
Red Island
Lifeboat Stn
Pier Beacon
Harbour
Martello Tower
Cairn
Knockan
Kilmainhams
Br.Ht. 3.70m
Rosepark
Gibbonsmoor
Glebe South
Reservoir
Barnageeragh
Knock
Cullen Hill
Strifeland
Fulacht Fia
Br.Ht. 3.20m
Fire Stn
College
SKERRIES
Na Sceirí
Darcystown
Black Hills
Townparks
Baltrasna
Cross of The Cage
Ardla
Windmill
Church
Decoy Br
Hynespark
Dennis' Fields
Martello Tower
Shenick's Island
Margaretstown
Dennis' Br
Grange
Holy Well
Br Ht 3.51m
Milverton Demesne
Salmon
Killalane
Burial Ground
Milverton
R128
Holmpatrick
Courtlough
Loughbarn
Baldongan
Loughland
Hacketstown
Holy Well
Palmerstown
Piercetown
Ballyhavil
Lane
Malheney
Balcunnin
Drumlattery
Hedgestown
Church
Baldongan
Dellabrown
Popeshall
Reservoir
Thomastown
Heathtown
Ballykea
Loughshinny
The Five Roads
Jordanstown
Ballymaguire
Rallekaystown
Ballaghstown
Laytown
Blackland
Pier
Carnhill
Ballustree
Drumanagh
Collinstown Br
Collinstown
Martello Tower
Oberstown
Rathmooney
Castle
Promontory Fort
Tyrrelstown Little
Holy Well
Greatcommon
Tyrrelstown
Big
Rush Demesne
Causestown
Regeens
Haystown
Megalithic Tomb
Bullaun Stone
Lusk
Lusca
Mound
Ballough
Regles
Raheny
Horestown
RUSH
An Ros
Proposed
Kingstown
Rathartan
Breakwater
Church and Round Tower
Lough Common
Rush
Pier
Newtowncorduff
Holy Well
Effelstown
Martello Tower
Whitestown
Racecourse Common
Dunganstown
Beau
Bishopland
Holy Well
Motorway
Bridetree
Rogerstown
Staffordstown
Corduffhall
Corduff
Pier
Br.Ht. 4.29m
Balleally East
Balleally West
R129
Corduff Common
Corduff Br
Tip Head
Thomondtown
Burrow
Rogerstown Estuary

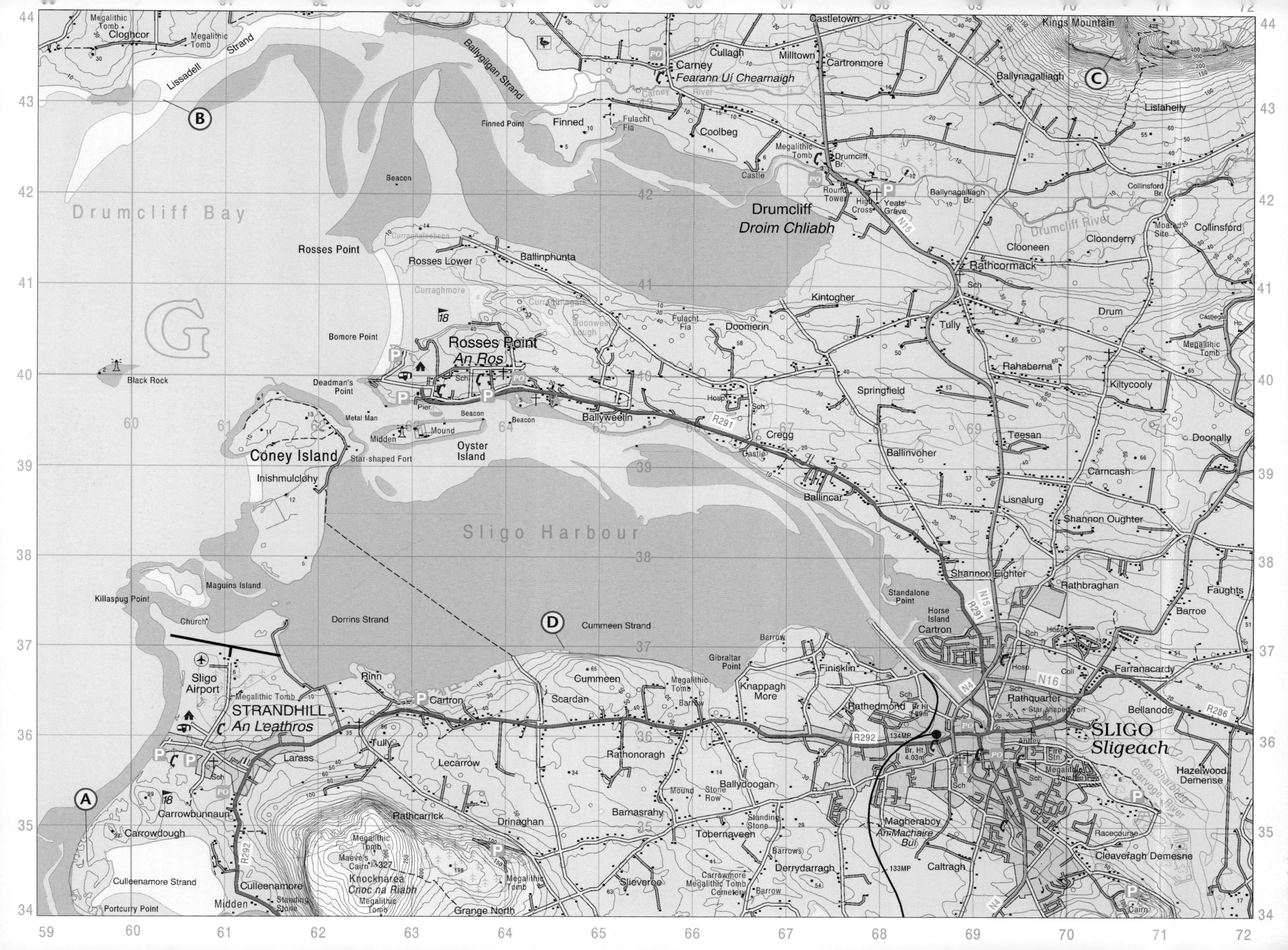
Drumcliff Bay
Lissadell Strand
Ballygilgan Strand
Cloghcor
Megalithic Tomb
Finned Point
Finned
Fulacht Fia
Coolbeg
Carney
Fearann Uí Chearnaigh
Carney River
Cullagh
Milltown
Castletown
Cartronmore
Ballynagalliagh
Kings Mountain
Lislahelly
Drumcliff
Droim Chliabh
Drumcliff Br.
Round Tower
High Cross
Yeats' Grave
Ballynagalliagh Br.
Collinsford Br.
Collinsford
Moated Site
Drumcliff River
Clooneen
Cloonderry
Rathcormack
Beacon
Rosses Point
Rosses Lower
Ballinphunta
Curraghmore
Doonweelin Lough
Kintogher
Doonierin
Tully
Drum
Bomore Point
Rosses Point
An Ros
Deadman's Point
Black Rock
Metal Man
Pier
Midden
Mound
Oyster Island
Star-shaped Fort
Coney Island
Inishmulclohy
Ballyweelin
Springfield
Rahaberna
Kiltycooly
Doonally
Cregg
Castle
Ballinvoher
Teesan
Carncash
Ballincar
Lisnalurg
Shannon Oughter
Shannon Eighter
Rathbraghan
Faughts
Barroe
Sligo Harbour
Maguins Island
Killaspug Point
Church
Dorrins Strand
Cummeen Strand
Standalone Point
Horse Island
Cartron
Gibraltar Point
Barrow
Finisklin
Knappagh More
Rathedmond
Sligo Airport
STRANDHILL
An Leathros
Megalithic Tomb
Rinn
Cartron
Cummeen
Scardan
Larass
Tully
Lecarrow
Rathonoragh
Ballydoogan
Mound
Stone Row
Standing Stone
Barnasrahy
Tobernaveen
Barrows
Derrydarragh
Carrowmore Megalithic Tomb Cemetery
Slieveroe
Magheraboy
An Machaire Buí
Caltragh
Farranacardy
Rathquarter
Star-shaped Fort
Bellanode
SLIGO
Sligeach
Abbey
Fire Stn.
Hazelwood Demense
Garvoge River
An Gharbhóg
Racecourse
Cleaveragh Demesne
Cairn
Rathcarrick
Drinaghan
Carrowbunnaun
Carrowdough
Culleenamore Strand
Culleenamore
Portcurry Point
Midden
Standing Stone
Maeve's Cairn
Knocknarea
Cnoc na Riabh
Grange North
R291
R292
R286
N15
N16
N4
Br. Ht 4.03m
134MP
133MP

◀ Fig. 16.12

Carefully study the Ordnance Survey map extract of the Sligo region, Fig. 16.12. Then do the following:

1. Draw a sketch map. On it mark and label
   - the main routeways
   - two beaches
   - one cliff area
   - two small seaside settlements
   - two inland pre-Christian settlement sites
   - two coastal pre-Christian settlement sites
2. Identify the physical landforms A to D.
   Explain fully the processes that have led to the formation of A and D only.
3. The Sligo region is an important tourist destination. Using evidence from the map, explain this statement fully, using the following headings:
   (a) tourist attractions
   (b) the natural landscape, ancient settlement
   (c) transport facilities
   (d) services
4. Explain the meaning of the following terms on the map: Midden G 612 342; Megalithic Tomb G 625 347; Fulacht Fia G 659 407; High Cross G 677 418.

Fig. 16.13

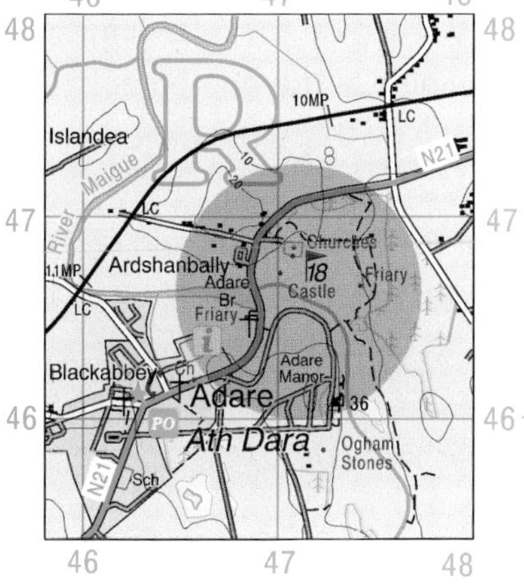


Fig. 16.14 The words Friary, Abbey and Priory regularly appear near to castles

## Norman Settlements

### Norman Ireland and the Defence Settlement

In earlier times, security from attack by enemy neighbours, invaders and pirates was often of great importance in siting nucleated settlements. In the late twelfth and thirteenth centuries in Ireland, the Normans conquered and occupied Viking settlements along the east and south coasts. As they advanced throughout the country, they built castles as fortresses in enemy territory to protect their captured lands. The Normans regularly chose sites which had already been established by earlier monastic communities. Norman knights required a fortified residence which could serve both as a dwelling and as a stronghold from which they could control the area of the country allotted to them. The earliest of these Norman defensive settlements was the **motte and bailey** type. This defence consisted of a **motte** or **mound** on which a wooden tower was built. It was connected by means of a bridge to a **courtyard** or **bailey**. In the courtyard stood a collection of sheds used as living quarters for the lord's soldiers, shelters for their horses and storehouses for grain and wine. The whole area was surrounded by a deep trench or ditch which was filled with water. This was called a **moat**, e.g. Naas in Co. Kildare and Granard in Longford. After AD 1200, the wooden towers were replaced by stone buildings.

Fig. 16.15

As the country came under Norman control, castles appeared at river crossings, river loops, islands, hilltops, peninsulas and rocky outcrops.

Norman settlers built their houses near the castles for security, fearing attack from the dispossessed Irish. The Middle Ages was a period of war and disturbance, and most towns were fortified with defensive walls, their guarded gateways being the only means of entering or leaving the town.

Anglo-**Norman settlements** are still our most important towns today. They are mainly located in the east and south of the country in areas of well-drained fertile land. Their original layout is often recognisable from the air. The portions of these towns in which the medieval plan has been preserved usually exhibit a rather **irregular plan** of **winding streets**, **narrow lanes** and a mixture of land uses, e.g. Navan in Co. Meath and Athlone in Co. Westmeath.

As towns grew, **charters** were granted which allowed townspeople to govern their own urban area. Once a week, a market was held in the town, so people focused on the settlement to sell their agricultural produce and to buy articles from local craftspeople. The more important and centrally located towns became centres for fairs to which merchants, traders and business people would come from a distance to buy and sell their goods.

From the fifteenth century onward, the Irish clan leaders erected castles based on Norman models. The fortified fourteenth-century tower house, a square building with massive walls and as many as four or five storeys, is the most characteristic Irish form of this style. Wealthy landowners and merchants erected tower houses for their own personal safety in the towns and countryside.

Apart from Athenry in Co. Galway where the medieval town walls are practically intact, only odd fragments of medieval fortifications remain today.

The ruins of the substantial medieval stone buildings — the castles, tower houses, churches and monasteries — have sometimes survived.

Carefully examine the photograph of Clonmacnoise, Fig. 16.16. Then do the following:

1. (a) Classify the two ancient settlement sites in the right middle of the photograph.
   (b) For each settlement, give three well-developed reasons why this settlement developed at this site.
2. Classify the major glacial landform in the centre background. Then with the aid of a labelled diagram explain fully the processes involved in its formation.
3. Give two well-developed reasons why this region is prone to flooding. In your answer, refer to landscape and settlement.
4. Explain fully how the landform in the right background was formed.

Fig. 16.16

## The European Abbeys

By the twelfth century, Irish monastic life had reached a very low ebb. In some cases, laypeople had taken over as abbots and abuses were rife in the monasteries. To help remedy this situation, St Malachy encouraged the Cistercian order of monks to come to Ireland and set up monasteries similar to those in Europe at this time. The Anglo-Normans, who were Christian, also encouraged religious orders such as the Augustinians and Franciscans to establish monasteries here. These religious centres are therefore often found

**Ennis**
River-meadow.

**Enniscrone**
River-meadow of the esker.

**Feeard**
High wood.

**Feenagh**
Place of woods.

**Garryduff**
Black garden.

**Garrymore**
Large garden.

**Glenaan**
Little glen.

**Inch**
Island or river-meadow.

**Inishbeg**
Little island.

in close association with Anglo-Norman castles. They are more common throughout the south and east, where Norman influence was strongest and where land was fertile with a high lime content. Examples include Adare in Co. Limerick, Killarney in Co. Kerry and Swords in Co. Dublin.

Riverside sites were often chosen for the availability of fresh water as well as for the food supply, as fish formed a large part of the monks' diet.

During the Middle Ages, ecclesiastical centres tended to play more important social functions than they do today. They frequently **provided alms for the poor, education for the young, accommodation for travellers and hospital services for the ill**. Abbeys therefore played a dominant role in encouraging settlement and urban growth, as at Ennis in Co. Clare and Kilmallock in Co. Limerick.

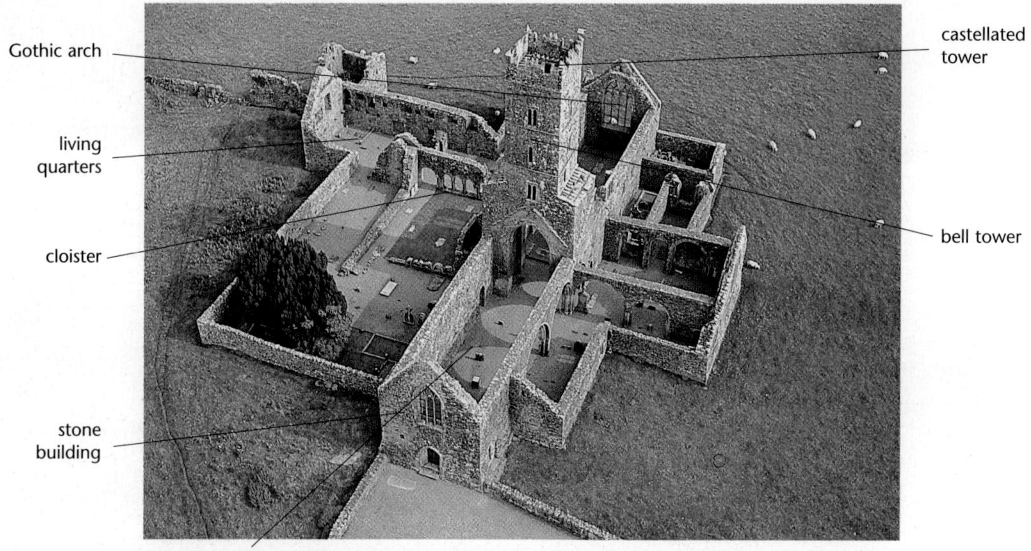


Fig. 16.17
Kilconnell Friary in Co. Galway

The medieval abbey was a specially designed and self-contained complex of buildings where monks could live as a group in the service of God. The abbeys differed from the Irish monasteries in architectural style, size and overall appearance. These ecclesiastical sites contained not only a large church, but a number of associated buildings such as dormitories, chapter house, kitchen and stores, all set around a square called a **cloister**.

The Cistercians played a key role in agricultural development, including the creation of markets within Ireland and abroad for cattle, horses and wool. The monks relied on their own labour for farming. Monastic lands were divided into farms or **'granges'**. The location of granges can often be identified now only by place name evidence.

Gothic architecture allowed larger and more graceful churches to be built. Pointed arches, large glass windows and tall thin walls often supported with buttresses became standard features. The largest window was generally on the east-facing gable located behind the altar.

## Sixteenth-century Change

The **ruins of monasteries**, such as those of Kilconnell Friary in Co. Galway, are **reminders of the closure of the monasteries** and the **introduction** of the **Protestant faiths** in Ireland (Reformation) in the sixteenth century. It was not until the relaxation of the Penal Laws and the granting of Catholic Emancipation in 1829 that we find a renaissance of Roman Catholic church architecture (see 'Larger Planned Towns — Eighteenth and Nineteenth Centuries', chapter 17, p. 349). Protestantism was introduced at this time and for the first time Protestant churches were constructed in various locations throughout the country.

**Class Activity**

Study the Ordnance Survey map extract of Buttevant in Co. Cork, in Ireland, Fig. 16.18. Then use the words within the circles to help you write three fully developed points on how you know this town began as a Norman settlement.

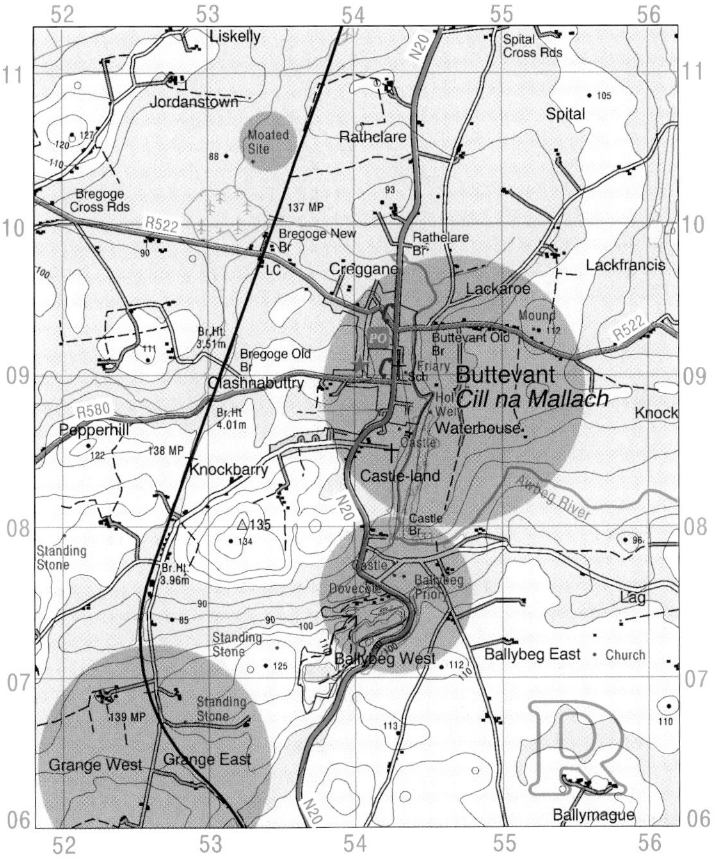

Fig. 16.18

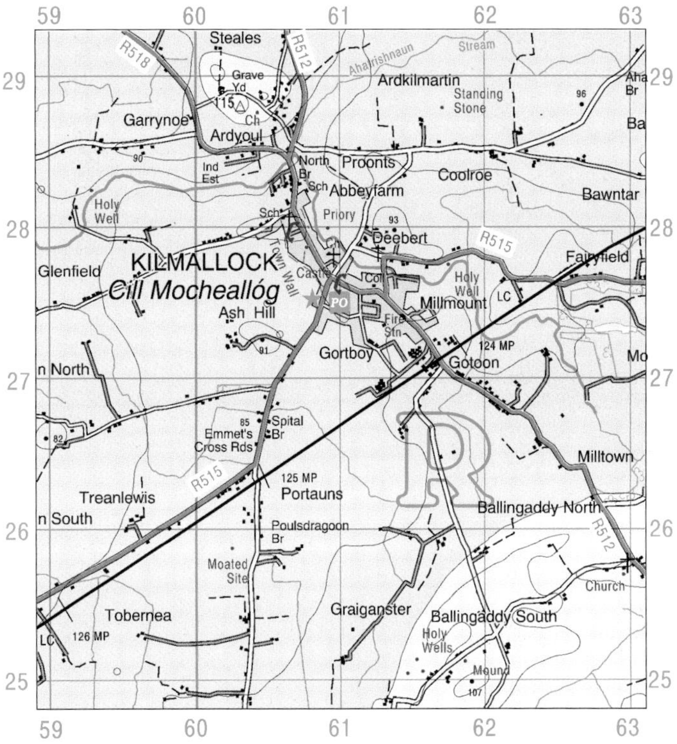

Fig. 16.19

**Class Activity**

Study the Ordnance Survey map extract, Fig. 16.19. Then give three fully developed points explaining why the town of Kilmallock developed at this location.

**Class Activity**

Study the Ordnance Survey map extracts, Figs. 16.20 and 16.21. Then using evidence from the maps explain the origin and subsequent development of Kells town and Kilfinnane village.

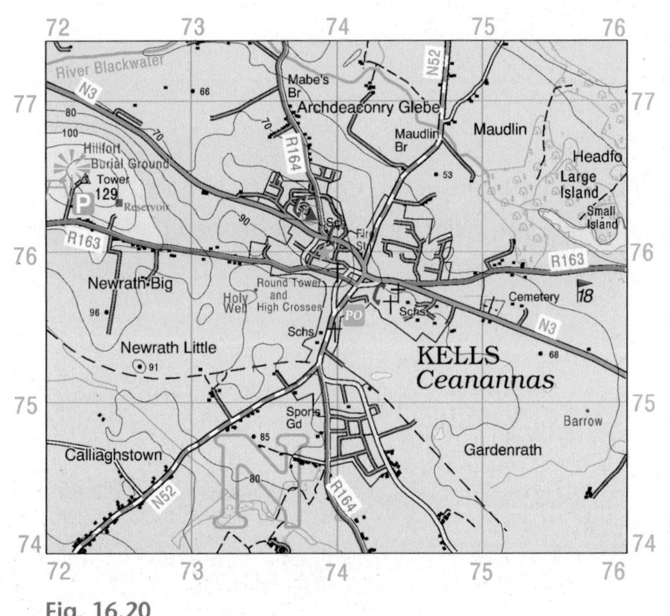

Fig. 16.20

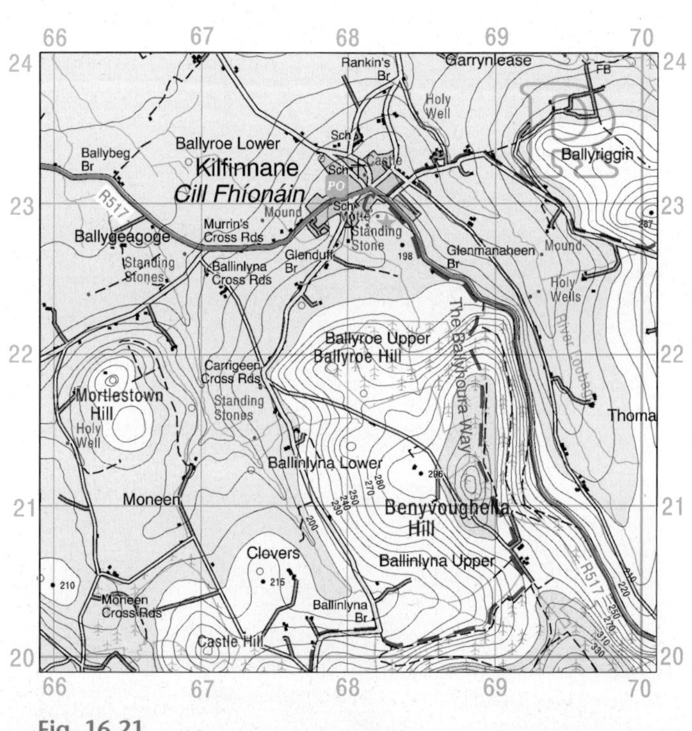

Fig. 16.21

Study the photograph of the centre of Kinsale town in Co. Cork, Fig. 16.22. Then answer the following:

1. What evidence on the photograph suggests that this town has a medieval origin? Explain fully.
2. What characteristics does this particular town display that makes it one of Ireland's most attractive settlements? Only use evidence from the photograph to explain your answer.

Fig. 16.22

Fig. 16.23

Study the photograph of Ardmore, in Co. Waterford, Fig. 16.23. Then answer the following:

1. What evidence on the photograph suggests that this settlement was founded in early Christian times?
2. Field markings, trenches, ditches, rivers and buildings etc. sometimes provide clues relating to the origin and development of some settlements. Does some evidence exist on this photograph that such is the case with Ardmore?
3. The layout and street pattern of some settlements suggests that they did not just develop piecemeal over time and that some thought was given to their layout. With reference to the photograph discuss this statement.

P.T.O. for Q. 4

4. Tourism in Ireland has experienced slow but continuous growth over past decades.
   (a) What evidence on the photograph suggests that Ardmore in some way has partaken in this growth?
   (b) (i) If you were a wealthy business person and you wished to build a hotel which would provide a lot of recreation facilities and related services in the area shown in the photograph, choose a site for this proposed building and give three reasons why this site and this region would be a suitable place for such a venture.
   (ii) Outline three arguments which local people could use to oppose such a venture.

Study the photograph of Carlow, Fig. 16.24. Then answer the following:

1. What evidence on the photograph suggests that this was a Norman town? Explain fully.
2. (a) Draw a sketch map of the area shown in the photograph and on it mark and name five different land use areas, and the street pattern.
   (b) What evidence on the photograph suggests that urban development and renewal in Carlow is in progress?

Fig. 16.24
Carlow town in Co. Carlow

# Field Patterns and Plantation Towns

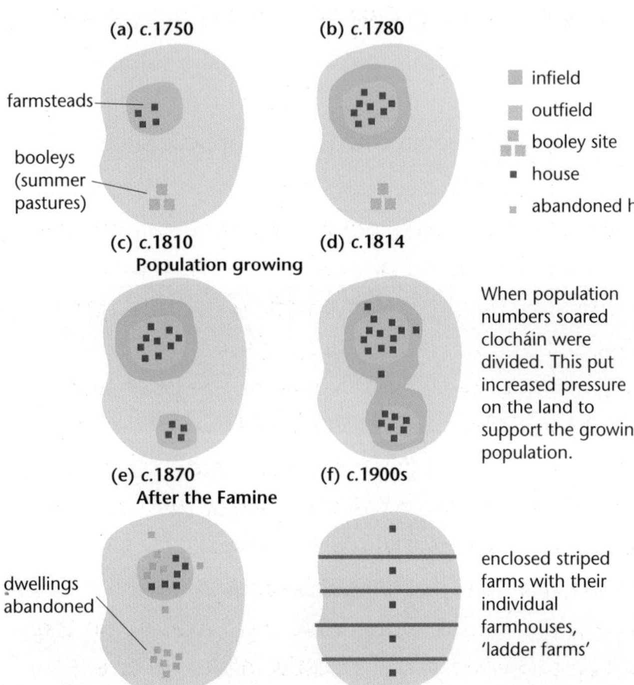


Fig. 17.1

Fig. 17.2
This is one of the very few remaining infields in Ireland on Tory Island off the Donegal coastline. Within the infield are unenclosed gardens laid out in a fan shape. The dark patches are vegetable gardens near the farmhouses.

## The Rundale System

Today, the most common feature of the social landscape is the farm, giving us a **bocage landscape** (fields enclosed by clay or stone fences with hedging on top). Eighteenth-century farms, with their fields and fences, were called **enclosures**. This name is derived from the fact that, prior to the eighteenth century, all land was **commonage** and the rundale and the open-field systems of farming were practised. The **rundale system** was practised in the Gaelic parts of the country such as the western counties of Mayo, Galway, Donegal and Kerry. Central to this system was the clochán or unplanned cluster of farmsteads (Fig. 17.2). The **infield** (a large field enclosed with a rail or wattle fence) was situated near the **clochán** and was divided into strips and tilled by the families. Oats, barley, wheat and rye were grown. Potatoes and turnips were introduced later. Surrounding the infield and beyond lay the **outfield**. The outfield was generally poorer, more marginal, hilly or boggy ground. This was grazed in common by all the farmers. **Booleying** or **transhumance** was practised on the upland slopes. Remnants of this farming system can still be found in remote parts of the west. (The word clochán has been anglicised variously as *cloghaun*, *cloghane* and *claughaun*.)

The rundale system was a sophisticated agricultural method which was in close touch with poor-quality land and social problems. This co-operative management of limited tillage land and poor grazing maximised output by shared labour and agreed land use.

The western small farms were a consequence and a cause of increasing population. Ireland's population soared from three to eight and a half million people between 1700 and 1845. This growth demanded reclamation, subdivision and expansion into previously unsettled marginal land. This population growth and farm expansion was aided by the potato crop, which could flourish in wet, thin, nutrient-poor acid soils.

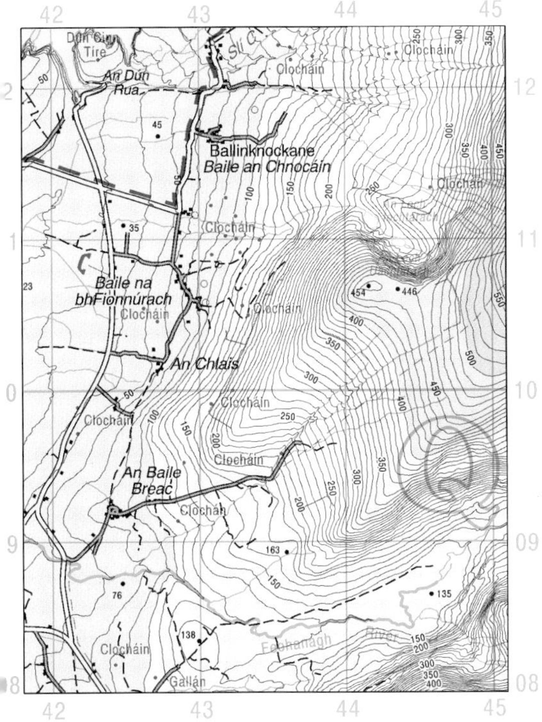

As a result of the weight of population, land subdivision and the damp foggy atmosphere, this social environment collapsed, aided by potato blight and the subsequent famine.

Emigration and death caused abandonment of land and contraction of clocháin to pre-Famine size. Later, the landlords or the tenants themselves reorganised the land by dividing it into individual 'ladder' farms.

Fig. 17.3
The Irish word clocháin refers to rural farming settlements where land was shared by neighbouring farmers

## The Open-field System

The **open-field system** which was practised in the rest of the country consisted of a village, more formal in plan than the clochán, which was located near the manor. Surrounding the settlement were three large open fields. One of these was generally left fallow while the others were divided into long, scattered and unfenced strips of land for each farmer. This ensured an equal distribution of land of varying quality. Each open field could be hundreds of acres in area. Landless peasants supplemented their incomes by grazing one or two cows on the commons which surrounded the open fields. When land was enclosed, peasants were unable to continue this practice.

In some western areas, enclosure fences were superimposed on the old tillage patterns (lazy beds) which are indicative of the rundale system.

Carefully study the photograph of Tory Island in Co. Donegal, Fig. 17.5. Then answer the following:

1. The settlement in the foreground displays evidence of an ancient Irish settlement pattern. Identify this pattern and then with the aid of a sketch map and evidence from the photograph describe this settlement pattern fully.
2. The northern coastline of Tory Island displays evidence of severe coastal erosion. Identify the landforms of coastal erosion at A, B, C, D, E and F. Then with the aid of labelled diagrams, explain how they were formed.

Fig. 17.4
Squaring pattern in Co. Kilkenny

### The Enclosures

By the early eighteenth century, the rundale and open-field systems had started to disappear. Commonage was enclosed with fences and hedges and the land was redistributed among the strip-holders. The enclosed farms were better managed. They used new farming methods such as crop rotation and newly invented machines such as seed drills which led to much heavier cropping and healthier, better-fed animals.

Each tenant built a house on his newly enclosed farm, and old villages or clocháin were pulled down. Thus a new landholding system evolved and a change came about in the settlement pattern of the Irish landscape.

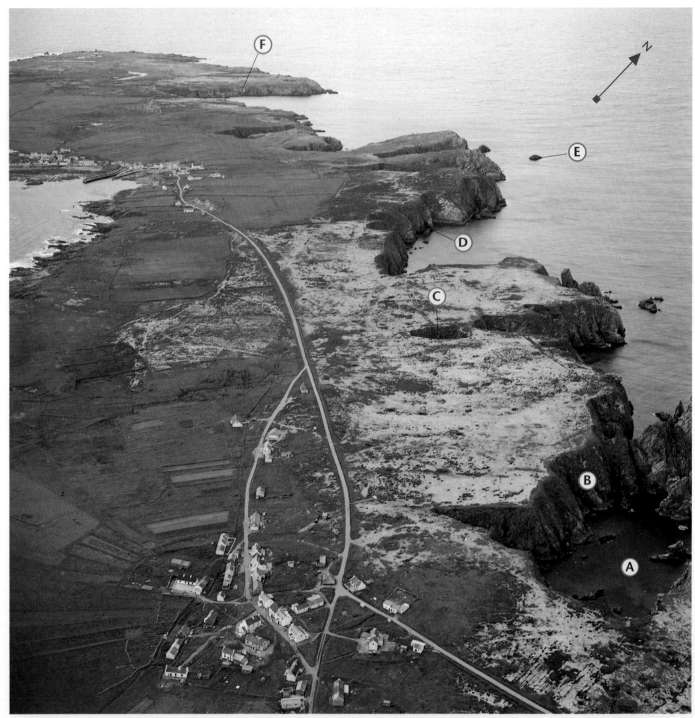


Fig. 17.5
Tory Island in Co. Donegal

Each individual farmer now held a farm of his own and no longer held a farm in common with others. Many nucleated clusters of villages in some areas were eliminated and replaced with a dispersed rural settlement pattern similar to that of today's landscape.

Each enclosure is divided into fields which are separated by fences or ditches of stone or clay, or both, upon which hedges of whitethorn grow (Fig. 17.4). Farms are separated from each other by **bounds ditches**, which are generally wider and thicker in their hedging than fences. However, in places such as the 'Ormond Lands' of south Kilkenny

Fig. 17.6
Ladder farms. Striping pattern in Co. Down.

and Waterford, tenant dwellings were arranged in clusters or 'farm villages' which were evenly spaced throughout the area. These were one type of post-feudal choice of the rural settlement pattern (Fig. 17.7).

On **lowland plains**, where land was of uniform quality, a system known as **squaring** was arranged. This was a grid pattern of farms connected by a complicated pattern of tracks or bohereens (Fig. 17.4).

**In hilly areas or places where land was of varying quality**, such as sloping land leading down to a river, improvers of the eighteenth and nineteenth centuries used a system of enclosure known as **striping**. In such cases, the farmland was arranged in single narrow strips or stripes arranged in parallel rows. They formed 'ladder farms' (Fig. 17.6).

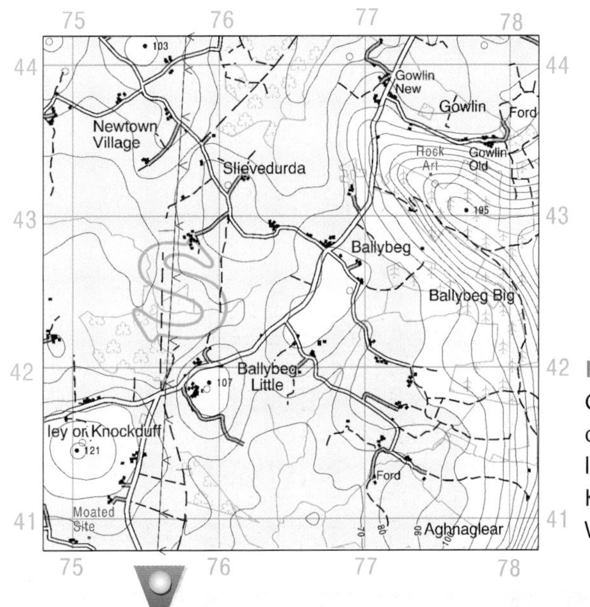


Fig. 17.7
Clusters of farm dwellings on enclosed land in the south Kilkenny/Waterford/ Wexford region

**Kan/Ken/Kin**
*Ceann* = head, e.g. Kanturk.

**Knockmore**
Big hill.

**Kyle**
*Coill* = wood, e.g. Kilduff = Black Wood.

**Letter**
*Leitir* = Wet Hillside.

**Lis/Lios**
A circular earthen fort.

**Lough Conn**
Lake of the hound.

**Lug/Lag/Leg**
A hollow.

**Maum/Madhm**
A high mountain pass.

**Mona/Mon**
*Móin* = a bog.

**Monasterevin**
*Mainistir Eimhín* = Monastery of Eimhín.

**Money**
Grove.

**Mull**
A summit; *Mullach* = Great Summit.

**Mullagh**
Hilltop.

**Mullen**
A mill.

**Owen**
*Abhainn* = river.

**Park**
*Páirc* = a field, e.g. Parkmore.

**Port Ballintrae**
Harbour of the town of the strand.

**Rath**
Circular fort.

**Ross**
Headland.

**Scart**
A thicket or cluster.

**Shan/Sean**
Old.

**Shanagarry**
Old garden.

**Slieve**
Mountain.

**Termon**
Sanctuary, church land.

**Termonbarry**
Church-land of St Bearach.

**Tobermore**
Large well.

**Turlough**
A seasonal lake which dries up in summer.

**Urris**
Peninsula.

**Ventry**
*Fionn Trá* = White Strand.

LANDSCAPES OF THE WORLD

## Case Study: Enclosure of Rundale Farms

**Ladder farms** generally represent the reorganisation of the older rundale system. It was preferred by the rundale farmers because they retained the rundale principle of a range of land-use types from lowland arable to rough grazing and bog.

These ladder farms indicate that this farmland was of varying quality. This long and narrow type of farm pattern is called striping.

The very narrow and parallel boundaries represent the old stripes of the ploughed infield. They are referred to as **'fossil strips'**.

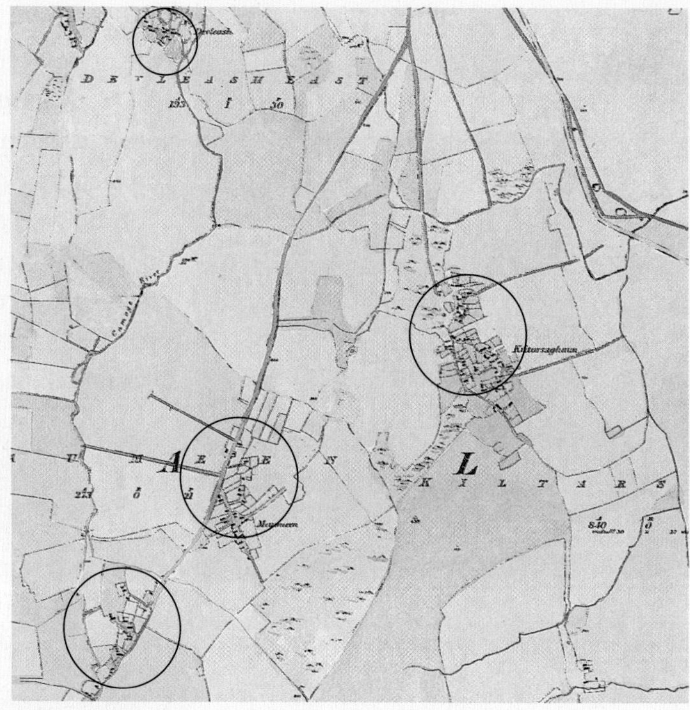

Fig. 17.8
Clochán and rundale of 1840.
Kiltarsaghaun townland in Co. Mayo.

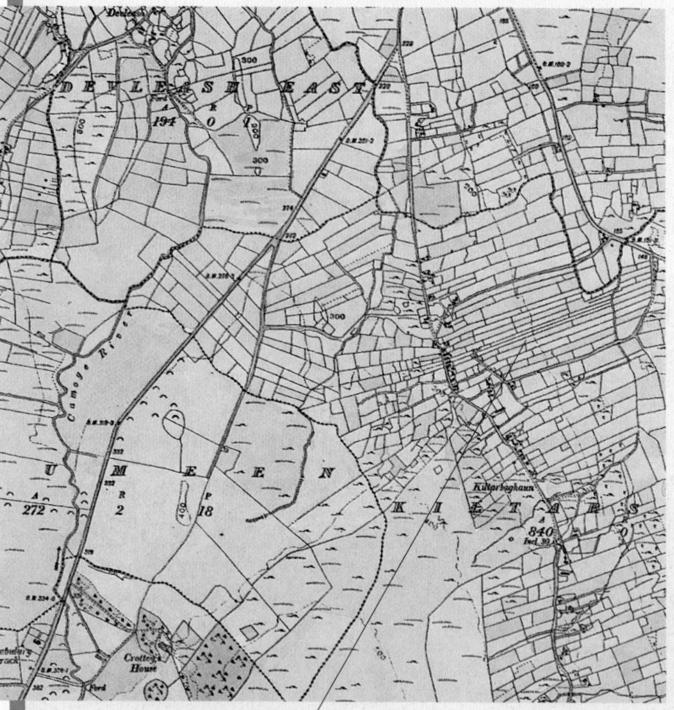

narrow fossil strips

Fig. 17.9
Reorganised enclosures in 1899

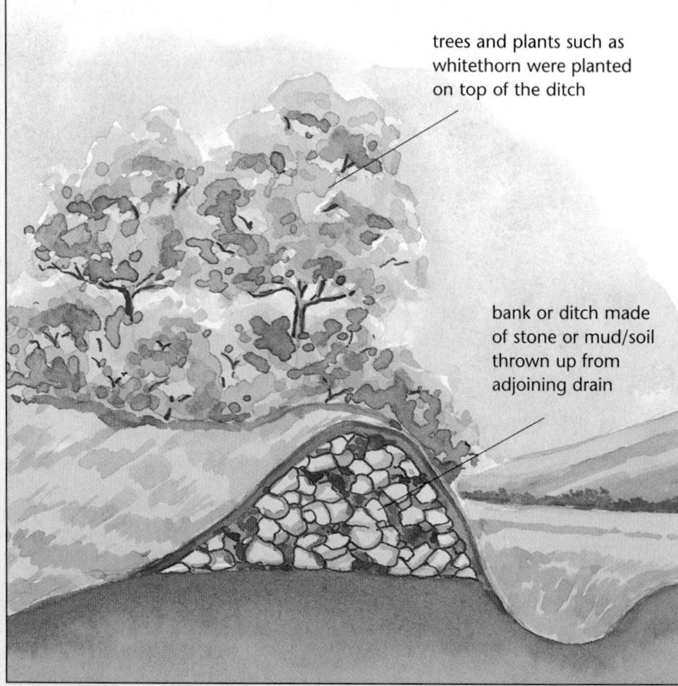
trees and plants such as whitethorn were planted on top of the ditch
bank or ditch made of stone or mud/soil thrown up from adjoining drain

Fig. 17.10
Bounds ditches were wider, higher and 'thicker' with whitethorn, blackthorn and trees than other field fencing

**Class Activity**

Carefully examine the maps of Kiltarsaghaun townland in 1840–99 in Co. Mayo, in Ireland, Figs. 17.8 and 17.9. Then answer the following:

1. What changes have taken place in this area with regard to farm structure?
2. How were routeways affected by the new land division patterns?
3. In what way did rivers and streams affect the new farm structures?

Fig. 17.11

Study the photograph, Fig. 17.11. Then answer the following questions:

1. (a) Name the type/category of farms according to their layout (see Fig. 17.6, p. 335 and p. 336).
   (b) What information does the layout of these farms give about the quality of the land? Explain fully, using evidence from both sources.
2. What other primary economic activities are carried on in this area? Explain fully using evidence from the photograph.
3. Describe the course of the river in the foreground and in the background of the photograph. Use evidence to support your answers.
4. Account for the pattern in the distribution of the various types of vegetation shown on the photograph. Explain fully.

**Class Activity**

Study the photograph of Glenelly Valley on p. 438. In what way does the slope of the land influence the shape of the fields/farms?

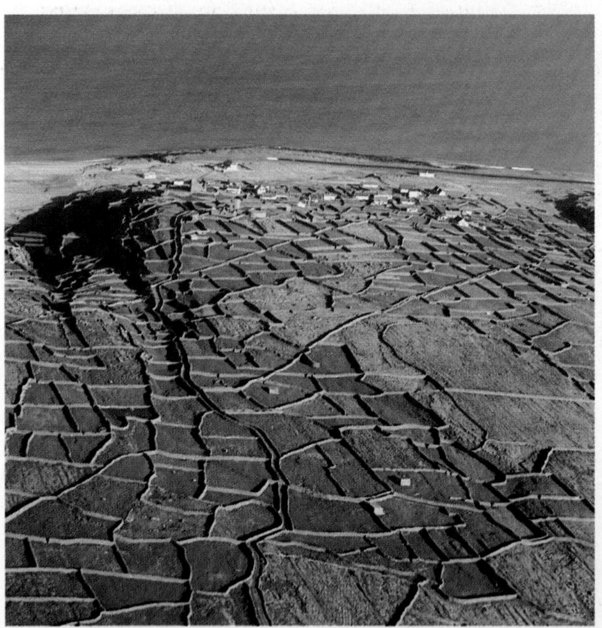

Fig. 17.12
Stone walls between fields in the west of Ireland

## Changing Field Patterns

Field patterns in Ireland may change from region to region, some, such as in mid-Galway, with their own unique characteristics.

### Munster, Midlands and Eastern Counties

In the Golden Vale in Munster, fields are generally regular and reasonably large. Intensive farming methods, such as strip grazing, in this area have led to the removal of some ditches, especially on the land of progressive farmers. However, hedges are needed for shelter, especially from cold northern winds in winter and early spring. Strong ditches with hawthorn, whitethorn and ash dominate. In the wetter areas in river flood plains, ditches can be overgrown due to little care as a consequence of rural depopulation and an ageing of the existing farming population. Some ditches are fenced off from grazing land as part of 'set aside', where funds are given to farmers who protect wildlife and their habitat.

### West of Ireland

Here fields are generally smaller than those in the Golden Vale. They are enclosed with hedges or walls and can be colourful in summer with fuchsia and other flowering plants. The farms are laid out in stripes, the 'ladder farms', with the parallel bounds ditches clearly stronger and bigger than the internal field ditches.

In Counties Clare and Galway, the stone walls are a characteristic feature of the landscape. It may be interesting to note that the very first farmers in the west some thousands of years ago used stone walls as field boundaries. What is also interesting is that some fields appear to have no gateways for access or exit of animals. What happens in such cases is that a portion of the loose stone wall is removed and then rebuilt when the animals are moved.

## Demesnes of the Eighteenth and Nineteenth Centuries

Fig. 17.13
A nineteenth-century demesne with its ornamental woodland adds great character to the Irish countryside

The late eighteenth and early nineteenth centuries introduced a new type of farming landscape, that of the demesne. These large landed estates embraced the whole country. Central to this development was the '**big house**', usually a large Georgian mansion, which represented the self-confidence of the landed class. Improvements in agricultural output, created by the introduction of new crops, crop rotation, liming, draining wet land, the building of outhouses and improved breeds of animals, all provided increased rental income. These great houses represented a stable and improving economy and a peaceful countryside which was to reach its peak in the 1780s.

In rural areas, many estate workers tended to locate their dwellings near the entrance leading up to the big house. Landlords wanted such houses to look well, so it was common for entire villages to be built as part of a landlord's overall plan for the estate, which was often improved further by his descendants.

The estate house therefore plays an integral part in the physical structure of many towns and villages. The landlord laid down the standards which were to apply and only the best materials were used, such as dressed stone for public buildings, high slated roofs and small windows. Such villages were formally arranged and carefully designed around a central square or green, with a wide main street leading up to the big house or estate entrance. The design often included a variety of houses, both large and small, a church, a school and a few shops.

The landlord often took a personal interest in the appearance of individual houses, as can be seen at Adare in Co. Limerick, Belmullet in Co. Mayo and Ardara and Glenties in Co. Donegal, all of which were built during the nineteenth century.

## Planned Plantation Towns and Villages

**Plantation towns** are those which were built during the various plantations in Ireland. Towns and villages which were built by landlords (who were generally of English ancestry) are also referred to as **plantation settlements**.

In Ireland, a Gaelic revival in the late fifteenth and sixteenth centuries caused towns located outside of Leinster to be isolated and dependent for their safety on their fortified walls. At this time town walls were in poor condition. To help remedy this situation, generous charters were granted to some towns such as Galway which were freed from all government taxes and liabilities. A town's trading profits could then be used to preserve loyalty to Britain and to maintain fortifications to protect its inhabitants.

### Plantation Towns of the Sixteenth and Seventeenth Centuries

Planned towns were developed or remodelled as part of the plantations of the late sixteenth and seventeenth centuries in Ireland. The first plantation towns developed in Laois/Offaly, e.g. Birr, Portlaoise. Other plantation towns developed in Munster including Youghal, Mallow and Bandon in Co. Cork, Killarney and Thomondgate (The Island, Englishtown) in Limerick. All these Munster towns have a few parallel narrow main streets.

They differ from the earlier planned towns of Laois/Offaly because they were established long before the plantations and were only reconstructed and reorganised during the Munster Plantation.

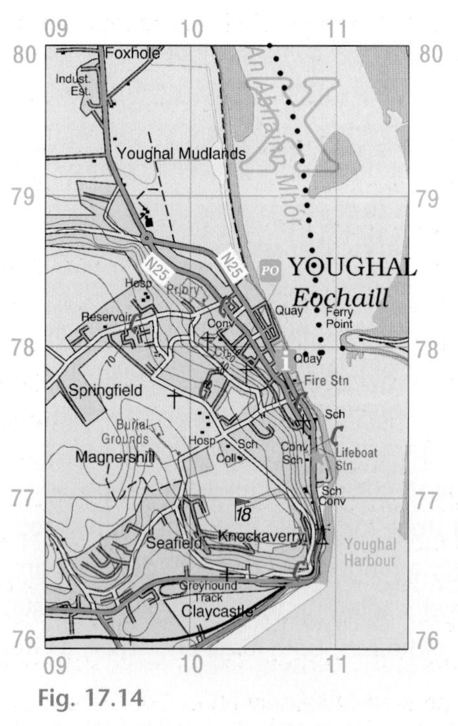


Fig. 17.14

Fig. 17.15
Youghal in Co. Cork — a late sixteenth-century 'remodelled' planned town in Munster

## New Planned Towns and Villages of the Seventeenth and Eighteenth Centuries

Fig. 17.16
Kiltinan in Co. Tipperary is one of the best-known of Irish deserted villages. Many of these early villages were sited within the boundaries of a planted demesne.

The coming of planned towns of the seventeenth and eighteenth centuries saw the destruction or desertion of earlier settlements. The centre of these deserted villages was the tower house and church. These open-field villages based on tillage gave way to newer settlements based on pastoral (grazing) farming. All that remains of these deserted settlements are subsoil markings sometimes visible only from the air as crop marks or dark lines in fields representing house foundations or streets. Most evidence of these 'fossil villages' are found in two zones, one between Roscommon and Louth and the other between Kilkenny and Limerick.

The new planters preferred to build new planned settlements on green-field sites around their new Anglican churches with houses set around a triangular fair green and, as in Ulster, the Diamond.

Ireland's deserted villages are not confined to this time. Deserted settlements are common to all ages, even in the twentieth century. Famine villages are especially interesting as many of their houses still stand. In the Burren near Fanore, in the valley of the River Caragh, lies a small cluster of unroofed dwellings. Their occupants fled, together, from the poverty and hunger of the west to the newer lands of America.

During the seventeenth century, much of southern Ireland was enjoying a time of relative peace when landowners built country seats and manor houses which were not primarily designed for defence or protection. These structures were more elegant and spacious homes than the tower houses.

Fig. 17.17
Coppinger's Court, Co. Cork, built during the seventeenth century

These seventeenth-century mansions (often called Jacobean, after the style of architecture which was common during the reign of James I) were frequently built as additions to earlier tower houses. They all have similar architectural details such as square stone-cut **window divisions**, **steep-gabled roofs** and chimney stacks incorporated into the gable walls. The plan consisted of a rectangular building. Some form of defence was generally incorporated into the building such as towers at the corners. Examples of such buildings can be seen at Leamaneh in Co. Clare, Donegal Castle and Kanturk Castle in Co. Cork, and the house of the Earl of Ormond at Carrick-on-Suir in Co. Tipperary.

**Case Study No. 1:** Birr in Co. Offaly — Seventeenth- to Eighteenth-century Town in the Midlands

Birr was founded in the sixth century at the confluence of two rivers, the Little Brosna and the Camcor, by St Brendan of Birr. But its neat and well-planned Georgian layout is due to the Parsons family, who were granted land in 1620. This 'recent' town began as a small village with a fair green and street, Castle Street, leading up to the large defensive castle. The town was later enhanced with the addition of a square and grid street plan in the eighteenth century.

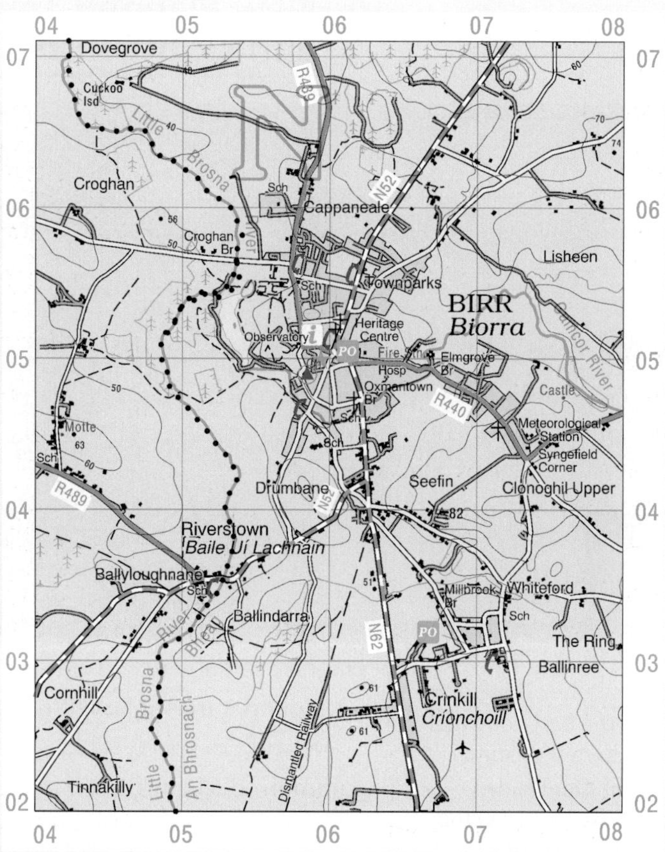


Fig. 17.18
Ordnance Survey map extract of Birr, Co. Offaly

Fig. 17.19
The main square in Birr, in Co. Offaly. Note the fine buildings of Georgian times, two to three storeys high, with simple classical features.

## Village to Town — Seventeenth and Eighteenth Centuries

Many inland towns started as villages, planned in the seventeenth century or even later, which prospered and outgrew their modest origins.

Study the Ordnance Survey map extract of Birr in Co. Offaly (Fig. 17.18) and the photograph (Fig. 17.19) and answer the following questions:

1. Suggest three reasons why Birr developed at its present location.
2. What evidence on (a) the map extract and (b) the photograph suggests that Birr was associated with the plantations? In your answer give two well-developed points.

## Case Study No. 2: Kenmare Town — Seventeenth- to Eighteenth-century Town in the South-west

The old name of Kenmare, Ceann Mara, means 'Head of the Sea'. This town was located on the south side of Kenmare River. The existing town is sited at the head of the Kenmare river inlet.

Kenmare is a fine example of a planned town in the latter half of the seventeenth and beginning of the eighteenth centuries. The town was designed like a giant 'X' with two wide main streets. The market house faces the town park. As with most other well-located towns it continued to grow, adding a church, a convent and other structures through the years, such as a cotton factory, a mill, market house, school house and an inn — the Lansdowne Arms.

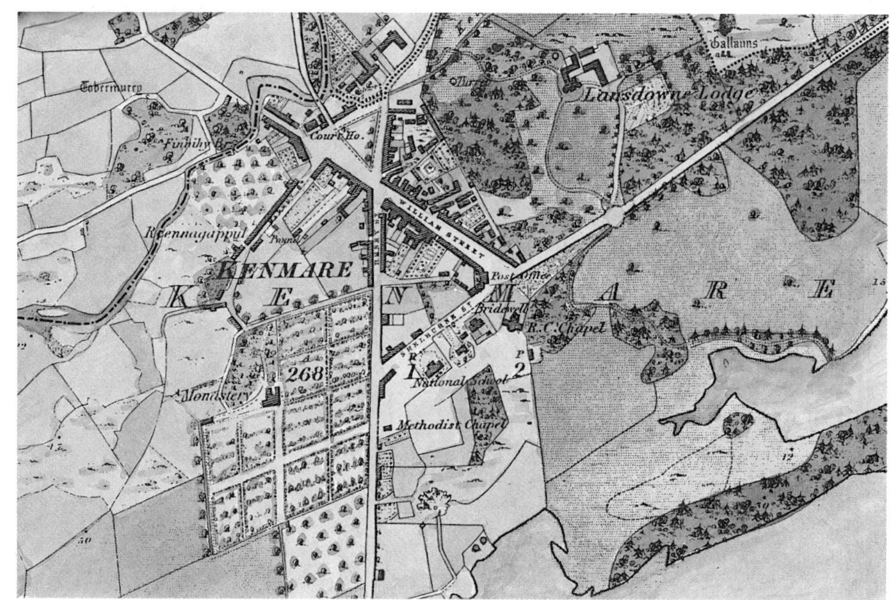


Fig. 17.20

Fig. 17.21

**Case Study:** **Sneem in Co. Kerry — A Planned Village of the Eighteenth Century**

Study the photographs of Sneem in Co. Kerry, Figs. 17.22 and 17.23. Then answer the following:

1. Describe the characteristics of Sneem that indicate it is a planned village.
2. This village is located on 'The Ring of Kerry'. What factors indicate that this village is an attractive settlement for tourists to visit?
3. There are two churches located in the village; one is a Roman Catholic church, the other a Church of Ireland Protestant church. Identify each church and list the characteristics of each structure.

Fig. 17.22

Fig. 17.23

## Seventeenth-century Planned Towns in Ulster

### Case Study: Donegal Town

Most plantation towns were built in Ulster after the Ulster Plantation of 1609. These included Donegal, Cavan, Monaghan and Clones. The principal result of the plantations was that they gave Ulster and Connacht urban centres which, up to then, were mainly rural and controlled by Gaelic chiefs. Establishing towns was an important part of encouraging settlers and creating economic development in Ireland. Old Norman walled towns in the south and east were revitalised and repaired at this time, e.g. Youghal in Co. Cork and Clonmel in Co. Tipperary.

In Ulster, where plantation was most successful, the main characteristics of planned towns were (Fig. 17.24)

1. a market square, green or **diamond** where two or more main streets converged
2. a fortified residence or castle belonging to the grantee
3. a Protestant church occupying a dominant position in the town
4. wide and regular streets

Fig. 17.24
The obelisk in the Diamond in Donegal commemorates the four Franciscans who wrote the Annals of the Four Masters, a history of Ireland which goes back to the time of Noah's grandmother, forty years before the flood

Planned towns were enclosed by strong defensive walls which incorporated the town gates, both for the protection of their citizens and for the control of traffic. The major streets of these planned towns focused on the central square or diamond which served primarily as a market place and encouraged trade.

The Ulster planters were very hard-working people who were conscious of the need for good communications, and the principal towns were linked together by a system of communications which helped trade to flourish. As towns grew, more routeways focused on each urban centre, thereby increasing trade, especially with ports through which goods were imported and exported. Towns prospered and many grew in size, often extending beyond their original walls.

Memorials to landlords started to appear at the centre of villages after 1850 during the post-Famine years. The emphasis on tenant loyalty in their inscriptions indicates many landlords' growing sense of isolation and insecurity, along with their increased awareness that tenants' allegiance could no longer be taken for granted.

All of these developments occurred at a time of great prosperity for the country estates. However, while medieval towns which had already withstood the test of centuries were successful and survived, some of these planned settlements declined once the enthusiasm of the founder was spent. Norman settlements were usually sheltered, sited on valley floors and in positions to defend river crossings, thus ensuring success, while some new planned settlements were off main routes or on exposed sites. These have since either deteriorated greatly or, on occasions, vanished completely.

With the development of routeways in the nineteenth century, numerous other planned settlements sprang up at road junctions throughout the country and gradually developed into hamlets and villages.

## Small Planned Towns — Eighteenth and Nineteenth Centuries

individual design
monument
eighteenth-century mall
octagonal square
regular (evenly wide) streets

Fig. 17.25
Westport in Co. Mayo — a small eighteenth-century town

A great sense of spaciousness was sought in eighteenth-century planning. Streets were laid out in a grid pattern, so buildings were planned in large regular blocks. In the grid pattern, the main streets are wide and run parallel to each other, while others cross these streets at right angles, thus creating a grid effect. Birr in Co. Offaly (Fig. 17.19) displays such a pattern, while Cookstown in Co. Tyrone has one long, wide and straight street with many regular intersections along its length.

The smaller planned towns usually have one central space — **a little street or square with buildings grouped around it**. As towns grew, the layout became more complex.

Most towns had **good houses** fronting the streets, a **regular square** and a **market house** in the square or on a wide street. These good houses were fine and graceful buildings which were erected as town dwellings for the country gentry or as homes of substance for the prosperous inhabitants such as professional people, merchants and army officers. These buildings were always **at least two storeys high** and are recognised by their **large chimney stacks** and **elegant doorways**. The square served as a market centre and the venue for fairs, which were allowed by licence acquired by the local landlord.

Fig. 17.26
Two house designs that were common to eighteenth-century towns

Many towns throughout the country have elegant houses set around late seventeenth- and eighteenth-century squares. Examples include Durrow and Ballinahill in Co. Laois, Bunclody and Newtownbarry in Co. Wexford, and Thurles in Co. Tipperary. Others have market houses centrally placed in their squares, such as Abbeyleix and Stradbally in Co. Laois, Kildare town in Co. Kildare, Sixmilebridge in Co. Clare and Kinsale in Co. Cork. Some of these centrally located buildings were courthouses.

### Case Study: Westport in Co. Mayo — A Small Eighteenth-century Town

Some towns had wide, tree-lined malls and often added an extra square as the town prospered and increased its range of services. The influence of the local landlord is clearly seen in some towns, e.g. Westport (see Fig. 17.25). The layout of the town is typical of the eighteenth century, with its wide and regular streets, an octagonal square, a mall and a canalised river. It displays local landlord influence, however, in that the street plan avoids the monotony of the severe grid pattern of many planned towns. (The streets are wide in keeping with eighteenth-century development but they do not run parallel.)

Many of these new settlements flourished as a result of the widespread improvement of roads throughout the country and the building of canals. Together, these increased trade and brought a new prosperity to the country (see 'Canals— Eighteenth and Nineteenth Centuries', p. 353).

### Case Study: Moy in Co. Tyrone— A Small Estate Town

The immaculate village of Moy is a fine example of a small estate town. The centrepiece is an elongated piazza. It could easily be used as a market or fair site. (The village was designed by James Caulfeild, Earl of Charlemont, who had been impressed by Marengo Square in Italy.)

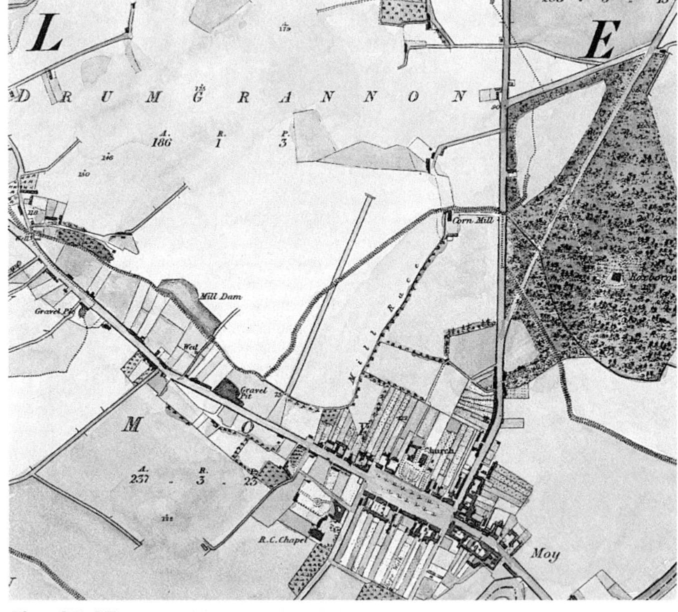

Fig. 17.27

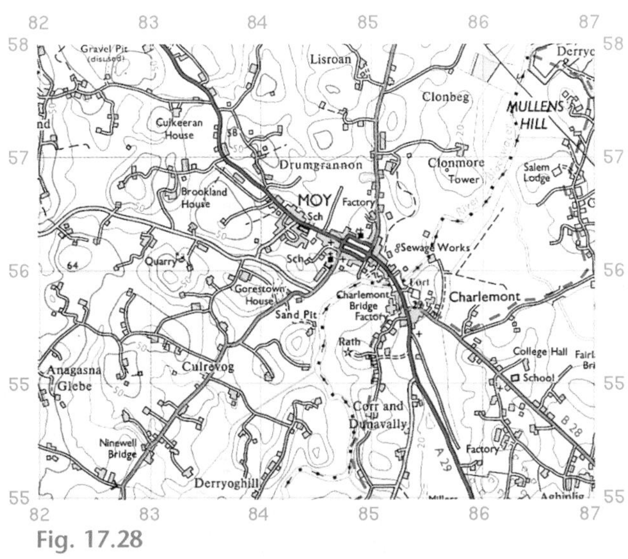

Fig. 17.28

Fig. 17.29
Moy village in Co. Tyrone

Study both Ordnance Survey map extracts and the photograph of Moy in Co. Tyrone, Figs. 17.27, 17.28 and 17.29. Then do the following:

1. Draw a sketch map of the town. On it mark and label
   - the street pattern
   - the river
   - the piazza
   - the original Protestant church
   - the Roman Catholic church
   - five areas of different land use
2. What settlements are visible on the photograph? Name them.
3. What changes have occurred to Moy since its beginning? Explain fully.
4. Describe the piazza/square.
5. (a) Classify the type of physical landscape in this region. Explain fully, using evidence from the given sources.

   (b) Examine the glacial formation distribution map of Ireland, Fig. 12.43 (a), p. 217. What is the name given to this physical region?

## Larger Planned Towns — Eighteenth and Nineteenth Centuries

The bulk of urban rebuilding and development took place in the eighteenth century and continued into the first two or three decades of the nineteenth century. This is called the **Georgian period** (1714 to 1830). The architectural style takes its name from the four Georges, the English kings who ruled during this period.

Dublin was a hive of activity when this Georgian construction was in progress. Blocks of stone were transported from quarries by horse-drawn carts. The construction of Dublin's Custom House, Dáil Éireann and other buildings created huge employment at this time.

In Dublin, Cork and Limerick, new Georgian suburbs were added to the city. As the wealthier families moved out into the new suburbs, the old town core or **Irish town** often deteriorated and became a slum area.

Fig. 17.30
A view overlooking Iveagh Gardens and Stephen's Green in Dublin. Parks such as these on the photograph formed an integral part of eighteenth-century urban planning.

These new Georgian suburbs had wide streets and squares or parks surrounded by large, four-storey, red-brick terraced mansions (Fig. 17.30). The rich made their wealth from property, trade, commerce and professions such as medicine and law. In Dublin many Irish lords and members of parliament added an extra touch of elegance to the town's social life. This ended with the Act of Union of 1800 when many of these lords and MPs moved to Westminster in London. Their houses were then reoccupied by merchants, businessmen and professionals.

In Georgian times, the street became a symbol of power and majesty. The spaciousness of such eighteenth-century urban developments is in striking contrast to medieval town areas (Irish towns) with their narrow streets.

Town houses, which were vacated for new homes in the Georgian suburbs or for life as country gentlemen in rural mansions, had now degenerated into use as warehouses or tenements. Today the Georgian suburbs form part of the new town centres' commercial districts (CBD — Central Business Districts). In many cases, the Georgian buildings have been changed into flats. Others have been renovated and are used as offices for insurance companies, solicitors, private businesses, commercial banks or local government.

### Architectural Styles

Towns and cities serve as administrative, social, religious and financial centres both for their own inhabitants and for people living in their hinterlands. The buildings in such centres of population which provided these services reflected the pride of various establishments and the status which the town held in the region. The influence of law and

## Some Variations of Neo-Gothic Style in Ireland

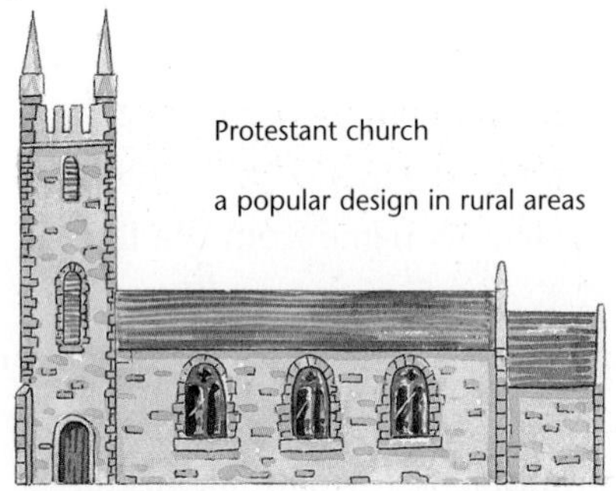


**Fig. 17.31 (a)**
Protestant church,
a popular design in rural areas

high spires
high pitch on roofs
tracery on windows
Gothic pointed arch in windows

**Fig. 17.31 (b)**
Gothic style for churches

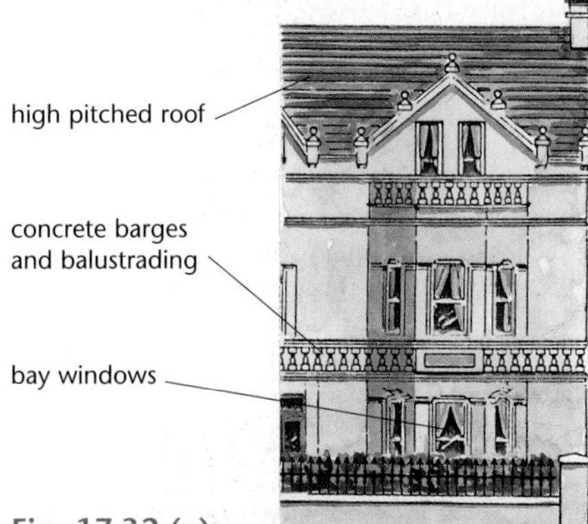


**Fig. 17.32 (a)**
Gothic domestic, common in seaside towns

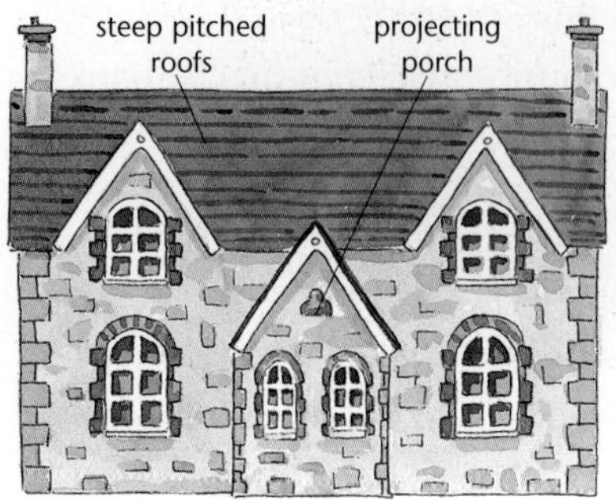


**Fig. 17.32 (b)**
Popular with estates and railways

administration was judged by the courthouse and police barracks. The landlord was represented by the development of fairs and markets, the religions by their churches and schools, and the financiers by their banks. All of these new buildings represented a new involvement in community affairs, adding to the heritage of local architecture. The style of architecture chosen for various buildings reflected the prevailing trends of that time.

It is interesting to note that in Norman times, abbey bell towers had a building style similar to their nearby castles. Both had merlons and crenelles along the parapet walls.

During the eighteenth century, at the height of the Georgian period, **classical architecture** flourished in Ireland. This style included not only the grid street plan and Georgian architecture, but the **Palladian** style (Italian Renaissance architecture which was used in the construction of rural mansions), **neo-classical** (resembling earlier architecture of France, Greece and Rome) and **neo-Gothic** (resembling the Gothic style). Some of the most attractive public buildings appeared in Ireland's towns and villages at this time.

In Dublin, buildings such as the Four Courts and the GPO were built in typical neo-classical style. Neo-classicism resembles Grecian and Roman architecture. Triangular pediments supported by Doric or Ionic columns are of Grecian origin, as is the use of simple geometric shapes of cut-stone buildings. Roman influence is displayed by domed roofs, similar to many in Italian cities.

**Class Activity**

Study the Ordnance Survey town plan of Dublin, Fig. 17.34 on p. 352. Then use the boxes to answer the following questions.

| When built: period | Buildings/street characteristics |
|---|---|
| A | |
| B | |
| C | |

By the mid-eighteenth century, the Penal Laws had been relaxed and Catholics began building churches once again. These early churches were barn-like or T-plan buildings. They were built of stone with lime-plastered walls and slated roofs. In the late eighteenth and early nineteenth centuries, the Catholic Church grew more confident. The professional and merchant classes became prosperous at this time and contributed to church funds, thus more impressive churches were erected. These classical or Gothic-style churches were to reach extremely high standards after Catholic Emancipation in 1829, as can be seen in Cobh Cathedral, Co. Cork and St John's Cathedral

Four Courts
GPO

triangular pediment supported by classical columns

parapet wall

columns

classical (public)

classical ( domestic)

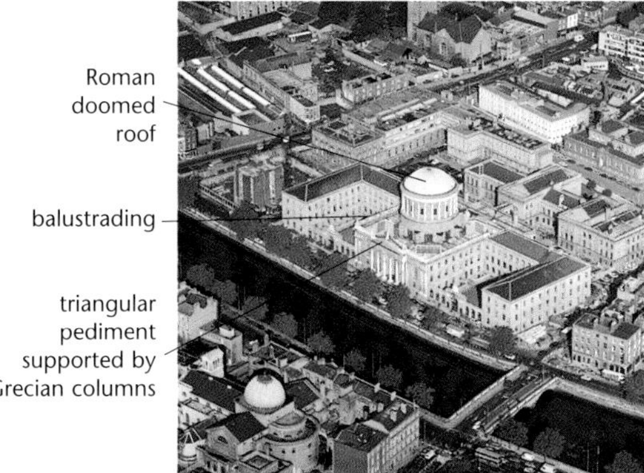


simple geometric style

railing around basement

classical doorway and overhead fanlight

terraced houses with slightly arched windows in classical style

shopfront in classical style in plaster or timber finish

Fig. 17.33
Neo-classical features

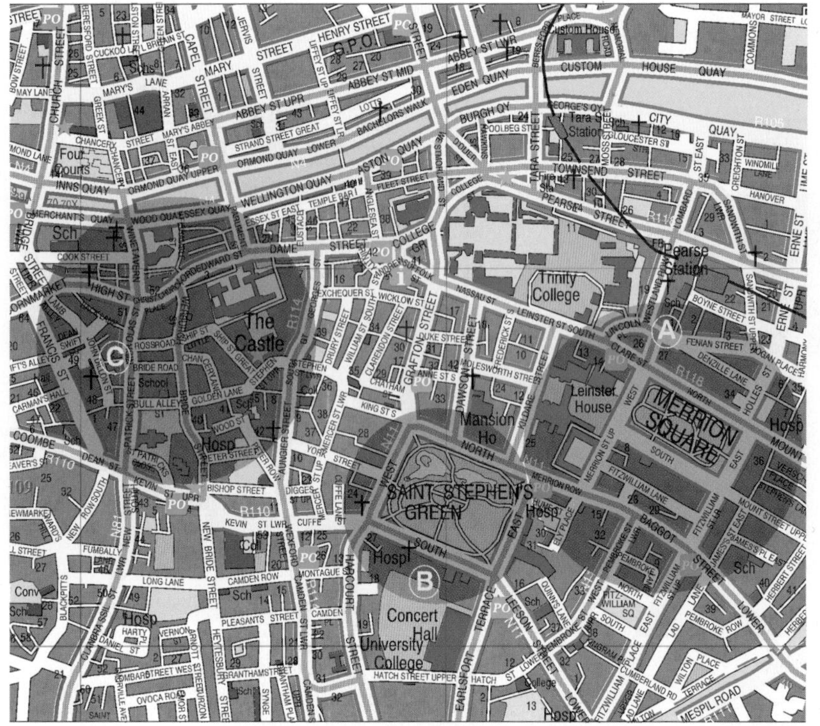


Fig. 17.34

in Limerick. Such churches are easily recognised by their high spires, their pointed windows with tracery patterns (as in the medieval abbeys), and their enormous size which dominates all other buildings in the neighbourhood. The building generally reflects the size and prosperity of the local community. In many Ulster towns, Gothic-style Church of Ireland churches command a dominant position in the town centres. In southern Ireland many of these churches are in a ruinous condition due to declining congregations, or have been sold to be reoccupied as homes or restaurants.

In or about 1845, Sir Robert Peel founded three university colleges. One was sited in Belfast (Queen's University), one in Cork (University College Cork) and another in Galway (University College Galway). These also display Gothic-style architecture. They were referred to as the 'Godless colleges' by Daniel O'Connell because they were interdenominational.

Fig. 17.35

Study the photograph of central Dublin. Then in class discuss the following:

1. Identify the following buildings and their era of construction: the Bank of Ireland, Trinity College (in this case the buildings represent different styles of architecture), Liberty Hall.
2. Parts of the city north of the river display characteristics of recent construction. From your knowledge of urban redevelopment and urban renewal, locate these buildings and suggest reasons why this particular area has undergone change.
3. Buildings and land use layout to the south of the river seem to portray a sense of prosperity when compared to those to the north of the river. Do you agree or disagree with this statement? Explain your choice.
4. Assume you are a developer who had purchased the triangular area in the centre of the photograph from the various owners and you wished to build an ultra-modern multi-storeyed hotel on this site.
   (a) Outline some of the concerns that urban planners might have concerning your proposal.
   (b) Outline three points to counteract these concerns.
   (c) Then, 'change coats' and give three conditions that you would place, as chief planning engineer for Dublin city, on a planning application similar to the above.

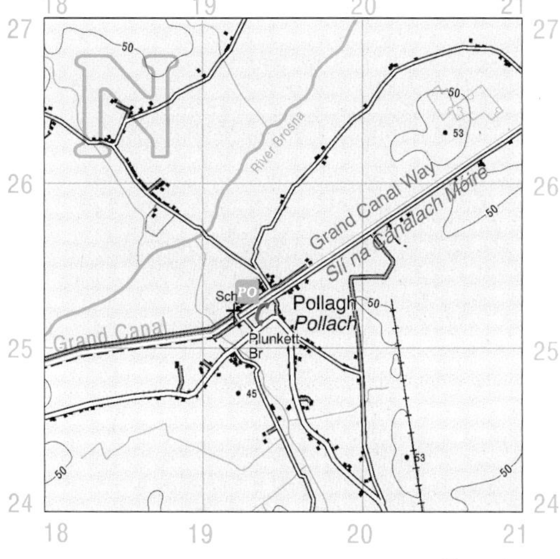


Fig. 17.36
The Grand Canal at Pollagh, Co. Offaly

## Canals — Eighteenth and Nineteenth Centuries

In 1715 a government act was passed to enable the main Irish rivers to be deepened and widened and a network of canals to be constructed. Many physical factors such as the following favoured such a network.

1. The **low-lying central plain** was crossed by important rivers.
2. **Low watersheds** existed between the rivers which enabled them to be easily connected to a canal system.
3. Many large towns with **established trade links** were already situated on the rivers.

Canals carried heavy and bulky materials such as timber, coal, wool, corn and flour. The important buildings associated with canals such as **warehouses**, **harbour workers' housing** and **hotels** were located in towns near to the canals. Dressed stone was widely used for many features such as doorways and windows, while quaysides were lined with stone. Many of these stone structures have stood the test of time and may still be seen in some Irish settlements.

Canal transport was considered more comfortable, more relaxed and more reliable than coaches at this time. Canals had the drawback, of course, that they serviced only a limited number of settlements.

Hotels such as Portobello House in Dublin which resembled the smaller country houses were built in towns to accommodate overnight travellers or short-term visitors.

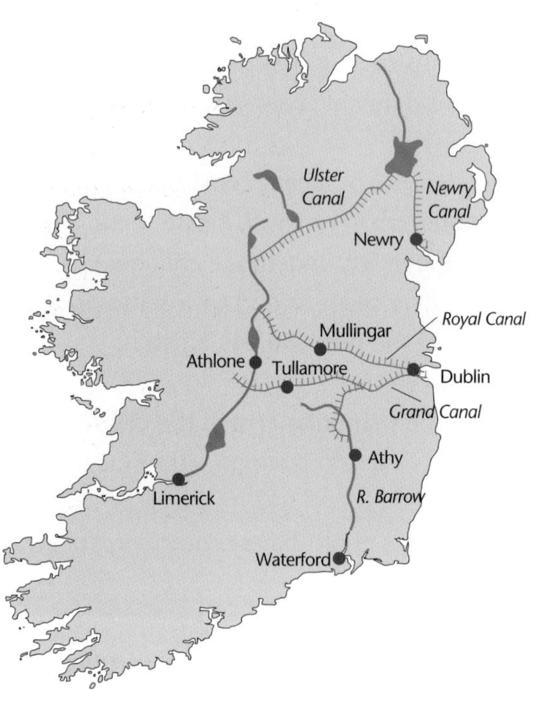


Fig. 17.37
Major canals of Ireland

However, no sooner were the canals constructed than the railways began to provide an alternative transport service. In any case, Ireland was not entirely suited to canal transport. This mode of transport is best for the carriage of heavy or bulky goods over long distances, and Ireland had no large quantities of either coal or iron ore. Canals were also unable to compete with the railway for the carriage of lighter goods over short distances. Ireland's main export outlet was Britain, but the River Shannon, which was the meeting point of the Grand and Royal Canals, opened to the west rather than to the east.

The belief that the canals would help drain the midland bogs and encourage industry and agriculture did not materialise.

At this time, the rapid fall in population (due to the Famine), the poverty of the people and the restricting effects of the land system upon farmers' ability to accumulate savings, provided for neither industrial development nor a market for industrial products. In addition, the cost of transporting goods to Britain and continental Europe, and the difficulty of establishing industry under a free trade system as existed at the time, exposed Ireland to European competition, which discouraged investors in an unpromising economy. Thus the canals were doomed to failure.

## Case Study: Monasterevin — A Canal Town

The word 'Monasterevin' means 'Mainistir Eimhín' — the monastery of Eimhín, which obviously suggests that the settlement has its origin in early Christian times. This canal town had a drawbridge which controlled traffic on the canal in the eighteenth century.

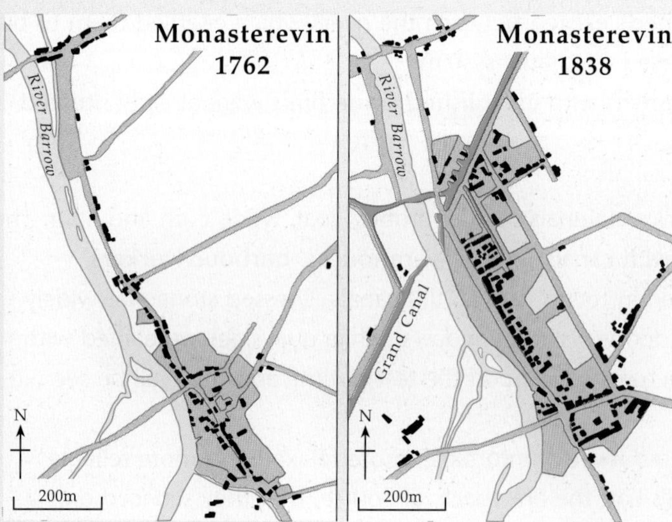


Fig. 17.38
Monasterevin, Co. Kildare was completely rebuilt between the 1760s and 1780s

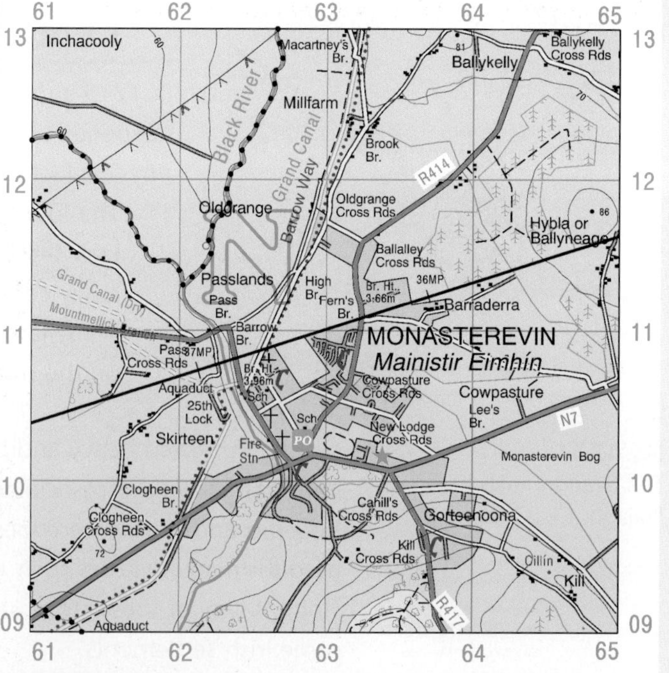


Fig. 17.39
Monasterevin, 1999

Fig. 17.40
Monasterevin, a canal town of the eighteenth century

## Home Activity

Carefully study the maps, Fig. 17.38, the Ordnance Survey map extract, Fig. 17.39 and the photograph, Fig. 17.40. Then explain the origin and subsequent development of Monasterevin using evidence from these figures to support your answer.

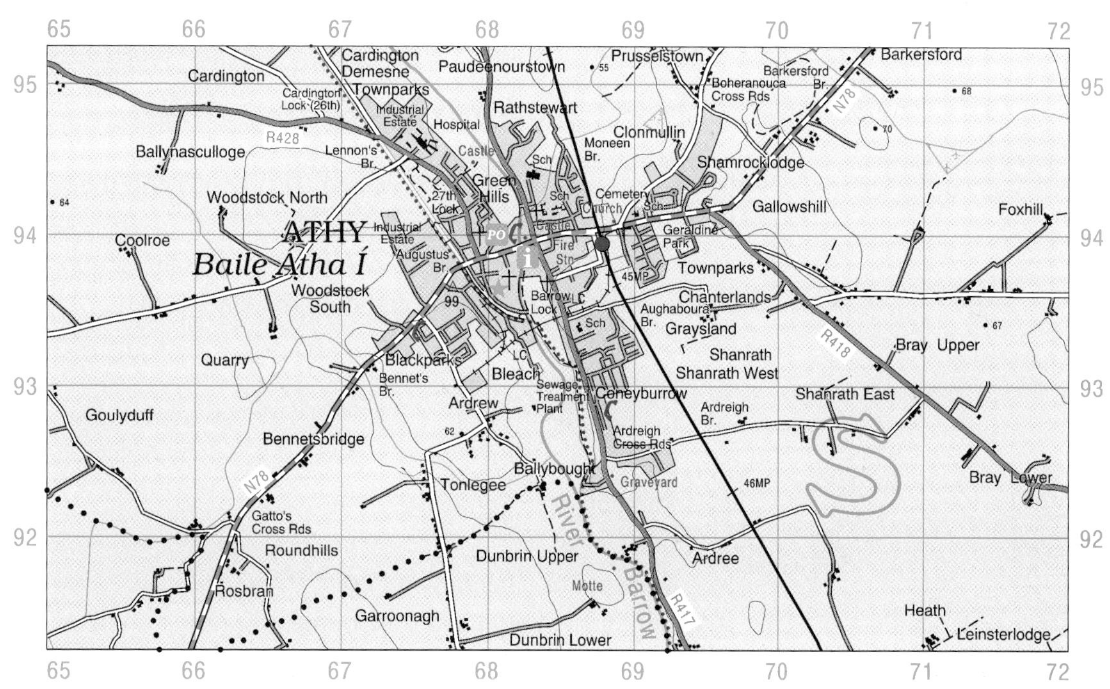


Fig. 17.41

Fig. 17.42
Athy in Co. Kildare

Study the Ordnance Survey map extract (Fig. 17.41) and the photograph (Fig. 17.42) of Athy in Co. Kildare and do the following:

1. Account for the development of Athy at its present location.
2. Draw a sketch map of Athy and on it mark and label
   - the River Barrow
   - Grand Canal
   - the bridges
   - the castle
   - warehouses
   - canal locks
   - five areas of different land use
3. Account for the origin and development of Athy from a study of its buildings, streets and waterways.

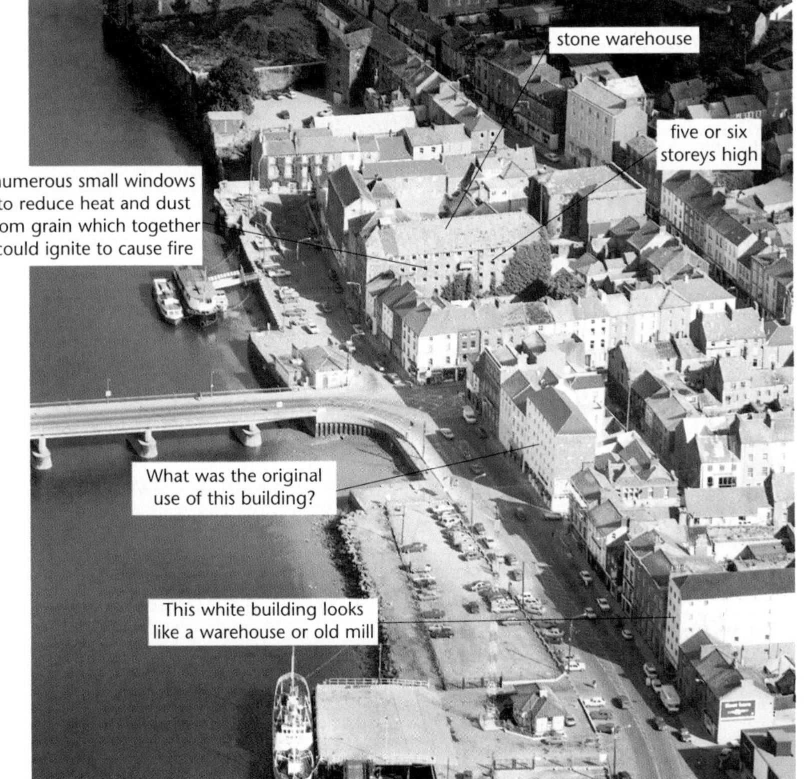
stone warehouse
five or six storeys high
numerous small windows to reduce heat and dust from grain which together could ignite to cause fire
What was the original use of this building?
This white building looks like a warehouse or old mill

Fig. 17.43
Typical warehouses or mills of the eighteenth and nineteenth centuries at New Ross, on the quayside. Why were these mills/warehouses located near to the waterfront?

## Industrial Growth

During the eighteenth century, urban areas were favoured for the building of factories. Industries in the south of Ireland were based mainly on agricultural produce such as grain for milling and wool for textiles. The milling industry continued to grow, especially in coastal ports as Britain became our most important export outlet, while smaller mills were often located in inland towns throughout the country. The demand for grain and wool continued up to the 1800s due to an increasing Irish population and a demand for grain in the fast-growing industrial cities of Britain. **In inland areas, many of these mills still held the waterside locations which had been so vital in earlier times** as water was needed to turn the mill-wheels. A sufficient flow of water was generated by building a weir or dam across a river in order to divert some of the flow into a **race** which rotated the mill-wheel (Fig. 17.53 on p. 364). Some of these dams are still visible today (see Carlow, Fig. 16.24, p. 331) near modern factories which are located on the sites of the older mills. Grain and other commodities, which were transported in barges on the canals, were stored in buildings or warehouses located near the canal quayside.

Some older mills in inland areas were powered by wind. These grain windmills were used up to the early 1800s. Some mills were powered by both water and wind in areas where river levels fell too low in summer. Once cheaper imported grain was available these inland and sometimes coastal windmills fell into disuse.

## Industrial Decline in the South of Ireland

There was a severe industrial decline in Ireland during the 1830s, largely due to the ending of the Napoleonic wars in Europe. The woollen and milling industries especially stagnated, and there was also a decline in the demand for foodstuffs. Many urban centres, especially the towns in the south of Ireland, were thrown into deep poverty and their economic activities were increasingly restricted to the provision of services for their rural hinterlands.

In addition, the Great Famine of the 1840s resulted in a rapid and continuous decline in rural populations. Thus, throughout the nineteenth century and part of the twentieth century, many towns either stagnated or declined. This is often reflected in their buildings which in some towns still stand dilapidated, or in their squares which lie empty of activity. Some of these old mills and factories have either been renovated or demolished entirely as a consequence of urban renewal and redevelopment. The railways, which will be discussed later, facilitated industrial growth in some centres. However, they could not generate growth and often served only to remove the isolation of remote towns and rural areas. Nevertheless, some industries still prospered during the nineteenth century. These included:

1. **the brewing industry** in such centres as Dublin, Midleton and many other towns
2. **shipbuilding** in Belfast
3. **linen manufacture** such as in the Lagan Valley
4. **local craft industries** which provided for the agricultural sector

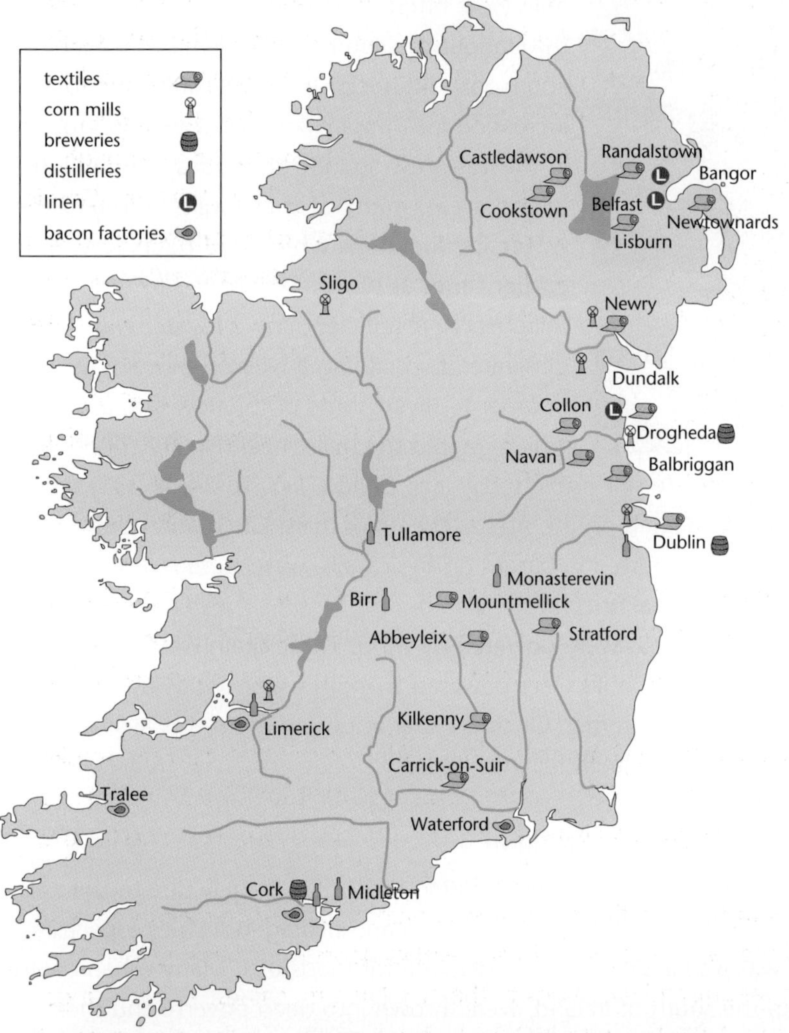


**Fig. 17.44**
Pre-Famine industry in Ireland

The new industrial centres were concentrated into well-defined areas in the cities, generally near the river or in sea ports near the harbour for the easy import and storage of raw materials such as coal and the export of products such as linen.

Large warehouses called **granaries** were also associated with the growing and storage of grain. These granaries were often similar in design to other industrial buildings such as mills. They were built of stone, four or five storeys high, with lines of small and slightly arched windows on each floor level. During the 1830s there were almost 2,000 mills in Ireland owned and run by landlords or merchants mainly for the export market. Today, many are in poor condition and may appear as drab and dark structures in aerial photographs. Seen on the ground, they are museums of another time both in their architecture and in their use.

Throughout the eighteenth and nineteenth centuries, the export of goods became centralised in the ports of the east and south which were nearest to Britain, Ireland's most important export outlet. These ports had wide and spacious quaysides for the easy movement of horse-drawn transport and the many dock workers that thronged the harbours loading and unloading both sailing ships and steamships.

Today, these spacious quaysides are less active in many smaller coastal towns where cranes have replaced manual labour. Also, while many western coastal towns retain their stone harbours, they have lost their trade to the eastern and southern ports. Large quaysides lie empty of ships, the mooring posts acting as reminders of busier times during the last century in places such as Limerick city and Galway city.

# Industrial Development in Villages and Towns in North-east Ulster

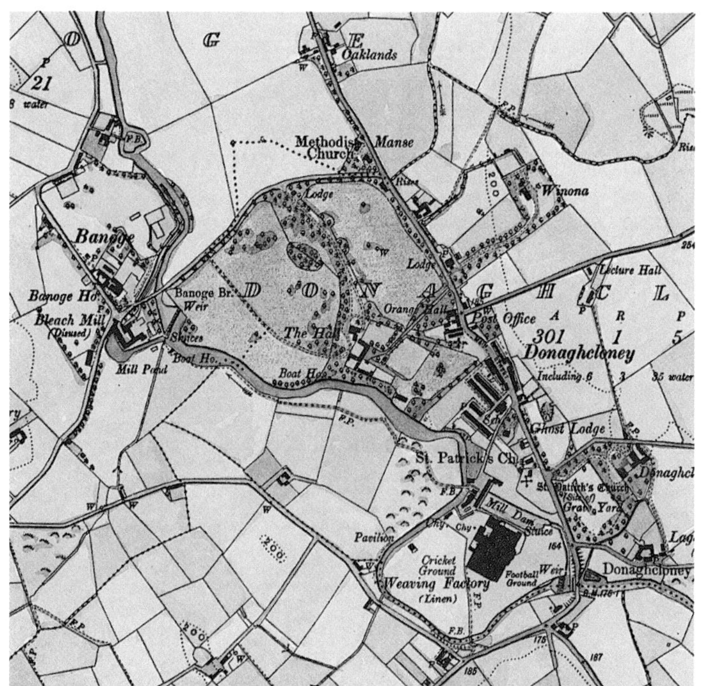


Fig. 17.45
Donaghcloney in Co. Down — a linen village. Associated with these settlements are the soccer and cricket grounds and Orange Halls.

## Ulster's Linen Towns and Villages

In contrast to the agriculturally supported towns of southern Ireland, the towns of north-east Ulster prospered in the nineteenth century. This was due to the close association of the linen industry and the Protestant faith. These earlier plantation villages and towns of the north-east adopted the linen industry of the eighteenth and nineteenth centuries once the milling and woollen industries declined, to secure their future prosperity. Many of these industrial villages are found in Armagh, Antrim, Down and Tyrone. Sion Mills with its neo-Gothic cottages, its black and white half-timbered buildings and chestnut trees is a typical mid-nineteenth-century linen village. This linen industry is continued today in a yellow-brick factory and old spinning mills in the village.

Indeed in this century the hard-working ethic of Ulster's rural population also ensured successful farming in the difficult drumlin landscape of its southern borders. In some cases this was accomplished by the additional investment in pig, poultry and mushroom industries on their farms.

In the 1820s the introduction of power spinning (the use of steam to work machines) to the Irish linen industry marked the end of the homespun (domestic) industry. The industry was concentrated in the north-east of Ireland, especially the Lagan Valley in Co. Down.

Associated with each linen factory was the nearby chimney stack that billowed smoke into the air in cities such as Belfast, which saw its population rise from 20,000 in 1801 to 100,000 by 1851. Many earlier grain mills were converted to suit power spinning for the linen industry.

New multi-storey buildings, often five or six storeys high, were constructed in order to carry great loads on each level. The stone walls and timber floors were massively constructed to carry the weight and vibrations of the new linen machines. The buildings were generally fairly narrow. Until the introduction of structural steel in the late nineteenth century, it was difficult to roof over wide spans.

The linen industry grew and became dependent on world markets. Thus it became sensitive to price fluctuations and changes in demand. Cotton, linen's old rival, was cheaper to make at this stage and could be finished in a variety of ways, as can modern fabrics. So Ireland's linen industry declined.

Today, while many industrial buildings are still in good condition, some are underused or empty, while others have been renovated and adapted to other uses such as community centres and commercial buildings (see Fig. 17.43, p. 357).

### Push-pull Factors

The availability of employment in industrial cities such as Belfast, along with the difficult working conditions in the surrounding landscape, caused many people to flock to the towns for work. Workers' houses, laid out in lines back to back and grossly overcrowded, were found close to linen factories.

## Railways — Nineteenth- and Twentieth-century Developments

### The Development and Decline of the Railway

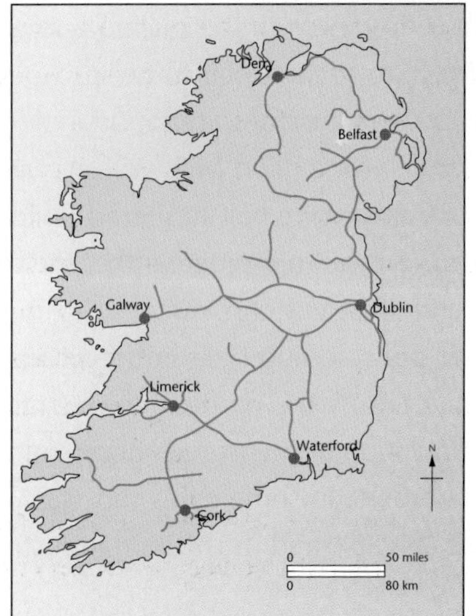


Fig. 17.46
Railways in 1860

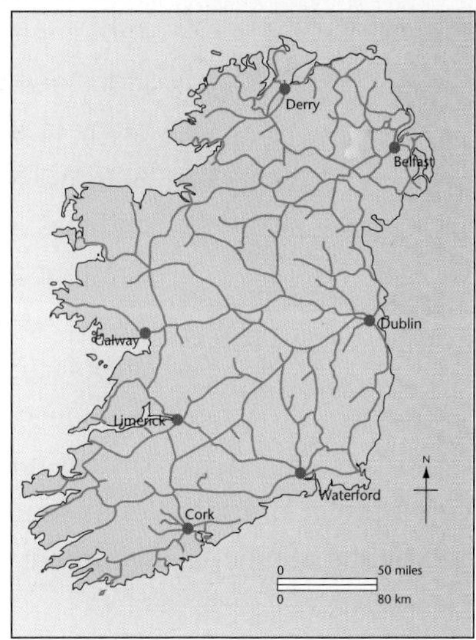


Fig. 17.47
Railways in 1923

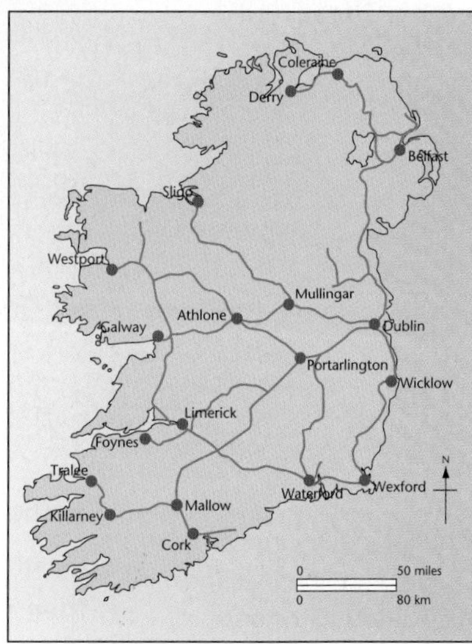


Fig. 17.48
Railways in 1983

Fig. 17.49
Bangor, Co. Down. A terrace of houses, with bay windows and plaster details, which is typical of many seaside towns

The development of the railway greatly influenced urban development in Ireland during the late nineteenth and early twentieth centuries. The first railway in the country was laid in 1834 and by 1912 a total of 5,500 kilometres of track had been laid. This new form of transport became an overnight success as it was cheaper, faster and more comfortable than either canal barges or road coaches. The British government encouraged investment in the railways and more than twenty companies came into existence. The most important of these were:

1. the great Southern and Western Railways, connecting Dublin, Cork and Limerick
2. the Midland Great Western Railway, connecting Dublin and Galway
3. the South Eastern Railway, connecting Dublin and Wexford
4. the Great Northern, connecting Belfast, Drogheda and Dublin

Towns which were fortunate enough to be located on railway lines grew. Other, less fortunate settlements declined, including canal towns at Daingean in Co. Offaly, Killaloe in Co. Clare (Fig. 17.52, p. 363) and Graiguenamanagh in Co. Kilkenny. The arrival of the

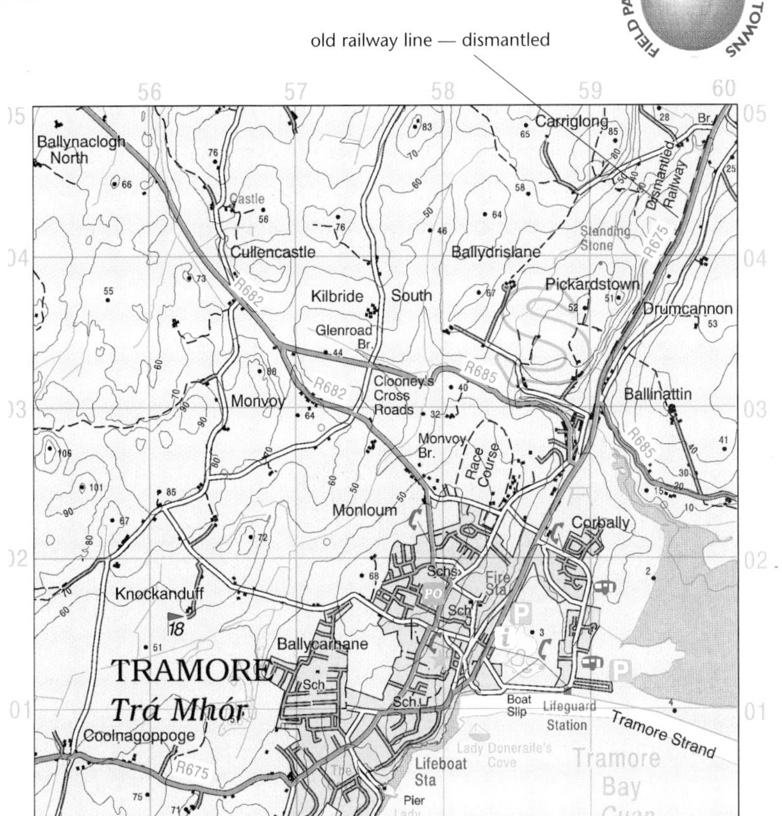


Fig. 17.50
Tramore, Co. Waterford

railway in small Irish towns added to a settlement's prosperity and increased its nodality for such events as fairs, which were held annually in market towns throughout the country. Railways were used to carry cattle from market centres in such areas as the Golden Vale to the fattening lands of Co. Meath, to the factories for processing, or to the ports for exporting, thus boosting Irish exports. Nearly all the new lines that were built in the late 1800s were intended to improve access to the ports. For years the local railway station was the hub of activity for reasons which include the following:

1. **social** — for holidays abroad, to the seaside or local town
2. **emigrants** from rural Ireland flocked to ports such as Cobh (Queenstown) for passage to America, escaping from the famine and poverty of the Irish countryside, especially in the west
3. **commercial** — for transport of goods. Cargoes which, until then, took three days for delivery by canal to the south and west could now be received by steam engine in ten hours.

The railways took thousands of tourists, from both the upper and middle classes, out of urban centres such as Dublin and Limerick to the west and south of Ireland. This in turn encouraged development of seaside resorts such as Kilkee in Co. Clare and Youghal in Co. Cork (Fig. 17.15, p. 340).

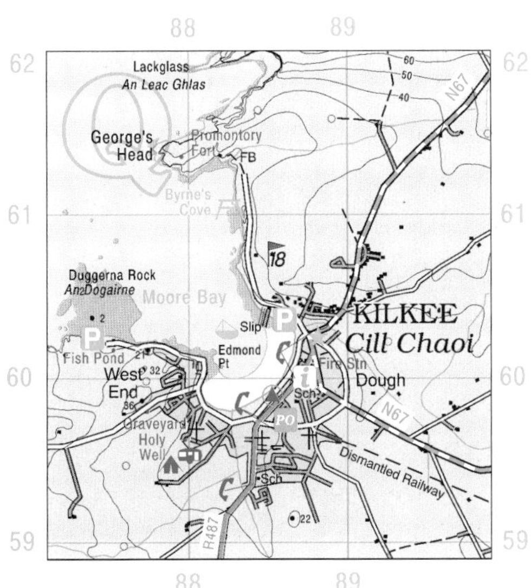


Fig. 17.51 (a)

Study the photograph of Kilkee in Co. Clare, Fig. 17.51 (b) and the Ordnance Survey map extract, Fig. 17.51 (a). Then answer the following:

1. In which direction was the camera pointing when the photograph was taken?
2. (a) What evidence suggests that the railway played an active part in the development of the town?
   (b) What does this suggest about the development of Kilkee?
3. Tourism is an important industry for the survival of Kilkee as a major seaside resort. As a consequence of this the human landscape needs to be carefully managed to protect the natural and cultural landscape. Discuss.
4. Identify three landforms of coastal erosion and one landform of coastal deposition. Then with the aid of labelled diagrams describe their formation.

The layout of such seaside resorts was often determined by the shape of the coastline. Straight coastal beaches were backed by a linear pattern of tall, terraced houses which often availed of a high vantage point for a view out to sea, or by low Italianate villas along the seafront. Crescentic or bayhead beaches had dwellings which followed the shape of the bay, as at Kilkee (Fig. 17.51).

**Fig. 17.51 (b)**
Kilkee in Co. Clare

Fig. 17.52

Study the photograph of Killaloe, in Co. Clare, Fig. 17.52. Then answer the following:

1. Evidence on the photograph shows how this settlement has prospered as a consequence of its location on the Shannon, both now and in the past. Discuss.
2. River edges offer many advantages for wildlife and for people. Discuss this statement with reference to the photograph.
3. Part of the deciduous woodland forms a circular cluster in the north-west corner of the photograph. Suggest what this cluster represents and give two fully developed reasons why it occurs at this site. (See Shannon basin video by ESB.)
4. The river has had an impact on the development of this settlement. Discuss this statement using the following headings: canal development, recreation, bridging point, scenic value.

Fig. 17.53
Bagenalstown in
Co. Carlow

Fig. 17.54

Study the photograph of Bagenalstown in Co. Carlow and the Ordnance Survey map extract, Figs. 17.53 and 17.54. Then answer the following:

1. In which direction is the river flowing? Are you certain?
2. (a) The land between the curved line of trees on the north side and the river appears to be summer grazing for farm animals. Do you agree?
   (b) If so, what kind of animals?
   (c) What time of year was the photograph taken? Explain.
3. A ruin stands between both waterways in the photograph. Suggest its original use. Explain your choice. Did you purchase your magnifying glass? If so, use it now.
4. A stone building is located 'on the water' in the north-western (right background) portion of the photograph. What was the original purpose of this building? Explain fully using evidence from the photograph.
5. The town appears to lack a focus. Discuss this statement with reference to the photograph.
6. There is a building with classical characteristics in the southern (left middle) portion of the photograph. Suggest a purpose for this structure.

Fig. 17.55
Bangor in Co. Down. Bangor experienced two eras of urban development. Of St Comgall's Abbey, founded in AD 558 and once the premier seat of Christian academic education in all of Europe, there is not a stone left. The second era came with the arrival of the Belfast, Holywood and Bangor Railway Company, which gave free first-class ten-year season tickets to new holders.

Study the photograph of Bangor in Co. Down, Fig. 17.55, and the sketch of seaside terraced houses, p. 360. Then answer the following:

1. Examine the terrace of houses in the centre of the photograph (just above the marina) and in the right foreground (facing the sea). What do these houses tell you about
   (a) the houses themselves
   (b) the town of Bangor?
2. What evidence indicates that this town is part of a prosperous region? Explain fully.
3. Would you think that this town experienced some effects of the industrial development, associated with Northern Ireland, of the nineteenth and twentieth centuries? Use evidence from the photograph and the headings: street plan, suburban development, house size.
4. Explain the purpose of the large, flat-topped building in the centre foreground of the photograph. Does this design seem to add to or take away from the character of the settlement?
5. Draw a sketch map of the photograph and on it mark and label
   - the coastline
   - five areas of different land use
   - the street pattern of the town
6. Describe and explain the structure and layout of the port area.

Fig. 17.56 Athlone in Co. Westmeath

Carefully study the photograph of Athlone in Co. Westmeath, Fig. 17.56. Then answer the following:

1. Draw a sketch map of the area shown in the photograph and on it mark and label
   - the river
   - the streets
   - the castle
   - a new shopping centre
   - three churches
   - an area of urban renewal
2. Account for the origin and subsequent development of the town from a study of its buildings, streets and waterway.
3. Which part of the town is oldest? Support your answer with evidence from the photograph.
4. The construction of the new shopping centre located in the right background above the river could have negative implications for that part of the town in the foreground of the photograph. Discuss.
5. Rivers can have both a positive impact and a negative impact on settlements sited along their banks. Discuss this statement, using evidence from the photograph.
6. What purpose does the weir across the river serve?

Fig. 17.57
Galway city



Study the photograph of Galway city, Fig. 17.57. Then answer the following:

1. Water has played and continues to play a key role in the life of Galway city. Discuss this statement with reference to the photograph.
2. With Galway, as with many other successful Irish towns and cities, change has almost always played a major role in its success. With reference to the photograph, discuss this statement.
3. Draw a sketch map, and on it mark and label
   - the coastline
   - five areas of different land use
4. The roadway in the foreground of the photograph is a recent addition to the Galway area. Identify its main purpose and discuss some effects, at least one positive and one negative, which this roadway may have on the city.
5. A number of landmarks which are named on the photograph have played an important role in the history and development of the city. With reference to the photograph and your knowledge of the Galway region, discuss this statement.

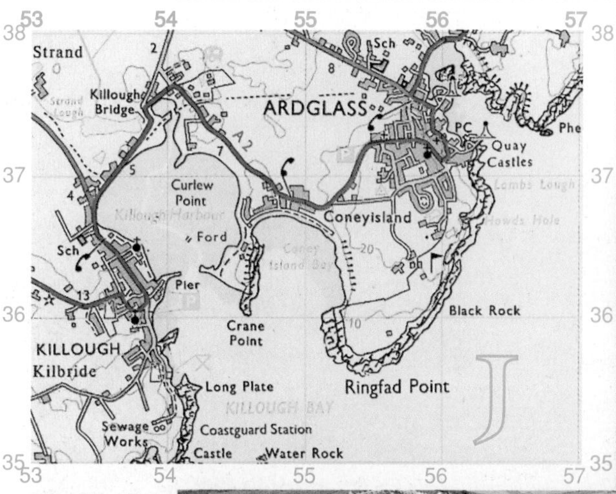

Fig. 17.58

**Fig. 17.59**
Ardglass, the main port of Lecale in the medieval period, with defended warehouses and tower houses, revived in the nineteenth century to become an important fishing port

Study the photograph and the Ordnance Survey map extract of Northern Ireland, Figs. 17.58 and 17.59. Then do the following:

1. Identify the settlement in the photograph.
2. What evidence exists to show that this settlement has had a varied function over a long period of time? Explain fully.
3. In which direction was the camera pointing when the photograph was taken?
4. Carefully examine the varied slopes of the land surrounding the bay. What evidence, if any, exists to prove that sea level has changed in this area?
5. What type of farming is carried on in this area? Use evidence from the photograph to support your answer.

**Fig. 17.60** ▶

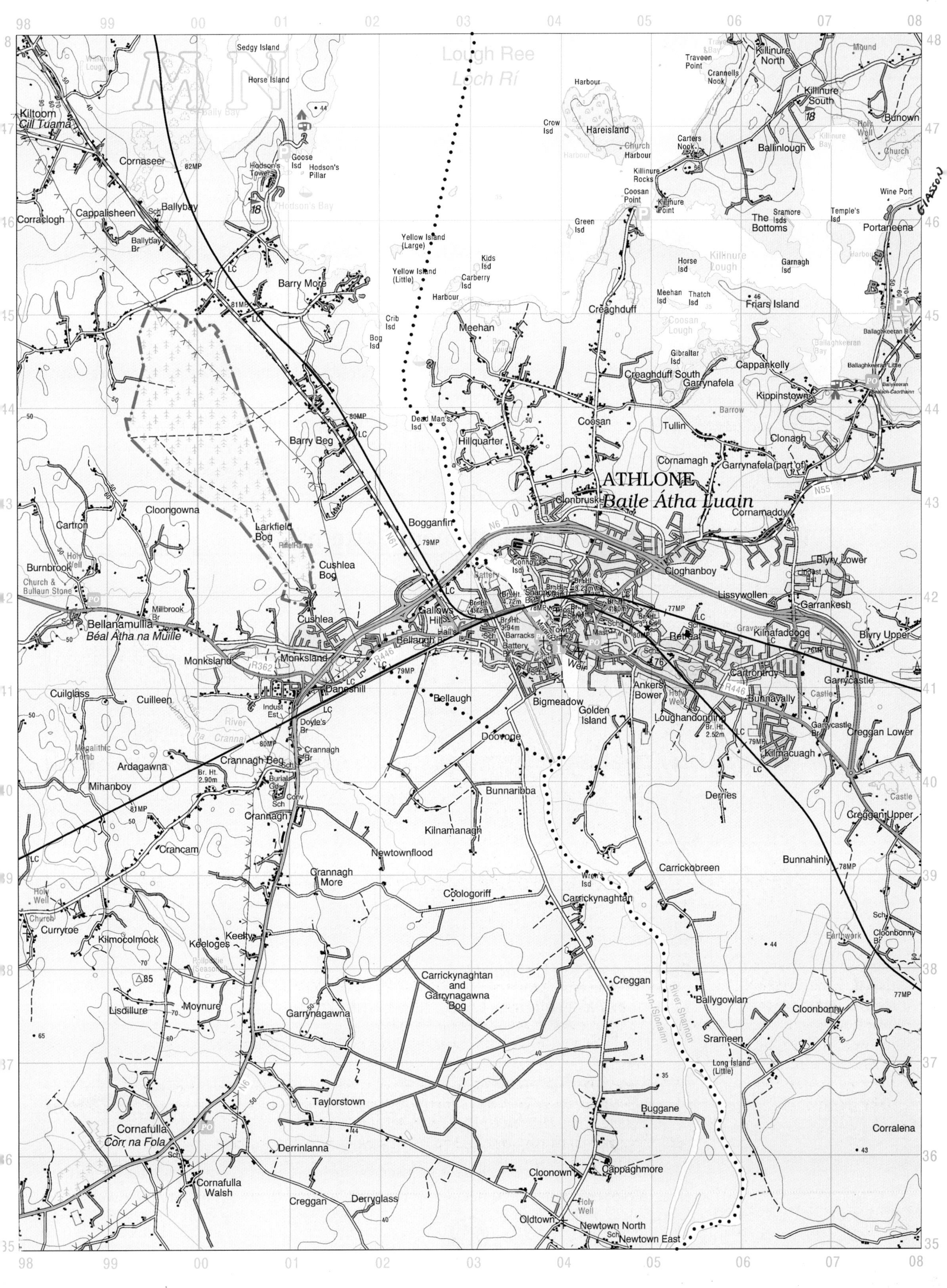
Lough Ree
Loch Rí
Sedgy Island
Horse Island
Kiltoom
Cill Tuama
Cornaseer
Goose Isd
Hodson's Tower
Hodson's Pillar
Hodson's Bay
Corraclogh
Cappalisheen
Ballybay
Ballybay Br
Barry More
Barry Beg
Yellow Island (Large)
Yellow Island (Little)
Kids Isd
Carberry Isd
Harbour
Crib Isd
Bog Isd
Dead Man's Isd
Meehan
Hillquarter
Crow Isd
Hareisland
Church Harbour
Killinure Rocks
Coosan Point
Killinure Point
Green Isd
Carters Nook
Traveen Point
Crannells Nook
Killinure North
Killinure South
Ballinlough
Bunown
Mound
Wine Port
Temple's Isd
Portaneena
Sramore Isds
The Bottoms
Killinure Lough
Horse Isd
Garnagh Isd
Meehan Isd
Thatch Isd
Friars Island
Creaghduff
Coosan Lough
Gibraltar Isd
Creaghduff South
Garrynafela
Cappankelly
Kippinstown
Ballaghkeeran Bay
Ballaghkeeran Little
Coosan
Tullin
Barrow
Clonagh
Cornamagh
Garrynafela (part of)
ATHLONE
Baile Átha Luain
Clonbrusk
Cornamaddy
N55
N6
N61
Bogganfin
Cloongowna
Larkfield Bog
Rifle Range
Cushlea Bog
Cartron
Burnbrook
Church & Bullaun Stone
Millbrook Br
Bellanamullia
Béal Átha na Muille
Cushlea
Monksland
R362
R446
Gallows Hill
Bellaugh Br
Battery
Barracks
Connolly Isd
Cloghanboy
Lissywollen
Blyry Lower
Garrankesh
Kilnafaddoge
Blyry Upper
Retreat
Weir
Cartrontroy
Garrycastle
Bunnavally
Ankers Bower
Golden Island
Bigmeadow
Loughandonning
Kilmacuagh
Creggan Lower
Creggan Upper
Castle
Cuilglass
Cuilleen
Daneshill
Indust Est
Doyle's Br
Bellaugh
Doovoge
River Cross
Crannagh Br
Crannagh Beg
Megalithic Tomb
Ardagawna
Mihanboy
Burial Gd
Conv
Crannagh
Bunnaribba
Kilnamanagh
Derries
Crancam
Newtownflood
Carrickobreen
Bunnahinly
Grannagh More
Coologoriff
Wren's Isd
Carrickynaghtan
Holy Well
Church
Curryroe
Kilmocolmock
Keeloges
Keelty
Earthwork
Cloonbonny Br
Cloonbonny
Carrickynaghtan and Garrynagawna Bog
Creggan
Ballygowlan
River Shannon
An tSionainn
Lisdillure
Moynure
Garrynagawna
Srameen
Long Island (Little)
Taylorstown
Buggane
Corralena
Cornafulla
Corr na Fola
Derrinlanna
Cornafulla Walsh
Cloonown
Cappaghmore
Creggan
Derryglass
Oldtown
Newtown North
Newtown East
GLASSON

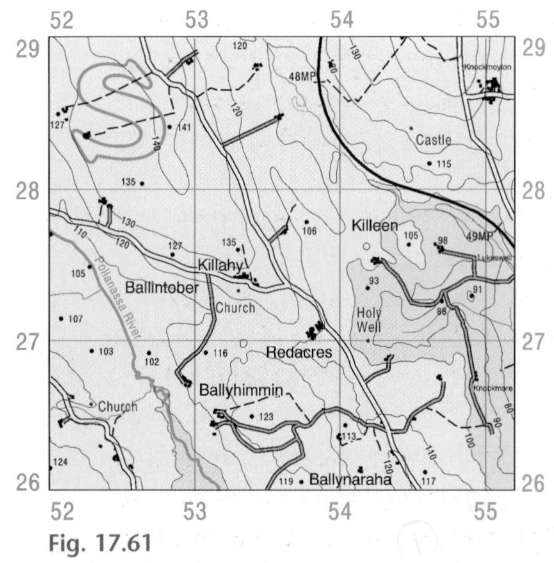


**Fig. 17.61**
Nucleated pattern (cluster)

Carefully study the Ordnance Survey map extract of the Athlone region, Fig. 17.60. Then do the following:

1. Account for the development of Athlone town at this location. In your answer, one developed point should refer to its linear shape (the irregular contours and the map extract of Clonmacnoise, Fig. 16.10, p. 319, should be of some assistance).
2. This area has many attractions for tourists. Describe the main attractions with reference to specific locations on the map extract.
3. Account for the pattern in the distribution of settlement/population displayed on the map extract. See 'Present-day Settlement Patterns' below.
4. With the aid of a sketch map, describe how the physical landscape has affected communications in this area.
5. Examine in detail how the map provides evidence of the evolution of Athlone's transport infrastructure over time.

## Present-day Settlement Patterns

### Rural Settlement Patterns

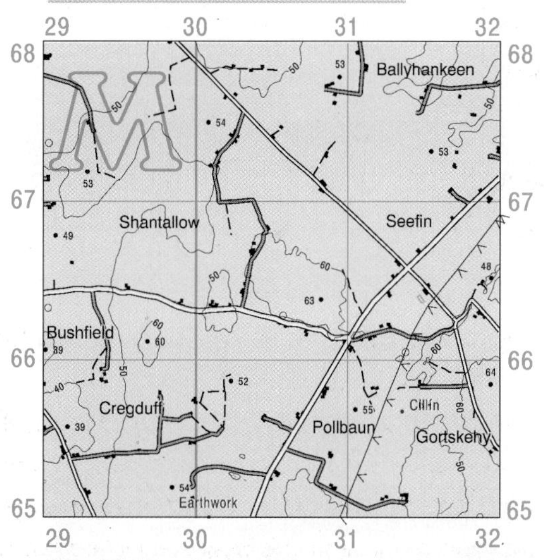


**Fig. 17.62**
Dispersed pattern (scattered)

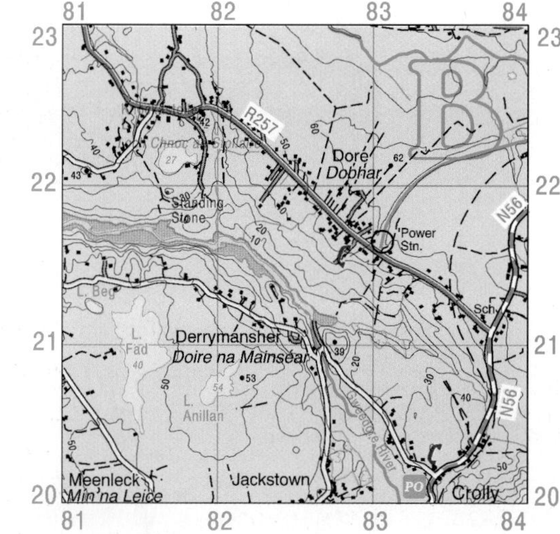


**Fig. 17.63**
Linear pattern (in a line)

## Patterns of Settlement

### 1. Linear Pattern (Figs. 17.63, 17.68 and 17.70)

Fig. 17.64

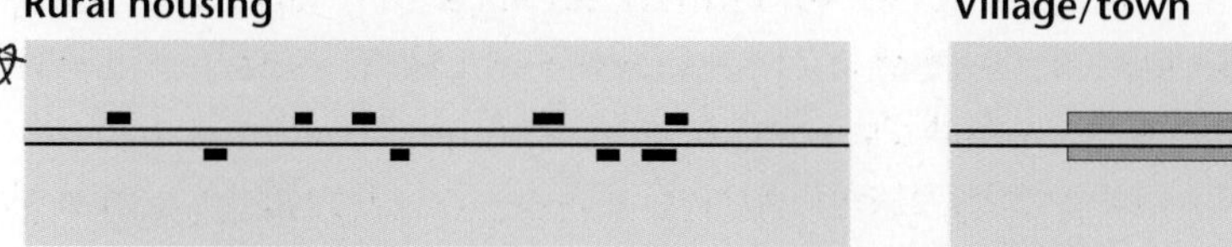

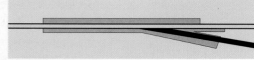

**When buildings occur in a line, they are said to form a linear pattern.**

Modern linear settlement is characterised by bungalows and two-storey houses. In rural areas, a plot of roughly 0.20 hectares is required in order to accommodate the dwelling

Follow this answer layout for each pattern in examination.

1. Name of pattern
2. Diagram of pattern or patterns from map extract
3. Definition of pattern
4. Description of pattern
5. **Refer to the individual pattern displayed in the photograph or Ordnance Survey map.**

and sewerage system before planning permission is granted. This land is generally desired along roadways because services such as telephone cables and water supply pipes are limited to roadside margins. Also, land is usually available only along roadways because farmers generally sell those narrow strips of land which will fetch higher retail prices.

The earliest linear pattern of dwelling houses existing today originated at the end of the nineteenth and beginning of the twentieth centuries. Up until then, the government-supported reorganisation of farm ownership had benefited only tenant farmers. Agricultural labourers were now to be granted half-acre sites along routeways with local government houses (council cottages). These were most common in large farm and tillage areas.

Where towns or villages develop on an important routeway, demand for frontage on such a routeway is high. Shops, garages and filling stations, pubs and so on need passing traffic to fulfil their **threshold needs**. Buildings therefore tend to form a line on both sides of the routeway. This pattern, which originated in western Europe, is now also found in North America and in the former colonies of some European countries.

## 2. Dispersed Housing (Scattered) (Figs. 17.62 and 17.65)

Fig. 17.65
Dispersed housing

**Dispersed housing**

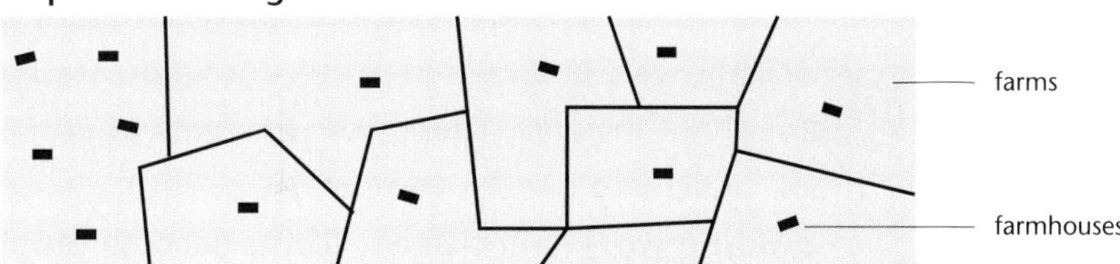


Follow this answer layout for each pattern in examination.

1. Name of pattern
2. Diagram of pattern or patterns from map extract
3. Definition of pattern
4. Description of pattern
5. **Refer to the individual pattern displayed in the photograph or Ordnance Survey map.**

When numerous buildings are dotted over an area such as on a fertile plain, they form a **dispersed pattern**.

By the early eighteenth century, the Gaelic and feudal land systems had started to disappear. Commonage was enclosed with fences and hedges, and the land was redistributed among the original strip-holders. Farms were arranged by squaring or striping and each tenant built a house on his newly enclosed farm. Each individual farmer now held a farm and farmhouse of his own rather than in common with others. A **dispersed** or **scattered rural settlement pattern** developed which persists to the present day. Such patterns are also found in Britain and in Brittany in north-west France.

Throughout Ireland, farmhouses are widely separated from each other. Some are sited at the end of long passageways far from routeways, while others have roadside sites. An overall dispersed pattern emerges.

The density of this distribution may vary from region to region. For example, in areas where farms are large, such as in the eastern part of Ireland, houses may be widely scattered. In areas where farms are small, as in the western part of Ireland, houses may be more closely spaced.

## 3. Nucleated Pattern (Figs. 17.61, 17.67 and 17.69)

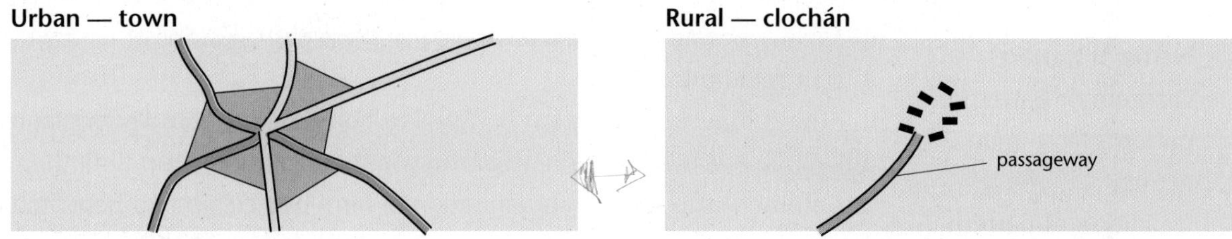


Fig. 17.66

Follow this answer layout for each pattern in examination.

1. Name of pattern
2. Diagram of pattern or patterns from map extract
3. Definition of pattern
4. Description of pattern
5. **Refer to the individual pattern displayed in the photograph or Ordnance Survey map.**

Nucleated settlements are those in which the buildings are grouped together. In rural areas, especially in parts of the west of Ireland, unplanned clusters of farmhouses provide examples of nucleated settlements. They are remnants of the **rundale** farming system which was practised in the **Gaelic parts** of the country (see the photograph of a clochán on Tory Island, Fig. 17.2, p. 332 and Fig. 17.3, p. 333).

Sometimes buildings are grouped together, such as at the foot of a hill or on a dry point in a marshy area, or at a wet point (near a spring) in an area of porous rock such as in the Burren in Co. Clare.

In places such as the 'Ormond Lands' of south Kilkenny and Waterford, tenant dwellings were arranged in clusters or 'farm villages' which were evenly spaced throughout the area. These were one type of post-feudal choice by the landlord of the rural settlement pattern (Fig. 17.61, p. 370).

Most towns are nucleated. Towns become the foci (singular: focus) of routes. Land at such a focus is in great demand for business, industry and housing. Thus the land at the junction of these routes, along these routes and in the sectors between these routes is built on to provide for these demands. Buildings are grouped together and so they are nucleated.

**Class Activity**

Figs. 17.67 to 17.70 display either nucleated or linear settlement patterns (shapes). For each settlement:

1. Describe its pattern (shape).
2. Suggest some reason/influence(s) why its settlement pattern (shape) developed.

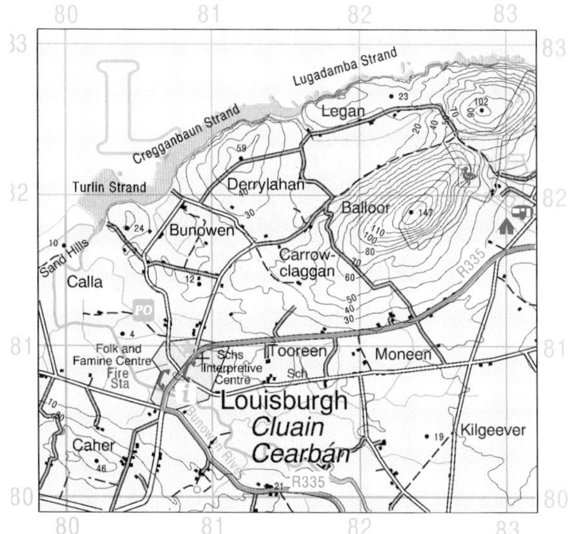


Fig. 17.67
Nucleated village

Fig. 17.68
Linear village

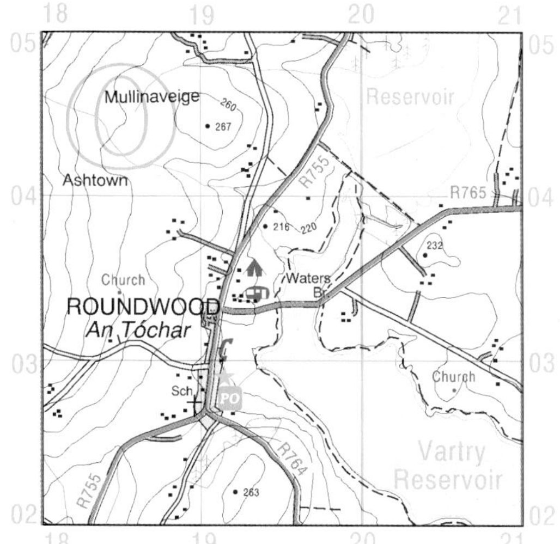


Fig. 17.69
Nucleated town

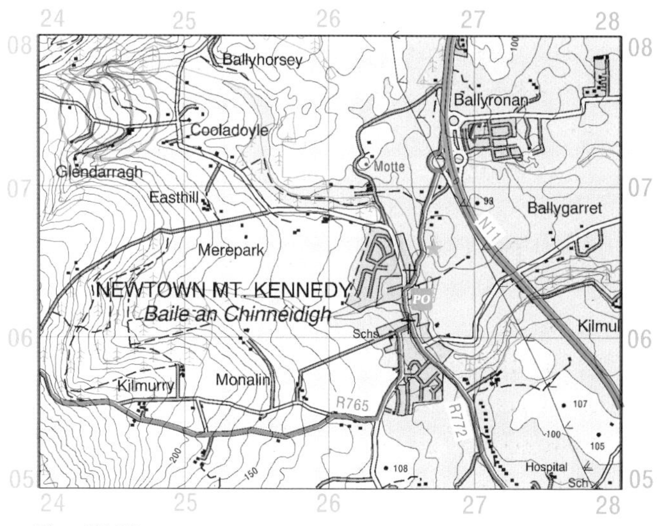


Fig. 17.70
Linear town

## 4. Absence Pattern (Figs. 17.71, 17.72 and 17.73)

Settlement is absent from elevated areas over 180 metres (600 feet approximately). Elevated areas are generally wet and exposed to strong winds, especially in winter. Temperatures are cooler than in lowland areas, so crop growth and crop variety are limited where soils are thin, acidic and unproductive. The factors which favour settlement are therefore absent and people avoid such areas as much as possible.

However, south-facing mountain slopes are warmer than north-facing slopes. A higher density of settlement may be found in these sunnier locations, especially on valley slopes in upland areas.

Settlement is also regularly absent from flat, low-lying areas.

1. Individual houses, villages and towns avoid the flood plains of rivers in their middle and lower courses. Such low-lying areas are prone to flooding during periods of heavy rainfall. At such times, rivers are no longer confined to their channels and they spread

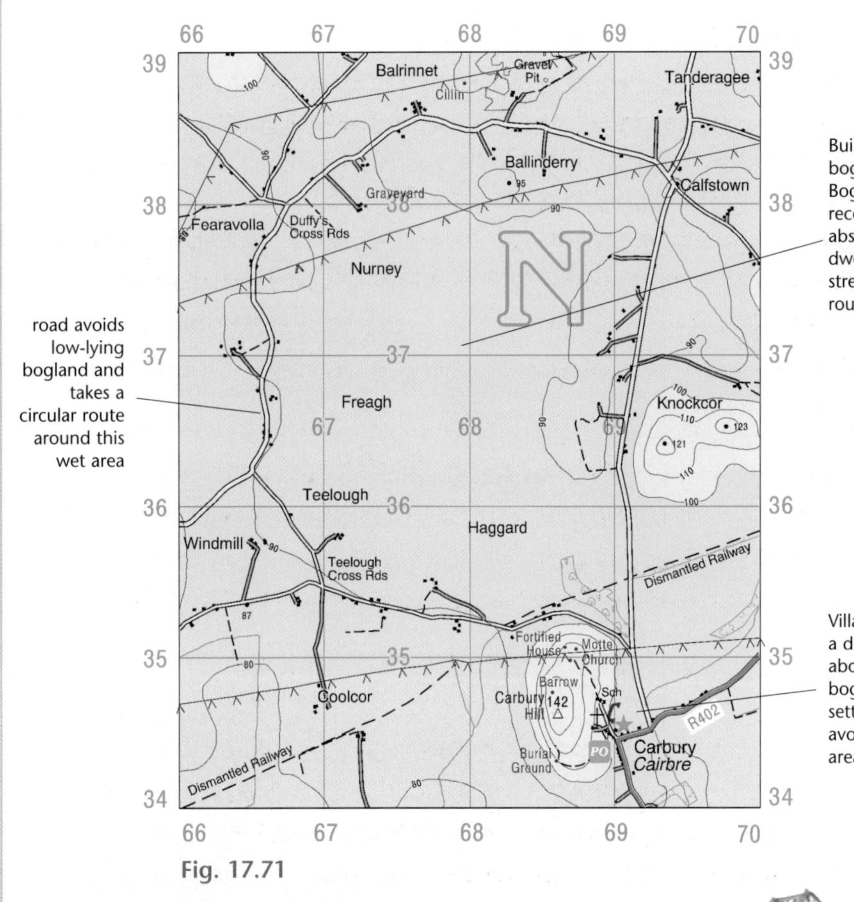


Fig. 17.71

across their flood plains, causing widespread flooding (Fig. 17.72). Where settlement does exist in such areas, it is confined to dry points on elevated patches of ground above the flood plain.

Routeways also avoid both steep, elevated areas and low-lying areas and those that are prone to flooding. Route construction and maintenance in such areas is expensive.

2. Raised bog areas in low-lying parts of the central plain in Ireland are also avoided by settlement. Such soft surfaces are not suited to road or building construction. These areas are recognisable on Ordnance Survey maps by the absence of streams and rivers and the absence of contours (see Figs. 17.71 and 17.73).

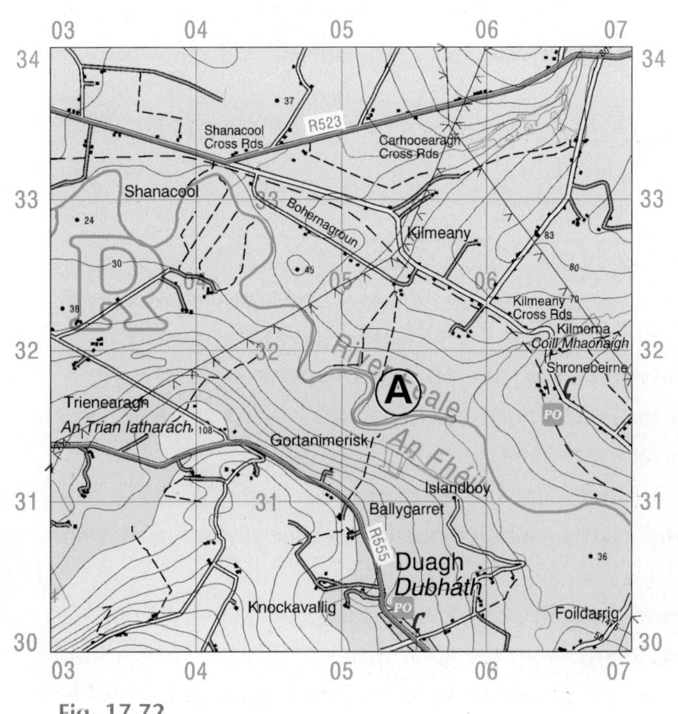


Fig. 17.72

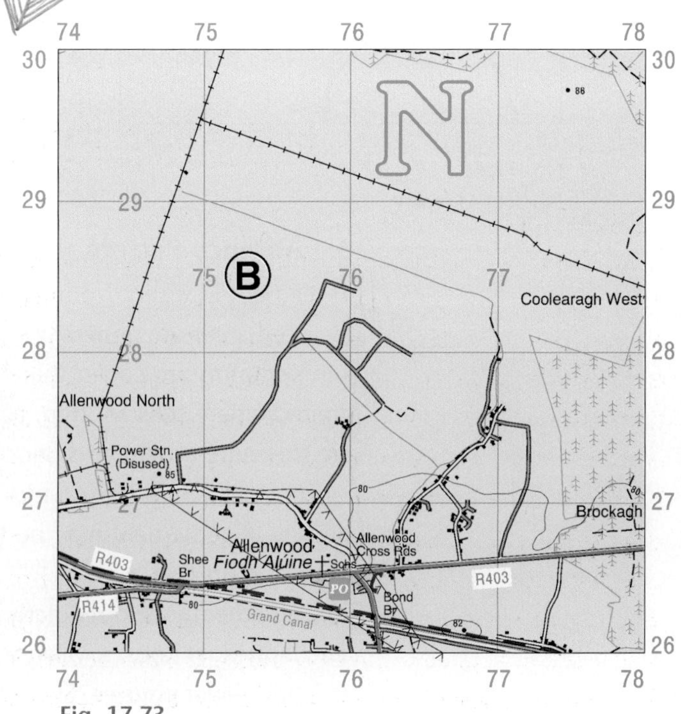


Fig. 17.73

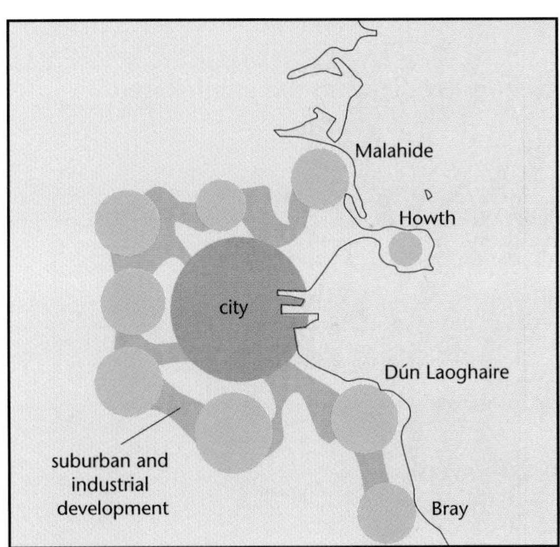


Fig. 17.74
Dormitory/satellite towns expand on the outskirts of cities. Such growth reduces agricultural land (green belts) between the towns and the city.

### 5. Clustered Pattern (Fig. 17.74)

Towns which develop around a larger settlement form a **cluster of nucleated settlements**. Malahide, Blanchardstown, Tallaght, Chapelizod, Howth, Dún Laoghaire and Bray, with Dublin at their centre, form a clustered pattern.

Suburbs grow around the main city and around the satellite towns. This growth reduces green belt areas when land use rezoning occurs, such as in Dublin.

**Home Activity**

Examine the Ordnance Survey map extracts, Figs. 17.72 and 17.73. Then suggest reasons why settlement has avoided locations A and B as shown.

Examine the Ordnance Survey map of Clonmacnoise, Fig. 16.10, p. 319. Then do the following:

1. Describe the patterns in the distribution of settlement displayed on the map (see 'Present-day Settlement Patterns', p. 370). Use the following headings as an aid in the layout of your answer: name pattern, locate pattern, sketch of pattern, definition of pattern, description of pattern, **refer to the particular pattern displayed on the map**.
2. (a) What type of power station is located at Shannonbridge? Be careful in your reply.
   (b) Give two well-developed reasons why this type of power station was built at this location.

## Patterns of Small Farms in the Twentieth Century

### Case Study: Golden in Co. Tipperary

Towards the end of the nineteenth century and the beginning of the twentieth century the wealthy large estates and their big houses fell into decline. The land acts and the political unrest of the time led many of the gentry to abandon these properties. The Land Commission, a government body in the new Irish Free State, purchased these estates and redistributed their lands to landless families who had supported the fight for Irish independence and to families from the west to relieve congestion of the countryside. The foundation of the new small family farm pattern and a population rooted in the land was now laid — the basis of Eamon de Valera's 'frugal comfort' policy. This new **policy of land ownership rather than** the provision of **economic units** led to the failure of these

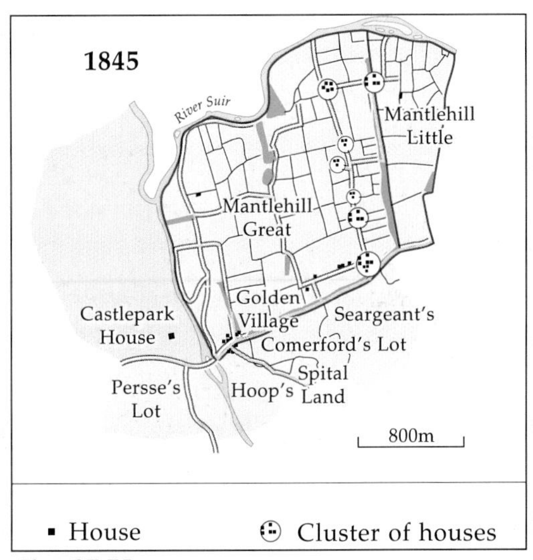


Figs 17.75

small farms to support their families. Mass emigration from the countryside resulted, leaving the uninhabited farmhouses to crumble on the overgrown passages through the fields.

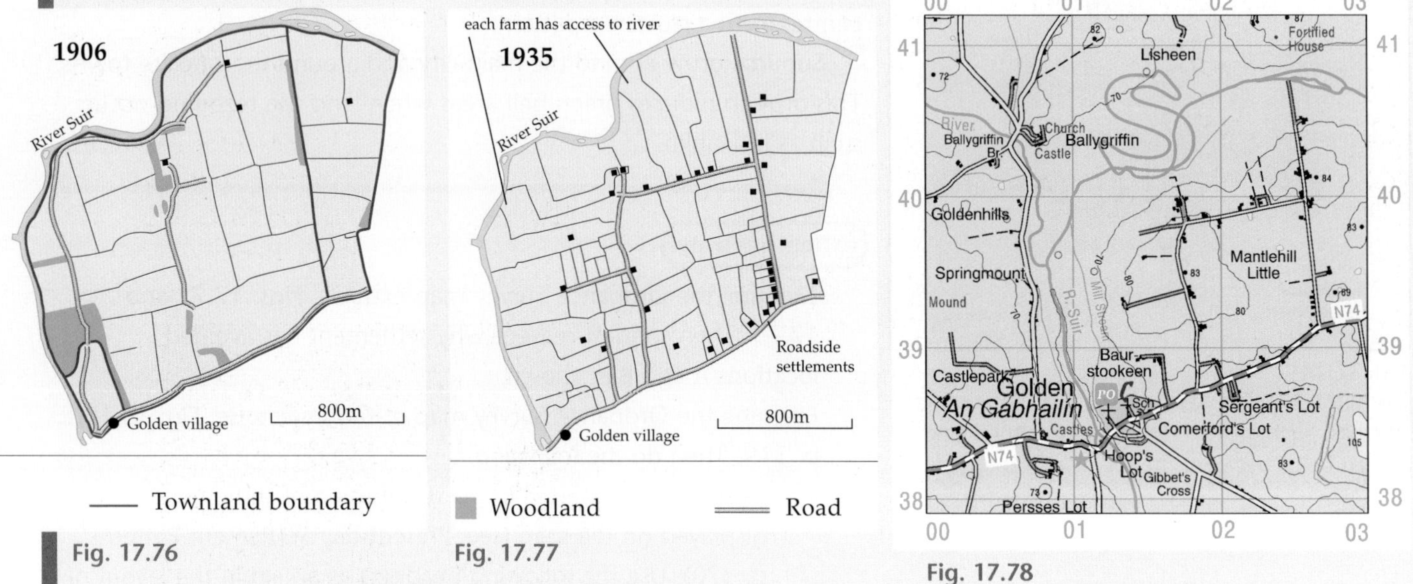


Fig. 17.76

Fig. 17.77

Fig. 17.78

## Activities to Describe the Location of Houses, Villages, Towns

**Do All the Following Activities Now**

### Location of Houses

#### How to Describe the Location of an Individual House or Group of Houses

**Site** — the land on which the house/s is/are built

**Situation** — the site in relation to the surrounding area

**Site**

1. What is the altitude of the land?
2. Is it flat, gently sloping or steeply sloping?
3. Is the slope facing south or south-east?
4. Is it a dry site or a raised site?
5. Is it a sheltered site?
6. Is it near a spring, well or stream for a water supply?
7. Is it on a road or passageway?

**Situation**

1. Is it an upland or lowland area?
2. Is it in a valley or plain or island or on the coast?
3. Is it on the edge of a town or village?

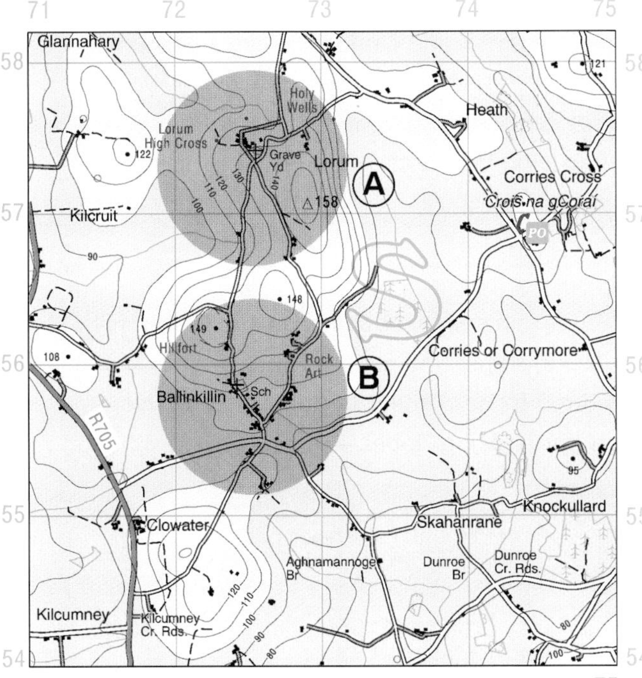


Fig. 17.79

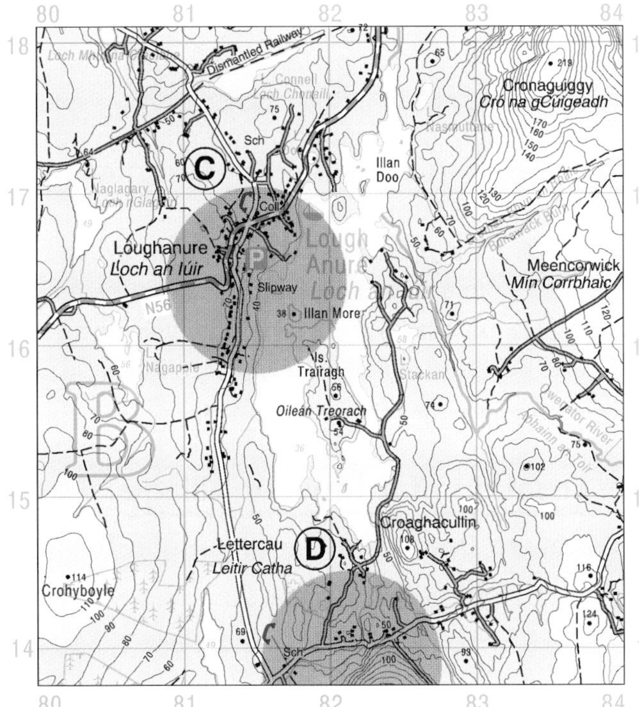


Fig. 17.80

4. Does the house(s) form a pattern with other house(s)?
5. Has a route(s) influenced the location of the house(s)?
6. Have mineral deposits influenced the location of the house(s)?

**Tip**

Once you have the answers to the questions above completed, create a heading indicating a suitable theme. Then rewrite the answers to these points making the various statements jell together.

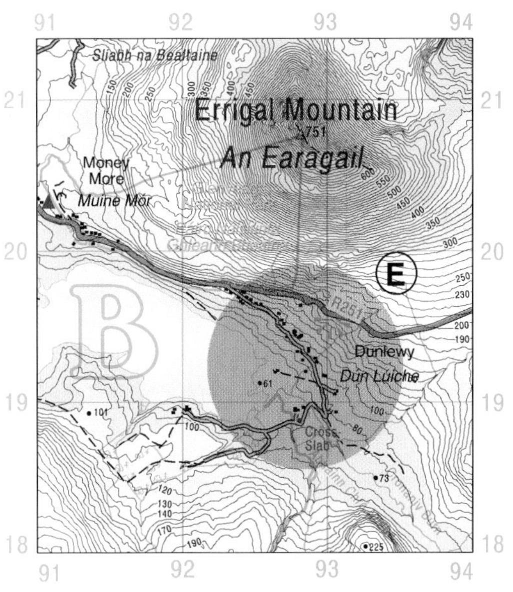


Fig. 17.81

**Home Activity**

Describe the location and the patterns of the houses within the circles A to E.

## Influences on House Location

### 1. Altitude

- The higher the altitude, the greater degree of exposure to climatic elements. Throughout the country there is a noticeable absence of settlement above 180 metres (600 feet approximately). This is partly due to exposure and partly because of the inhospitable nature of the terrain, the thin, unproductive soils, or a covering of blanket peat. Some settlements may be found above 180 metres,

but these are generally shelters for herdsmen and cattle (booleys, transhumance), which are probably now deserted.

- Low-lying land may be prone to flooding. Rivers rise and cover their flood plains in times of heavy rainfall. Other low-lying land such as bogland or marsh is avoided by people and routeways.

## 2. Aspect

- In upland and highland areas, houses generally seek shelter at the foot of hills.
- They generally avoid cold north-facing slopes and exposed sites.
- South-facing slopes are often favoured for settlement because they receive greater amounts of sunshine than north-facing slopes — explain why this is so.

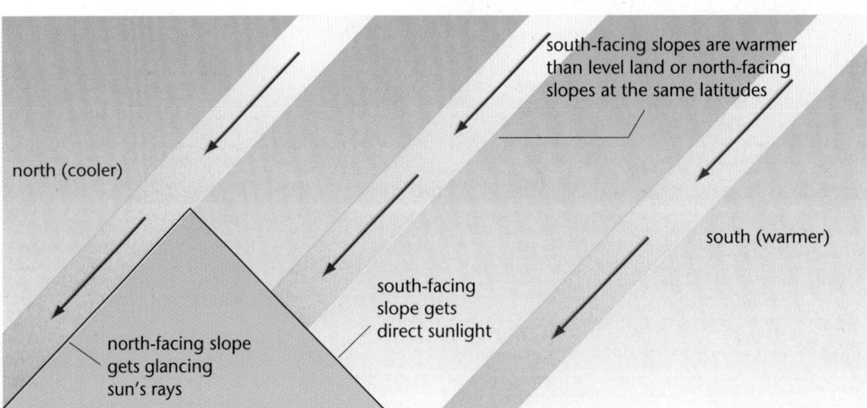


**Fig. 17.82**
The influence of aspect on temperature

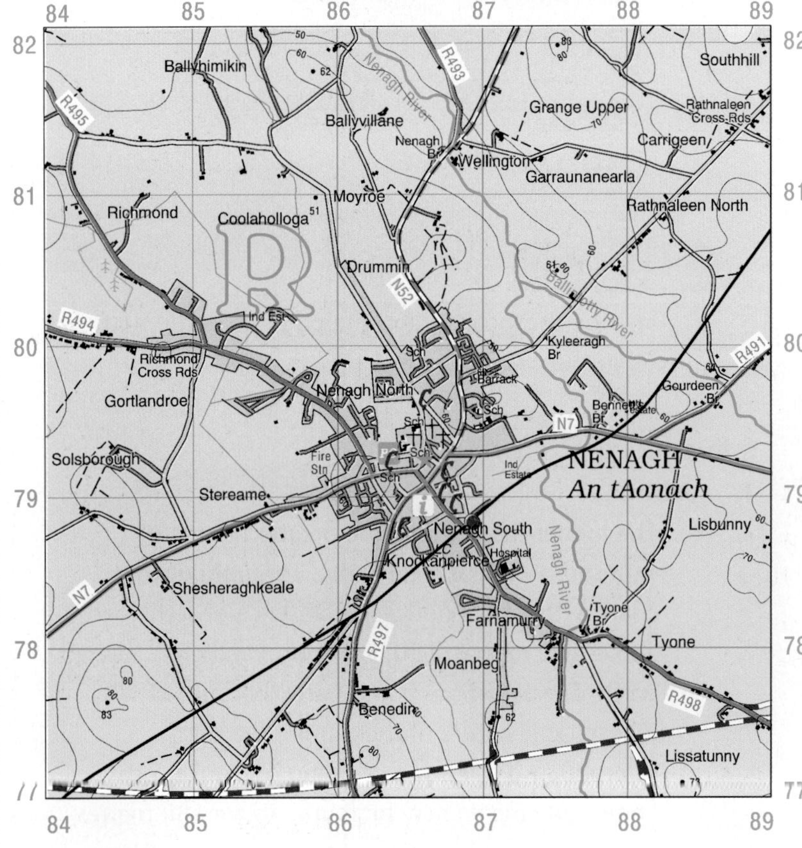


**Fig. 17.83**

## 3. Water Supply

- The presence of a spring may give rise to a **spring line settlement** at the base of a hill. Fast-flowing mountain or hill streams may have been used as a source of power for mills.

# Location of a Town or Village

### How to Describe the Location of a Town or Village

**(Location = site + situation)**

Site — the actual land on which the settlement is built

1. What is the altitude of the land?
2. Is it flat or gently sloping?
3. Is it sited on a river? If it is, which bank?
4. Is it a defensive site? (Is there a castle in the settlement or nearby?) If so, why?

**Tip**
The top of a map is always north

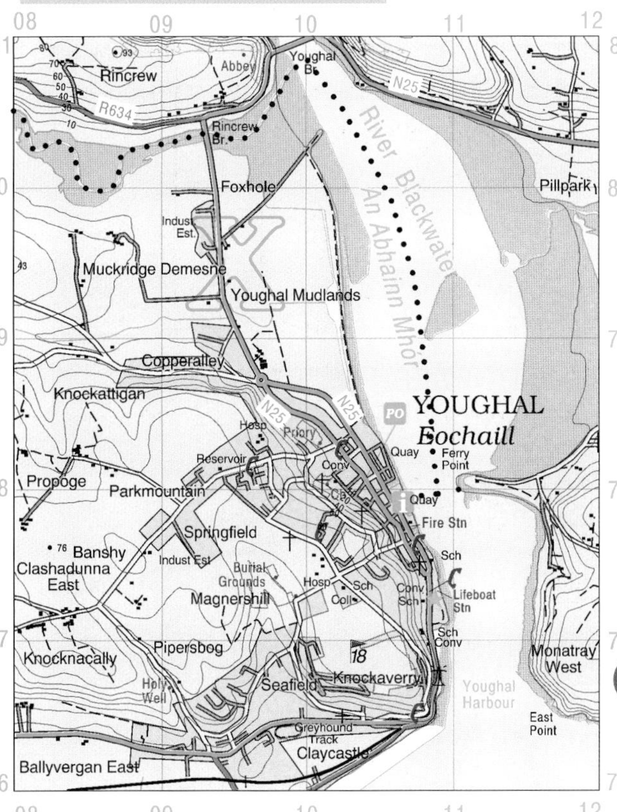


Fig. 17.84

5. Is it a bridging point of a river? If it is, name the river.
6. Is it the lowest bridging point of the river? If so, why?
7. Is it a confluence town? (Is it at or near the confluence of two rivers?)
8. Does the place name suggest anything about the settlement? If so, what?

**Situation** — the site in relation to the surrounding area

1. Is it sited in a valley or on a lowland plain?
2. Is it near the coast? Is it at an estuary? Name the bay or inlet.
3. Where is the settlement in relation to some prominent relief feature?
4. Is it a focus of routes? Name the routes and classify them.
5. Have routes influenced the development of the settlement?
6. Has the presence of mineral deposits influenced the location/development of the settlement?

**Class Activity**

Examine the Ordnance Survey map extracts, Figs. 17.83 and 17.84. Then describe the location of each of the nucleated settlements Nenagh and Youghal.

## Case Study: Location of Lismore

### Site

Lismore is sited on the south side (right bank) of the Blackwater River. The town is sited on land varying between ten metres on the north side and fifty metres on the south side. The northern part slopes steeply to the river while the southern side slopes gently away from the river. There is a castle in the town, Lismore Castle. This suggests that the site for the town was chosen because of its elevated defensive position overlooking the crossing point of the River Blackwater.

### Situation

Lismore is situated in the river valley of the Blackwater, which runs east–west. It is bounded by steep slopes (bluffs) to the north and south, represented by contours which are close together. The present valley floor is flat and this is represented by the absence of contours to the north of the river.

Lismore is a bridging point of the River Blackwater. A national secondary road, N72, regional roads R668 and R666 and some third-class roads all meet at the town. Thus it is a major focus of routes.

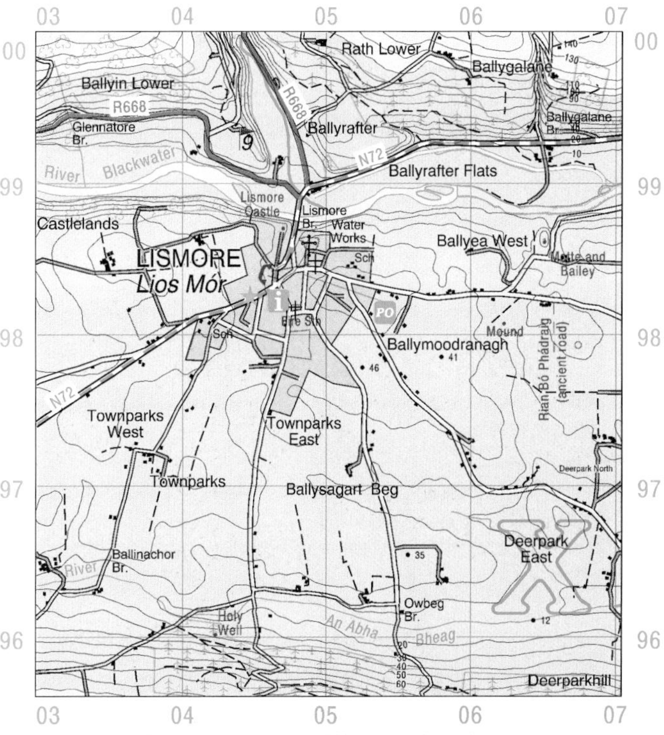


Fig. 17.85
Lismore, on the River Blackwater

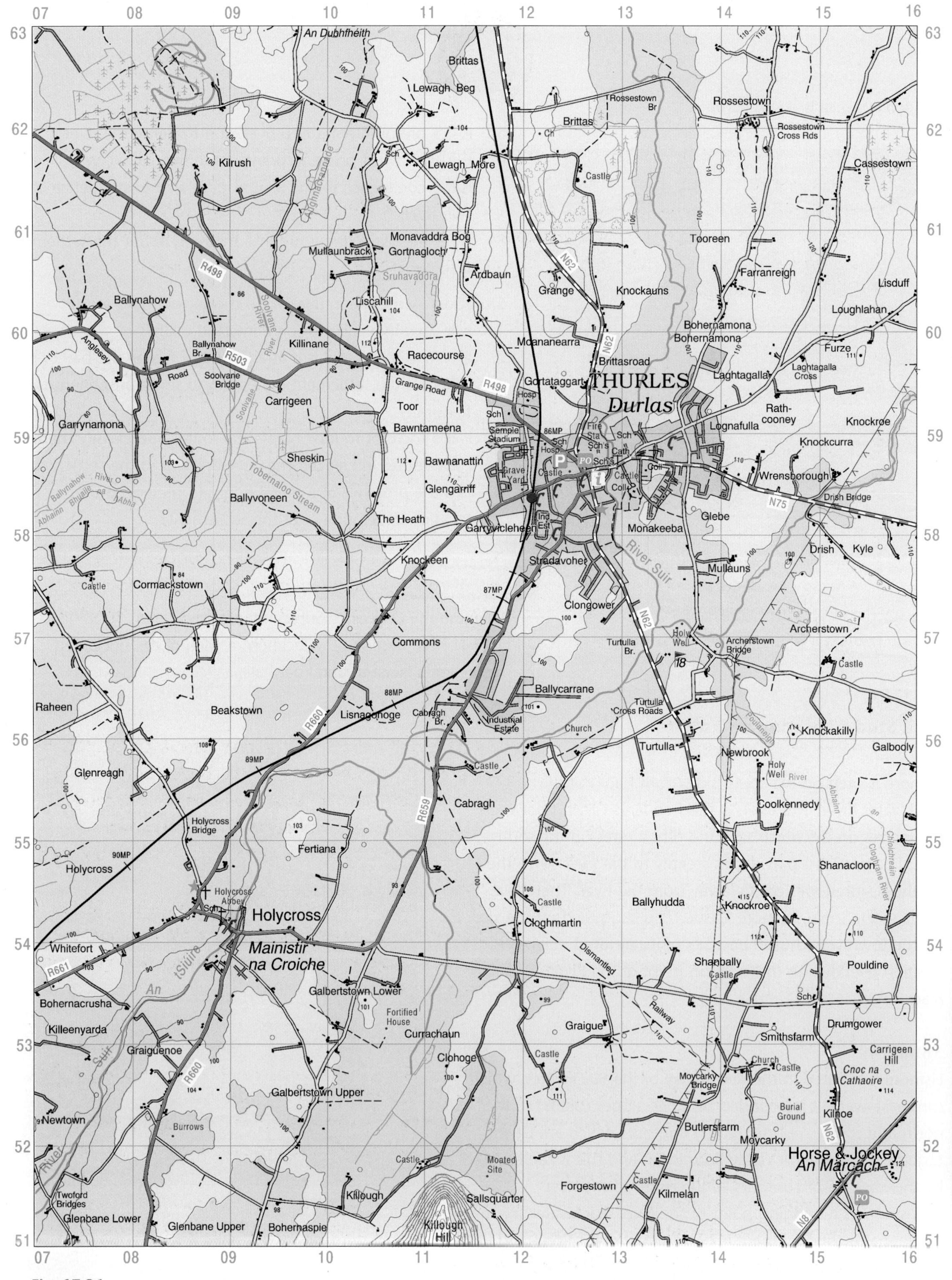


**Fig. 17.86**
Thurles region

## Case Study: Account for the Development of Thurles at Its Present Location

### Route Focus

Thurles is situated at a bridging point of the River Suir. As a result, it has become a route focus. Two national secondary routes, N75 and N62, three regional roads, R498, R659 and R660, third-class roads and a railway line all meet here (Fig. 17.86). These routes meet here because Thurles is located in the heart of a lowland plain.

### Market Town

The word 'grange' to the north of the town suggests monastic lands associated with the Cistercian order. Cistercians played a key role in the creation of markets for cattle within Ireland and abroad. Their lands were divided into farms or 'granges' and relied on their own labour. This land at grid reference S 124 604 is raised slightly above the flood plain of the River Suir. This would suggest that much of the monastic land was quite level and dry for cattle and tillage. Some of the lands probably included portion of the Suir flood plain S 135 605 which gave access to vital water supplies and summer meadows for fodder.

The land in and around Thurles is low-lying and level as the 100-metre contours are widely spaced. The area is well drained by the River Suir and its tributaries, which seem to flow in a southerly direction. This level or slightly undulating land therefore appears to be a fertile farming region. Thus Thurles is a market town. In addition, the presence of a railway station suggests that at one time fairs may have been held in the town, with the railway carrying purchased cattle to other destinations, such as ports for export.

### Defensive Settlement

There are two castles in Thurles town, which suggests that it was a very important defensive town. One of these castles is near to the River Suir at a bridging point. This castle would have controlled/protected this point as all goods and vehicles would have crossed the river here. The castle was probably built, therefore, by the Normans, who erected these castles in the twelfth and thirteenth centuries to protect their captured lands from the dispossessed Irish farmers. At this time the town was probably defended also by a town wall which would have enclosed the Norman settlement.

The second castle to the west of the town may have also been Norman, or it could have belonged to an Irish chieftain. The Irish chieftains in the fifteenth century built castles similar to the Norman ones but larger in size. In any case, Thurles probably developed as a consequence of its location at a strategic river crossing and route focus which was centrally located on a fertile plain.

### Services

Thurles has many services. The fire station, the hospitals, graveyard, schools and churches suggest that the town is well established and recognised as a major centre for the surrounding region.

The industrial estate in the south-western part of the town would influence many people to locate nearby to be near their place of work. The estate would also create spin-off services which in turn would increase the prosperity of the settlement.

Semple Stadium and the racecourse to the west of the town suggest its focus for sporting events, which again would create extra income for local businesses. The railway services the town and the local area for the transport of goods and people.

Fig. 17.87

KILKENNY
CILL CHAINNIGH
Troyswood
Church and Castle
Longsgate
Burial Ground
Coolbricken Br.
Kyleroe
Sandfordscourt
Loughmerans
Crannóg
Baun
Bonnetsrath
Water Works
Castle
Burntfurze
Hosp
Greenridge
Oldpark
Glendine
Brownstown
Coolgrange
Neworchard
Newtown
Fulacht Fia
Dunmore
Friarsinch
Raheennagun
Talbotsinch
Sch
Seixeslough
Charterschool Land
Lousybush
Kingsland
Newpark
Ch
Garrincreen
Cellarstown
Jamespark
Lough Macask
Ayresfields
Cathedral and Round Tower
Barracks
Pennefatherslot
Palmerstown
Sports Ground
Archersrath
Bishops Demesne
Blanchfieldsland
Indust. Est.
MacDonagh Station
Kilcreen
Hospital
Crokershill
Rothe House
Leggetsrath
Cross-Inscribed Stone
MP29
Collegepark
Lacken
Aughmalogue Br.
Lyrath
Robertshill
Dukesmeadows
Hosp
Middleknock
Templemartin
Castle
Walkinlough
Schs
Maudlinsland
MP30
Drakeland
Deansground
Sionhermitage
Indust. Est.
Poulgour
Maidenhill
Cashel
Wetland
Gallowshill
Archersgrove
Mortgagefields
Sch
Smithsland
Loughboy
Danville
MP31
Reviewfields
Warrington
Archersleas
Burial Ground
Shellumsrath
Joinersfolly
Maddockstown
Baile Mhadóg
Springhill
Donaghmore
Birchfield
Raggetsland
Castle
Derdimus
Goslingstown
Kilferagh
Burial Ground
Foulkstown
Holy Well
Parkmore
Outrath
Furzehouse
Raheenapisha
Glebe
Ballynalina
Knocklegan
Standing Stone
Sheastown
Rathaleek
Wallslough
Bullaun Stone
Grevine East
Bodalmore
Grave Yard
River Nore
R693
N77
R695
R700
N10
N76
R697
48
49
50
51
52
53
54
55
60
59
58
57
56
55
54
53
52
51

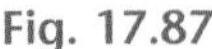

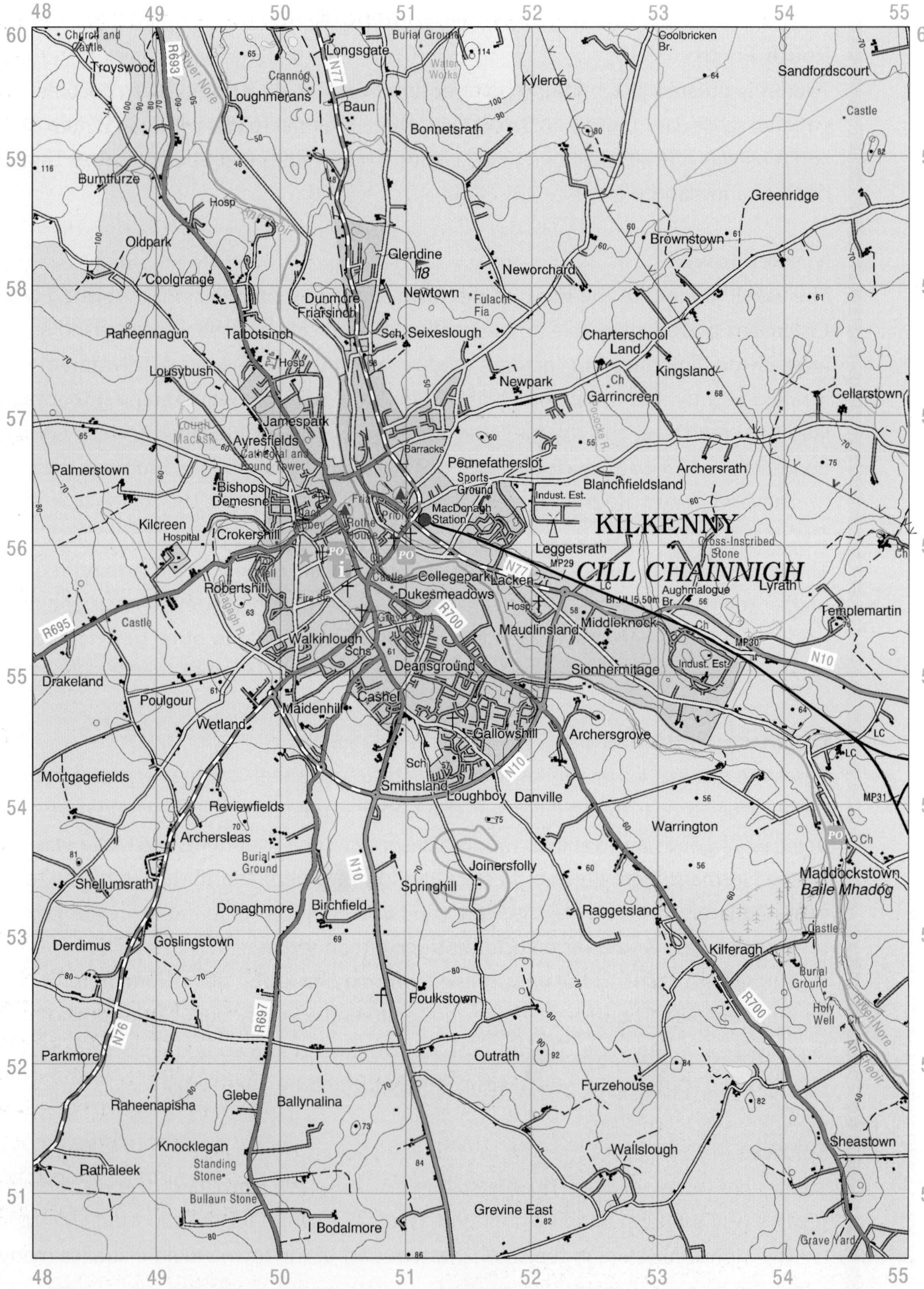

FIELD PATTERNS AND PLANTATION TOWNS

Study the map extract of the Thurles region, Fig. 17.86, and then do the following:

1. Account for the origin and development of Holycross village.
2. Describe the course of the River Suir from where it enters the map at grid reference S 131 630 to where it exits the map at grid reference S 070 513.
3. Account for the distribution of settlement within the area enclosed by eastings 12 and 15 and northings 60 and 63.
4. Account for the patterns of drainage shown on the map.
5. What evidence suggests that this area was settled in pre-Christian and Norman times? Use evidence from the map to support your answer.

Look at the 1:50,000 Ordnance Survey map extract of the Kilkenny region, Fig. 17.87. 'This map shows evidence of the development of human settlement AND of land use over a long period.'

Discuss this statement, with reference to the map. **(35 marks)**

'The Kilkenny region is well served by various means of transport.'

With the aid of a sketch map, examine this statement, referring to map evidence. **(35 marks)**

The map provides evidence that Kilkenny city offers a variety of attractions for tourists. Discuss this statement. **(30 marks)**

(1996 Leaving Certificate)

radial routeways focus on the original monastic settlement
round tower
semicircular street represents the boundary wall of the early monastery

Fig. 17.88
A curved street in association with a round tower or early church suggests an early Christian origin in Clondalkin in Co. Dublin

## Street Patterns in Irish Towns

### Radial Pattern

In some instances, towns developed around monastic sites. Routes focused on the monastery and formed **a radial pattern**. The town which grew up retained these routeways and a road developed around the enclosing stone wall. A circular street with roads radiating outward from the town centre is typical of a monastic settlement. Few of these remain in Ireland today, although Kells in Co. Meath and Lusk in Co. Dublin are two such towns.

### Planned Pattern

Planned towns of the eighteenth century are laid out with rows of streets at right angles to each other. They form 'blocks' just as in New York City except on a smaller scale. This net or mesh layout is called grid pattern (Fig. 17.89).

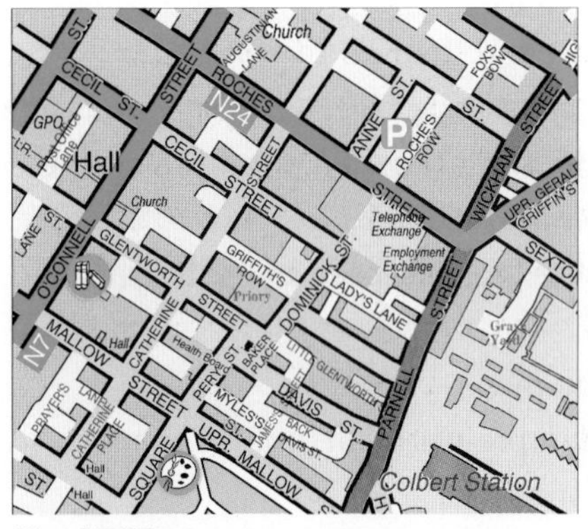

Fig. 17.89
Planned grid pattern of streets in Limerick city

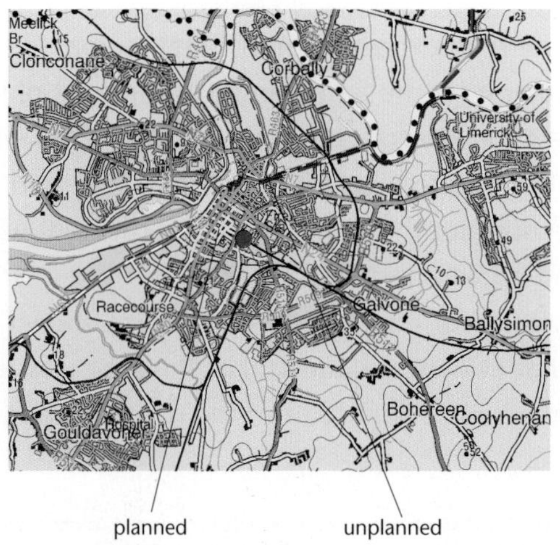

Fig. 17.90

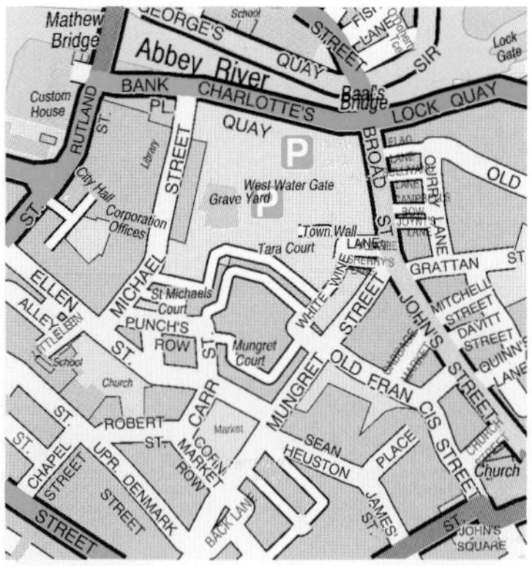

Fig. 17.91
Unplanned streets in Limerick city

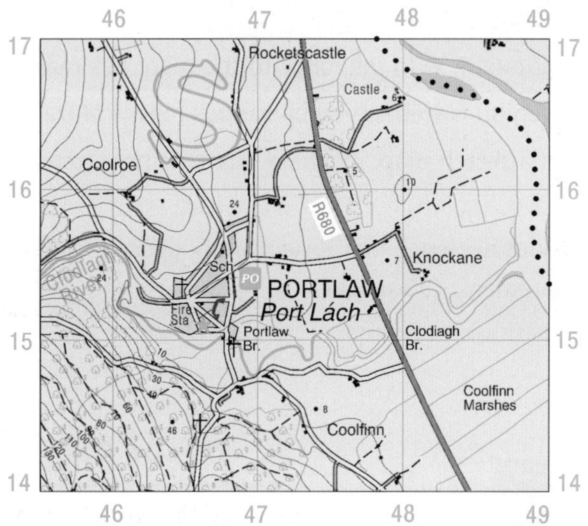

Fig. 17.92
Planned industrial town, Portlaw in Co. Waterford

Some exceptions to the rule are to be found, however. For example, Portlaw in Co. Waterford, a planned industrial town, was laid out with a plan similar to that of a human hand (Fig. 17.92).

## Unplanned Pattern

An **unplanned street pattern** with narrow streets suggests an old or medieval origin to the town. Where streets in a town form a Y-shape, it suggests that markets were held at the junction of such streets. Medieval towns had many market areas, such as the potato market and fish market. Identify the market area type on Fig. 17.91.

Where a town is not obviously planned, the presence of a castle, in association with an abbey or monastery, may suggest a medieval origin.

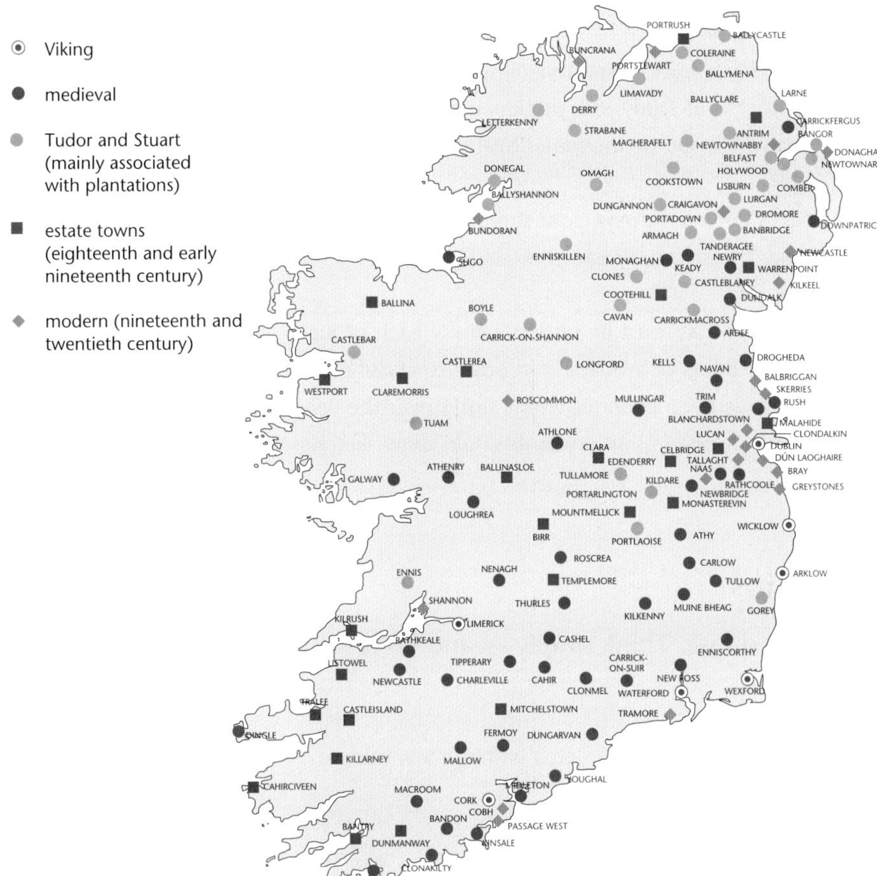


Study Fig. 17.93 and see if your town or village is mentioned or marked. If it is, and using what you have studied in previous chapters, organise a fieldwork exercise on the historical or economic development of your settlement under the headings below.

Fig. 17.93
Irish historical settlements

## A Fieldwork Exercise in Your Own Town

*Title:* An Urban Land Use Survey in . . ., Co. . . .

*Objectives*

1. To show how land use varies with accessibility.
2. To map the various land uses in a given area.
3. To show how to present information in a number of ways, e.g. statistics, sketches, charts, etc.
4. To measure area on a map.
5. To work with maps of various scales.
6. To use computers to design charts.

Choose one block of buildings near the town centre, or an area defined by certain streets in any part of the town.

Write an account of your survey bearing the following points in mind.

1. Describe clearly the title and purpose of the exercise.
2. Explain how you decided on the following: (a) the type of information needed; (b) where this information was gathered; (c) what methods were used in gathering the information.
3. Describe clearly both the gathering and analysis of the information.
4. Explain the results of that analysis and the methods used to present these results.
5. Explain how this experience would alter your approach to any similar fieldwork exercise in the future.

*Other suggested project titles*

1. 'My village since AD 1900 — a change over time'
2. 'Buildings and their relevance to the growth of my town'
3. 'Industry and the development of my town'

**Useful Tips in Doing Your Fieldwork**

1. Be precise in your title. For instance, you may limit the period for development within certain dates, or you may restrict your study to a particular street.
2. Use a large-scale map which can be increased further by photocopying for easy recording of data.
3. Always orientate your map when doing your survey.
4. Because your time may be limited, divide your class into small groups so that each group has a specific task.
5. Be specific about the work that must be undertaken by each group.
6. Always look at the upper floors of a building in order to classify it into a particular category, as the ground floor may have been changed many times over the years.
7. Use all the skills of your group, e.g. art students could draw sketches of building façades.

## Other Fieldwork Tips

**NB: Only choose a limited task. Think simply — and do it well.**

**Fieldwork Tips**

It may help to limit a fieldwork exercise to an area within a few minutes' walking distance of the school, so that

- facts can be rechecked
- the area can be visited a number of times if needed
- class time is gainfully used to a maximum
- students can develop a sense of responsibility towards working without supervision, earning trust, and learning the importance of time management, i.e. returning to class on time.

### Title of Survey

Choose a title that gives a clear idea of the work to be carried out. This will help you to focus on the key ideas of the fieldwork, and it will help to outline clearly the type of information that is required for completion of the survey.

### Aims of Fieldwork

Give at least five aims, e.g.

- to identify land uses in . . .
- to record data accurately
- to learn to work with maps of various scales in the field and in the classroom — orient, update, amend and colour, etc.

### Careful Gathering of Information

- What steps were taken in the classroom to gather the information before the exercise and during the fieldwork?
- State how it was to be gathered, e.g. orientation of map, key for land uses, how the

key was to be applied during the fieldwork; questionnaire: its layout, who was to be asked and why, how this was to be carried out during the exercise.

- Two groups should do the same section, to compare for accuracy of each group.
- Each group should give a description of how their various tasks were to be gathered carefully.

### Careful Recording of Information

- Box options for questionnaire; give sketch.
- Letters used instead of colours for land uses; give sketch of method.
- Cross-section, e.g. of beach material — measurement and material at various measured intervals; give sketch.
- Photographs — explain why they were needed.
- Sketches, e.g. of building types and characteristic features.

### Results

What were the fieldwork findings? — discovered facts, e.g.

- How many people live in the settlement?
- How many derelict houses in the settlement?
- Numbers of vehicles and types in the traffic survey.
- Fall or slope of the beach.
- State the facts of the gathered information.

### Conclusions

What conclusions were drawn from the facts mentioned above? Develop each conclusion fully and in detail.

### Presentation of Findings

- Draw sketches of facts, using various charts, e.g. pie charts, bar charts, line graphs, to display clearly. **Use colour** where possible.
- Coloured Ordnance Survey maps with colour key marked in.
- Freehand sketches, photographic material.
- Headings and written script on display board.
- Group plenary session.

**Examination Preparation Tips**

- Limit final account so that it can be handwritten within forty minutes (Higher Level) or thirty minutes (Ordinary Level) in an exam.
- Write neatly.
- Lay out your survey under the headings given in the previous Leaving Certificate examination, and for each heading —
  1. give a written account in sufficient detail to gain maximum marks
  2. use sketches and diagrams when possible to explain points made
  3. practise this many times within the allocated time, and compare it with your own fieldwork account to see if any important item was omitted

**Tips**
Use these questions as point developers. Expand each point fully. Constantly refer to places on the map or photograph.

## Settlement Distribution

### How to Describe the Distribution of Settlement and How to Account for that Distribution

### High Density — Where There is Much Settlement

Villages, towns and individual settlements

1. In which portion of the map/photo is there most settlement? Why?
2. What pattern(s), if any, do the villages and towns form?
3. Why do they form this/these pattern(s)?
4. What pattern(s) do the individual houses form?
5. Why do they form this/these pattern(s)?
6. Is this high density of settlement located in a valley or plain?
7. On which side of the valley/plain are these settlements located? Explain.
8. Is settlement concentrated in a few larger villages/towns? If so, is it because they are on a lowland/coastal plain or is it due to some other influence?
9. Is this land well drained and level and suited to construction?
10. Is there a high density of routeways? Were these settlements influenced by the route network or were the routes influenced by the settlements, which in turn drew more settlement to the area?

**Case Study:** Barrow and Nore Valleys — Settlement Distribution Sketch Map

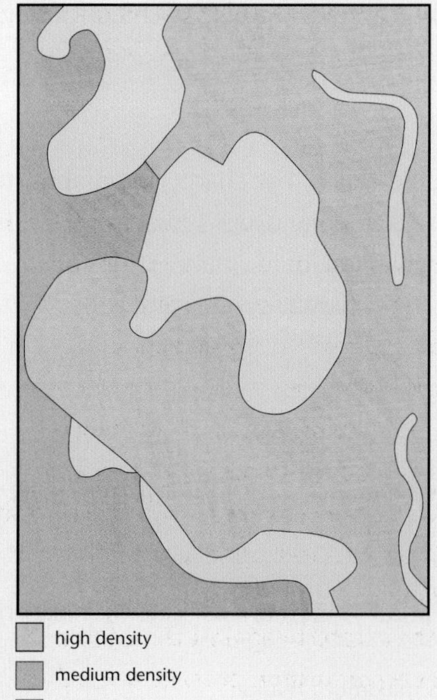


Fig. 17.94
Settlement distribution sketch map

### Low Density — Where There is Less Settlement

Use these questions as point developers. Expand each point fully. Constantly refer to places on the map or photograph.

1. In which portion of the map/photo is there less settlement?
2. What pattern, if any, do these settlements form?
3. Why do they form this pattern?
4. Is this low density of settlement confined to an upland valley(s)?
5. On which side(s) of the valley(s) is this settlement located?
6. Is settlement confined to a particular slope(s)? If so, explain why?
7. Is this land low-lying and prone to flooding and are the settlements avoiding these flood plains?
8. Is this settlement confined to routeway sites and/or avoiding rugged lowland?

### Absence — Where There is No Settlement

1. How high is the land? Is it highland, upland or lowland?
2. Is the land steep and unsuited to construction of buildings?
3. Is the land too high for settlement purposes?
4. Has the land gentle slopes or is it flat?
5. If it is flat, is it on a flood plain and prone to flooding?
6. Has the area few routes or no routes? Explain.
7. Is the area exposed to cold winds and unsuited to settlement?
8. Is there an absence of rivers or small streams? If so, is it bogland?

9. Is much of the land facing north or north-west and steep, so that it would discourage settlement?

Use these questions as point developers. Expand each point fully. Constantly refer to places on the map or photograph.

**Home Activity**

Study the following:

1. the Ordnance Survey map extract, Fig. 7.40, p. 102
2. the settlement distribution sketch map, Fig. 17.94 and
3. the questions relating to high, low and absence settlement patterns

With the aid of the sketch map and the questions provided, describe the distribution of settlement in the Barrow and Nore basins.

# Patterns in the Distribution of Woodland

## Distribution of Woodland in Elevated Areas

### The Price of Land

Land in elevated areas in Ireland is sometimes purchased at a low price per hectare by government bodies such as **Coillte**. Much of it is commonage, and due to rural depopulation, many farmers are willing to sell their land in order to invest their money elsewhere. As Ireland is the least forested country in western Europe, it is government policy to continue its commitment to extensive planting in the future, especially on poor-quality soils to the west of the River Shannon.

### Climate

Trees on mountains or high upland areas must withstand hard climatic conditions — extremely cold winters and short cool summers. They must also withstand a continuously damp climate and, at times, winter precipitation in the form of snow.

Conifer trees such as Sitka spruce, Norway spruce, *Contorta* pine and Scots pine are generally grown in Irish coniferous forests because they can cope with these harsh climatic conditions. Their needle-shaped leaves reduce transpiration to a minimum. Their compact conical structure both helps their stability against the wind and prevents too heavy an accumulation of snow upon the branches, as in the colder northern lands of the boreal region. Conifers will grow quickly in exposed mountain areas. They produce a large amount of soft wood in relation

slender shape offers little resistance to wind and so reduces the risk of trees being blown over during storms

downward sloping branches in some species allow snow to slide off before it becomes heavy enough to break the branches

thin, needle-like foliage is not withered by cold, biting winds

thick bark protects the conifer from the cold, dry wind which might deprive the tree of the moisture it needs

shallow roots can develop in the thin layers of soil above the bedrock

**Fig. 17.95**
Coniferous/softwoods. How conifers are adapted to elevated areas.

LANDSCAPES OF THE WORLD

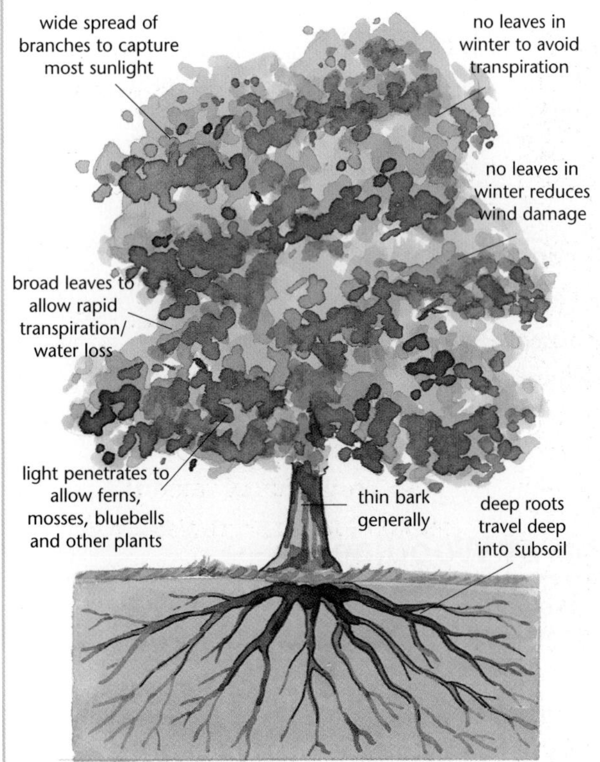
wide spread of branches to capture most sunlight
no leaves in winter to avoid transpiration
no leaves in winter reduces wind damage
broad leaves to allow rapid transpiration/ water loss
light penetrates to allow ferns, mosses, bluebells and other plants
thin bark generally
deep roots travel deep into subsoil

**Fig. 17.96**
Deciduous/hardwoods. How deciduous hardwoods are adapted to deep soils in lowland areas.

to land occupied in a relatively short period of time, while their foliage allows for the maximum utilisation of sunlight throughout the growing season. South-facing mountain slopes capture warm, high-angle rays of the sun. These slopes, which are sheltered from cold northerly winds, produce high growth rates in species such as Sitka spruce. Forestry is a valuable alternative land use in areas that would otherwise be unproductive. It also provides a supplementary income for subsistence farmers in areas of heavy outmigration.

### Soil Type

Conifer trees are frugal in their needs. They will thrive in soils such as **leached soil** or **bog** where deciduous trees may fail to grow.

High upland and mountain areas in Ireland receive between 1,500 and 2,000 mm of rainfall annually. This gives rise to leaching. As percolating water drains downward, iron hydroxides and the humus from rotting vegetation near the surface are carried in solution to the lower soil layer, thus leaching the top soil or upper horizon. The top zone or **A horizon** is bleached (changed in colour) to a predominantly greyish tint. Such a soil is called **podsol**.

The leached ferro-humus material may accumulate at a depth of a few centimetres. There, mixed with particles of clay and silt, it forms a hard cemented band (hard pan) which prevents further drainage and leads to waterlogging. So there is a marked tendency for peat bogs to develop.

Trees may be planted on steep mountain slopes. The tree roots help bind soil particles together, thereby preventing erosion. The trees also prevent rapid surface water run-off. Thus areas that might otherwise be non-productive agriculturally are now put to a commercial use.

## Distribution of Woodland in Lowland Areas

### Demesne Woodland/Parkland

In the eighteenth and nineteenth centuries, there was a system of large landed estates throughout the country. Their boundaries were often defined by high stone walls. The main characteristics of these demesnes were great houses or castles with numerous outbuildings set in parkland with ornamental trees, gardens and lakes. On the inside, and running parallel to the perimeter wall, was a narrow strip of deciduous trees such as beech, oak and chestnut. Other parts of the estates may have been cordoned off by walls for the production of commercial lumber. Such woodland areas may be identified on maps by noting plantations near to castles or large rural residences.

Deciduous trees such as beech, oak and chestnut thrive on lowland soils. Deep soil is needed for deciduous trees. Their roots grow deep into the soil as well as spreading laterally. Deciduous woodland gives rise to 'brown earth' soils which are rich in humus. They form most of the soils in mid-latitude lowland areas.

Fig. 17.97
Demesne woodland outside Kenmare, now used as hotel and golf course

## Mixed Soils

Some lowland areas have soils which are not desired for immediate agricultural use. Such soils may be too wet or too dry for high agricultural yields. These may, however, be put to use for long-term investment, and so conifer trees may be planted for the following reasons.

1. Coniferous trees produce a larger proportion of wood in relation to space occupied when compared with deciduous trees.
2. Conifer trees grow very quickly in lowland areas and are ready for felling in forty to sixty years.
3. Pines thrive on dry, sandy soils. Spruce thrive on damper soils such as flood plains, while larch thrive on soils of fair quality.

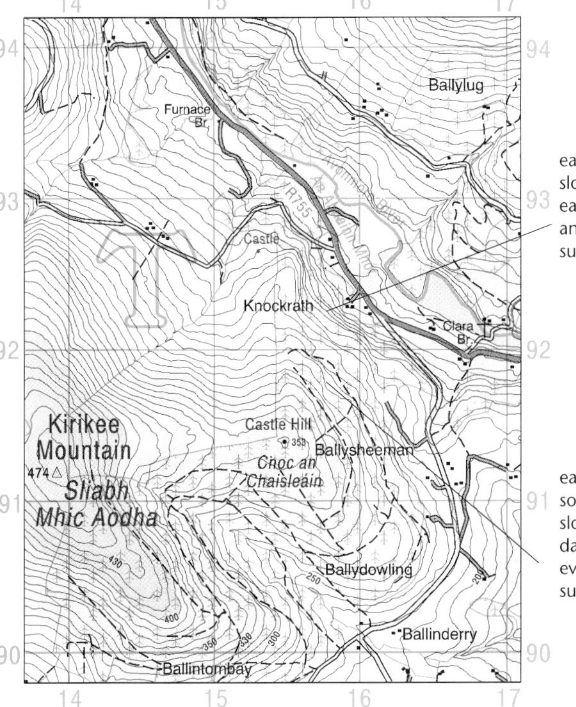


Fig. 17.98

# Absence of Woodland

## Elevated Areas

High upland and mountain land is often devoid of forestry. This may occur for a number of reasons.

1. Some mountain areas are too high (greater than 600 metres) for the growth of trees. In elevated areas such as these, insufficient heat and exposure to strong winds limit growth to grasses, mosses and lichens.
2. Most high mountain areas in Ireland have little or no soil cover. This absence of soil occurs because the agents of denudation (weathering and erosion) constantly remove weathered material (regolith), thus restricting the growth of soil.

## Lowland Areas

3. Most Irish lowland areas have no forest cover. Lowland is generally used for agricultural purposes. Farming activities such as tillage or dairying are intensive forms of land use and so produce high yields each year.

Forestry, on the other hand, does not produce a return for at least forty years. It is seen as a long-term investment and is thus restricted to marginal lands.

## Green Belts

Green belts are open spaces of agricultural land, parks or areas of woodland within large towns and cities. They are specific designated areas for the purpose of breaking up the monotony of urban development such as housing estates and industrial zones. They are designed by urban planners and become permanent features when legislated by local government.

Green belts and individual trees in urban areas have many purposes:

1. As mentioned above they prevent continuous urban development in large towns and cities, which improves the quality of urban environments.
2. They provide shade on streets and in urban outdoor restaurant areas in hot countries such as Italy, France and Spain in Mediterranean areas.
3. Urban recreation areas such as Central Park in New York City, St Stephen's Green and the Phoenix Park in Dublin, and Eyre Square in Galway provide a welcome relief from an office environment at lunch hour.
4. They provide sports fields, for leisure walks and runs, golf courses, etc.
5. Some trees are chosen for their ability to survive in heavily populated environments, such as cities, while at the same time provide a wildlife sanctuary for birds such as sparrows. In Mediterranean areas these trees hold hundreds of birds whose sounds create a welcome chorus at evening time.
6. Parts of green belts may be farmland whose owners wish to retain their family farm rather than sell out for a high price and purchase elsewhere.

**Home Activity**

Look at the 1:50,000 Ordnance Survey extract, Fig. 17.84, p. 379 and answer the following questions.

(a) This segment of the river Blackwater acts as a sheltered harbour that serves a variety of human activities. Justify this statement using evidence from the map only. **(50 marks)**

(b) As a Geographer you have been asked to draw up a feasibility study to consider building a bridge across the Blackwater between County Cork and County Waterford.
Describe, using map evidence only, the positive **and** negative aspects of such a proposal. In your answer identify and justify an exact location for the bridge. **(50 marks)**

(1999)

# Modern Ireland — Late Twentieth-century Developments

## Urban Development in Recent Years

Fig. 18.1
Refurbishment of old buildings in Dublin; one of the city's renewal schemes

Irish towns and cities such as Cork, Limerick and Dublin have expanded enormously since the early 1960s. Large tracts of fertile farmland have been encroached upon by new developments such as housing estates and industrial sites. The earliest of these estates, e.g. Southhill in Limerick and Tallaght in Dublin, were planned with little concern for the social needs of their inhabitants. Playground areas, community facilities and imaginative landscaping were often omitted, while the houses were laid out in monotonous rows. Tallaght, for instance, has grown into a large residential settlement with a population in excess of 70,000 people. Hospitals (Limerick, with a similar population, has five active hospitals), local shopping centres (The Square) and other necessary community facilities and services have only recently been provided, while all its heating systems were initially based on the burning of bituminous coal.

In recent years, much of this type of planning has changed. More creativity and imagination in design as well as a strict balance on land use are demanded.

### Inner-city Development

Fig. 18.2
New office blocks beside the River Liffey in Dublin

Many of our inner-city areas had deteriorated by the 1970s. Buildings were derelict or underused and so contributed little to the commercial life of the Central Business Districts. A recent surge of urban renewal in cities such as Cork, Limerick, Waterford and Galway has brought new life into places which were derelict some ten years ago. This has occurred in almost all of Ireland's older established settlements. Old buildings such as granaries, factories, warehouses and deteriorated Georgian commercial units have been either refurbished or demolished. New buildings have replaced them, with façades which are often in keeping with the traditional character of their areas. These new developments have created office blocks and shopping plazas, such as the Stephen's Green Centre in Dublin, Arthur's Quay Centre in Limerick and Merchants' Quay Centre in Cork.

Shopping complexes may play both a social and commercial role in some towns, e.g. in the planned Georgian settlements of Dublin and Limerick which otherwise lack a central focus point such as a square. Other improvements which have occurred within towns are areas where the pedestrian has precedence over traffic. These pedestrianised streets, such

as Grafton Street in Dublin, Princes Street in Cork and Cruises Street in Limerick, provide natural foci within their respective settlements.

Towns which have central spaces within their Central Business Districts, such as Eyre Square in Galway, have ensured that they will become genuine foci of community activity and interest, with improvements such as bandstands, paving and landscaping (decorative use of trees and flowers), which may have a special attraction for tourists as well as local residents. Many English, European and American cities have been successful with such developments, e.g. New York City with its café and ice rink at Channel Promenade. The contrast between long narrow streets and the openness of a square, regardless of how small the square may be, always adds character and interest to a settlement. Broad Street in Waterford is such an example.

Up until recently, old inner-city residential areas were allowed to deteriorate, with no provision for community facilities or vital services. Sean McDermott Street in Dublin is one such example. Such areas have received some funds for renewal and long-established communities have lately been rehoused in the inner city.

Trying to strike a balance between increasing traffic demands, restoration of traditional buildings' façades and the needs of shoppers and residents is a major task. The widespread use of the motor car has, up to now, put the pedestrian in second place within the Central Business Districts. Today this is changing. The street has an integral part to play in the aesthetic quality and social activity of all our towns. The retention and, more recently, the addition of traditional Irish shop façades has once again added great character to streets which had been dominated by Italian tiles and American-style neon signs.

The rapid growth in the supply of domestic accommodation in seaside towns and cities continues unabated. While this recent development has created some local employment, it has also brought some problems. Urban sewerage systems such as that in Kilkee, Co. Clare have created pollution of the coastal waters. Sewage is pumped into the sea to the south of Kilkee (Moore) bay. From here it is carried by longshore drift, aided by the south-westerlies, to the beach. Other seaside towns may well be affected in time as services are rarely installed in advance of development.

Dublin, Galway, Limerick and Waterford have all renewed part of their cities, the provision of domestic accommodation being an integral part of this process. The expansion and upgrading of some of Ireland's third-level colleges has increased accommodation demands within their respective vicinities, leading to the development of new housing estates for rental purposes.

However, in some instances, even though this renewal has taken place, many of the midtown apartments lie empty as a consequence of their purchase by large firms only for tax rebate allowances.

## Roads and Streets

Traffic flow has improved within Irish towns. The addition of one-way streets, traffic islands, roundabouts, freeways, ring roads, central parking spaces, new streets, new bridges and the electrified railway (DART system in Dublin) have all helped to speed up the movement of traffic within towns and cities.

Fig. 18.3
An interchange on Dublin's new ring road

In Dublin, however, traffic congestion is still a major problem. Public transport, particularly the bus system, is unable to maintain schedules. The expansion of the electrified rail system as well as bus lanes, bicycle lanes, reduced private parking areas and increased parking fees are just some of the measures proposed for improving traffic flow. The reintroduction of tramways is one suggestion which may help to alleviate congestion within a city. The recent construction of ring roads helps to divert some traffic travelling across cities.

Many of Ireland's arterial routeways have been upgraded from trunk roads to national primary and secondary status. Money from the European Regional Development Fund has helped to widen and resurface many roads, while those connecting important centres, e.g. Dublin city and Dublin Airport or Limerick city and Shannon New Town, have become either motorways or dual carriageways.

Bypasses on national primary roads, such as at Naas and Newbridge in Co. Kildare, have reduced travelling time on long journeys, while ring roads around cities such as Limerick will alleviate much urban congestion within their Central Business Districts.

port-side loading facilities

container cargo on an ocean carrier

Fig. 18.4
Container cargo on an ocean carrier

Pfizer chemical plant

deep-water container port

queuing area for ferry on outward journey — cars and juggernauts

incoming and outgoing foot-passenger area

Fig. 18.5
Tivoli container terminal, Cork

## Water Transport

Many of Ireland's ports have been improved to cater for increased traffic such as cargo and passengers. At some ports, quaysides have been extended to cater for containers (Dublin), oil storage (Dublin, Foynes and Cork), and warehousing and manufacturing (Waterford, Cork and Dublin).

Passenger and roll-on roll-off services have been provided at Dún Laoghaire, Ringaskiddy and Rosslare. Queuing facilities for both of these services are provided at such places, thus reducing congestion and confusion.

The most modern coal trans-shipment facilities in Europe are located at Moneypoint on the Shannon estuary. Its marine terminal can handle bulk carriers of 180,000 dwt (dead-weight tonnage) and can be upgraded to cater for 250,000-dwt vessels.

## Traffic Management in the Town of Cashel in Co. Tipperary

Fig. 18.6

yellow mesh-box to identify no parking at any time for fire engine exit
off-street parking spaces identified by white rectangular spaces sufficiently 'deep' for car parking
road markings indicate location of vehicle park
white arrows and white writing identify direction of traffic flow and route choice
traffic lanes for regulating flow of vehicles
large and wide white markings indicate bus parking spaces
large car park with spaces marked with white lines

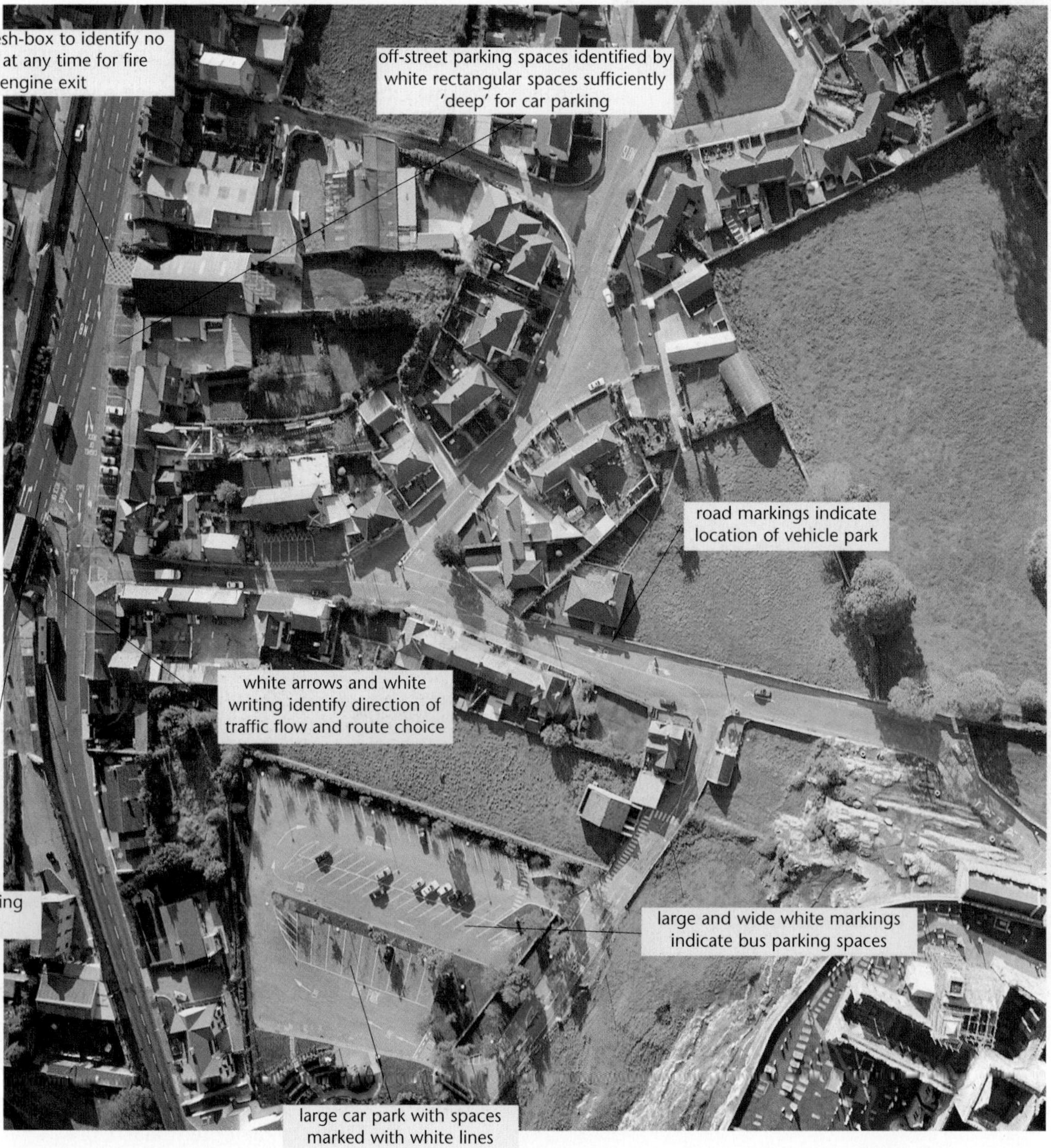

Examine, with reference to evidence from the photograph, how the civic authorities in Cashel manage motor traffic in the area. (1997)

In an examination question such as this, locate, identify and describe road/street markings, which indicate some measures used by a town to regulate traffic.

## Some Other Traffic Management Systems

Fig. 18.7
Special lane

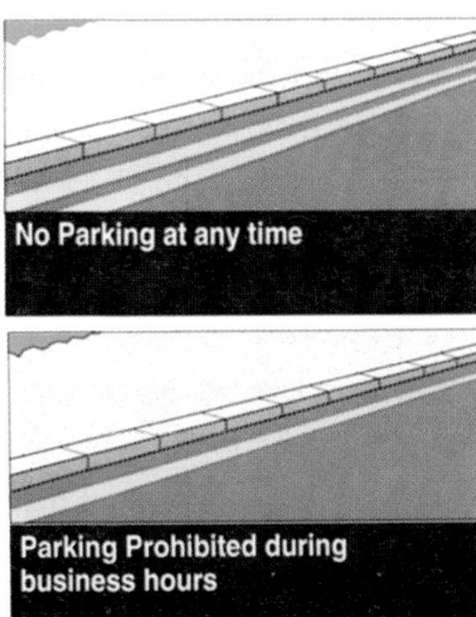

No Parking at any time
Parking Prohibited during business hours

Fig. 18.8

Fig. 18.9
Special lanes

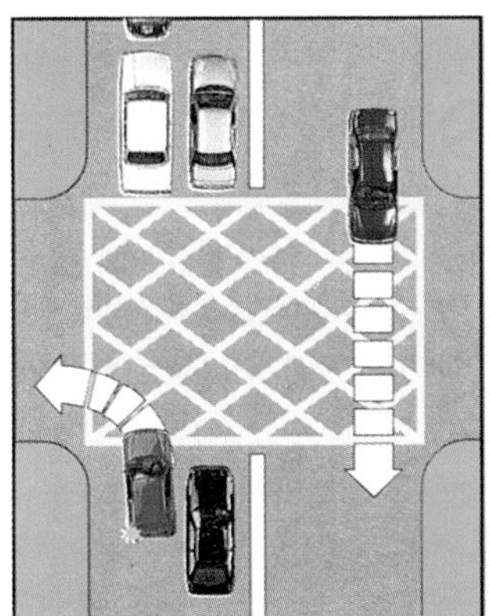

◀ DON'T ENTER THE "BOX" UNLESS YOUR EXIT AHEAD IS CLEAR.

There is one ▶ exception – if you are turning right you may enter the "box" and wait until your way is clear.

Fig. 18.10

MAKING A LEFT TURN

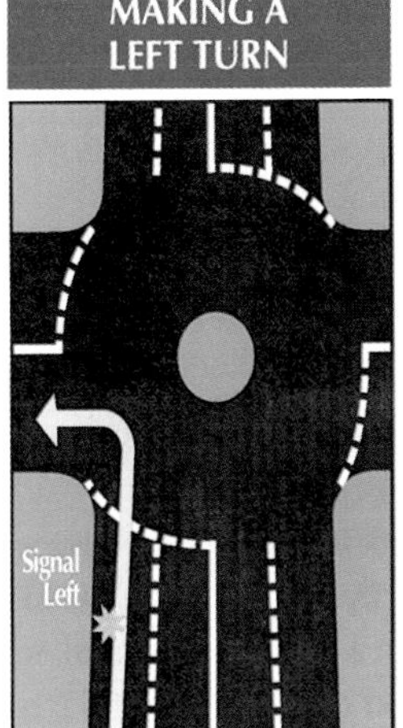

Stay in the left-hand lane

TRAVELLING STRAIGHT AHEAD

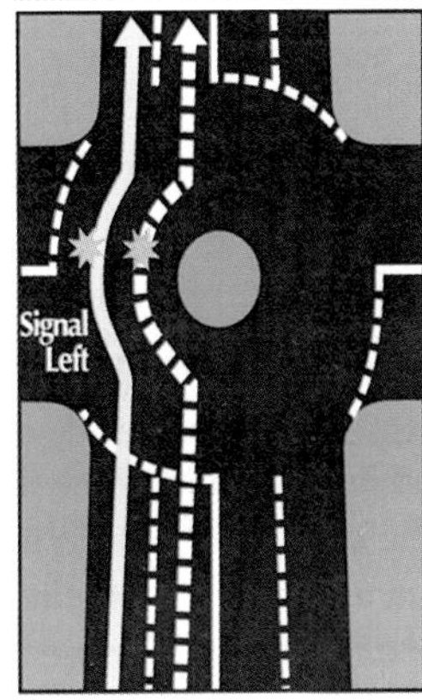

Stay in the left-hand lane, but take second exit

MAKING A RIGHT TURN

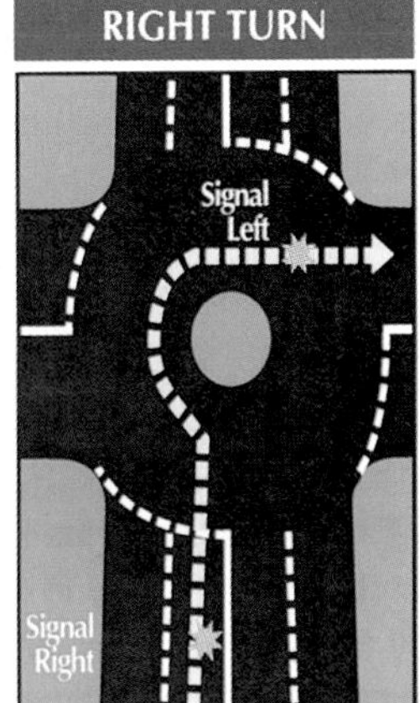

Stay in the right-hand lane

Fig. 18.11

Fig. 18.12
Safe crossing places: zebra crossings, pedestrian lights

Outline the advantages of each of the above and suggest how each adds to a town's transport management systems.

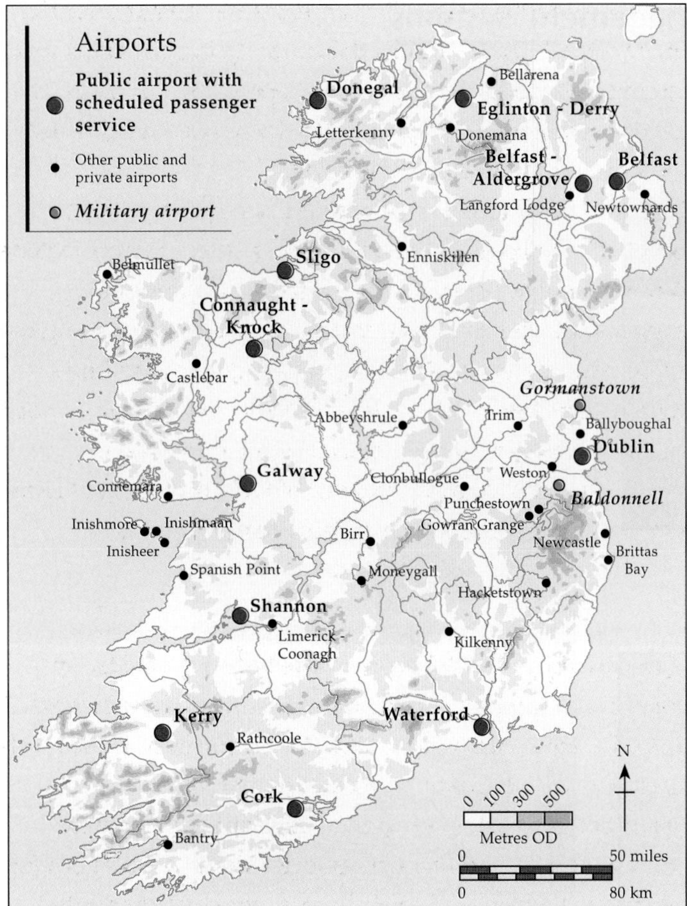


Fig. 18.13
Airports and landing strips in Ireland

## Air Transport

The development of air transport in Ireland since the 1950s has led to the creation of new growth centres such as Shannon New Town in Co. Clare. Shannon New Town is linked to Limerick city by means of a dual carriageway. This caters for very heavy commuter traffic to the Industrial Free Zone in Shannon as well as tourist traffic, especially in summer.

Until the early 1960s, tourism was mainly confined to **intranational** (within a nation) travel. Recent advances in aircraft technology, including the development of the jet engine and improved passenger accommodation in large aircraft such as the Boeing 747, have boosted international travel, so that holidays abroad are more the norm today.

The development of a new international airport at Knock in Co. Mayo and the expansion of others such as at Farranfore in Co. Kerry have helped to promote the west of Ireland for tourists and returning emigrants. Knock Shrine is becoming an important pilgrimage centre for British and American tourists, while Lahinch in Co. Clare and Killarney and Ballybunion in Co. Kerry have become golfing centres for Europeans, Americans and Japanese. Private aviation companies provide shuttle services for business executives by helicopter throughout Ireland, while the use of light aircraft for leisure has become popular in recent times.

## Urban and Rural Land Use

Town planners have divided settlements into existing and future land use zones. This ensures that residential needs and industrial wants do not come into conflict. For instance, industrial estates are generally located on national primary routes on the outskirts of towns and away from residential areas. These industrial estates are designed and serviced for industry only. As they are not in keeping with the desired aesthetic quality of housing estates (unsightly factory buildings, heavy articulated vehicles and, on occasions, unacceptable odours), they are zoned apart.

On the other hand, zones such as recreational areas (parks) and educational areas (schools) are located in close association with residential zones as they complement each other.

Residential zones differ from one area to the next. The high cost of some private housing creates varying zones of residential accommodation, which are defined by the value of land. The availability of open space for new construction is limited in some well-

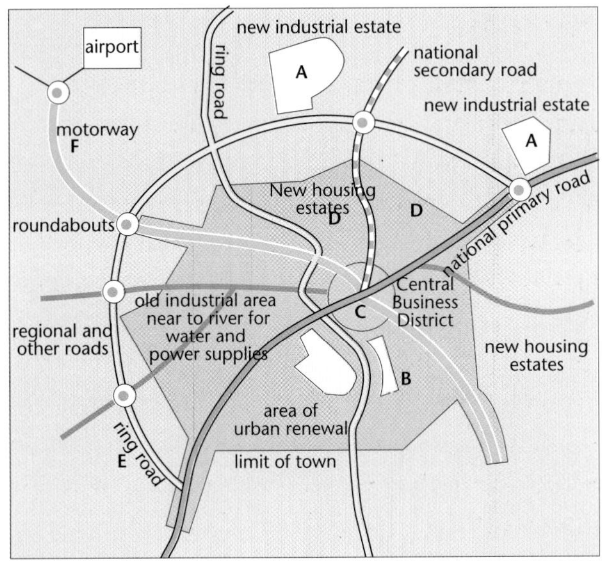


Fig. 18.14
Land use near a city

established residential areas. This increases the demand for the remaining available land and so the property in such areas soars during times of economic boom. This segregation by value creates different zones of high and low income earners, which may be reflected by local crime rates.

The future of all towns depends on a series of factors in addition to the influence of town planners. These factors include: (a) planning legislation; (b) land ownership; (c) private developers; and (d) economic forces. The likelihood of all these factors coming under the influence of one developing body is very slim so that future urban developments are difficult, if not impossible, to predict. Some trends, however, may be easy to forecast, such as the outward growth of Dublin's satellite towns to form one large **conurbation** called Greater Dublin.

## Rural Housing

Throughout the countryside, especially since the 1960s, new dwellings are constantly being built. In the early stages of modern housing development in Ireland, homes were freely built along what were then our major trunk roads. Since then, however, these roads have been upgraded and renamed and are now referred to as national primary and national secondary routes. Planning permission for all types of development on these routes, outside speed limit areas, is strictly limited since numerous exits and entrances for vehicles are regarded as potential traffic hazards and so likely to cause road accidents.

Permission for housing is therefore generally restricted to areas within speed limits, thus increasing village sizes, and to areas along less important roads such as regional and minor roads. Patterns of settlement are therefore:

1. nucleated in villages and towns
2. linear along roads
3. dispersed in farming areas or, as on occasions, in clusters in parts of the west of Ireland where remnants of the rundale system still exist

Bungalow dwellings were the preferred style of family homes during the years 1970–94. These structures offered the largest habitable floor area for a government grant-sized house. Also at this time mortgage institutions were limited to building societies and local government bodies, such as county councils. This restricted the availability of finance for housing.

Since then all banks have ventured into mortgage lending. In addition, recent higher incomes and increased numbers of third-level graduates as well as low interest rates have created a surge of very large detached family homes in suburban housing schemes and in rural areas.

In recent years, planning permission for linear settlement is also being restricted as a consequence of contamination of local water supplies by ineffective septic tank percolation areas.

Fig. 18.15
Clondalkin in Co. Dublin

The settlement in Fig. 18.15 originated in early Christian times. It was initially a monastic settlement.

With this information in mind:

1. (a) **Carefully** examine the street pattern and identify, if you can, the location of that early monastery around which this settlement has developed.
   (b) Identify the building(s) associated with this early settlement. Can you find the round tower?
   (c) Draw a sketch map to help explain your answers to (a) and (b) above.
2. Identify the function/land uses at A, B, C and D. Describe the buildings and associated characteristics that are typical of such land uses at A and B. In each case use evidence from the photograph to explain your answer.
3. This type of photograph is generally classified as a vertical photograph. Can you develop any argument to cast doubt on this classification? (Examine the edges of the photograph.)

Fig. 18.16
Birr in Co. Offaly

Study the photograph of Birr in Co. Offaly (Fig. 18.16). Then do the following:

1. Account for the origin and subsequent development of the town shown in the photograph. Use evidence in the photograph to support your answer.
2. The landscape in the southern half seems to display an undulating appearance on this vertical photograph. Give a reasoned explanation for this landform, using evidence in the photograph to support your answer. See glacial deposition, chapter 12.
3. Account for the field pattern shown in the photograph.
4. Account for the distribution of woodland shown in the photograph.

Fig. 18.17

Study the photograph, Fig. 18.17. Then do the following:

1. Identify the town (this town has already been dealt with).
2. What characteristics of the settlement have helped you in your identification?
3. What era of development has been instrumental in the success of this town?
4. Identify the largest area of industrial development on the photograph. With reference to the photograph, give reasons for its location.
5. The pattern of recent settlement has been influenced by two main factors. Identify each influence and account for the pattern.
6. Explain the street pattern and its association with the development of the town.

Fig. 18.18

Study the photograph of Clara in Co. Offaly, Fig. 18.18. Then do the following:

1. (a) Account for the origins of the varying surface materials (soils) in this area. Use evidence from the photograph to support your answer.
   (b) How have these materials influenced the economic development of the area?
   (c) How have these materials influenced the pattern and shape of the fields?
2. Draw a sketch map and on it mark and label
   - the transport network
   - sites of extractive primary industry
   - two factories
   - one housing estate
   - one area of woodland
3. Account for the location and subsequent development of the town from a study of its buildings, street plan and transport system.
4. Suppose that you are moving with your family to this region. Give reasons why this area is attractive for your family home.
5. Account for the patterns in the distribution of population.

Fig. 18.19

Study the photograph of Downpatrick in Co. Down, Fig. 18.19. Then do the following:

1. Draw a sketch map of the area shown on the photograph. On it mark and label
   - five areas of different land uses
   - the main roadways and streets
   - the Quoile River
   - an early Norman settlement
2. Account for the field pattern and agricultural land use in the photograph (see glacial landforms, p. 215).
3. (a) Discuss and account for the pattern in the distribution of woodland.
   (b) (i) What species of tree dominates in this area?
   (ii) How would this tree type influence the local flora and fauna? (Tip — deciduous/woodland plant and animal life.)
4. Account for the origin of this town (examine the town's name).
5. (a) What type of early Norman settlement is located in the south-west corner of the photograph?
   (b) What does this tell you about the early stages of this settlement?
6. Describe and explain the course of the Quoile River. (Locate Downpatrick on your atlas to help you with this question.)

Fig. 18.20

Carefully study the photograph of Graiguenamanagh, on the River Barrow, Fig. 18.20. Then do the following:

1. Draw a sketch map of the area shown on the photograph. On it mark and label
   - the River Barrow
   - the street pattern
   - the canal section
   - an old castle or millhouse
   - an old river cliff
   - a disused pathway near the canal
   - four areas of different land use
2. What was the purpose of the pathway near the canal? Explain.
3. Explain using evidence from the photograph the main function of this settlement in the eighteenth century.
4. What material do you think was quarried in the north-east of the photograph. Support your answer using evidence on the photograph.
5. Account for the pattern in the distribution of woodland.

# Environmental Studies

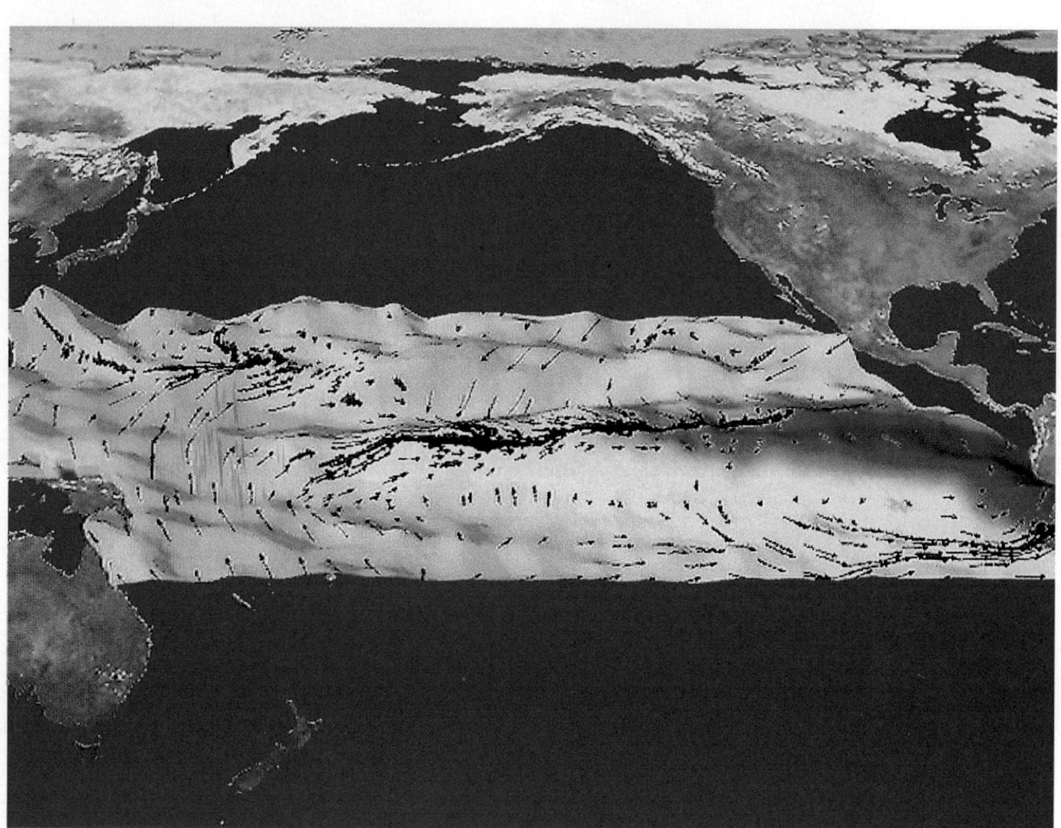

Fig. 19.1
Conditions in the Pacific Ocean leading to El Niño

## El Niño

El Niño means 'the little boy' or 'the Christ child' in Spanish. This name is given to a warm body of ocean water that occurs every three to seven years in the tropical Pacific Ocean. The 'El Niño effect' within this Pacific region is based on changes in wind patterns and the way heat is exchanged between ocean and air when El Niño moves away from its 'normal' position. When it does occur, its climatic effects are wide-ranging, triggering an intricate set of changes in the ocean and the atmosphere across the 'tropical Pacific', which covers a third of the earth's circumference.

### Normal Years — Characteristics of the East Pacific and West Pacific Regions

1. **East Pacific Region**
   (a) Trade winds blow from east to west across the Pacific Ocean. These winds push the cold east Pacific water westward across the equator. As it travels, the water is heated by the equatorial sunshine and by the time it reaches Indonesia and south-east Asia it has become a vast warm ocean body.

**Fig. 19.2**
Normal

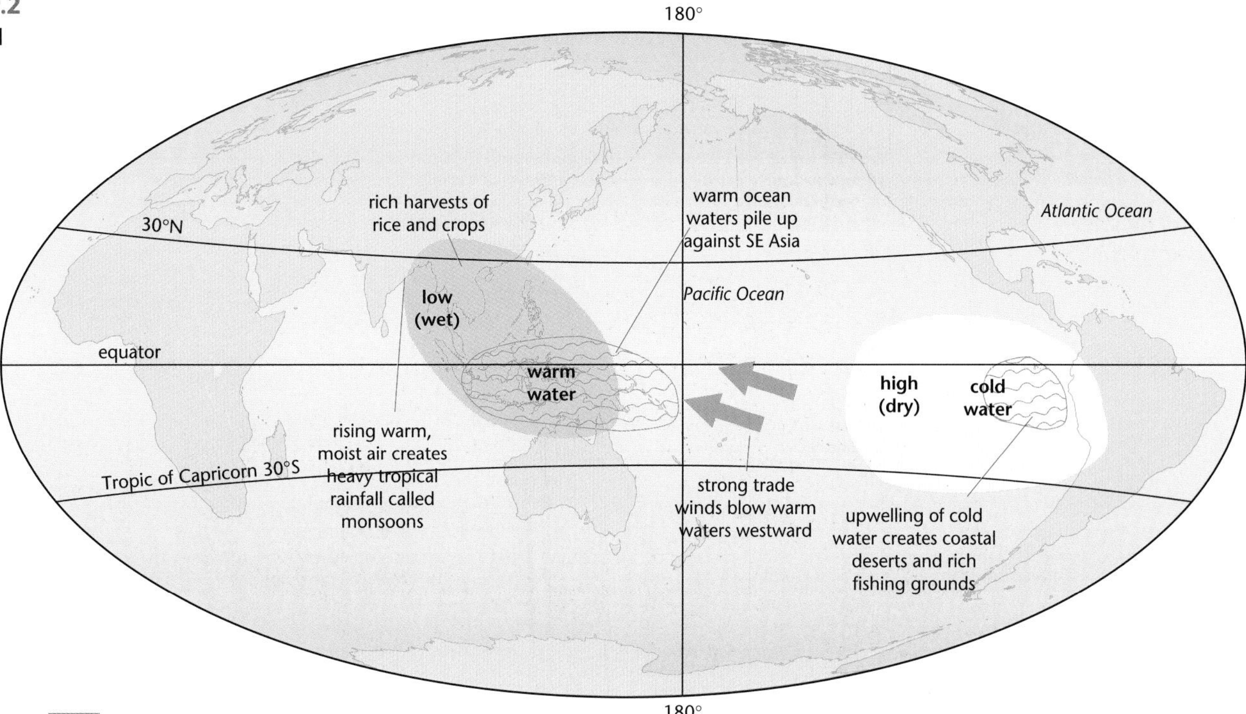
180°
rich harvests of rice and crops
warm ocean waters pile up against SE Asia
Atlantic Ocean
30°N
low (wet)
Pacific Ocean
equator
warm water
high (dry)
cold water
rising warm, moist air creates heavy tropical rainfall called monsoons
Tropic of Capricorn 30°S
strong trade winds blow warm waters westward
upwelling of cold water creates coastal deserts and rich fishing grounds
180°

The torrential rain that falls during the summers in south-east Asia is called the monsoons.

(b) Winds blowing across the cold Peru current are cooled and cause fog at sea. However, when they move inland they are heated and become very dry, creating clear skies. These movements are a major factor in the creation of the Atacama Desert in coastal Peru and northern Chile. This is, normally, one of the driest areas of the world (see deserts, Figs. 14.3 and 14.4, p. 286).

(c) Upwelling of cold deep ocean water off Peru and Ecuador brings nutrient-rich water to the surface. This encourages the growth of plankton, which in turn attracts vast numbers of fish, such as anchovies, to the region.

### 2. West Pacific Region

(a) Warm ocean surface water heats the air above it. This warm, moist air rises quickly, creating an extensive area of low pressure over south-east Asia in July and over northern Australia in January. This causes monsoon rain.

(b) Heavy rainfall and high temperatures allow many crops to be harvested each year in Indonesia, the Philippines and India (see Fig. 19.2).

## El Niño

El Niño, when it does occur, reverses the 'normal' weather systems over the west and eastern Pacific Ocean. The cold waters off Peru and Ecuador are replaced with warm waters. The high pressure system over coastal Peru and Ecuador is replaced with a low pressure system. The dry coastal deserts suffer torrential rainstorms, mudflows and road and bridge destruction. The normally rain-drenched west Pacific suffers drought, creating forest fires and crop failure.

LANDSCAPES OF THE WORLD

Fig. 19.3
El Niño

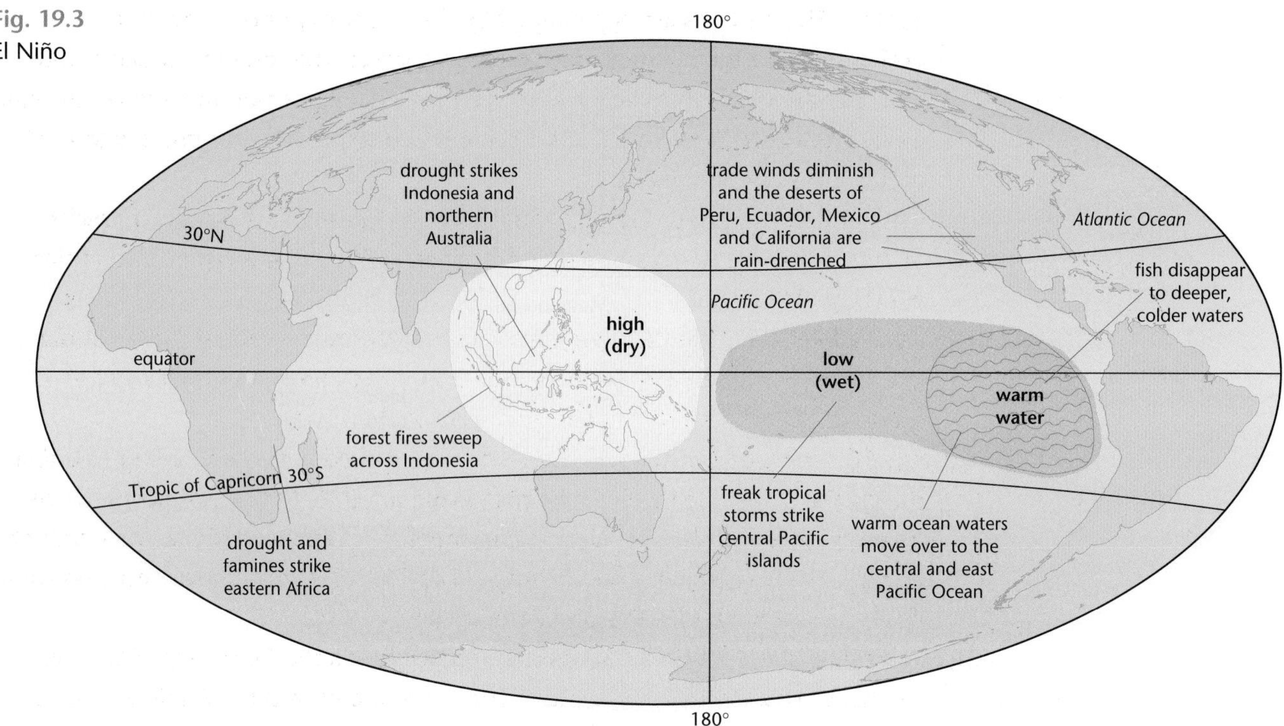
180°
drought strikes Indonesia and northern Australia
trade winds diminish and the deserts of Peru, Ecuador, Mexico and California are rain-drenched
Atlantic Ocean
30°N
fish disappear to deeper, colder waters
Pacific Ocean
high (dry)
equator
low (wet)
warm water
forest fires sweep across Indonesia
Tropic of Capricorn 30°S
freak tropical storms strike central Pacific islands
warm ocean waters move over to the central and east Pacific Ocean
drought and famines strike eastern Africa
180°

## 1. East Pacific Region

(a) The tropical low pressure system that normally builds up over south-east Asia and northern Australia overshoots and centres itself over the central tropical Pacific. This patch of the ocean, the size of the USA, becomes the centre of a tropical storm system that can change atmospheric circulation around the globe.

(b) The trade winds reduce in strength. This creates a chain reaction.

 (i) The cold tongue of ocean water in the east retreats towards South America.

 (ii) This in turn allows the warm west Pacific water to surge eastward towards Peru and Ecuador.

 (iii) As the east Pacific warms, the trades die down and as they die down, the water warms. This becomes a feedback loop in a chain reaction. The east Pacific sea surface temperatures rise from one to five degrees centigrade.

(c) The fish off Ecuador, Peru and northern Chile dive for or move to cooler waters rich in nutrients and oxygen and become out of reach for the Peruvian and Ecuadorean fishing vessels. This destroys the region's fishing industry and the local impact is devastating.

(d) The deserts of Peru, Ecuador, Mexico and California are rain-drenched, creating mudslides and landslides, killing hundreds of people at a time, destroying homes and cars, eroding beaches and closing roads.

## 2. West Pacific Region

(a) In the western Pacific, forest fires started by farmers (slash and burn farming), and logging companies, generally kept in check by the tropical rains, go out of

control. These blazes are encouraged by the tinder-dry undergrowth, and firefighters find them almost impossible to control. The resultant choking and blinding smog carries toxic fumes of sulphur dioxide, carbon monoxide, nitrogen oxide and ozone. This smog closes schools and airports and at times is so thick that it turns day into night.

(b) Parts of Indonesia and Australia and eastern Africa, such as Zimbabwe, suffer extreme drought. This causes severe famines in the underdeveloped countries which are affected.

(c) Meanwhile heavy rainfall causes freak tropical storms on central Pacific islands. Typhoons hit Hawaii and Tahiti. Island bird life is destroyed as their nests in sandbars are flooded by the rain and waves.

(d) Intense monsoon rainfall, well above normal, caused extreme flooding in China on the flood plain of the River Yangtze Kiang in 1998. Millions of people were made homeless, thousands died in the flooding and billions of pounds worth of damage was inflicted on the landscape. The 1998 El Niño was the most severe to date and the biggest in the twentieth century.

The most baffling development in recent years is the frequency of El Niño events. Over a period of four years, 1994–8, there were three El Niños. This is causing concern. Does it mean that El Niños are no longer confined to the three- to seven-year cycle? Does it mean that El Niños are going to be the norm? Are El Niños being triggered by global warming? Will all unusual climatic phenomena be blamed on El Niño? As yet there are no definitive answers to these questions.

What we do know is that climates have changed before. Is the climate changing now, and if it is will it change as rapidly as that being proposed for the Greenhouse Effect? At present, all we can do is speculate.

## Ozone in the Atmosphere

### Formation of Ozone

Ozone is created when ordinary oxygen molecules $O_2$ are bombarded with ultraviolet rays high up in the atmosphere. This radiation shatters the oxygen ($O_2$) molecules into unstable single atoms, O, which readily combine with other $O_2$ molecules to form ozone ($O_3$).

The newly formed **ozone** ($O_3$) **has a property that ordinary oxygen does not have: it can efficiently absorb ultraviolet light**. In doing so, ozone protects oxygen at lower altitudes from being broken up and keeps most of these harmful ultraviolet rays from penetrating to the earth's surface.

### Ozone Destruction

Ozone is a form of oxygen, $O_3$, that has three atoms instead of the normal two in each molecule.

At ground level, ozone can be dangerous to humans, but high up in the atmosphere the maintenance of a thin layer of ozone is essential for human, animal and plant life.

This ozone layer absorbs most of the harmful ultraviolet rays of the sun and thus prevents serious damage to life on earth.

In modern times, human activity has introduced chemicals that destroy this ozone layer. The most destructive chemicals contain chlorine (e.g. CFCs) and bromine (e.g. halons). Chlorine and bromine steal oxygen atoms from the ozone molecule and change it into ordinary oxygen. Each chlorine and bromine atom can do this many thousands of times (see Fig. 19.4 (c)).

Chlorofluorocarbons (CFCs) were developed as cooling agents for fridges and air-conditioning systems. Since the 1950s they have been used as propellants in aerosol cans. They are also used to blow foam for furniture, packaging and insulation, and to clean delicate computer circuits.

(a)

(c)

$$CL + O_3 \rightarrow CLO + O_2$$
$$CLO + O \rightarrow CL + O_2$$

Ultraviolet rays split CFCs, releasing chlorine. This creates a chain reaction which depletes the ozone layer.

**Fig 19.4 (a) and (b)**
Destruction of El Doñana National Park in Andalusia, Spain in 1996 is an example of soil pollution. The wall of a reservoir holding mining residues (top photo), 40 km upriver from the park, collapsed. Acidic water swept down the valley into the river system and poisoned the park (bottom photo).

(b)

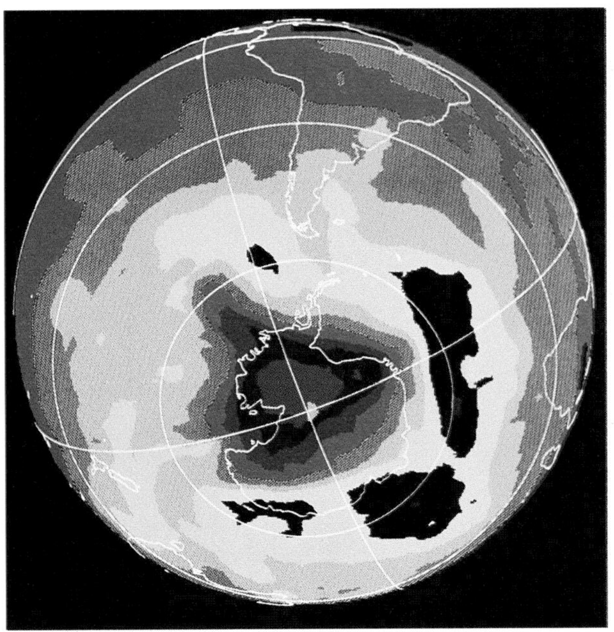

Fig. 19.5
Satellite image showing ozone hole over Antarctica

Near the earth's surface, CFCs are immune to destruction. But high in the stratosphere they are affected by ultraviolet light. They break apart easily and a chlorine atom is released from a chlorofluorocarbon molecule (CFC). This chlorine atom (CL), a man-made gas, attacks ozone ($O_3$) and steals an atom of oxygen (O) to form chlorine monoxide (CLO) and an oxygen atom ($O_2$).

Then chlorine monoxide combines with a free oxygen atom to form oxygen ($O_2$) and a chlorine atom (CL). This chain reaction repeats itself indefinitely. For every chlorine atom you release, says F. Sherwood Rowland, the person who initially highlighted ozone destruction, 100,000 molecules of ozone are destroyed. Thus the threat to life is enormous.

## Solutions to Ozone Destruction

In 1985 a convention was concluded under the supervision of the United Nations Environment Programme (UNEP). This led to a meeting at which countries agreed that the production and consumption of all CFCs should be phased out by the year 2000. Intermediate targets were set to achieve a 50 per cent reduction by 1995 and an 85 per cent reduction by 1997. The EU countries promised to phase CFCs out by 1997. These agreements also supported the phasing out of halons (chemical compounds which contain chlorine and bromine) by 2000 and recognised that other substances that contribute significantly to ozone depletion (e.g. chlorine) should be controlled. Global action is the only way of tackling the depletion of the ozone layer. Another convention was held in December 1997 in Kyoto in Japan, where new targets were set to improve the atmosphere.

### Research Activity

1. Electricity/energy/power generation in recent decades has had a dramatic effect on
   (a) the atmosphere
   (b) the waters of the earth.
   Discuss this statement fully. Use case studies to develop your answer.

## The Greenhouse Effect and Global Warming

# CLIMATE CHANGE TREATY

Global warming or just hot air?

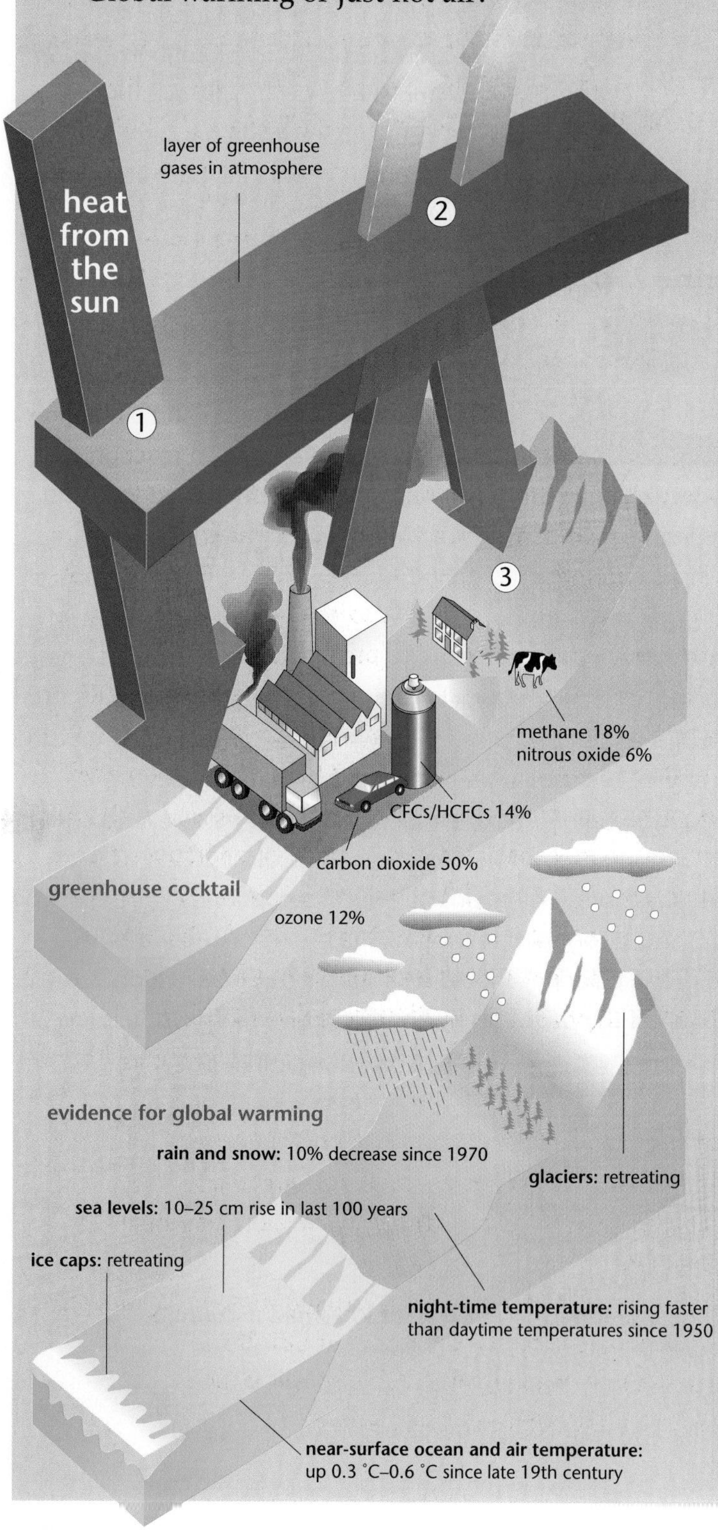

heat from the sun

layer of greenhouse gases in atmosphere

① ② ③

methane 18%
nitrous oxide 6%

CFCs/HCFCs 14%

carbon dioxide 50%

greenhouse cocktail

ozone 12%

evidence for global warming

**rain and snow:** 10% decrease since 1970

**glaciers:** retreating

**sea levels:** 10–25 cm rise in last 100 years

**ice caps:** retreating

**night-time temperature:** rising faster than daytime temperatures since 1950

**near-surface ocean and air temperature:** up 0.3 °C–0.6 °C since late 19th century

### UNITED NATIONS CLIMATE CONFERENCE

**More than 150 signatories of the 1992 UN climate change convention met in Kyoto in December 1997 to hammer out binding targets for industrialised nations to cut emissions of 'greenhouse gases' in the next century.**

Targets vary from Japan's average of 5% from 1990 levels by 2012 to the EU's 15% by 2010. A UN panel of 2,000 scientists predicts unchecked greenhouse gas emissions will warm the planet by up to 3°C over the next 100 years. This compares to a 1° rise since the start of the Industrial Revolution and a 4° rise since the last Ice Age, 18,000 years ago.

### The greenhouse effect

**1. Sunlight:** Earth's atmospheric blanket of gases allows most of the sun's radiation to pass through, warming up both the atmosphere and the land and sea beneath.

**2. Radiated heat:** The warmed earth radiates solar energy, as heat, back into space.

**3. Trapped radiation:** The cocktail of gases – mostly carbon dioxide – acts like the glass in a greenhouse, reflecting much of the solar energy back into the atmosphere.

### Sources of greenhouse gases

| | |
|---|---|
| Carbon dioxide | From burning fossil fuels, especially coal, oil and wood |
| CFCs/HCFCs | Used in aerosols and older refrigerators |
| Methane | Produced by livestock and landfill rubbish tips |
| Nitrous oxide | Use of agricultural fertilisers |
| Ozone | Caused by effect of sunlight on atmospheric |

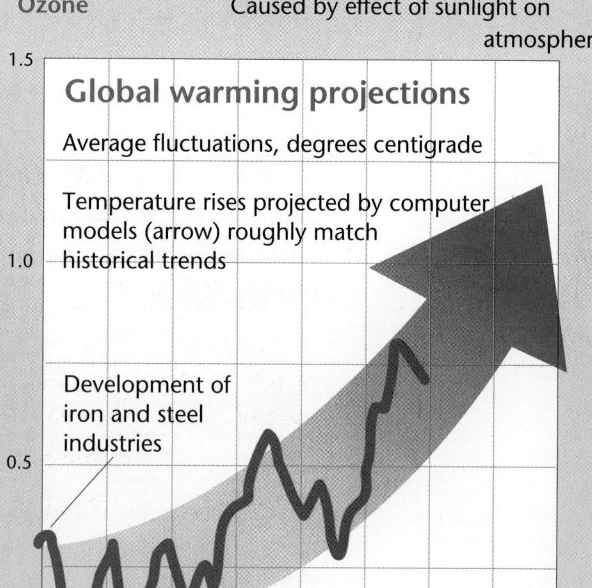


**Sources:** IPCC, US National Climatic Data Center, Reuter, Associated Press

Fig. 19.6

## What is Meant by the Greenhouse Effect

The earth's land surface and oceans are warmed during the day by incoming light from the sun. This light/energy is called **insolation** and it travels through our atmosphere in the form of short-wave radiation. The warm earth's surface in turn heats the air above it and at night this heat radiates outward in the form of longer-wave radiation. Normally, water vapour in the atmosphere is the most important gas for trapping this outgoing heat, but trace gases such as carbon dioxide, methane, nitrous oxide and ozone also retain some of this heat.

Because these gases trap heat much as a greenhouse does, they are referred to as greenhouse gases. Without these naturally occurring greenhouse gases the earth's average temperature would be 33 °C colder than it is today — far too cold for life as we know it. During the last ice age, average temperatures were only 4 °C colder. This balance of heating and cooling allows the earth's natural environment to remain as it is.

## Global Warming

Over the past 150 years a change in the natural balance of the earth's atmosphere has been occurring. As **industrialisation** has increased from modest beginnings in AD 1750 in Britain to vast proportions worldwide in AD 2000, so too have the emissions of carbon dioxide and nitrous and sulphur oxides.

In 1850 the world's **population** was approximately 1,200 million people. Its population today is approximately **6,000 million**. Most of these people use fossil fuels in their domestic fires for heating and cooking, a practice which adds greatly to the amounts of heat-retaining gases. These greenhouse gases allow the sun's rays through but will not allow excess heat from the earth's surface out into space. Any increase in these gases should, theoretically, create a corresponding increase in atmospheric temperatures, leading to global warming.

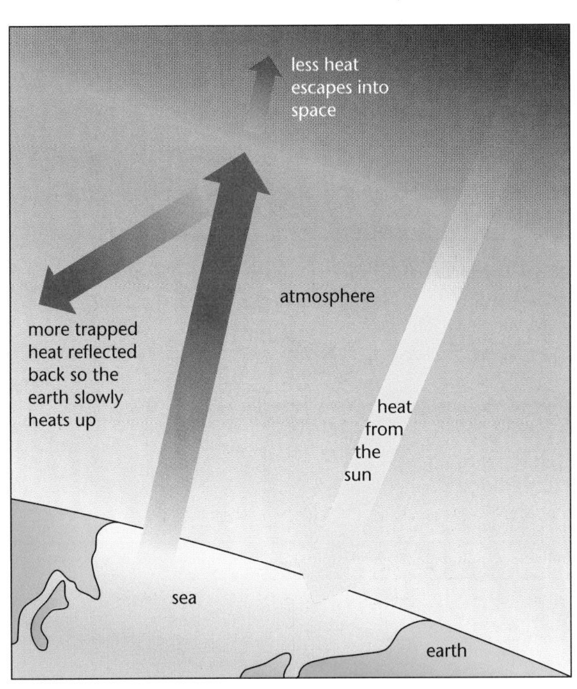


Fig. 19.7
The Greenhouse Effect

### Sources of Greenhouse Gases

| | **Gas Sources (Natural and Man-made)** |
|---|---|
| **water vapour** | evaporation from the ocean, evapotranspiration from land |
| **carbon dioxide** | burning of fossil fuels (power houses, industry, transport), burning rainforests, respiration |
| **methane** | decaying vegetation (peat and in swamps), farming (fermenting animal dung and rice growing), sewage disposal and landfill sites |
| **nitrous oxide** | vehicle exhausts, fertiliser, nylon manufacture, power stations |
| **CFCs** | refrigerators, aerosol sprays, solvents and foams |

**Did You Know?**

To go from an ice age to a period of warmth requires a change of just 5 °C to 6 °C. The rise in the sea level over the same period will be half a metre, enough to flood Bangladesh and pose a real threat to the Nile delta, parts of China and much of the South Seas coastlines in south-east Asia. In Europe, coastal stretches from Biarritz in south-west France to the Baltic Sea would be likely to be submerged.

Floods and droughts would occur in different parts of the world to existing deserts due to atmospheric changes, such as wind patterns and associated water cycle.

Class Activity

1. Explain what is meant by acid rain.
2. What effects has acid rain on
   (a) forests
   (b) lakes and fish in Finland and Sweden
   (c) crops?

See P.E.F. O'Dwyer, *Geography Revision Notes for Leaving Certificate* (new edition, 1999), Gill & Macmillan.

## Proof of Greenhouse Gas Increase

The ice caps of Greenland and Antarctica have built up over hundreds of thousands of years. Trapped within these annual layers of ice are bubbles of 'fossil air' — tiny samples of the atmosphere that existed at the time that each layer formed when on the surface.

Ice cores have been drilled in the Antarctic and Greenland glaciers. By analysis of the trapped air bubbles and the content of the ice cores, layer by layer, scientists are able to trace the rise and fall of atmospheric carbon dioxide over the time-span covered by the core.

The fossil air bubbles indicated that over the past 250 years, there has been a steady growth in the percentage of carbon dioxide in the atmosphere. This growth rate is identical to the growth in the amount of fossil fuels being burned since the beginning of the Industrial Revolution.

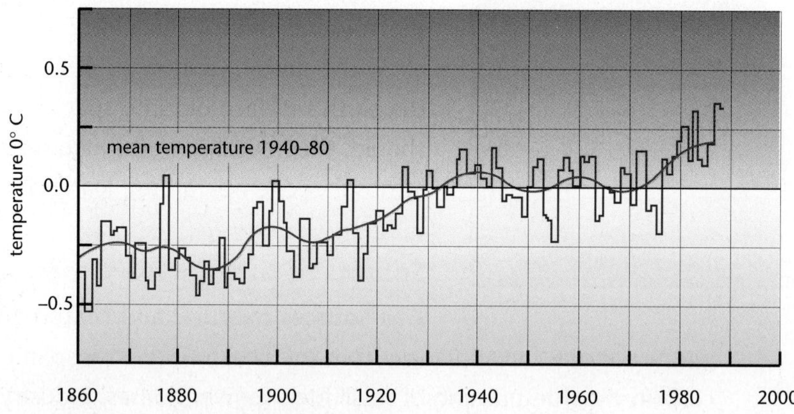


Fig. 19.8
Average global temperatures since 1860

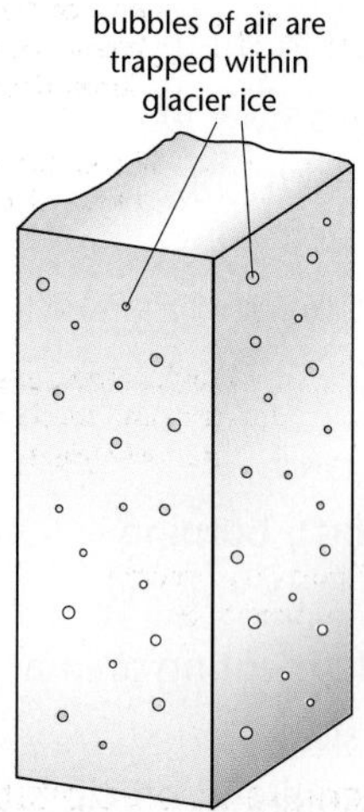


Fig. 19.9
Arctic ice core showing fossil air in glacier ice

Scientists believe that the increase of greenhouse gases will lead to an increase in worldwide temperatures. This process is called **global warming**.

## Effects of Global Warming

Present evidence suggests that global warming is having the following effects on the earth's systems.

### Glacier Melt

1. The glaciers of Mount Kenya and the Andes are in retreat. So also is the Arctic sea ice, which has shrunk 2 per cent in the past thirty years.

### Famines

2. Increased drought and subsequent famines in marginal tropical areas, such as the Sahel and African east coast countries.

### Tropical Storms

3. Increased frequency and severity of tropical storms such as tornadoes and hurricanes.
4. Increased frequency of El Niño in the Pacific Ocean (see 'El Niño', p. 407).

### Ice Cap Melting

5. Scientists at the British Antarctic Survey who identified the ozone hole in 1985 now believe that 'a dangerous warming' is first becoming evident in the frozen continent. Flowers and grass are spreading, and more than two-thirds of the 2,000-square-kilometre Wordie Ice Sheet has melted away.

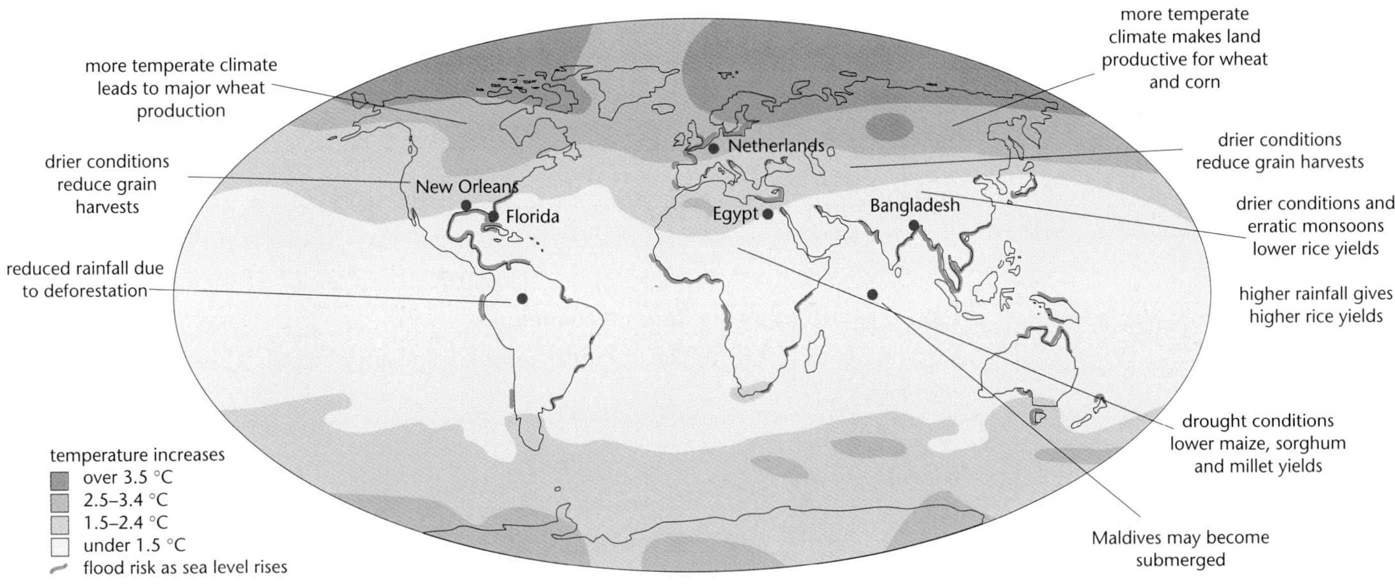


**Fig. 19.10**
Weather forecast for the year 2100: assuming, by then, $CO_2$ content in the atmosphere has doubled

### Water Supply

6. Computer projections suggest that there will be a significant increase in precipitation in some areas and a decrease in others. This in turn could lead to great shortages of water supply in places such as Los Angeles, which is only barely able to cope with present demands. Increased hot spells and extreme cold spells may become commonplace.

## Rise in Sea Levels

7. Higher temperatures will cause sea levels to rise because the water in the oceans will expand and the polar ice sheets will get smaller. It has been calculated that levels will rise by about 1.5 metres in the next forty years.

   The predicted rise in sea levels is such that it could devastate south-east Asian rice-growing delta areas and flood many of the world's major coastal cities.

### Island-nation Disappearance

In 1998 an alliance of the world's small island nations — several of which are due to disappear altogether as sea levels rise — formally drafted a treaty which would bind industrialised countries to cut carbon dioxide emissions by 20 per cent by the year 2005. As it lacks the backing of powerful countries such as Britain, it stands little chance of success.

## Effects of Proposed Sea Level Rise in Ireland

8. In Ireland there are many places that are vulnerable to flooding from rising sea levels, including the following:
   - The southern coast is likely to suffer first. Cork Harbour has some 18,000 hectares at risk by 2030. The harbour is a drowned river valley and any further rise in sea level is likely to reinforce a tendency to flood its low-lying margin. Other parts of Co. Cork likely to be affected include river estuaries such as Kinsale and Clonakilty.
   - Dublin Bay areas such as Sandymount and Sandycove in the past have been prone to flooding from easterly winds and storm surges.
   - Other areas:

     — The impact on the Shannon estuary is likely to be extensive with some 19,000 hectares at risk.

     — Further south, reclaimed polderland in Wexford may be affected.
   - Co. Mayo includes more land at risk than any other county. This includes parts of Clew Bay and other sheltered estuaries.
   - Large parts of the Belmullet peninsula are composed of sand dunes and may simply disappear.
   - Sligo town may be vulnerable to high tides and surge floods.
   - The northern half of Ireland has been rising slowly ever since the last ice age, when the ice cap crushed down the north but not the south. This reduces the risk of flooding in many areas. Co. Donegal is the least likely area to be flooded by AD 2030.

**Home Activity**

Research and explain how the following may influence global warming:

- farming
- aerosols

## Plate Tectonics and Global Warming

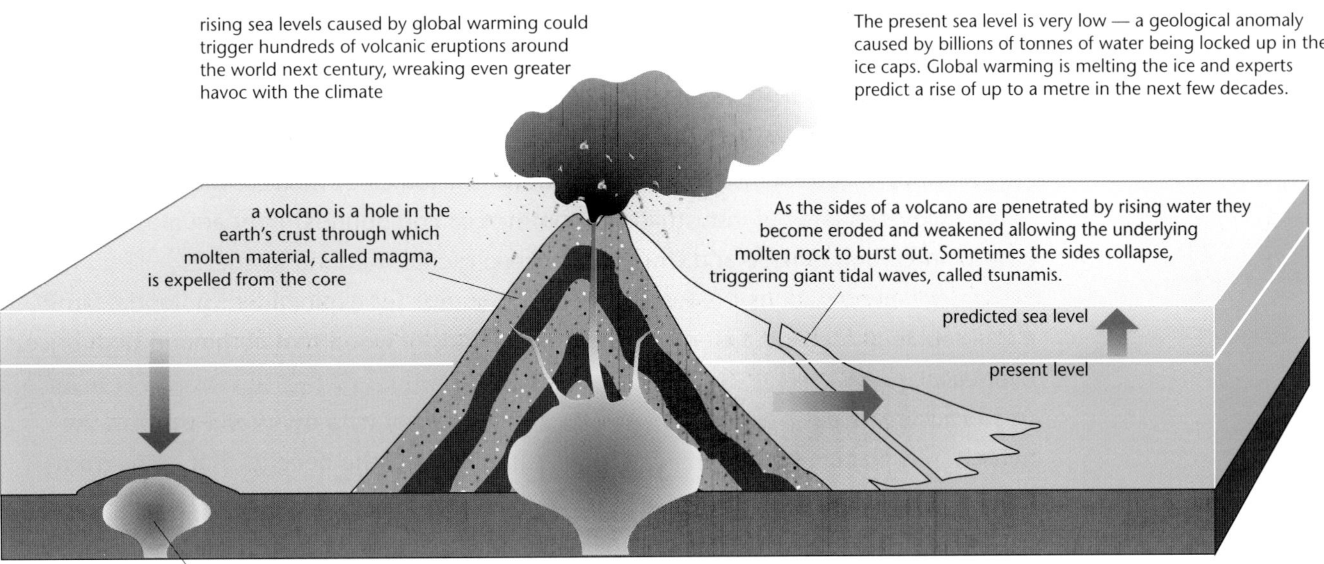

rising sea levels caused by global warming could trigger hundreds of volcanic eruptions around the world next century, wreaking even greater havoc with the climate
The present sea level is very low — a geological anomaly caused by billions of tonnes of water being locked up in the ice caps. Global warming is melting the ice and experts predict a rise of up to a metre in the next few decades.
a volcano is a hole in the earth's crust through which molten material, called magma, is expelled from the core
As the sides of a volcano are penetrated by rising water they become eroded and weakened allowing the underlying molten rock to burst out. Sometimes the sides collapse, triggering giant tidal waves, called tsunamis.
predicted sea level
present level
huge magma reservoirs often lie close to the seabed, held in check by the water pressure above — but if levels change the seabed may crack, allowing the magma out and forming new volcanoes


**Fig. 19.11**
The volcanic century

Sea level changes caused by global warming could trigger the eruption of hundreds of new or dormant volcanoes, turning the twenty-first century into the fieriest on record, according to a study by British scientists.

Deciding that the best way to predict the future behaviour of volcanoes was to study their distant past, they found that rises in sea levels caused by periods of global warming have almost all been followed by a surge in volcanic activity. An article published in *Nature* magazine described how the crucial geological evidence dated back 12,000 years to when sea levels surged by 12 metres in two centuries — immediately followed by intense volcanic activity across the world. By contrast, there were far fewer eruptions when sea levels were stable, as during the ice age that ended 18,000 years ago.

### Class Activity

1. (a) What is smog?
   (b) How is it formed?
   (c) How does it affect people and the environment?
2. (a) Identify three sources of water pollution.
   (b) Explain fully how each of these sources pollutes water. In your answer refer to fresh- and salt-water bodies.

See P.E.F. O'Dwyer, *Geography Revision Notes for Leaving Certificate* (**new 1999 edition**), Gill & Macmillan.

### Causes

The team observed that 90 per cent of volcanoes are close to or surrounded by sea. As water rises it begins to erode the lava. The rising water also causes landslides and weakens the outside rock. Eventually the mountain becomes unable to withstand the internal pressure of the molten rock and explodes.

The long-term consequences of a surge in eruptions remain unclear, however, because different types of eruption have different products. One possibility is that the millions of tonnes of ash thrown into the atmosphere could reflect enough sunlight to cool the earth, reverse global warming or even prompt a new ice age. Alternatively, if the explosions released huge volumes of greenhouse gases, the earth could heat up even faster.

## Deforestation in the Tropics

The world's tropical forests are being cleared at the rate of 14 hectares per minute or 7.3 million hectares a year. As a result, fragile tropical soils are being turned into wastelands, many tribal peoples are being wiped out, and thousands of unique plant and animal species are being destroyed.

To industrial countries, tropical forests are a vast source of medical vaccines, as well as an undiscovered array of industrial and commercial compounds. They are also a means of absorbing carbon dioxide and thus help to keep global warming in check.

To developing nations the forests are resources ripe for exploitation: potential farmland, a free source of fuel and a storehouse of exotic kinds of wood that command high prices overseas.

Much debate on these forests centres on the issues of who owns and controls the genetic information stored in these species. Traditionally, the benefits that come from genetic materials — seeds, specimens or drugs derived from plants and animals — go to whoever finds a way to exploit them.

Vanilla, for example, was a biological resource found only in Central America. It later became an important cash crop in Madagascar. Now a US biotechnology company has developed a process to clone the vanilla flavour in a cell culture. If the firm sells the bioengineered versions for less than natural vanilla and takes some of the market share, who will compensate the Madagascar farmers, or the Central American Indians from whose lands the genetic material originated?

Recently it was hoped that agreement could be reached that would include a provision making genetic materials of all kinds the sovereign resource of the originating country. Nations would have control over who had access to their genetic resources, and if someone else found a way to make money from them, the originating country would collect royalties on each sale.

**Research Activity**

1. Outline three reasons why tropical forest environments are being destroyed in developing countries.
2. What effect has this deforestation on
   (a) soil (c) water
   (b) air (d) indigenous (native) people(s)?

# The Expanding Deserts

### Desertification in Africa

Desertification is the spreading of deserts into the surrounding regions due to

(a) human interference
(b) changing weather patterns
(c) other influences or a combination of (a) and (b) above

## Case Study: The Sahel Region of North Africa

The Sahel region of Africa — a dry land sixty-three times the size of Ireland, straddling parts of six of the world's poorest countries — was at one time a prosperous empire where trading, herding and farming tribes interacted well and kept their environment in balance. They exchanged meat and milk for cereals and vegetables. In the dry season, herds grazed the leftover cereal plants and manured the land in doing so. Rotational grazing (grazing cereal stubble) and farming patterns placed no strain on the soil.

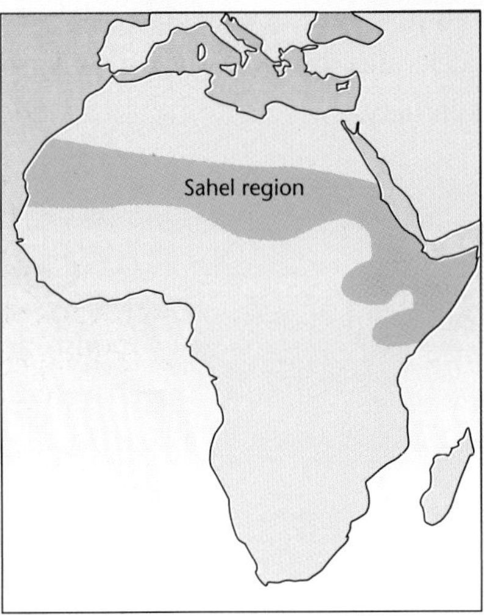


Fig. 19.12
Sahel region

Natural vegetation is destroyed because of:

- demand for wood in growing urban centres
- trees and natural vegetation destroyed to make way for cash crops
- overgrazing leading to exposure of soil and cutting of trees for income

### Human Interference

Colonisation (by France, Spain and Portugal), private land rights and cash cropping have damaged the delicate ecological balance in the Sahel.

In colonial times, administrative centres grew up across the Sahel. This urban development continued into recent times, creating a huge demand for wood for fuel and construction.

This deforestation in association with global warming (the 'Greenhouse Effect') has increased the water supply problem in marginal areas and has led to desertification. Climate change is brought about by the interaction between the forest and the atmosphere. Of the rainfall, 25 per cent drains into the rivers, 25 per cent evaporates off the leaves, and 50 per cent returns to the atmosphere through transpiration and makes clouds. When trees are absent, so are clouds. With no clouds, the land becomes dried up by the equatorial sun.

### Changing Weather Patterns

The Sahel region of north Africa is located between the Sahara Desert and the equatorial rainforest. The natural vegetation of the Sahel region is savannah grassland and scrub. Traditionally it is a cattle grazing region on a nomadic basis. This means that traditional herdsmen such as the Fulani, the Tuareg and the Masai moved within their own traditional areas with their herds of cattle in search of fresh pasture.
For some of these tribes cattle numbers are a status symbol, and therefore numbers rather than cattle quality are the priority. From 1950 to 1970 the annual rainfall on the southern fringe of the Sahara was greater than average and the corresponding increase in the supply of pasture allowed for:

(a) a substantial increase in numbers of cattle owned by these herdsmen. This led to livestock overgrazing these marginal lands. Land that used to be green pasture during short wet spells turned into desert after four years without rain.

(b) an invasion of tillage farmers and the attendant land clearance in what had been a pastoral area. In Niger, for example, farmers pushed northward, planting fields of millet and groundnuts in the area traditionally occupied by nomadic herdsmen.

Cash cropping was imposed through taxes. Agricultural demands led to the clearance of natural vegetation. Over the past thirty years the woods that covered the Sahel have been reduced by felling. The

disappearance of tree cover has had serious consequences for the whole of the environment. Run-off accelerates, groundwater is no longer replaced, water-tables drop, and erosion attacks the soil, robbing it of its fertility.

## Coasts, Coastal Tourism, People and the Environment

Tourism is the world's second fastest industry in terms of money generated and is expected to be the European Union's major source of employment in the year AD 2000.

As a consequence of improved living standards and education, higher incomes and an increasing and pressurised workload, many families and individuals feel they deserve at least one holiday abroad or within their own country each year. Most countries of the world offer various attractions to entice people to visit. The combination of these factors as well as competitive transport rates results in a myriad of choices for the eager tourist. Income from tourists helps to develop particular regions that are attractive, especially their roads, services and accommodation, which all leads to improved facilities for the local population.

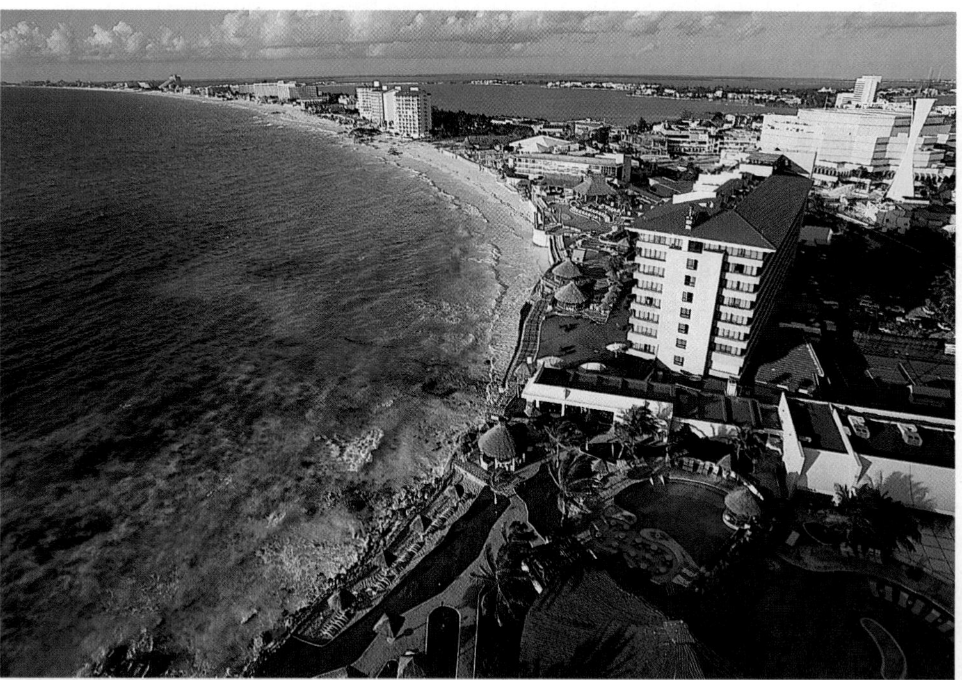

Fig. 19.13
Development such as this at Cancun, Mexico is now being carefully controlled

However, the perils of success are also threatening the most picturesque attractions in these countries. Britain attracts about 30 million tourists each year. The cathedral town of Canterbury, for instance, has a population of only 36,000 people but attracts 2.5 million tourists annually, an invasion rising by 100,000 each year. The Cathedral of Notre-Dame in Paris has so many visitors each day that the moisture of people's breath is damaging parts of the building's structure.

Famous places also suffer in other ways. Malibu Beach, one of the best-known beaches in the world, is no longer as accessible as it once was. Motorways, private homes and hotels line the beach head and are so close to each other that access to the beach is extremely limited. As in Mediterranean Europe, those that own property on the backshore of a beach generally own that part of the beach between their building and the shore. Sometimes this can be identified by a rope railing or single-coloured umbrellas on the beach in front of the building. Hotels generally allow their patrons free access, but charge a fee to others who wish to sit on their sandy patch, while private homes erect 'private property' signs.

## Rivers, Tourism, People and the Environment

### Case Study 1: The Victoria Falls in Africa

The Victoria Falls is locally known as 'Mosioatunya', the smoke that thunders. Upstream of the falls, the Zambezi River, which divides Zimbabwe and Zambia, is nearly 2 kilometres wide, and its foaming 100-metre descent into white-water gorges below makes Victoria Falls one of the natural wonders of the world and the biggest scenic attraction in sub-saharan (south of the Sahara) Africa.

Since peace and stability came to Zimbabwe in 1980, tourism has soared. Over 1.5 million visitors are expected to visit the falls within the next decade. Hotel and lodge construction there is booming. There are now ten major hotels on the more popular Zimbabwe side, two more are under construction and it is planned to construct two more within the next year. Nearby is the township of Chinotimba. It is overcrowded and classed by Stanley Katsenga, a well-known local environmentalist, as an environmental time bomb.

Fig. 19.14 (a)
Too many trips to the falls. The battle is on to save Africa's great scenic wonder from the ravages of a tourist boom.

As with the Grand Canyon, white-water rafters are creating a rash of river-bank campsites and causing damage to the native bush by the supply and recovery vehicles of the rafting companies. Helicopter flights are the norm, creating much local hostility.

Most agree that a joint action by Zimbabwe and Zambia is required to protect this fragile environment.

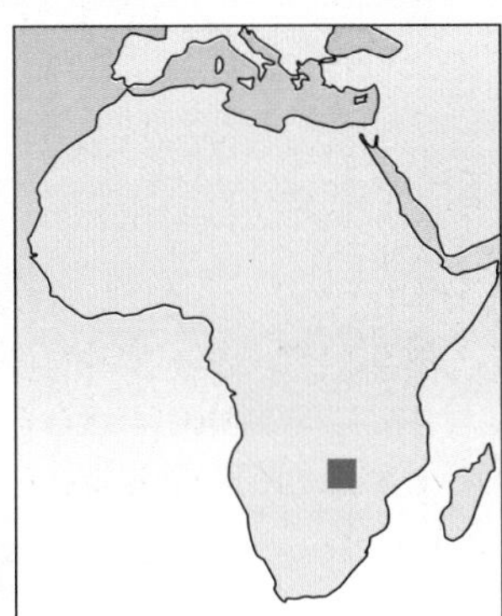

Fig. 19.14 (b)

## Case Study 2: The Grand Canyon in North America

### The Grand Canyon National Park

The Grand Canyon National Park is the most popular tourist spot in the United States. Some 18,000 people visit the park each day. This multiplies to a daily flood of 28,000 during July and August. In the midsummer heat, long waits for shuttle buses and queues for parking spaces regularly lead to fist fights, exhaustion and heart-related problems.

The number of park visitors has doubled in a decade and a half, from two million in 1984 to five million in 1998. If present trends continue, it is estimated that this figure will rise to seven million visitors by 2010.

The Grand Canyon National Park is 500,000 hectares in size. However, most visitors are concentrated on narrow trails along the 90 kilometres of the 446-kilometre canyon edge. This high-density policy and overcrowded visitor centre facilities do not appear to be changing. Noise levels and visual pollution are increasing. Some 43 aircraft services provide 10,000 plane and helicopter flights over the canyon in the peak summer months alone.

White-water rafting has become a major tourist attraction. Average waiting time for a private rafting permit on the Colorado River, which has created the canyon, is nine years. Intense pressure is being put on park officials to allow more private rafting and kayaks.

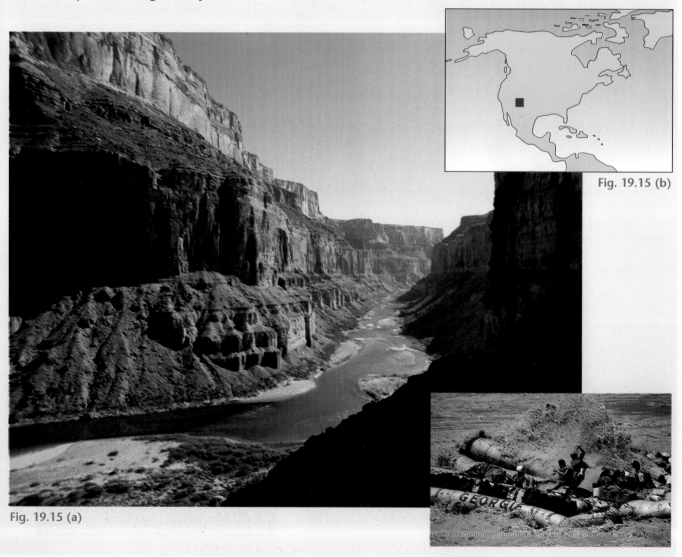

Fig. 19.15 (b)

Fig. 19.15 (a)

# Rivers, Industry, People and the Environment

## Case Study: The Three Gorges Hydroelectric Scheme on the Yangtze Kiang in China

### An Environmental Dilemma

Midway between its icy source in Tibet and the fertile delta at its mouth in Shanghai, 6,300 kilometres to the east, China's Yangtze River rushes through a series of vertical-sided channels, known as the Three Gorges. The Chinese government intends to use these gorges to build the world's largest hydroelectric dam. The Chinese leaders argue that the Three Gorges scheme is vital to their country's future and will be good for the environment as a whole. They say it will prevent the periodic flooding of the Yangtze Kiang that has claimed 500,000 lives in this century. More importantly, its production of clean hydroelectric power will reduce China's reliance on coal, the dirtiest of all fossil fuels, which now supplies 75 per cent of the country's needs. At present, the burning of coal has helped make lung disease the nation's leading cause of death.

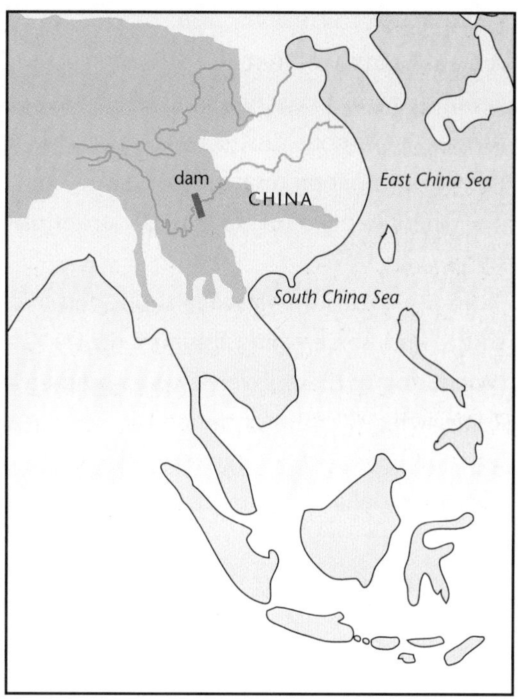


Fig. 19.16
Three Gorges dam on the Yangtze Kiang

Many opponents of the Three Gorges plan have no quarrel with the effort to move away from coal towards hydropower. But they argue that for a lower price, numerous smaller dams could produce more power and greater flood control benefits. They fear that a dam so large on the muddy Yangtze River will lead to dangerous build-ups of silts in some parts of the river, creating new obstacles to navigation and causing floods upstream. Chinese officials respond that both big and small dams are needed. In fact ten projects smaller than 'Three Gorges' are under construction on the upper reaches of the Yangtze and its tributaries.

Fig. 19.17
Construction of the Three Gorges dam on the Yangtze Kiang

# Rivers, Farming, People and the Environment

## Case Study: The Aral Sea

### A Sea Turns to Dust

Grounded ships beached in a desert now some eighty kilometres from the sea, poisonous dust storms, salt-laden soils and sick children . . . this environmental disaster area is the Aral Sea basin, in central Asia. The Aral, once one of the world's largest lakes, has been shrinking for thirty years.

In the 1950s the Aral Sea supported mixed agriculture around its shores and a flourishing fishing industry. Of the twenty-four fish species found there, twelve were commercially valuable, and catches approached 50,000 tonnes a year. The surrounding river deltas were fertile regions, rich in animals and plants. Fish used to breed in the abundant delta lakes, and the reeds that surrounded them were harvested to make paper. The deltas were fringed with forests several kilometres wide where 173 species of animals and birds lived.

The Aral Sea has no outlet. The rivers' water simply evaporated and kept a local ecosystem and its dependent population of thirty million people in balance with the second largest inland body of water in Asia.

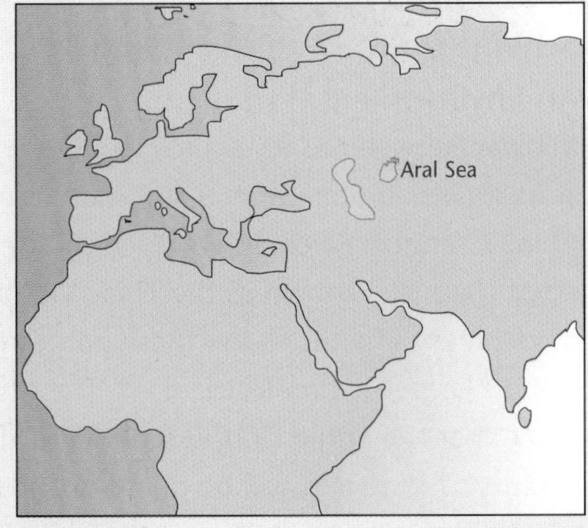


Fig. 19.18
The location of the Aral Sea in Asia

### The Aral Sea Project and its Original Purpose

The waters of the two largest rivers, the Amu and Syr, which flow through a desert region and feed the Aral Sea, were diverted through canals to irrigate seven million hectares of cotton, rice and melon fields for the markets of the former Soviet Union. Though the water diversion was carried out to increase agricultural production, it has only partially done so, but at extreme cost to the local physical and human environment.

Prior to its break-up, the Soviet Union, along with United Nations environmentalists/scientists, had devised a rescue programme for the region. The break-up has shifted responsibility from Moscow to the five newly independent central Asian republics of the Aral basin. Now progress depends on whether Uzbekistan, Kazakhstan, Turkmenistan, Kirghizia and Tajikistan can co-ordinate their policies and resources to solve a problem they did not create.

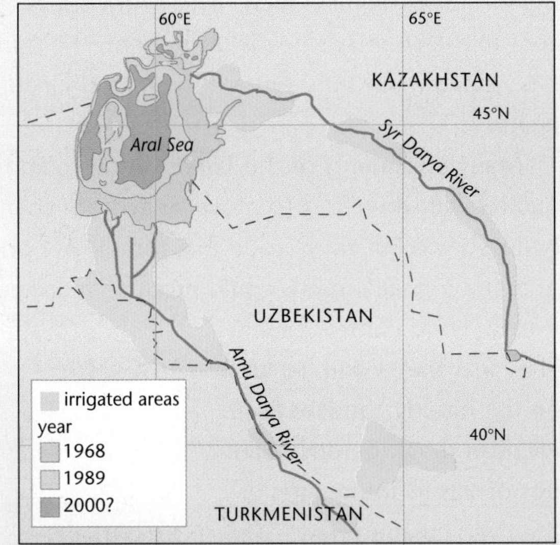


Fig. 19.19
This sketch clearly shows a shrinking Aral Sea. The original coastline can be up to eighty kilometres from the reduced shoreline. Glasnost brought the problem to the world's attention, but with the break-up of the Soviet Union rescue plans have fallen apart.

### Effects of Shrinking Aral Sea

1. Salinity has increased from 10 per cent to 40 per cent due to evaporation of sea water.
2. Delta lakes have dried up and delta forests have been reduced.

3. Dust storms from exposed lake-bed sediments increase dust and salt amounts in the air.
4. Canals are leaking, creating waterlogging in some areas.
5. Summers are hotter (up to 45 °C). Winters are colder as sea is much reduced.
6. Cancer and other diseases, e.g. hepatitis A and typhoid, are five times the national average.
7. Many fish species have been wiped out.

Study the photograph and the Ordnance Survey map extract of Sandycove in Dublin, Figs. 19.20 and 19.21. Then answer the following:

1. What is meant by global warming?
2. Outline the main causes and effects of global warming.
3. What effect(s) could global warming have on this coastal area in the year 2030, approximately thirty years from now?
4. What efforts could be made by
   (a) the local council
   (b) the residents of this area, to cope with such environmental changes as outlined in question 3 above?
5. In what direction was the camera pointing when the photograph was taken?
6. What is the name given to the low, circular building in the right background? Use your encyclopedia or other source to explain the purpose of this structure and the era when it was built.
   In relation to compass direction, what is the correct location of this structure on the photograph?
7. A flat stone surface is exposed on the right margin of the photograph. Name this coastal landform and explain the processes involved in its formation.
8. What evidence on the photograph suggests that this is a prosperous part of Greater Dublin?

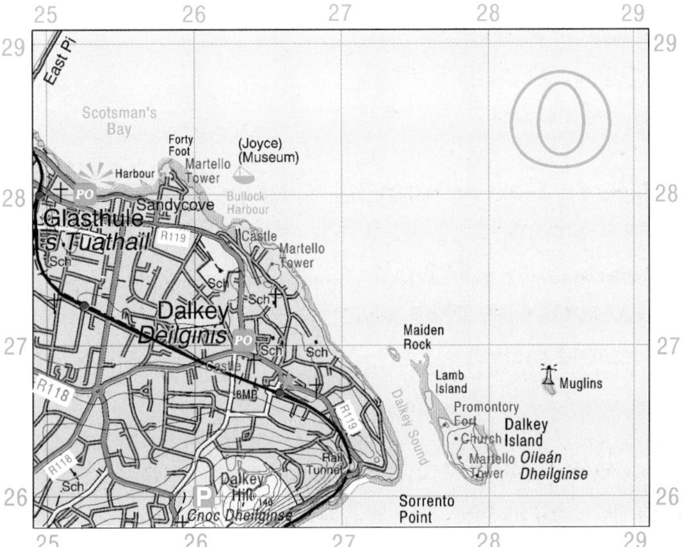


Fig. 19.20
Nuclear waste in the Irish Sea (see chapter 2)

Fig. 19.21

Ordinary Level

**Environmental Issues**

**1997**

Ozone damage, Acid rain, Greenhouse effect.

(i) Select any **TWO** of the above issues and describe their impact on the environment. **(40 marks)**

(ii) For **EACH OF THE TWO** selected, explain how Man can reduce the negative impact. **(40 marks)**

**1995**

(i) In the case of **EACH** of the following environmental issues, explain how it occurs AND describe its effect on the environment.

Acid Rain, Greenhouse Effect. **(48 marks)**

(ii) The seas around Ireland have suffered pollution from a number of sources, both local AND international. Examine this statement, referring to examples you have studied. **(32 marks)**

**1991**

(i) With reference to **BOTH** air quality **AND** water quality, describe examples of the damage which has been caused to the European environment. **(60 marks)**

(ii) 'Protecting the environment should not interfere with economic development.' Examine this statement. **(20 marks)**

**1990**

Smog, Acid Rain, Greenhouse Effect.

(i) Select any **TWO** of the above environmental problems and for **EACH** one you choose explain why geographers and other scientists regard it as a problem and how human activities contribute to it. **(40 marks)**

(ii) Suggest and explain **TWO** changes which modern society must make in our way of life, if these problems are to be controlled. **(40 marks)**

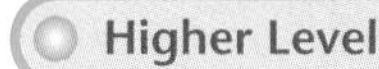


**1998**

'The human species is becoming an increasingly significant agent of change in the environment.'

Discuss this statement. **(100 marks)**

**1996**

'Humankind has become an increasingly important agent of environmental change.'

Examine this statement, with reference to (i) weather and (ii) climate. **(100 marks)**

**1989**

'The potential for human activities to change the environment has grown to the stage where today we may be witnessing irreversible damage to the planet's natural systems.'

Examine this statement, with reference to (i) soils, (ii) forests, (iii) waters. **(100 marks)**

**1988**

'Mankind is becoming an increasingly important agent of environmental change.'

Examine this statement, with reference to (i) the atmosphere and (ii) the waters of the earth. **(100 marks)**

# Exam-type Questions in Maps and Photographs

Fig. 20.1

Study the photograph of the Great Sugar Loaf mountain in Co. Wicklow, Fig. 20.1. Then do the following:

The Great Sugar Loaf is composed of quartzite, a hard rock which formed as a consequence of intense heat and pressure associated with the colliding of continental plates. Sandstone, which once formed beach sediments, was compressed and heated and was metamorphosed to form quartzite.

Quartzite is a hard rock because it is composed of quartz particles which are bonded together by a strong quartz cement. As a result, it is very resistant to weathering and so this mountain has a steep, cone-shaped appearance.

1. (a) What evidence on the photograph suggests that the Sugar Loaf is composed of hard rock?
   (b) Explain the processes involved in the formation of the Sugar Loaf. Use information from your study of plate tectonics — focus on Ireland to explain your answer.
2. What evidence suggests that it develops a poor soil cover when weathered?
3. What industrial raw material in this area developed as a consequence of denudation (weathering and erosion)? Use evidence on the photograph to support your answer.
4. (a) Account for the pattern in the distribution of woodland.
   (b) What species of trees are found in this area? Explain fully.
5. Describe and account for the distribution and density of population displayed on the photograph.
6. Account for the field pattern displayed in the photograph.

Carefully examine the Ordnance Survey map of western Galway, Fig. 20.2. Then do the following:

1. This part of the Connemara region in western Galway has a long tradition of settlement, from pre-Christian times to the present day. Discuss this statement using evidence from the map to support your answer.
2. Glacial action in this region has greatly influenced both the topography and settlement patterns. Discuss this statement fully, using evidence from the map to support your answer.
3. (a) With reference to processes of erosion and to processes of deposition, examine three ways in which the sea shapes the Irish coastline in this area.
   (b) Areas such as this have been used for the successful production of seafood. Discuss this statement with reference to the map.
4. With the aid of a sketch map, examine how the relief and drainage of this area have influenced the development of communications.
5. Tourism is the world's second fastest growing industry. With reference to the map discuss the potential of this region as a tourist destination.
6. Explain the following references to ancient settlement on this map: crannóg, oratory, midden, cillín, castle.
7. There is an Atmospheric Research Station located at grid reference L 735 324. Suggest reasons why this site was chosen for such a purpose.

ROUNDSTONE
Cloch na Rón
Illaunacorka
An tOileán
Oileán an Choirce
Quay
Pier
PO
Sch
Craft Centre
Cregduff Br.
R341
Cregduff Lough
Lough na gCreag Dubh
Ervallagh
Oirbhealach
Pier
Green Isds
Inishnee
Inis Ní
Church
Roundstone Bay
Cuan Chloch na Rón
Beacon
Illaunnaguroge
Oileán na nGiúróg
Oghly Island
Oileán an Chlaí
Rosroe Lough
Loch an Rosa Rua
Canower
Ceannúir
Seal Isd
Carraig an Róin
Cashel Bay
Oileán an Ghiolcaigh
Cuan an Chaisil
Doolick
Dúleac
Illaunnamrogue
Oileán na mBróg
Illaundoon
Oileán Domhain
Illaungorm North
An tOileán Gorm Thuaidh
Illaungorm South
An tOileán Gorm Theas
Illaunacroghnut
Oileán Cruachnaite
Holy Well
Croghnut
Cruachnait
Illaunaknock
Oileán an Chnoic
Illaunfadda Islands
BERTRAGHBOY BAY
Cuan na Beirtrí Buí
Carrickleagh
An Charraig Liath
R340
L Chaleb an Chnoic
L na Caileí
Lough Naneeve or Saints Lough
Loch na nAobh
Lough Na
Loch na
Inishlackan
Inis Leacan
Midden
Pier
Holy Well
Inishtreh
Inis Troighe
Holy Well
Quay
Pier
Cillín
Holy Well
Lough Aheeran
Loch an Roisín
Glinsk
Glinsce
Sch
Droichead Ghlinsce
Creenaghmore Lough
Callancruck
Caladh an Chnoic
Lough Nagraigue
Loch na Gráige
Letterard
Leitreach Ard
Lough Nacalaga
Loch na Calgaí
Loch Chnoc na gCapall
Lough Cam
Loch Doirín an Iarla
Tullynaman L
Holy Well
Lough Derreenanearla
Loch Doirín an Iarla
Saineaghmore Lough
L an Ghiolcaigh
Shallow Lough
Loch Tan
Lough Cam
Loch Cam
Mering L
Black Pool
Freaghillaun
Fraochoileán
Sch
Holy Well
Quay
Loch na Scinín
Moyrus
Mairos
Holy Well
Inishbigger
Inis Bigir
Lough Bola
Crannóg
Loch Buaile
Knockboy
An Cnoc Buí
L Black
Loch Dhúleitir Beag
Lough Nadullagh
Loch na Dúloc
Lough Truskeena
Glennaun L
Quay
Church
Owenglin River
Sruthán na mBreac
Lough an tSeanmharai
Loch na nEan
Lough Tuaí Buí
Black Lough South
Dooletter
Dúleitir
Holy Well
Dooyeher
Dumhaigh Ithir
Mill Lough
Loch an Mhuilinn
Culleen
An Coillín
Culleen
L Dohola
L Dúthulach
Grannóg
L Keamnacally
L na bhFaoileán
Colt Island
Oileán an Bhromaigh
Atmospheric Research Station
Mace Head
Halfmace
An Leathmhás
Mace
An Más
Quay
Sch
Ard West
An Aird Thiar
Ard East
An Aird Thoir
Holy Well
Letterdeskert
Leitir Deiscirt
Loch na hAbhann Bige
Crannóg
Lough Skann
Yellow Rock
An Charraig Bhuí
Yellow Isds
Glasoileán an Táilliúra
Ard Bay
Cuan na hAirde
PO
Carna
Sch
Lough Aghara
Lough Keeraun
Loch an Chaoráin
Crannógs
Lough Sleedagh
Loch Sc
Quays
Lough Shemus
Holy Well
Quay
Castle
Quay
Lough Nageeron
Holy Well
Inishtroghenmore
Inis Srathair
Lough Ateesky
Loch an Taosca
Rusheenyvulligan
Roisín an Bholgáin
Rusheennamanagh
Roisín na Mainiach
St Macdara's Isd
Cruach na Cara
St Mac Dara's Early Christian Monastery
Wherroon Isd
Na Foiriúin
Rough Rock
Avery Island
Oileán Aimhréidh
Quay
Carra Isd
Oileán na Cara
Rusheenacholla
Roisín an Chalaidh
Rusheenyvulligan Rock's
Holy Well
Quay
Masson Island
Oileán Máisean
Oratory
Straddle Rock
Mullán an Bhealaigh
Coarse Rock
Oileán an Chartha
Ardnacross Isd
Aill na Croise
Mweenish Island
Iron Rock
Maínis
Burial Ground
Midden
Sch
Holy Well
Holy Well
Holy Well
Mweenish Bay
Cuan Mhaínse
Joyce's Rock
Carraig na bPortán
Finish Rocks
Quay
Holy Well
Cillín
Sheep Rock
Midden
Finish Island
Fínis
Midden
Calf Island
Bound Rock
Eagle Rock North
Carrickaview

◀ Fig. 20.2
Roundstone region

Study the photograph and the Ordnance Survey map extract of Carlow town, Figs. 20.3 and 20.4. Then answer the following:

1. Describe the origin and development of Carlow using evidence from the map extract and photograph.
2. There is a dam built across the River Barrow which is visible in the photograph. Suggest
   (a) when this dam/weir was built
   (b) why this dam/weir was built
3. If you were searching for a house site in order to build your own home, outline three reasons why you might purchase some of the unoccupied land to the east of the river in the foreground and middle of the photograph.
4. Now see the photograph of Carlow on p. 193. Having examined this other photograph:
   (a) state why you might not build on this land to the east of the river
   (b) state why you may or may not wish to purchase one of the newly constructed apartments in the centre of this photograph
5. Carefully examine the light green line in the south-west corner of this photograph.
   (a) Suggest a reason for this light-coloured marking.
   (b) New buildings are being constructed on the west side of the River Barrow in the centre of the photograph. Give reasons why these buildings could be less susceptible to flooding than areas on the west side of the river.
6. The photograph and the Ordnance Survey map extract, Figs. 20.3 and 20.4, display a variety of transport methods, some of which are generally profitable over both long and short distances and continue to be used, while others are generally profitable over long distances only. Discuss this statement.
7. A light black line with V-shapes on its west side runs in a north-west/south-east direction in Fig. 20.4.
   (a) Identify what this line represents.
   (b) In what ways might such a structure in a scenic area or an urban area
      (i) affect the landscape
      (ii) affect people?
   (c) If planning permission was sought for this structure in a scenic rural area, what kind of conflicting conditions could arise as a consequence of local objections to the proposal? Tip — 'energy fields', carcinogenic, unsightly, ugly, power supply, industrial estate jobs.
8. As a geographer you have been asked to draw up a feasibility study to consider building another bridge across the River Shannon.
   Describe the positive **and** negative aspects of such a proposal. In your answer identify and justify an exact location for the bridge.

Fig. 20.3
Carlow region

Fig. 20.4 ▶
Carlow region

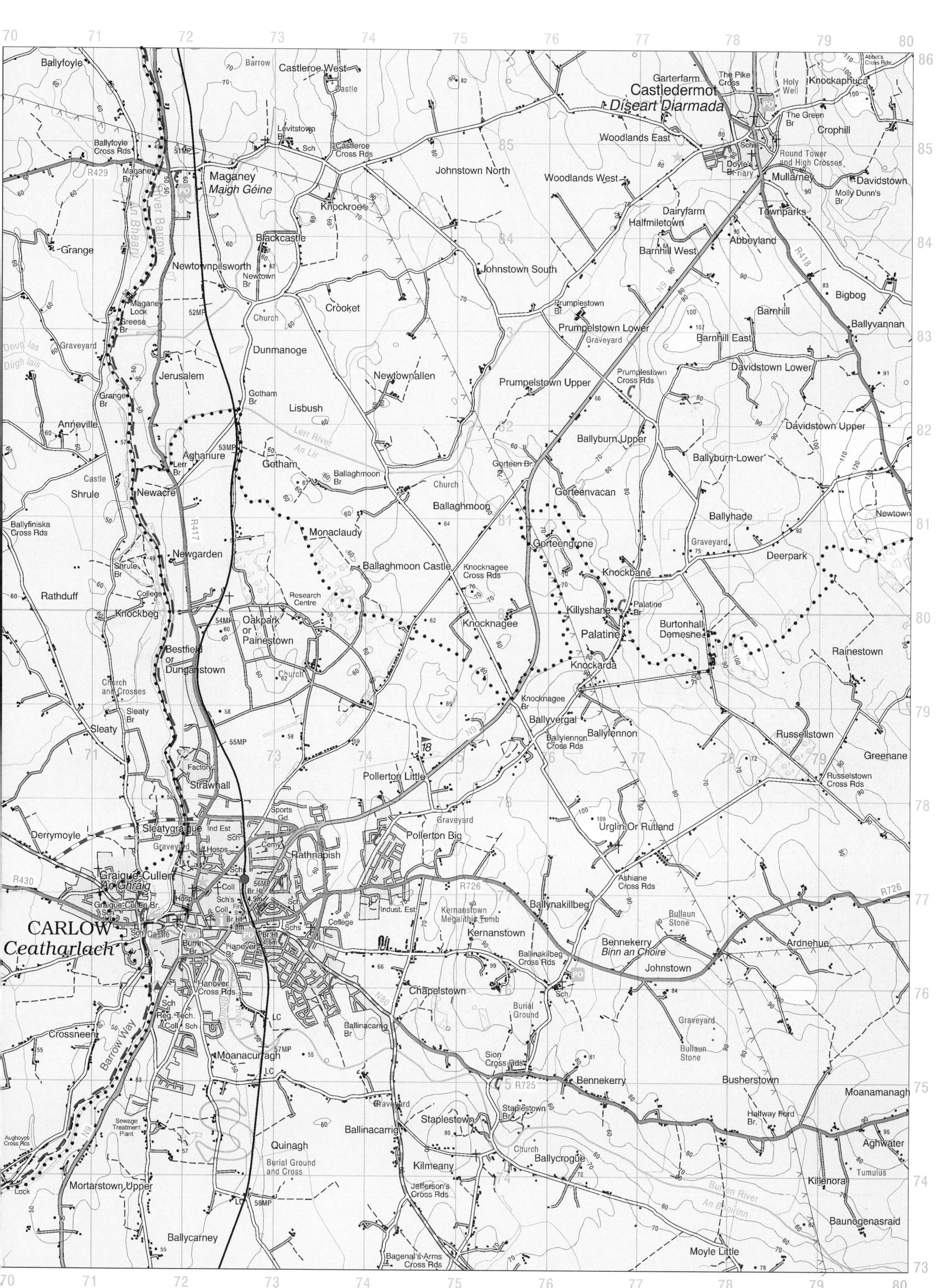

Castledermot
Díseart Diarmada
Garterfarm
The Pike Cross
Holy Well
Knockaphuca
Abbot's Cross Rds
The Green Br
Crophill
Woodlands East
Round Tower and High Crosses
Doyle's Br
Friary
Mullarney
Davidstown
Molly Dunn's Br
Townparks
Woodlands West
Johnstown North
Dairyfarm
Halfmiletown
Abbeyland
Barnhill West
Johnstown South
Bigbog
Barnhill
Ballyvannan
Barnhill East
Prumplestown Br
Prumpelstown Lower
Graveyard
Prumplestown Cross Rds
Prumpelstown Upper
Davidstown Lower
Davidstown Upper
Ballyburn Upper
Ballyburn Lower
Newtown
Gorteen Br
Gorteenvacan
Ballyhade
Deerpark
Gorteengrone
Knockbane
Palatine Br
Killyshane
Palatine
Burtonhall Demesne
Rainestown
Knockarda
Knocknagee Br
Ballyvergal
Ballylennon
Ballylennon Cross Rds
Russellstown
Greenane
Russelstown Cross Rds
Urglin Or Rutland
Ashlane Cross Rds
Ballynakillbeg
Bullaun Stone
Ardnehue
Bennekerry
Binn an Choire
Johnstown
Ballinakilbeg Cross Rds
Kernanstown
Kernanstown Megalithic Tomb
Chapelstown
Burial Ground
Sion Cross Rds
Bennekerry
Busherstown
Moanamanagh
Halfway Ford Br.
Aghwater
Tumulus
Killenora
Baunogenasraid
Moyle Little
Burren River
An Bhóirinn
Staplestown
Staplestown Br
Church
Ballycrogue
Kilmeany
Jefferson's Cross Rds
Bagenal's Arms Cross Rds
Ballinacarrig
Ballinacarrig Br
Quinagh
Burial Ground and Cross
Ballycarney
Mortarstown Upper
Moanacurragh
Sewage Treatment Plant
Aughoyle Cross Rds
Lock
Crossneen
Barrow Way
Hanover Cross Rds
Reg. Tech. Coll
Hanover Br
Burrin Br
CARLOW
Ceatharlach
Castle
Graigue Cullen Br.
Graigue Cullen
An Ghráig
R430
Derrymoyle
Sleatygraigue
Ind Est
Graveyard
Hosps
Cemy
Rathnapish
Sports Gd.
Pollerton Big
Pollerton Little
Indust. Est.
College
R726
R725
N80
N9
Strawhall
Factory
Sleaty
Sleaty Br
Church and Crosses
Bestfield or Dunganstown
Oakpark or Painestown
Research Centre
Knockbeg
College
Rathduff
Shrule Br
Newgarden
Ballyfiniska Cross Rds
Shrule
Castle
Newacre
Aghanure
Lerr Br
Gotham
Ballaghmoon Br
Church
Ballaghmoon
Monaclaudy
Ballaghmoon Castle
Knocknagee Cross Rds
Knocknagee
Anneville
Lerr River
An Lir
Gotham Br
Lisbush
Jerusalem
Grange Br
Dunmanoge
Newtownallen
Church
Crooket
Graveyard
Douglas
Dúghlais
Greese
Maganey Lock
Newtownpilsworth
Newtown Br
Blackcastle
Grange
River Barrow
An Bhearú
Maganey
Maigh Géine
Maganey Br
Ballyfoyle Cross Rds
R429
Knockroe
Levitstown Br
Castleroe Cross Rds
Castleroe West
Castle
Barrow
Ballyfoyle
R417
R418

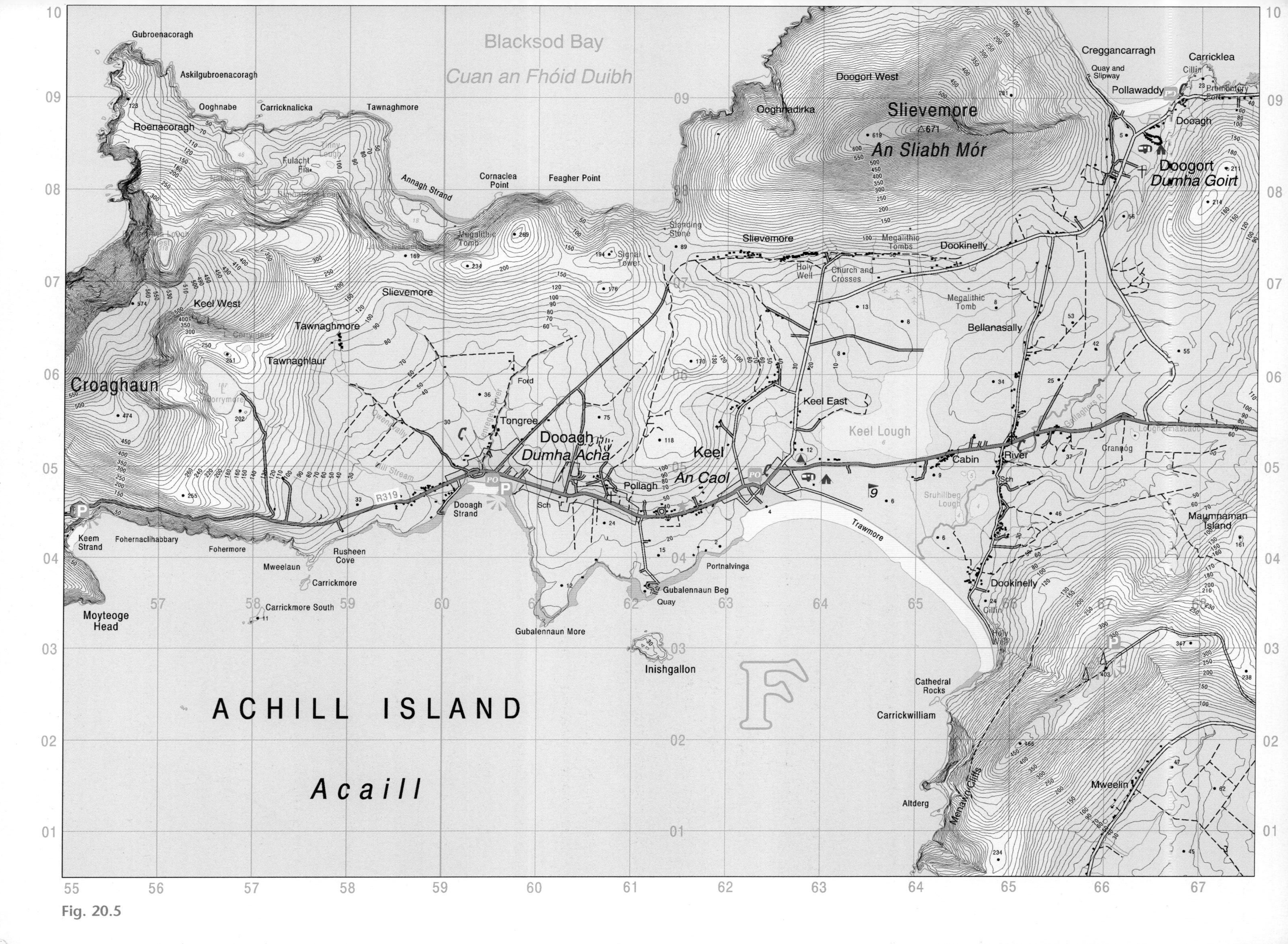

ACHILL ISLAND
Acaill
Blacksod Bay
Cuan an Fhóid Duibh
Croaghaun
Slievemore
An Sliabh Mór
Keel
An Caol
Dooagh
Dumha Acha
Doogort
Dumha Goirt
Keel West
Keel East
Keel Lough
Tawnaghmore
Tawnaghlaur
Dookinelly
Bellanasally
Pollagh
Tongree
Cabin
River
Keem Strand
Moyteoge Head
Fohernaclihabbary
Fohermore
Mweelaun
Carrickmore South
Carrickmore
Rusheen Cove
Gubalennaun More
Gubalennaun Beg
Portnalvinga
Inishgallon
Quay
Trawmore
Menawn Cliffs
Cathedral Rocks
Carrickwilliam
Altderg
Mweelin
Maumnaman Island
Cranóg
Annagh Strand
Cornaclea Point
Feagher Point
Signal Tower
Standing Stone
Megalithic Tomb
Megalithic Tombs
Holy Well
Church and Crosses
Gubroenacoragh
Askilgubroenacoragh
Roenacoragh
Ooghnabe
Carricknalicka
Fulacht Fia
Ooghnadirka
Doogort West
Creggancarragh
Quay and Slipway
Pollawaddy
Carricklea
Dooagh Strand
Ford
R319
Sch
PO


Fig. 20.5

Carefully study the Ordnance Survey map extract of Achill Island, Fig. 20.5. Then do the following:

1. (a) Draw a sketch map of the region.
   On it mark and label
   - the coastline
   - the main physical regions

   (b) With regard to one upland region and one lowland region describe each area under the headings: relief, drainage, settlement.
2. With reference to specific landforms explain how the processes of sea and ice have shaped this scenic landscape.
3. Explain how recent settlement trends in coastal areas in Ireland could have a negative influence on this natural landscape.
4. Explain how the settlement pattern has been influenced by the physical landscape.

## Map Questions

1. This map extract, Fig. 20.6, indicates that there is a long history of settlement in this area, even before early Christian times. Discuss this statement. Use evidence from the map to support your answer.
2. With the aid of a sketch map, describe how the relief of the area has influenced the communications network.
3. Describe the course of the River Laney from Clonavrick Bridge, grid reference W 346 783, to its confluence with the River Sullane, grid reference W 353 726.
4. (a) Account for the presence of the reservoir in the south of the map.

   (b) Large water areas are of great benefit to local populations. Discuss.
5. Measure the length of the River Laney, from where it enters the map at grid reference W 389 826 to its confluence with the River Sullane, grid reference W 353 726.
6. Account for the pattern in the distribution of woodland.
7. What steps have been taken to preserve the natural wildlife of this area?

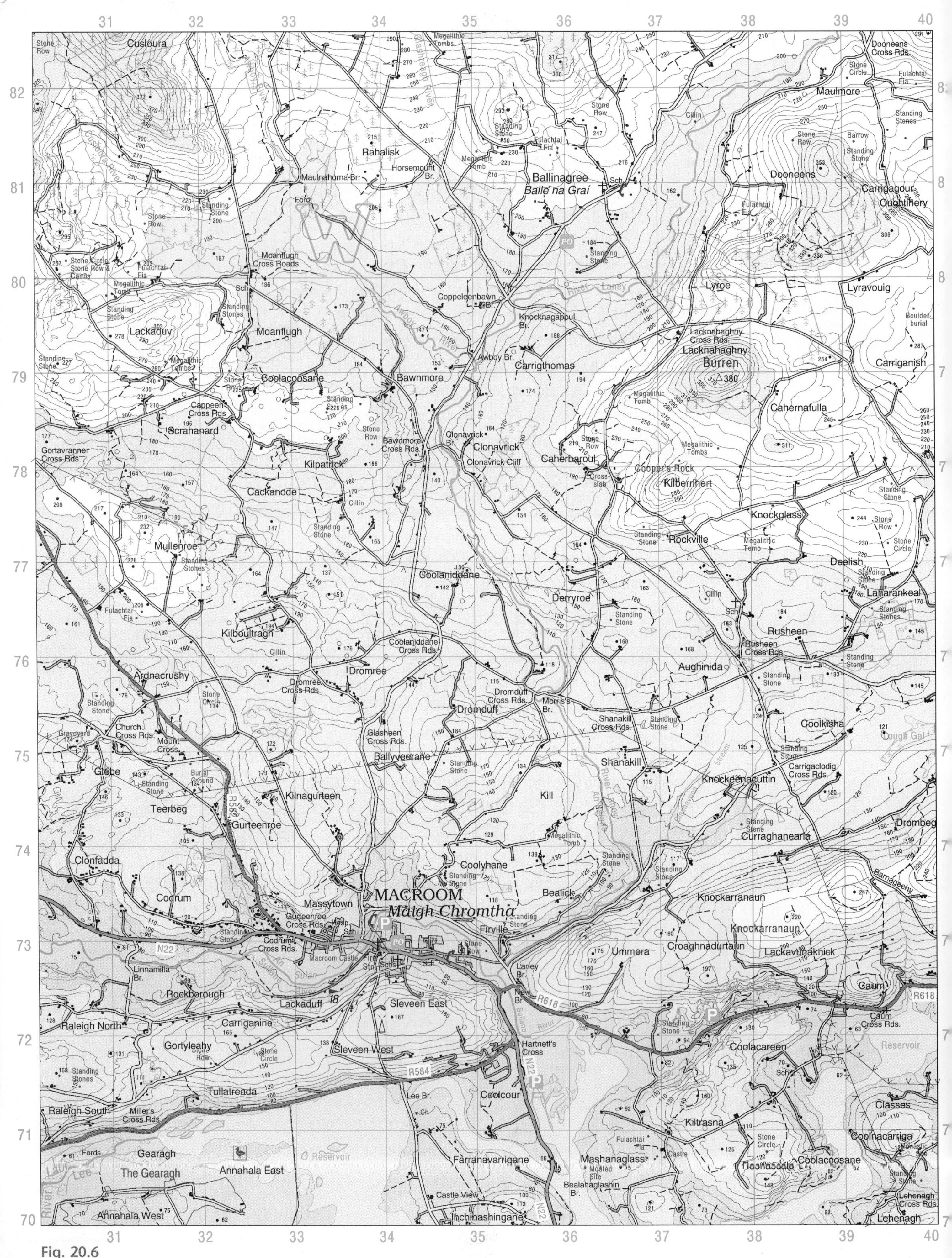


**Fig. 20.6**
Macroom region

Fig. 20.7

## Photograph Questions

1. Explain the processes involved in the creation of the course of the river. Use a sketch map to explain your answer.
2. Account for the field patterns shown on the photograph, Fig. 20.7. In your answer refer to (a) old patterns and (b) new patterns.
3. Account for the origin and development of Macroom. Use photographic evidence ONLY to support your answer.
4. Draw a sketch map of the photograph. On it mark and label
   - the rivers
   - the streets of the settlement
   - the roadways
   - a low ridge
   - five areas of different land use
5. What effort, if any, has been made to reduce pollution in the area?

## Photograph and Map Questions

1. Describe the origin and subsequent development of Macroom at its location. Use evidence from both sources.
2. What side of the photograph is to the north of the settlement? Explain.
3. There is a large ring-fort on the photograph. Identify this landform and give a grid reference for this feature on the map.
4. Account for the origin and present shape of the low ridge. Use evidence from both sources.
5. Account for the settlement pattern on the third-class road to the west of Macroom.
6. Rivers and roadways seem to separate many small hills of more or less similar altitude. Account for this pattern.

**Fig. 20.8**
Glenelly Valley

Examine the photograph of the Glenelly Valley in Co. Tyrone in Northern Ireland, Fig. 20.8. Then do the following:

1. What major erosive force created this valley? Explain fully.
2. Describe the course of the Glenelly River and use evidence from the photograph to support your answer.
3. Describe and account for the pattern in the layout of fields and farms in this area.
4. Account for the pattern in the distribution of settlement in the photograph.

Study the Ordnance Survey map of Dublin, Fig. 20.9. Then do the following:

1. The expansion of Dublin out into the countryside over the past thirty years has increased fears that the few existing open spaces (green belts) within Greater Dublin may not exist in the near future, thus adding to the city's problems. Discuss this statement with reference to the map.
2. (a) Identify the coastal landforms at A, B, C and D.
   (b) Explain the processes involved in the formation of B and C.
3. (a) Explain one example of how people have always attempted to manage or control the natural processes that operate along coastlines.
   (b) Attempts to control coastal processes have often led to erosion of or deposition along neighbouring places. With reference to one example on the map, explain this statement fully.
4. Explain the efforts that the Dublin Harbour Board has made in order to create access to and development of its docking facilities.
5. The disposal of urban waste has created major difficulties for urban councils. Explain fully how waste disposal in Dublin could interfere with human activity and natural processes in the area shown on the map.

Study the photograph of Ballyleeson in Co. Down, Fig. 20.10, p. 441. Then do the following:

1. There is a large circular structure in the centre of the photograph. What is this structure and what was its purpose?
2. Explain the pattern in the distribution of woodland shown in the photograph. Refer to urban and rural areas.
3. (a) Identify the function of a large structure to the south of the river, in the north-western corner of the photograph.
   (b) Describe and explain its location.
4. Describe the farming activities shown on the photograph and justify your choice using evidence from the photograph.
5. Suppose you wish to build a large footloose industry in this area.
   (a) State the products you wish to manufacture.
   (b) Choose a suitable site for your factory and justify your choice using evidence from the photograph.
6. Suppose you are an urban planner with the local council and you intend to draw up a land use plan for the future development of the area. With the aid of a sketch map, outline your land use zones and justify your layout using evidence from the photograph to support your answer.

Swords
Sord
Seatown East
Malahide Point
Mountgorry
Barrack Br.
Br. HT. 2.21m
Yellow Walls
Commons East
Drinan
Holy Well
Crowscastle
Barrysparks
School
Malahide
Mullach Íde
Martello Tower
Castle
Church
Malahide Demesne
Robswalls
Holy Well
Nevinstown East
Auburn
Marshallstown
Broomfield
Motte
Sainthelens
Martello Tower
Kinsaley
Feltrim Hill
Faoldroim
Grange
Rahulk
Feltrim
Greenwood
Beechwood
Abbeyville
Br. Ht. 3.25m
Hazelbrook
Glebe
Ballymacartle
Kinsaley
Cionn Sáile
Kinsaley Bridge
Church
Portmarnock
Port Mearnóg
Baskin
Stockhole
Agricultural Institute
Portmarnock Bridge
Sluice River
Bohammer
Middletown
Springhill
Portmarnock Bridge
Velvet Strand
Mound
Drumnigh
Holy Well
Church & Cross
Clonshagh
St. Doolagh's Burgage
St. Doolagh's Bridge
Br. Ht. 4.01m
Maynetown
Mayne Bridge
Clonshagh Br.
Wellfield Bridge
Snugborough
Belcamp
Balgriffin
Baile Ghrífín
Mayne River
Grange
Belcamp Park
College
Father Collins Park
Stapolin
Darndale
Portmarnock Point
Donaghmede
Domhnach Míde
Baldoyle
Baile Dúill
Hosp
Cush Point
Kilmore
Sutton
Cill Fhionntain
Mound
Castle
Beaumont Hospital
Coolock Br.
Edenmore
Éadan Mór
Br. Ht. 4.90m
Kilbarrack
Cill Bharróg
Church
Sutton Strand
Church
Howth Castle
Beaumont
Coolock
An Chúlóg
College
Santry River
Megalithic Tomb
Hospital
Raheny
Ráth Eanaigh
Donnycarney
Artane
Árd Aidhin
Cemetery
Church
Muck Rock
Holy Well
Killester
College
Cairn
Casino
St. Annes Park
North Bull Island
Oileán an Bhulla
Cliff Walk
Martello Tower
Visitor and Interpretive Centre
Br. Ht. 4.65m
Hosp
Dollymount
Cnocán Doirinne
Church
Br. Ht. 5.33m
Clontarf
Cluain Tarbh
Bull Wall
To Isle of Man Ferry (V) Seasonal
The Point Depot
Alexandra Basin
Ferry Terminal
DUBLIN HARBOUR
Breakwater
North Bull Lighthouse
North Bank Lighthouse
Toll Bridge
Poolbeg Lighthouse

◄ Fig. 20.9
Dublin region

Fig. 20.10

## Map Questions

1. Examine the course of the River Suir south of Caher town in the map, Fig. 20.11. Then do the following:
   (a) State the stage of maturity of the river.
   (b) Give reasons for your choice and support it with accurate detail from the map.
2. The map displays a long history of settlement in the area. Discuss.
3. Tourism is forming a larger portion of foreign earnings each year. Discuss why this region is an important stop in tourist itineraries.
4. Carefully describe the location of Ardfinnan.
5. Describe and explain the pattern in the distribution of farm houses/buildings.

## Photograph Questions

1. What major factor has influenced the road network pattern in the photograph, Fig. 20.12? Explain fully with exact references to the photograph.
2. Draw a sketch map and on it mark and label
   - the River Suir
   - the road network
   - the flood plain of the river
   - three areas of different rural land use
   - three areas of different urban land use
3. Explain how aerial photographs can sometimes provide us with information concerning our culture that no other related method can offer.
4. Suppose you are moving house from (a) Dublin, (b) Garryroe, grid reference S 055 165, near Ardfinnan, to live in this area. Choose a place on the photograph where you would like to build/purchase your home. Give reasons for your choice.

## Map and Photograph Questions

1. Account for the historical development of Caher from its earliest origins. Use evidence from the map and photograph to support your answer.
2. Explain how the River Suir has influenced the development of the town over time.
3. The combination of maps and photographs is necessary to study the landscape accurately. Discuss.
4. There is a large earth-moving activity in progress in the north-east corner of the photograph. Suggest a likely purpose for this excavation and give reasons for your choice.
5. (a) Describe and explain the origin of the field pattern(s).
   (b) State and explain the type of farming activity(ies) carried out in this region.
6. Account for the pattern in the distribution of woodland.

Fig. 20.11 ▶
Caher region

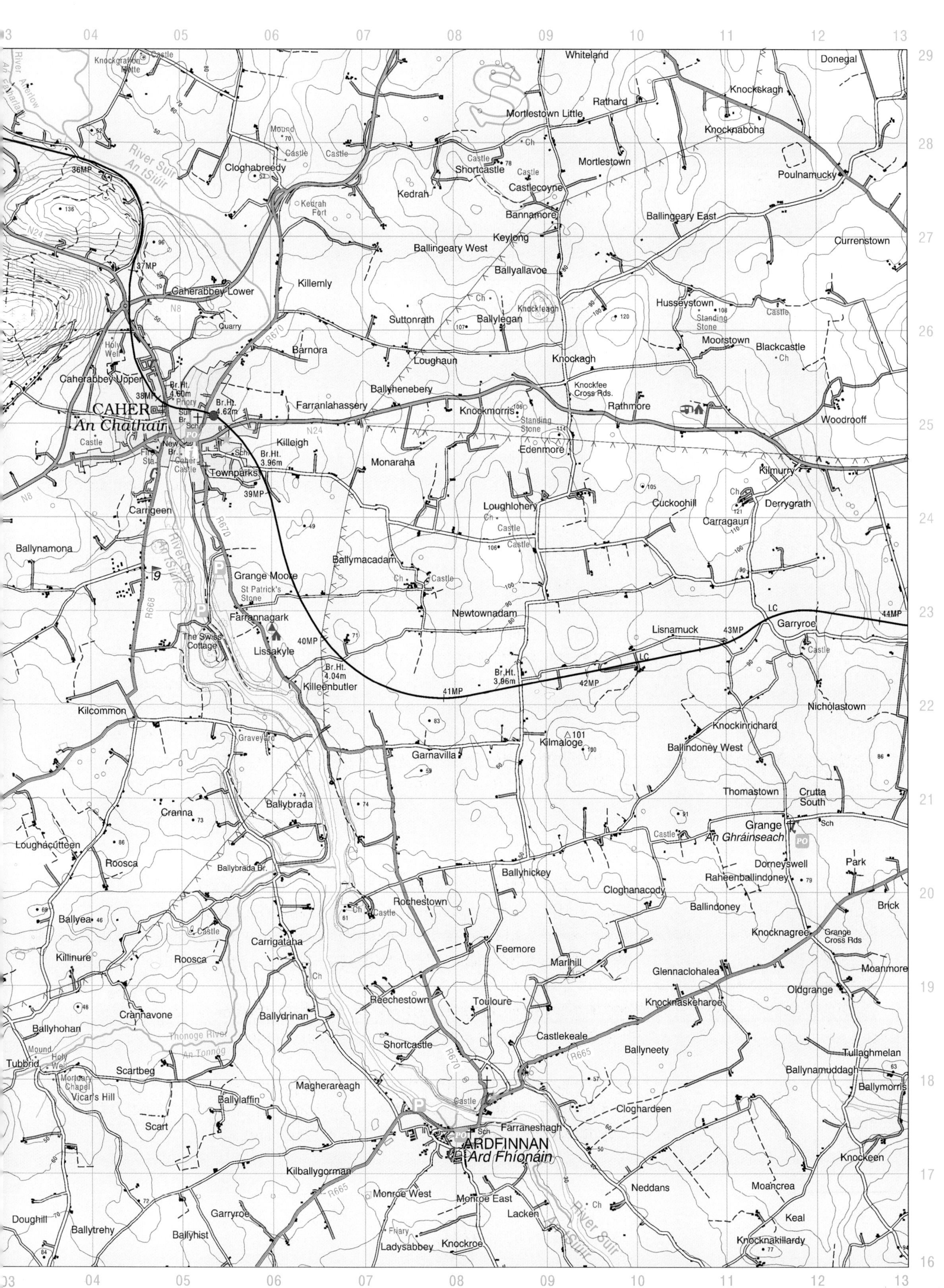
CAHER
An Chathair
ARDFINNAN
Ard Fhionáin
Grange
An Ghráinseach
River Suir
An tSiúir
Whiteland
Donegal
Rathard
Knockskagh
Mortlestown Little
Knocknaboha
Cloghabreedy
Shortcastle
Mortlestown
Poulnamucky
Kedrah
Castlecoyne
Bannamore
Ballingeary East
Keylong
Ballingeary West
Currenstown
Ballyallavoe
Killemly
Caherabbey Lower
Husseystown
Suttonrath
Ballylegan
Knockfeagh
Moorstown
Blackcastle
Barnora
Loughaun
Knockagh
Caherabbey Upper
Ballyhenebery
Knockfee Cross Rds.
Farranlahassery
Knockmorris
Rathmore
Woodrooff
Killeigh
Edenmore
Monaraha
Townparks
Kilmurry
Carrigeen
Loughlohery
Cuckoohill
Derrygrath
Carragaun
Ballynamona
Ballymacadam
Grange Moore
St Patrick's Stone
Farrannagark
Newtownadam
Garryroe
Lisnamuck
The Swiss Cottage
Lissakyle
Killeenbutler
Kilcommon
Nicholastown
Knockinrichard
Kilmaloge
Ballindoney West
Garnavilla
Thomastown
Crutta South
Cranna
Ballybrada
Loughacutteen
Roosca
Dorneyswell
Park
Ballyhickey
Raheenballindoney
Ballybrada Br.
Cloghanacody
Rochestown
Ballindoney
Brick
Ballyea
Carrigataha
Knocknagree
Grange Cross Rds
Killinure
Roosca
Feemore
Marlhill
Glennaclohalea
Moanmore
Reechestown
Touloure
Oldgrange
Crannavone
Ballydrinan
Knocknaskeharoe
Ballyhohan
Thonoge River
An Tonnóg
Castlekeale
Shortcastle
Ballyneety
Tullaghmelan
Tubbrid
Scartbeg
Ballynamuddagh
Vicar's Hill
Magherareagh
Ballymorris
Ballylaffin
Cloghardeen
Farraneshagh
Scart
Kilballygorman
Knockeen
Monroe West
Monroe East
Neddans
Moancrea
Doughill
Lacken
Garryroe
Keal
Ballytrehy
Ballyhist
Ladysabbey
Knockroe
Knocknakillardy
River Aherlow
An Eatharla
Knockgraffon Motte
N24
N8
R670
R668
R665
Holy Well
Quarry
Kedrah Fort
Standing Stone
Graveyard
Mortuary Chapel
Mound
Friary
36MP
37MP
38MP
39MP
40MP
41MP
42MP
43MP
44MP
Br.Ht. 4.60m
Br.Ht. 4.62m
Br.Ht. 3.96m
Br.Ht. 4.04m
Br.Ht. 3.96m
04
05
06
07
08
09
10
11
12
13
29
28
27
26
25
24
23
22
21
20
19
18
17
16

Fig. 20.12

Fig. 20.13 ▶
Kenmare River

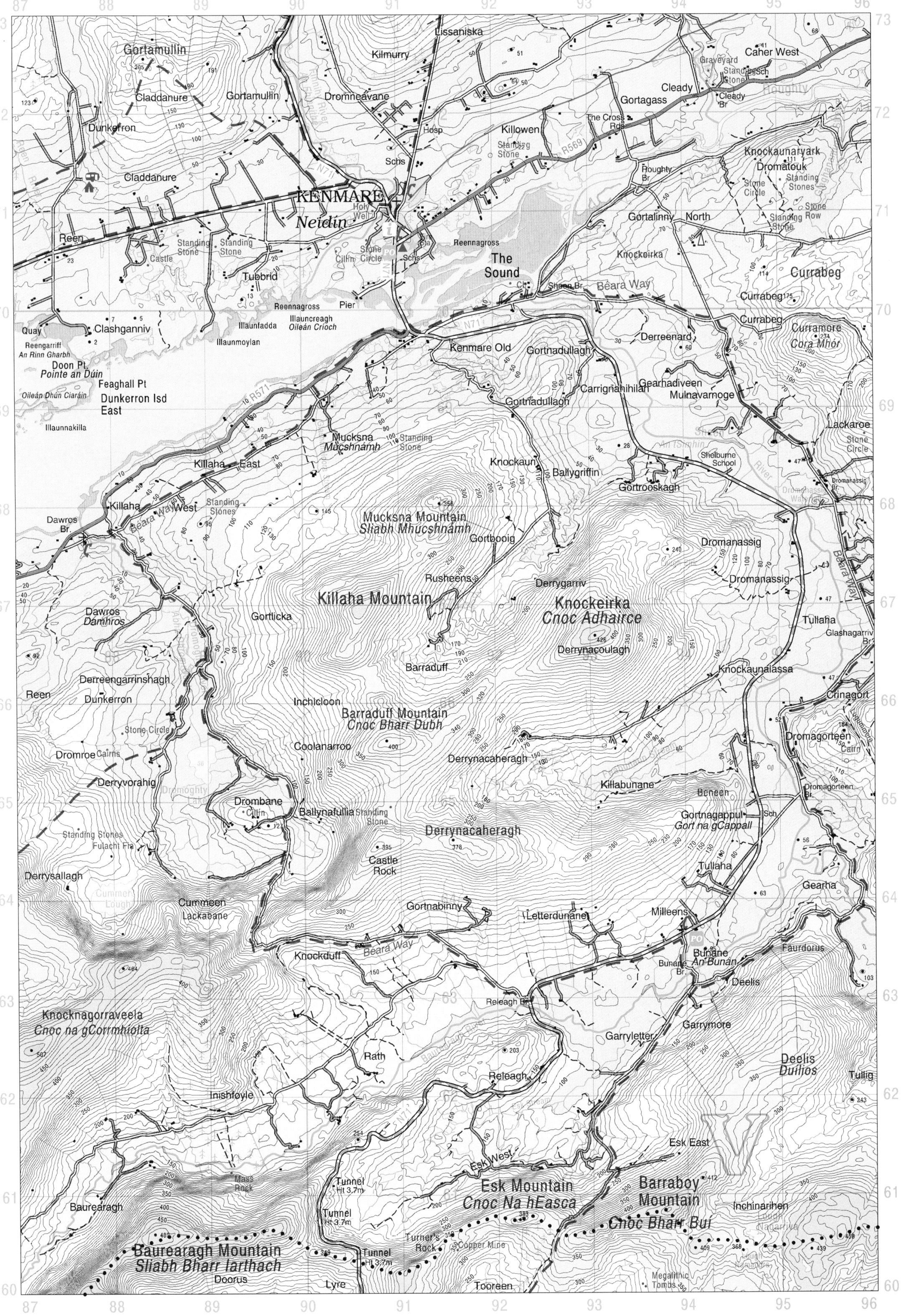

Gortamullin
Claddanure
Gortamullin
Dromneavane
Kilmurry
Lissaniska
Caher West
Graveyard
Standing Stone
Sch
Cleady
Cleady Br
Roughty
Gortagass
Dunkerron
The Cross Rds
Killowen
Hosp
Standing Stone
R569
Knockaunaryark
Dromatouk
Stone Circle
Standing Stones
Schs
Claddanure
Roughty Br
KENMARE
Neidín
Holy Well
Gortalinny
North
Stone Row
Standing Stone
Reen
Standing Stone
Standing Stone
Castle
Cillín
Stone Circle
Schs
Reennagross
The Sound
Knockeirka
Currabeg
Tubbrid
Sheen Br
Beara Way
Currabeg
Reennagross
Pier
Currabeg
Quay
Clashganniv
Illaunfadda
Illauncreagh
Oileán Críoch
Illaunmoylan
N71
Derreenard
Curramore
Cora Mhór
Reengarriff
An Rinn Gharbh
Kenmare Old
Gortnadullagh
Doon Pt
Pointe an Dúin
Feaghall Pt
Oileán Dhún Ciaráin
Dunkerron Isd East
R571
Carrignahihilan
Gearhadiveen
Mulnavarnoge
Gortnadullagh
Illaunnakilla
Mucksna
Múchsnámh
Standing Stone
An tSimhín
Sheen River
Shelburne School
Lackaroe
Stone Circle
Killaha
East
Knockaun
Ballygriffin
Dromanassig Br
Gortrooskagh
Dromanassig Waterfall
Killaha
West
Standing Stones
Dawros Br.
Beara Way
Mucksna Mountain
Sliabh Mhúcshnámh
Gortbooig
Dromanassig
Rusheens
Derrygarriv
Dromanassig
Killaha Mountain
Knockeirka
Cnoc Adhairce
Tullaha
Dawros
Damhros
Gortlicka
Glashagarriv Br
Derrynacoulagh
Barraduff
Knockaunalassa
Derreengarrinshagh
Reen
Dunkerron
Inchicloon
Crinagort
Barraduff Mountain
Cnoc Bharr Dubh
Stone Circle
Dromagorteen
Cairn
Dromroe
Cairns
Coolanarroo
Derrynacaheragh
Derryvorahig
Dromoghty Lake
Killabunane
Beneen
Dromagorteen
Drombane
Cillín
Ballynafullia
Standing Stone
Gortnagappul
Gort na gCappall
Sch
Standing Stones
Fulacht Fia
Derrynacaheragh
Castle Rock
Tullaha
Derrysallagh
Gearha
Cummeen Lough
Cummeen
Lackabane
Gortnabinny
Letterdunane
Milleens
Beara Way
Bunane
An Bunán
Faurdorus
Knockduff
Bunane Br
Deelis
Releagh Br
Knocknagorraveela
Cnoc na gCorrmhíolta
Garryletter
Garrymore
Rath
Deelis
Duilios
Tullig
Releagh
Inishfoyle
N71
Esk East
Esk West
Mass Rock
Tunnel Ht 3.7m
Esk Mountain
Cnoc Na hEasca
Barraboy Mountain
Cnoc Bharr Buí
Inchinarihen
Lough Nagarriva
Baurearagh
Tunnel Ht 3.7m
Turner's Rock
Copper Mine
Tunnel Ht 3.7m
Baurearagh Mountain
Sliabh Bharr Iarthach
Doorus
Lyre
Tooreen
Megalithic Tombs

## Map Questions

1. Describe the main tourist attractions of this area with reference to specific locations on the map extract, Fig. 20.13.
2. Using map evidence, describe how the physical landscape has affected communications in this area. Draw a sketch map to illustrate your answer.
3. Kenmare is a focal point of socio-activity in this area. Justify this statement with reference to past or present map features.
4. (a) There are many areas in Ireland today which show that erosion by moving ice played an important role in shaping our landscapes in the past. Identify three landforms typical of such areas and explain the processes that produced them.
   (b) Describe and explain the patterns which you can identify in the drainage system of this region. Use sketch maps to support your answer.
5. With reference to both upland and lowland areas, describe and explain the patterns in the distribution of woodland.
6. Suggest why the location V 890 650 would be an ideal site for an outdoor pursuits centre.
7. Suppose you are the chief planning engineer for this region and you have decided to create a 'superdump' at grid reference V 915 670.
   (a) Give three well-developed reasons for the advantages of your choice of site.
   (b) Give three well-developed arguments that may be offered against the site by local residents and environmentalists.

## Photograph Questions

1. Using evidence from the photograph, Fig. 20.14, describe why this town appears to have had a planned origin.
2. Draw a sketch map of the town. On it mark and label
   - the coastline
   - the street plan
   - the road network
   - the 'Big House'
   - five areas of different land uses
   - a river
3. Account for the distribution of woodland.

Fig. 20.14

## Map and Photograph Questions

1. (a) Account for the deposition of sediment in the area called the Sound.
   (b) How could this sediment affect water traffic?
2. Is the land on the far side of the estuary from the town on the photograph to the north or to the south of the settlement?
3. Name the estuary (water area), classify the inlet and then explain its formation.
4. Describe the location of Kenmare.
5. Assume that you are a government official responsible for industrial development in this region. You plan to create an industrial estate near to Kenmare, somewhere on the photograph. Identify this proposed site, and give specific reasons for your choice of location.
6. Give two fully developed reasons why the bridge over the river was chosen at this site.

1. Look at the 1:50,000 Ordnance Survey map extract and legend supplied, Fig. 20.15.
   (a) Examine in detail how the map provides evidence of the evolution of Dublin's transport infrastructure over time. **(35 marks)**
   (b) Dublin is Ireland's largest city and its capital. Examine, with reference to map evidence, how the range of services available reflects this statement. **(35 marks)**
   (c) Select a site which might be suitable for the location of a new national sports centre. Discuss fully *one* reason in favour of your selected site and *one* reason why local residents might object. **(30 marks)** 1998 Leaving Certificate examination (HL)
   (i) On a sketch map of the north side of Dublin mark and label
   - three types of roads
   - a hospital
   - a ferry terminal
   - two railway stations
   - a named river
   - a university
   - the Phoenix Park

   [Note: The sketch map should not be a tracing and should be smaller than the Ordnance Survey map.] **(30 marks)**
   (ii) The Phoenix Park is a major tourist attraction in Dublin. What evidence is there on the Ordnance Survey map to suggest this? **(20 marks)**
   (iii) The city centre area of Dublin is represented quite differently on the Ordnance Survey map than on the aerial photograph supplied (Fig. 20.16). Account for this difference. **(30 marks)**

*Please use evidence from both map and photograph in your answer to part (iii).
1998 Leaving Certificate Examination (OL)

Balcurris
Balseskin
Turnapin
Belcamp
Belcamp Park
Sillogue Park
Meakstown
Poppintree
Santry Demesne
Darndale
Kildonan
Santry Stadium
Santry
Seantrabh
Poppintree Park
Kilmore
Ballymun
Baile Munna
Finglas
Fionnghlas
Glasnevin North
St Mary's Hospital (Cappagh)
Beaumont
Beaumont Hospital
Coolock
An Chulóg
Dublin City University
Whitehall
Scribblestown
Donnycarney
Domhnach Cearna
Artane
Ard Aidhin
Killester
Ashtown
Baile an Asaigh
Cardiffs Br
Tolka Valley Park
Glasnevin
Glas Naíon
Prospect Cemetery
Botanic Gardens
Casino
Drumcondra
Droim Conrach
Marino
Cabra
An Chabrach
Ashtown Castle
Dalymount Park
Prison
Croke Park
Fairview Park
The Phoenix
Áras an Uachtaráin
Mater Hospital
Papal Cross
St Brendan's Hospital
PHOENIX PARK
Páirc an Fhionnuisce
Dublin Zoo
Connolly Station
St Mary's Hospital
People's Garden
Busáras
The Point Depot
Alexandra Basin
Ferry Terminal
Chapelizod
Séipéal Iosóid
Islandbridge
Droichead na hInse
Heuston Station
St James's Gate Brewery
Toll Bridge
War Memorial Park
Royal Hospital
Christchurch Cathedral
Dublin Castle
Trinity College
Pearse Station
Ringsend
An Rinn
Kilmainham Gaol
Kilmainham
Cill Mhaighneann
St James's Hospital
Merrion Square
Greyhound Race Track
Irishtown
Sean Moore Park
Irishtown Park
Inchicore
Inse Chór
St Stephen's Green
Grand Canal
Dolphins Barn
An Carnán
Coombe Hospital
National Concert Hall
Lansdowne Rugby Gd
Sandymount Strand
Drimnagh
Droimeanach
Brickfields Park
Ranelagh
Raghnallach
Ballsbridge
Droichead na Dothra
Sandymount
Dumhach Thrá
R.D.S.
Our Lady's Hospital
Mount Jerome Cemetery
Harolds Cross
Herbert Park
Simmonscourt
Martello Tower
Drimnagh Castle
Crumlin
Cromghlinn
Eamon Ceannt Park
Greyhound Race Track
Rathmines
Ráth Maonais
Donnybrook
Domhnach Broc
Merrion Gates
R.T.E. Studios
Clonskeagh
Cluain Sceach
St Vincent's Hospital
Merrion
Muirfín
Rathgar
Ráth Garbh
Milltown
Baile an Mhuilinn
Kimmage
Camaigh
Terenure
Tír an Iúir
St Lukes Hospital
Booterstown

Fig. 20.15

Fig. 20.16

# Index

LANDSCAPES OF THE WORLD

EXAM-TYPE QUESTIONS IN MAPS & PHOTOGRAPHS

LANDSCAPES OF THE WORLD